Research Methodology: A Guide for Researchers in Agricultural Science, Social Science and Other Related Fields

Pradip Kumar Sahu

Research Methodology: A Guide for Researchers in Agricultural Science, Social Science and Other Related Fields

Springer

Pradip Kumar Sahu
Department of Agricultural Statistics
Bidhan Chandra Krishi Viswavidyalaya
West Bengal
India

ISBN 978-81-322-1019-1 ISBN 978-81-322-1020-7 (eBook)
DOI 10.1007/978-81-322-1020-7
Springer New Delhi Heidelberg New York Dordrecht London

Library of Congress Control Number: 2013933139

Printed on acid-free paper

Springer is part of Springer Science+Business Media (www.springer.com)

To
My Parents

Preface

Starting from the initiation of the human civilization, all efforts have remained directed towards improving the quality of life. In the process, "Research," the search for knowledge, is ever increasing. Huge amount of resources in different forms is being channeled for this purpose. As such, the importance of research methodology is being felt day by day to have more and more successful research program. The voyage of discovery could be more meaningful if due attention is given to the art of scientific investigation, particularly in designing and meticulous implementation of a research program. Taking all these into consideration, this book has been written in such a way so that researchers in the field of agriculture, social science, and other fields could get a guideline about the appropriate methodology to be adopted and, in the process, the methods and techniques to be used so as to have successful research program.

This book contains 14 chapters altogether. The first two chapters ("Scientific Process and Research" and "Research Process" will help the research students, teachers, and researchers in different fields to get ideas/conception and also plan better research programs. Chapter 3 ("Research Problem") will guide researchers in formulating a research problem under the given situations. Chapter 4 ("Research Design") is aimed at imparting ideas of quality research to researchers so that these could be attributed to a research program. Discussions of the types of variables, their measurement, and scaling techniques have been made in Chap. 5. In Chap. 6, mainly the techniques of drawing appropriate sample have been discussed. This will help researchers working on heterogeneous population. Different types of data and their collection methodologies are discussed in Chap. 7. To interpret the data collected, processing and analysis are the most important aspects. Processing of raw data, its scrutiny, and preliminary analytical techniques have been discussed in Chap. 8. Many research programs deal with problems that involve hypothesis testing. Details of hypothesis testing techniques, both parametric and nonparametric, are discussed in Chap. 9. Analysis of variance, along with different experimental designs and their analysis, is discussed in Chap. 10. In Chap. 11, an attempt has been made to demonstrate the techniques of analysis of different genetics and breeding related problems. Multiple regression analysis, discriminant analysis, principal component analysis, cluster analysis, etc., have been taken up in Chap. 12. The emphasis has been on how to tackle multivariate problems with the help of computer programs and their interpretations. Instruments are nowadays an indispensible part of any research program. An attempt is made to discuss some of the instruments, precaution to be taken to use the laboratory safety measures, etc., in Chap. 13 . In this chapter, the utility of a computer in statistical analysis and the misuse of statistical theories are discussed in brief. In the last chapter, formulation of project proposal, interpretation, preparation of final report, etc., are discussed. In each chapter, theories are followed by examples from applied fields, which will help the readers of this book to understand the theories and applications of specific tools. Attempts have been made to familiarize the problems with examples on each topic in a lucid manner. During the preparation of this volume, a good number of books and articles in different national and international journals, have been consulted; efforts have been made to acknowledge and provide these in the reference section. An inquisitive reader may find more material from the literature cited.

To help the students, teachers and researchers in the field of agriculture and other allied fields remains the prime consideration in conceptualizing this book. Sincere efforts have been made to present the material in simplest and easy-to-understand form. Encouragements, suggestions and helps received from my teachers and my colleagues in the Department of Agricultural Statistics, Bidhan Chandra Krishi Viswavidyalaya are acknowledged sincerely. Their valuable suggestions towards improvement of the content helped a lot and are acknowledged sincerely. Dr. BK Senapati, Associate Professor, Department of Plant Breeding, and Dr Kusal Roy, Department of Agricultural Entomology, Bidhan Chandra Krishi Viswavidyalaya, have extended their full support towards the improvement of several chapters of this book. The author sincerely acknowledges the help received from Dr Senapati and Dr Roy. The author is happy to acknowledge the help received from Sri Pradeep Mishra, K Padmanabhan, and Ms Piyali Guha, the research students, during the preparation of this book. Thanks are due to my two daughters, Saswati and Jayati, and their mother Swati for their full cooperation and continuous encouragement during the preparation of this manuscript. The author thankfully acknowledge the constructive suggestions received from the reviewers towards improvement of the book. Statistical packages used to demonstrate the techniques of analyzing data are also gratefully acknowledged. Thanks are also due to M/S Springer Publishers for the publication of this book and continuous monitoring, help and suggestion during this book project. Help, co-operation, encouragement received from various corners, which are not mentioned above, the author thankfully acknowledge the same. Every effort will be successful if this book is well accepted by the students, teachers, researchers, and other users whom this book is aimed at. I am confident that like the other books written by me, this book will also receive huge appreciation from the readers. Every effort has been made to avoid errors. Constructive suggestions from the readers in improving the quality of this book will be highly appreciated.

Department of Agricultural Statistics P.K. Sahu
Bidhan Chandra Krishi Viswavidyalaya

About the Book

This book is the outcome of more than 20 years of the author's experience in teaching and in research field. The wider scope and coverage of the book will help not only the students/ researchers/professionals in the field of agriculture and allied disciplines but also the researchers and practitioners in other fields. Written in a simple and lucid language, the book would appeal to all those who are meant to be benefited out of it. All efforts have been made to present "research," and its meaning, intention, and usefulness. The book reflects current methodological techniques used in interdisciplinary research, as illustrated by many relevant worked out examples. Designing of research program, selection of variables, collection of data, and their analysis to interpret the data are discussed extensively. Statistical tools are complemented with real-life examples, making the otherwise complicated subject like statistics seem simpler. Attempts have been made to demonstrate how a user can solve the problems using simple a computer-oriented program. Emphasis is placed not only on solving the problems in various fields but also on drawing inferences from the problems. The importance of instruments and computers in research processes and statistical analyses along with their misuse/incorrect use is also discussed to make the user aware about the correct use of specific technique. In all chapters, theories are combined with examples, and steps are enumerated to follow the correct use of the available packages like MSEXCELL, SPSS, SPAR1, STATISTICA and SAS. Utmost care has been taken to present varied range of research problems along with their solutions in agriculture and allied fields which would be of immense use to readers.

Contents

1 Scientific Process and Research

Inquisitiveness is the mother of all inventions. Human being, by its instinct, is curious in nature; everywhere they want to know what is this, what is this for, why this is so, and what's next. This inquisitiveness has laid the foundations of many inventions. When they want to satisfy their inquisitiveness on various phenomena in a logical sequence of steps, they should take the role of the scientific process.

1.1 Scientific Methods

There are various ways or methods of knowing the unknowns, to answer to the inquisitiveness. Among the various methods, the scientific method is probably the most widely used method. The scientific process aims at describing explanation, and understating, of various known or unknown phenomena in nature. Thus, it increases the knowledge of human beings in multifarious ways. Any scientific process may have three basic steps: systematic observation, classification, and interpretation. Scientific methods are characterized by their objectivity, generality, verifiability, and creditability. Objectivity refers to procedures and findings not influenced by personal feelings, values, and beliefs of a researcher. Thus, objectivity in a scientific process ensures an unbiased, unprejudiced, and impersonal study. Generality refers to the power of generalization of the study from a particular phenomenon under study. Scientific studies should have general or universal applications. Universality refers to the fact that a study could provide a similar result under the same situations. Verifiability refers to the verifications of results or findings from subsequent studies. Not all scientific studies should lead to predictability, but when it predicts, it should predict with sufficient accuracy.

Scientific method is a pursuit of truth as determined by logical considerations. Logical sequence aims at formulating propositions/hypotheses explicitly and accurately so that these not only explain the phenomenon and unearth the undiscovered truth but also possible alternatives. All these are done through investigations/experimentations/observations. Thus, observation, investigation, and experimentation are the integral part of a scientific process either in isolation or in combination; through observation, one can describe and explain the phenomenon of interest. Investigations lead to provide a deeper insight into the phenomenon to unearth or expose the hidden truth. On the other hand, experimentations are mostly conducted under control conditions where most of the variables under considerations are allowed to remain constant, barring the variables of interest. The scientific method is mainly based on some basic postulates like:

1. It assumes the ethical neutrality of the problem under consideration, which leads to adequate and correct statement about the population objects.
2. It relies on empirical evidences.
3. It utilizes pertinent and relevant concepts.

P.K. Sahu, *Research Methodology: A Guide for Researchers in Agricultural Science, Social Science and Other Related Fields*, DOI 10.1007/978-81-322-1020-7_1, © Springer India 2013

4. It is committed to objective considerations.
5. It is committed to a methodology known to all concerned for critical scrutiny and possible replications.
6. In most of the cases, it is based on probabilistic approach.

1.2 Research

Research is "re-search," meaning a voyage of discovery. "Re" means again and again, and "search" means a voyage of knowledge. Research facilitates original contribution to the existing stock of knowledge, making for its advancement for the betterment of this universe. Research inculcates scientific and inductive thinking, and it promotes logical habits of thinking and organizations. The wider area of research and its application has laid to define research in various ways by various authors. By and large, research can be thought of as a scientific process by which new facts, ideas, and theories could be established or proved in any branch of knowledge.

Research is an art of scientific investigations; it leads to unearth the so-called hidden things of this universe. Research leads to enrichment of knowledge bank. Scientific research is entirely different from the application of scientific method towards harnessing of curiosity. Scientific information is put forward by a scientific research in explaining the nature and properties of nature.

1.2.1 Motivations of Research

Though inquisitiveness is the prime motivation for research, there are other motives or possible motives for carrying out a research project:

1. Joy of creativity
2. Desire to face challenges in solving the unsolved problems
3. Affinity to harness intellectual joy while doing some new work
4. Desire to serve the society
5. Desire to get respect
6. Desire to get a research degree
7. Desire to have means of livelihood

Neither the above list is exhaustive, nor the motives mentioned above operate in isolation; they may operate individually or in combinations. These motivations are from individual point of view. But motivations in research may come as directives from the government as part of their policy and may come under certain compelling situations and so on. Government may ask to undertake a research program so as to make the public distribution system more effective towards eradicating hunger, may direct to undertake research so that the productivity of food grain production may reach to the target within the stipulated time period, may ask researchers in economic policy so that the economic recession in different parts of the globe may not touch the economy of a particular country, etc.

Compelling situations like certain natural disasters in the form of tsunami, Aila, flood, and tornado, and outbreak of pests and diseases may lead to a research work towards finding and understanding causal relationships and overcoming approaches.

1.2.2 Objective of Research

The objective of any research is to unearth the answers to the questions an experimenter/researcher has in mind. The purpose of any research is to find out the truth which has so long remained undiscovered. Any research may have one or more objectives befitting to the purpose of the study. However, these can be categorized into different groups:

1. To gain familiarity of new things or to get an insight into any phenomenon
2. To picturize or to describe the characteristics of a particular situation, group, etc.
3. To work out the relative occurrence of some things and their associated things
4. To taste the hypotheses of relationship among the associated variables, to study the dynamics of relationships of different associated variables in this universe

As such a research may be exploratory or formulative, descriptive, diagnostic or hypothesis testing, etc., in nature. A research can be oriented to explore new things or new solutions and new insights in a particular phenomenon. This type of research can lead to the formulation of new things.

1.2.3 Research Methodology

The systematic process of solving a research problem is termed as research methodology. The science of studying how a research is carried out scientifically is known as research methodology. It generally encompasses various steps followed by researchers in studying research problems adopting logical sequences. In doing so, researchers should clearly understand the following questions in their mind: how to reach the answers to the questions, what could be the other steps, and what methods/techniques should be used? They should have ideas about a particular technique to be used among the available techniques, what are the assumptions and what are the merits and demerits. Thus, proper knowledge of research methodology enables a researcher to accomplish his or her research projects in a meticulous way. Research methodology helps a researcher in identifying the problems, formulating problems and hypotheses, gathering information, participating in the fieldwork, using appropriate statistical tools, considering evidences, drawing inferences from the collected information or experiment, etc. Research methodology has a great role to play in solving research problem in a holistic way by the researcher. Research methodology helps a researcher:

1. To carry out a research work more confidently
2. To inculcate his or her abilities/capabilities
3. To extract not only the undiscovered truth of the objective of the research but also his or her hidden talent
4. To better understand the society and its need

1.2.4 Research Method

Techniques or methods used in performing research operation known as research method. Research method is mainly concerned with the collection and analysis of information generated in answering the research problem that researchers have in mind. Thus, research method mostly includes analytical tools of research. The method of collecting information (from observation, experimentation, survey, etc.) using an appropriate survey/case study/experimentation method is included in a research method. The next group of a research method consists of statistical techniques to be used among the host of available statistical tools in establishing the characteristics and relationships of the knowns or unknowns. Research method also includes techniques for evaluating the accuracy of the result obtained.

1.2.5 Research Methodology vs. Research Method

From the above discussion, it is clear that research methodology is a multidimensional concept in which research method constitutes a part. As such research methodology has a wider scope and arena compared to research method. While discussing research methodology, a researcher talks not only about the techniques or collections of information and analysis of information but also on the logic behind the use of particular methods commensurating with the objective in mind and capability of generalizations of information generated from the research. Research methodology is also concerned with why the research problem has been chosen, how the specific problem has been defined, how the possible indicator has been identified, how the hypothesis is framed, what the data requirements for testing those hypotheses are, how to collect those data, why some particular methods of collections of analysis of data are against the other methods, and other similar questions. Thus, research methodology is concerned with the whole research problem or study, whereas research method is concerned with techniques, collections of information, its analysis, and validation. Research method can be regarded as a subset of research methodology.

1.2.6 Research Approach

Depending upon the inquisitiveness/problems the researcher has in mind, the approach to find a solution may broadly be categorized under two groups: qualitative and quantitative approach. In qualitative approach, research is mainly concerned with subjective assessment of the respondent. It is mainly concerned with attitudes, opinions, behaviors, impressions, etc. Thus, qualitative research is an approach to research to generate insights of the subject concerned in nonquantitative form or not subjected to rigorous quantitative analytical tools. In quantitative research approach, researchers undertake generations of information in quantitative form which are subjected to rigorous quantitative analysis subsequently. Generally, the quantitative approach has three different forms:

1. Inferential approach
2. Experimental approach
3. Simulation approach

1.2.6.1 Inferential Approach

In this approach, information is obtained to use or to draw inference about the population characteristics, their relationships, etc. Generally, survey or observations are taken from a studied sample to determine its characteristics and their relationships, and then, sample behavior is used to infer about the population behavior on the same characteristics and their relationships. Though the objective remains to study the population behaviors, characteristics, and interrelationships, because of constraints like time, money, resource, accessibility, and feasibility, it becomes difficult to study each and every unit of the population. Representative part of the population samples are drawn to study the behavior, characteristics, and interrelationships; these are again subjected to inferential tools to draw conclusions and to draw inference about the populations, behavior, characteristics, and interrelationships. In this approach, the researcher has no control over the characteristics or variables or respondent under study.

1.2.6.2 Experimental Approach

Experimental approach is characterized by control over a research environment by a researcher. An experiment is defined as a systematic process in which a researcher can have control over variables under considerations to fulfill the objective of his or her research process. Experiments are of two types: absolute experiment and comparative experiment. In absolute experiment, researches are in search of certain descriptive measures or characteristics and their relationships under control conditions, for example, how an average performance of a particular variety of paddy is changing over different nutrient regimes and how the nutrient regimes and average performance are associated. In comparative experiment, on the other hand, an experimenter is interested in comparing the effects of different treatments (control variables). For example, one may be interested in comparing the efficacy of different health drinks.

1.2.6.3 Simulation Approach

Stimulation means operations of numerical model that represents the structure of a dynamic process. In a simulation approach, artificial environment is created within which required information can be generated. Given the values for initial or ideal conditions, parameters, and exogenous variables, stimulation is run to represent and to regenerate the behavior of a process again and again so that it reaches to a stabilized condition providing consistent results. Future conditions can also be visualized under different varying conditions, parameters, and exogenous variables.

1.2.7 Criteria of Good Research

The ultimate objective of any research program should be oriented towards providing benefit to the society. The purpose of research, may it be in the field of life science, social science, business, economics, and others, will definitely facilitate better standard of living and conditions for the poorest of the poor. A good research should have the following criteria:

1. It should have clearly defined objectives
2. It should have described the research procedure sufficiently in detail and performed in a manner such that it could be repeated
3. It should have research design properly planned and formulated to provide results as far as possible to fulfill the objective of a research program
4. It should have adequate analytical work to reveal the significance in the tune of validity and reliability of the results for the betterment of the community
5. It should find out and report the flaws and lacunae of the research carried out so that it could be rectified in future program
6. Inference should be drawn only to the extent to which it is justified by the information support and validated by the statistical tool. Unnecessary generalization and extra variant comments not warranted by the observations should be avoided.

With the above guidelines, a good research should be (a) systematic, (b) logical, (c) empirical, (d) replicable, and, above all, (e) creative.

Systematic research is one which is structured with specific steps in proper sequence in accordance with well-defined set of rules. Specific steps in proper sequence never rule out creative thinking in changing the standard paths and sequence.

Any scientific process is guided by the rules of logical reasoning and logical process of intention and deduction. Induction is a process of reasoning from portion to the whole, contrary to deduction, meaning a process of reasoning from some premises to the conclusion which comes out from the premises; without having any logical reasoning, a research is meaningless.

A good research should take into consideration one or more aspects of real-life situations which could deal with concrete data that provides basis for validity of research results.

Replicability of any research loves research results that should be verified by replicating a study and thereby providing a sound basis in taking decision with respect to the research findings.

Creativity is the most important factor in research proposal. Ideally no two research proposals should be identical to each other. Research proposal should be designed meticulously so as to consider all factors relevant to the objective of the project. Difference in the formulation and structure of two research programs results in difference in creativity and also in findings. Any sorts of guessing or imagination should be avoided in arriving at conclusions of a research program.

1.2.8 Types of Research

Research is a journey towards the betterment of human lifestyle. It is innovative, intellectual, and fact-finding tools. Depending upon the nature, objectives, and other factors, research program has been defined or categorized under different types. The different types of research program mentioned below are to some extent overlapping, but each of them has some unique differentiable characteristic to put them under different categories:

1. Descriptive research
2. Analytical research
3. Fundamental/pure/basic research
4. Applied/empirical research
5. Qualitative research
6. Quantitative research
7. Conceptual research
8. Original research
9. Artistic research
10. Action research
11. Historical research
12. Laboratory research
13. Field research
14. Intervention research

15. Simulation research
16. Motivational research
17. One-time research
18. Longitudinal research
19. Clinical/diagnostic research
20. Conclusion-oriented research
21. Decision-oriented research
22. Exploratory research
23. Explanatory research
24. Evaluation research
25. Operation research
26. Market research
27. Dialectical research
28. Internet research
29. Participatory research

1.2.8.1 Descriptive Research

This research is sometimes known as ex post facto research. In this type of research, the objective is to describe a state of phenomenon that already exists. Generally the researchers attempt to trace probable causes of an effect which has already occurred even when a researcher doesn't have any control over the variables. The plight of human beings after tsunami may be the objective of research projects. In this type of project, the researcher's emphasis is on the causes of their plight so that appropriate measures could be taken at proper level. Ex post facto research may also be undertaken in business and industry, for example, reasons for changing behavior of consumer towards a particular commodity or group of commodity. In this type of studies, all measures to describe the characteristics as well as correlation measures are considered.

1.2.8.2 Analytical Research

Analytical research study is based on facts. A researcher has to use the facts or information available to them, analyzes them to critically evaluate the situation and followed by inferences. Thus, the difference between the descriptive and analytical research, though there is no silver lining, is that analytical research most likely goes deep inside the information for critical evaluations of the situations, whereas descriptive research may have the sole objective in describing the characteristics of the situations.

1.2.8.3 Fundamental/Pure/Basic Research

Fundamental or basic research, sometimes known as pure or exploratory research, is mostly related to the formulation of theory. Fundamental researches are concerned with the generalization of nature and human behavior at different situations. It may aim at gathering knowledge for knowledge's sake. Research findings which have resulted to the Newton's law of gravity, Newton's law of motion, etc., are examples of pure or fundamental research. Fundamental research is more often intellectual explorations arising out of intrinsic inquisitiveness of human beings. It is not associated with solving a particular problem, rather exploring the possibility of unearthing universal laws or theories.

1.2.8.4 Applied Research

Applied researches are mostly application-oriented research programs. This type of research aims at finding a solution for an immediate problem faced by a society, nation, business organization, etc. Market research is an example of applied research. Applied research is action oriented. Applied researches are often criticized by the nonacceptance or poor acceptance of their results by the people. Among the many reasons, one might be the fact that action research is conducted under controlled conditions which may not match entirely in reality with the people's living and working conditions. These problems of applied research have given rise to the concept of adaptive research. Thus, adaptive research should have emphasis on the usefulness of its results in the society and should be conducted under the prevailing situations of the targeted people.

1.2.8.5 Qualitative Research

Qualitative research is concerned with qualitative phenomenon. It is associated with phenomena like reasons of human behaviors. It aims at discovering the reasons of motivations, feelings of the public, etc. This type of research explores the psychological approach of human behavior and qualitative aspects of other areas of interest. Instead of analyzing data, based on observations, this depends on the help and guidance of psychologists, experts, etc. Why a group of people in a particular area prefers a particular type of tea may be one of the examples of qualitative research. Qualitative research uses techniques like word association test, sentence test, or story competition test. Exit or opinion poll is conducted during election in determining how people react to political manifestoes, candidatures, etc., and in deciding the outcomes of election.

1.2.8.6 Quantitative Research

Quantitative research is another kind of research in which systematic investigations having quantitative property and phenomenon are considered. Their relationships are worked out in this research. Quantitative research designs are experimental, correlational, and descriptive in nature. It has the ability to measure or quantify phenomena and analyze them numerically. Statistics derived from quantitative research can be used in establishing the associative or causal relationship among the variables. Fluctuations relating to performance of various business concerns, measured in terms of quantity or data, and agricultural experiments relating to measurement of quantitative characters and their correlational activities are examples of quantitative research. Quantitative research depends on the collection of data, the accuracy of data collection instruments, and the consistency and efficiency of the data. Utilization of proper statistical tool in testing of hypothesis or in measuring the estimate of the treatments is the prerequisite of the quality inference drawn from this type of research.

1.2.8.7 Conceptual Research

Conceptual research leads to an outline of conceptual framework to be used for a possible course of action in a research program. Thus, conceptual researches are aimed at formulating intermediate theory that is related to, or connected to, all aspects of inquiry. Conceptual research is related to the development of new concepts or innovations and interpretations of new ideas for existing methods. It is generally adopted by the philosopher and policy makers or policy thinkers.

1.2.8.8 Original Research

To some extent, original research is fundamental or basic in nature. This is not exclusively based on a summary, review, or synthesis of earlier publications on the subject of research. The objective of original research is to add new knowledge rather than to present the existing knowledge in a new form. An original work may be an experimental one, may be an exploratory one, and may be an analytical one. The originality of research is one of the major criteria for accessing a research program.

1.2.8.9 Artistic Research

One of the characteristics of artistic research is that it is subjective in nature, unlike those used in conventional scientific methods. Thus, artistic research, to some extent, may have similarity with qualitative research. Artistic research is mainly used to investigate and taste an artistic activity in gaining knowledge for artistic disciplines. These are based on artistic methods, practice, and criticality. The main emphasis is enriching the knowledge and understanding the field of arts.

1.2.8.10 Action Research

Action research is mostly specific problem oriented. The objective of this type of research is to find out reasons and to understand any situation to go deep into the problem so that an action-oriented solution could be attended. This is mostly used in social science researches; participation of local people is the main characteristic feature of this type of research. There are three main steps in action research: (1) examining and analyzing the problem with local people's participation, (2) deciding upon actions to be taken under the given situations in collaboration with active local people's participation, and (3) monitoring and evaluating so as to eradicate constraints and improve upon the action plans. Generally this type of research doesn't include strong theoretical basis with high complex method of analysis; rather, it is directed towards mitigating immediate problems under a given situation.

1.2.8.11 Historical Research

Historical research means an investigation of the past. In this type of research, past information is noted and analyzed to interpret the condition existed during the period of investigations. It may be a cross-sectional research or a time series research. This method utilizes sources like documents, remains, etc., to study past events or ideas, including the philosophies of people and groups at any remote point in time. In forecasting generally the past records are analyzed to find out the trends and behaviors of a particular subject of interest. These are sometimes mathematically modeled to extrapolate the future of a phenomenon with an assumption that past situations will likely continue in the future. The Delphi and expert opinion methods are examples of using such historical method of research in solving research problems.

1.2.8.12 Laboratory Research

Laboratory research is one of the most powerful techniques in getting answers a researcher has in his or her mind. Laboratory research is also one of the wings of experimental research. It may combine with field research or may work independently. Modern scientific research is heavily dependent on laboratory/instrumental facilities available to researchers. A better laboratory facility in conjunction with tremendous inquisitiveness of a researcher may lead to an icebreaking scientific achievement for the betterment of human life. There are so many good popular, renowned laboratories in the world which continue to contribute to the knowledge bank not only by genius scientists but also by the facilities available to them. Under laboratory condition, a research is being carried out to some extent under control conditions. In laboratory research, an experimenter has the freedom to manipulate some variables while keeping others constant. Laboratory experiments are characterized by its replicability.

1.2.8.13 Field Research

Field research is a type of research that is not necessarily confined to laboratories. By field research, we generally mean researches were carried out under field conditions; field conditions may be agricultural, industrial, social, psychological, and so on. Field experiments are the major tools in psychology, sociology, education, industry, business, agriculture, etc. An experimenter may or may not have control over the variables. Field research may be an ex post facto research. It may be an exploratory in nature and may be a survey type of research. Field research can be very well suited for testing hypothesis and theories. In order to conduct a field research, an experimenter is required to have high social experimental, psychological, and other knowledge.

1.2.8.14 Intervention Research

Interventional research mainly aimed at searching for the type of interventions required to improve upon the quality of life. Its main objective is to get an answer to the type of intermediaries and intervention (may be institutional, formal, informal, and others) required to solve a particular problem. Once the finding of an interventional research is obtained, the government and other agencies work out an intervention strategy. For example, it is decided that to ensure food and nutrition security by 2020, the productivity should increase at 3% per annum. The question is not only related to what should be the strategy to achieve the target but also related to what should be the appropriate steps; what should be the role of government, quasi-government, and nongovernment agencies; and what are technical, socioeconomic, and other interventions required at different levels—intervention research is the only solution. In business, let a particular house wants to popularized their products to the consumer. Now research is required to answer the questions what should be the marketing/promotional/intermediary strategies to reach to the consumer, in addition to motivational research about the consumers? At government levels, different health-improving programs are taken up based on intervention research findings on particular aspects. Thus, intervention research is most pertinent in social improvement perspective, having clear-cut objectives.

1.2.8.15 Simulation Research

Simulation means an artificial environment related to a particular process is framed with the help of numerical or other models to indicate the structure of the process. In this endeavor, the dynamic behavior of the process is tried to be replicated under controlled conditions. The objective of this type of research is to trace the future path under varying situations with the given site of conditions. Simulation research has wider applicability not only in business and commerce but also in other fields like agriculture, medical, space, and other research areas. It is useful in building models for understanding the future course.

1.2.8.16 Motivational Research

Motivational research is, by and large, a quality type of research carried out in order to investigate people's behavior. While investigating the reasons behind people's behavior, the motivation behind this behavior is required to be explored with the help of a research technique, known as motivational research. In this type of research, the main objective remains in discovering the underlying intention/desire/motive behind a particular human behavior. Using an in-depth interview/interactions, researchers try to investigate into the causes of such behavior. In the process, researchers may take help from an association test, word completion test, sentence completion test, story completion test, and some other objective test. Research is designed in such a way so as to investigate the attitude and opinion of people and/or to find how people feel or what they think about a particular subject. Thus, mostly these are qualitative in nature. Through such research, one which can motivate the likings or dislikings of people towards a particular phenomenon. There is little scope of quantitative analysis of such research investigation. But one thing is clear: while doing such research, one should have enough knowledge on human psychology or take help from a psychologist.

1.2.8.17 One-Time Research

A research program either can be carried out at a particular point of time or can be continued over the time course. Most of the survey types of research designs are one-time research. This type of research may be both qualitative or quantitative in nature and applied or basic in nature and can also be action research. Survey research can be taken out by the National Sample Survey Organization (NSSO) in India, and another similar type of organizations in different countries takes up the survey type of research to answer different research problems which come under mostly one-time research program. Though research programs are taken up at a particular point of time, these can be replicated or repeated to verify their results or study the variations of results at different points of time under the same or varying situations.

1.2.8.18 Longitudinal Research

Contrary to one-time research, longitudinal research is carried out over several time periods. In experimental research designs, particularly in the field of agriculture, there is a need for checking the consistency of the result immerged out from a particular research program; as such, these are coming under a longitudinal research program. Long-term experiments in the field of agriculture are coming under the purview of longitudinal research. This research program helps in getting not only the relationships among the factors or variables but also their pattern of change over time. Time series analysis in economics, business, and forecasting may be included under a longitudinal research program. Unearthing the complex interaction among variables or factors under long-term experimental setup is of much importance compared to a single experiment or one-time experiment.

1.2.8.19 Clinical/Diagnostic Research

In medical, biochemical, biomedical, biotechnological, and other fields of research, a clinical/diagnostic research forms an important part. Generally this type of research follows a case study method, in which in-depth study is made to reveal the causes of a particular phenomenon or to diagnose a particular phenomenon. This type of clinical/diagnostic research helps in knowing the causes and the relationship among the causes of a particular effect which has already occurred.

1.2.8.20 Conclusion-Oriented Research

In conclusion-oriented research, a researcher picks up his or her problem design and redesigns a research process as he or she proceeds to have a definite conclusion from the research. This type of research leads to yield definite conclusion.

1.2.8.21 Decision-Oriented Research

In decision-oriented type of research, a researcher needs to carry out a research process so as to facilitate the decision-making process. A researcher is not free to take up a research process or research program according to his or her own inclination.

1.2.8.22 Exploratory Research

Exploratory research is associated with the development of a hypothesis. It is the type of research in which a researcher has no idea about the possible outcome of the research. Research is an innovative activity. It is an intellectual process to provide more benefits to the community. Exploratory research is the tool to discover hidden facts underlying the universe. Exploratory research can, however, be conducted to enrich a researcher on a particular phenomenon he or she wishes to study. For example, if a researcher intends to study social interaction pattern against the polio vaccination program or polythene avoidance program but has little idea about human behavior, a preliminary interaction with persons related against the polio vaccinations or against the polythene avoidance may help the researcher in designing his or her research program.

1.2.8.23 Explanatory Research

The purpose of explanatory research design is to answer the question *why*. That means in explanatory research, the problem is known and descriptions of the problem are with the researcher but the causes or reasons or the description of the described findings is yet to be known. Thus, explanatory research starts working where descriptive research ends. For example, descriptive research has findings that more than 30% of people in a particular community suffer from a particular disease or abnormality. Now, an explanatory researcher starts working to search the reason for such results. He or she is interested in knowing why a particular phenomenon is happening, what are the reasons behind it, and how could it be improved upon.

1.2.8.24 Evaluation Research

In a broad sense, evolutionary research connotes the use of research methods to evaluate the program or services and determine how effectively they are achieving the goals. Evaluationary research contributes to a body of verifiable knowledge. The government or different organizations are taking up different developmental projects or programs. Evaluationary research develops a tool to evaluate the success/failure/lacunae of such research program. Evaluationary research program helps in decision-making process at the highest level of government departments or NGOs.

1.2.8.25 Operation Research

Operation research is a type of a decision-oriented research. It is a scientific method of providing quantitative basis for decision making by the decision makers under their control.

1.2.8.26 Market Research

Nowadays, market research is an important field of business or industry. This type of research generally falls under socioeconomic research pattern in which social economic behavior of the consumer, of the middlemen, of the business community, of the nation, and of the world is investigated in depth. This type of research is qualitative as well as quantitative in nature. Attitude, preference, liking, disliking, etc., towards adoption or non-adoption of a particular product are coming under qualitative research, whereas the area/quantum of consumers, possible turnover, possible profit or loss, etc., constitute the quantitative parts of a research. Market research is the basic and must-do type of research before releasing new product into the market. Market research also includes the types of marketing channels involved in the process, their modernization and upgradation, etc., so that the products are welcomed by ultimate users. No business unit can bypass the importance of market research. Market research investigates about the market structure, dreams, and desires of the customers. An effective marketing strategy should have clear idea about the nature of the targeted customer choices, the competitive goods, and the channels operative in the system.

1.2.8.27 Dialectical Research

This is a type of qualitative research mostly exploratory in nature. This type of research utilizes the method of dialectic. The aim is to discover the truth through examination and interrogation of competing ideas, arguments, and perfectives. Hypothesis is not tested in this type of research; on the other hand, the development of understanding takes place. Unlike empirical research, these are researches working with arguments and ideas instead of data.

1.2.8.28 Internet Research

If research is a journey towards knowledge, then Internet search can also be regarded as research. It is a widely used, readily accessible means of searching, enhancing, defining, and redefining the existing body of knowledge. Internet research helps to gather information for the purpose of furthering understanding. Internet research includes personal research on a particular subject, research for academic project and papers, and research for stories and write-up by writers and journalists. But there is an argument whether to recognize Internet search as research or not. Internet research is clearly different from the well-defined rigorous process of scientific research. One may or may not recognize Internet research, but it is emphatically clear that in modern scientific world, Internet has a great role to play in unearthing or discovering the so-called hidden truth for the betterment of the humanity.

1.2.8.29 Participatory Research

From the above discussion on the type of research, it can emphatically be noted that the type of research is based on its objectives, tools, nature, and perspectives/approaches. Different types of research are not exclusive to each other; rather, they are overlapping. Thus, the field experiment to compare the efficacy of yield of different varieties of rice may be put under the experimental research as well as applied research and also under field research. In fact, most of the research programs are following one or more of the research types mentioned above.

In social sciences, particularly when the research problem is aimed at solutions to the problems faced by a community or a group of people, participatory research plays an important role. In this type of research, people participation is ensured, leading to a better understanding of the problem, the need, and the solution needed. In finding out the solutions to social problems strategically, people perception about the solution is very much helpful, in particular during the implementation of the solution. This method is generally most effective for research programs involving relatively homogeneous community. The advantages of a participatory research method are as follows: (1) researchers (the outsiders) get an inner perspective of the community and (2) it is easier, (3) faster, (4) cheaper, and (5) more effective. Generally this type of research is used in intervention studies, though not solely.

1.2.9 Significance of Research

Research is a process to satisfy the inquisitiveness of human beings. It is a pursuit of the unknown, to know the things in a better way. Research is a systematic, formal, or informal process of knowing or explaining the so-called hidden truth for the betterment of mankind. It is gaining importance day by day. Though in its initial phase it was mostly individual efforts, but with passage of time, it has been realized that individual efforts could be streamlined in a systematic way for the benefit of this universe in a better way. As a result it has become an indispensable component for the development of the nation, for the development of the industry, and for the development of a particular group of people, to guide the solutions to particular problems, in addition to the development of theories and hypothesis. It is consistent and sincere endeavor towards enriching the knowledge bank. Today's progress is based on the past days' research. If inquiry is the base of a research, then all progress is born of inquiry. Research inculcates scientific and inductive thinking in human beings. It helps human beings to think and organize logically. Research is an integral part of the development of the nations. In the absence of any fruitful research outcome, government policies will have little or no effects. The economic systems of the nations and the whole world are entirely based on the research outcomes. All government policies and activities are guided by the findings of the research.

The importance of scientific research can very well be felt in government, business, and social science sectors. Decision making may not be a part of research, but research can facilitate the decision of policy makers. Government is by the people, for the people, and of the people. It is an official machinery to run the administration. In the context of good administration, research as a tool to economic policy has three faces of operation: (1) investigation of economic structure through continuous assessment of facts, (2) diagnostics of events and analysis of forces underlying the event, and (3) the prediction of future development–prognosis.

Research has a significant role in solving various operational and planning problems of business and industry. The existence of different new products in the market is the ultimate result of the research carried out in the development of that product, operation research, motivational research, and market research. While pure and/or applied research helps in the development of different new products, operation research is useful in optimizing the cost, namely, the profit maximization of the business concern. In a competitive market, any customer wants to have higher-quality products with the lowest price. Market research helps in unearthing the scenario in the market as well as the feelings and desires of the consumers and accordingly sets a marketing strategy. Motivational research is related with psychological behavior of the market. Motivational research helps in identifying the psychology or the motives of the consumers in favor of or against a particular product. Market analysis also plays a vital role in adjusting demand and supply for business houses. The researches in sales forecasting, price forecasting, and quality forecasting may help the business houses to develop effective business strategies.

Research is for the betterment of humanity. As a result, this is equally important in social science. Scientists of social science are to take up research programs so as to get solutions on different social problems. Social scientists are benefitted in two ways: social research provides them the intellectual satisfaction of gaining knowledge for the sake of enriching knowledge and also the satisfaction in solving practical social problems. The findings of social science research not only help to have better lifestyle for the community but also help in two other sectors—the business sector and the government. The results of social science are being effectively used by the business and industry for the effective expansion of business and marketing strategy. For the government, social science research provides the inputs in getting the impulse, feeling, likings, disliking, problems, constraints, etc., faced by the people, namely, the formulations of interventions or developmental strategy.

Keeping all these importances of research in view, both government and nongovernmental agencies are setting up the research wings for the purpose. Joining to a research laboratory, may be as a professional, may be as a means of livelihood, may be as a means for satisfying the inquisitiveness, and may be to flourish the intellectuality of human beings, has become prestigious to the society. By knowing the research methodology and the methods, one can accomplish research intention in a better way. The knowledge of methodology provides a very good idea to the new research workers to enable them to do better research. The knowledge of research methodology not only enriches the logical capability of the researcher but also helps to evaluate and use the research results with reasonable confidence. The intellectual joy of developing a new theory, idea, product, etc., can never be compared with any other thing.

2 Research Process

Research is a systematic process of knowing the unknown. A research process is a stepwise delineation of different activities to accomplish the objective of a researcher in a logical framework. It consists of a series of actions and/or steps for effective conduction of research. The main steps of a research process are given below. It may clearly be understood that the steps noted below are neither exhaustive nor mutually exclusive. These steps are not necessarily to be followed in the order these have been noted below. Depending upon the nature and objective of the process, more steps may come into the picture or some of the steps may not appear at all.

2.1 Steps in Research

1. Identification and conceptualization of problem
2. Purpose of study
3. Survey of literature
4. Selection of the problem
5. Objectives of the study
6. Variables/parameters to be included in the study to fulfill the objectives
7. Selection of hypothesis
8. Selection of samples
9. Operationalization of concepts and optimization/standardization of the research instruments
10. Collection of information
11. Processing, tabulation, and analysis of information
12. Interpretation of results
13. Verification of results
14. Conclusion
15. Future scope of the research
16. Bibliography
17. Appendix

2.1.1 Identification and Conceptualization of Problem

The inherent inquisitiveness of a researcher leads to an abstract idea about the type of research or area of research he/she is intending for. A newcomer in the field of research, like a PhD student or a Master's student doing a dissertation work under a supervisor/guide, may get ideas from the guide or the supervisor. The best way to understand the problem is to discuss with colleagues or experts in the relevant field. In private, government, and research organizations having definite research objectives, the research advisory committees and scientific advisory committees may also guide a researcher to the type of research work to be undertaken. A researcher may review conceptual and empirical type of literature to have a better understanding about the problem to be undertaken. There are mainly two types of researches that were carried out: *study of nature* and *relational studies*. The first one is concerned with the development of theories, laws, principles, etc., whereas the second one is related to working out the relationships among the variables under study. In the process of identification and

P.K. Sahu, *Research Methodology: A Guide for Researchers in Agricultural Science, Social Science and Other Related Fields*, DOI 10.1007/978-81-322-1020-7_2, © Springer India 2013

conceptualization of problem, a researcher frames and reframes his/her ideas so as to reach to a meaningful research program.

2.1.2 Purpose of Study

Any research program is guided by three "w's": what to do, why to do, where to do, and also how to do. All these are guided by the purpose of the study. The purpose of the study guides a researcher to the type of research problem(s) he/she should undertake, which means what to do, whether the study is basic in nature or an empirical one. Once the problem has crept into the mind of the researcher, the immediate question that rises into his/her mind is "why." That means what is the purpose of the study, why should one undertake this type of study, and what would be the benefit to human beings. Next, he/she should try to satisfy the query in his/her mind, that is, where to take up a research program, meaning the type of research program to be undertaken—experimental, social, etc. He/she should have a clear-cut understanding of the objective of the study. The problems to be investigated must be conceptualized unambiguously, because that will help in deciding whether further information are needed or not.

2.1.3 Survey of Literature

Once a problem is conceptualized, it should be checked and cross-checked for its possible acceptance or nonacceptance in the light of the scientific inquisitiveness. Extensive survey of literature should be done so as to have a comprehensive idea about not only the theoretical aspects but also the operational aspect of the program to be undertaken. A survey of research on theoretical aspects will help a researcher in fine-tuning a research program. This will also help in boosting the self-confidence of a researcher in his/her line of thinking. The operational part of the survey of the literature will give an idea to a researcher on how to accomplish a research program, what could be the possible problems in reaching the objective of the program, and other information on relevant field of studies. These may help a researcher in anticipating the outcome of the research also. When a researcher has a better access to literature, the better he understood a research program; as a result, each and every research organization must have a well-equipped reference library. With the advancement of computer and Internet technology, a researcher, in any part of the world, is now better placed than their counterparts a few years ago. Thus, not only having a good library but also selecting an appropriate library are important for the smooth conduction of a research program. The review of literature, thus, constitutes the integral part of a research program, in spite of its being time-consuming, daunting, and, to some extent, frustrating but rewarding. The review of literature provides clarity and enhances the knowledge of a researcher, improves methodology in the context of findings, etc.

2.1.4 Selection of the Problem

The selection and formulation of a research problem is the most important phase of a research program. A research must answer the questions "for whom," "what to do," etc. Once the conceptualization and purpose of the study are fixed along with an extensive literature survey, formulation and statement of the problem are next steps. The problem should examine the issues to be addressed and the path. While stating the problem unambiguously, care must be taken to verify the subjectivity and validity of the background facts and information of the problem. The path of formulation of the problem generally follows a sequential pattern in which the number of formulations is set up, each being more specific than the previous one and each being framed more analytically and realistically on the basis of the available information and resources. In selection and formulation of the problem, the subject area, the situations, the resources available particularly the time and monetary resources, etc., are required to be kept in mind. Any research program is usually associated with time consumption and possibility of unforeseen problems. One should select a problem of great interest which can sustain such problems. A too big or too extensive research program may be unmanageable within the given time frame and resource. So a researcher should take into cognizance these factors and narrow down the area of the topics up to that extent that it is already manageable. A researcher should make sure that the indicators by which research problems are going to be characterized/investigated are explained and measurements are followed. A researcher should select the indicators/measuring tools/analytical tools in which he/she has a good level of expertise, which are relevant in the context of the program, and in which the data would be available and, of course, the ethical issues are not compromised. Thus, in formulating a problem, a researcher should identify the broad field of a study or subject of interest, dissect those into subareas and select the most interested and pertinent subareas, and lastly double-check all of the above steps.

2.1.5 Objectives of the Study

The objectives of a study are the goals set to attend in the study. These objectives should clearly specify what a researcher wants through a study. An objective may be a main objective, and sub-objective, or specific objective. A main objective is a statement about the overall trust area of a study. In an empirical or applied research, these objectives may also set the relationship or associations a researcher wants to discover or establish. Sub-objectives are specific objectives of a particular topic a researcher wants to investigate within the main framework of a study. Generally, each objective of a study should be concerned with only one aspect of a study. The objectives of a study are generally written in words starting with "to determine," "to find out," "to ascertain," "to measure," "to describe," "to compare," "to demonstrate," etc. Thus, the objective should be framed in such a way that it gives a complete, clear, and specific idea about a study, identifying the main and the correlational variables.

2.1.6 Variables/Parameters to be Included in the Study to Fulfill the Objectives

In a research program, particularly the quantitative ones, it is important that the concepts used should be in measurable terms so that the variations in the respondent could be put under mathematical treatments. Thus, the knowledge of variables and measures plays a vital role in extracting the truth from the observation of a research program, so long remained hidden. The types and measurement of these variables have been discussed in Chap. 5. What is meant by a variable? What is the difference between a variable and a parameter? How are these measured? The variable is an entity which varies over various situations (time, space, both time and space, individuals or units). In other words, the antithesis of a constant is a variable. There are various types and concepts of variables and their measures/scales.

2.1.7 Selection of Hypothesis

At the very beginning, it may be clearly understood that not all researchers have a definite hypothesis or a hypothesis to be tested. A hypothesis is an assumption/assertion of an idea about a phenomenon under study. This hypothesis forms the base of the entire research program. Hypothesis brings clarity and specificity, focuses to a research problem, but as has already been mentioned, it is not essential for all research programs. Hypotheses are based on the main and specific objectives of studies, and they direct a researcher to the type of information to be collected or not to be collected. Hypothesis enhances the objectivity of a research program. There are different types of hypothesis like null hypothesis, alternative hypothesis, and parametric and non-parametric hypotheses.

2.1.8 Selection of Sample

In fact, most of the empirical or applied studies and, to some extent, some basic studies for their effective outcome heavily depend on the type of sample being used for the purpose. Any research program is directed towards enriching the lifestyle of the human being. Population can be studied by examining its individual unit. Because of constraints like time, resource, and feasibility, it may not be possible to study each and every individual unit of a targeted population. So a representative part of the populations—"sample"—is taken; studies or researches are conducted on the sample, using statistical techniques; it is verified whether the sample behavior would be taken as the population behavior or not. Information can be obtained from experimental fields/social field depending upon the nature of the research program. There are different methods of drawing samples (probability sampling, non-probability sampling) depending upon the nature of the population. Certainly not all the methods are equally welcome in all situations. Thus, the technique of drawing samples and the nature of the samples plays a vital role in analyzing the sample characteristics, namely, drawing inference about the population characteristics towards fulfilling the objective of a research program.

2.1.9 Operationalization of Concepts and Optimization/Standardization of the Research Instruments

Nowadays most of the research programs are characterized by the use of high-end sophisticated modern instruments/equipments. Instrumentation centers have become an integrated component of any research organization. Starting from a simple pH meter to the modern-day electronic scanning microscopes or other sophisticated equipments required to be standardized, depending upon the nature of the experiments and types of the instruments. A faulty or unstandardized instrument may lead to erroneous information and subsequently derail the entire research program from its ultimate objective. So, before taking up any information or recording data from any instruments, a researcher should be very careful about the authenticity and usefulness of the data in the light of the objective of a research program.

2.1.10 Collection of Data

Whatever may be the research type, qualitative or quantitative, information are required to be collected or collated in a most scientific manner to fulfill the objective of a study. It may emphatically be noted that though there exists no clear-cut silver lining in differentiating between information and data, by and large, by information we mean both qualitative and quantitative; on the other hand, by data one has an instinct to categorize them under quantitative form. But, quite frequently in a general sense, these two are used synonymously. There are different methods of collection of information/data depending upon the source, nature, etc. According to the source of information, data may be *primary* or *secondary*. Primary data are those data collected by the user from the field of investigations with specific objectives in mind. Secondary data, on the other hand, are collated from different sources by the user for his/her purpose. Thus, meteorological information is the primary data to the department of meteorology, but to a researcher who uses the information from the meteorological department, these are secondary data. There are different methods of collections of data: observation method, survey method, contact method, experimental method, etc.

2.1.11 Processing, Tabulation, and Analysis of Information

Having fixed the objective of a study and collected the required information, the first thing a researcher should attempt is to scrutinize or cross verify the information. Once the information is passed through the above process, processing and/or tabulation is required to be done to extract firsthand concise information from the raw data. With the help of the processing and tabulation of the information, a researcher will be in a position to understand to some extent whether the results are in possible direction with the objectives or not. At this juncture, a researcher may note the findings in brief in the form of text or paragraphs, tables, graphs, etc. Analysis of data is the most critical and vital part of a research program in unearthing the hidden truth from the area of research concerned. Various analytical (mathematical and statistical) tools are available, and with the advancement of computing facility, most of them are used nowadays. But the most important thing is that not all analytical tools are useful/applicable in extracting/discovering the hidden truth in all types of research problems or under all situations. The researcher should know what he/she wants to get from a particular analysis, whether the analysis taken up is applicable/useful under a particular situation, whether proper tools have been used in the most correct manner, and whether the findings from the analysis are being explained in a proper manner. Given a set of data, it can be analyzed in numerous ways using a number of techniques, but definitely all are not going to provide accurate result in line of the objective of a research program. The researcher should be well aware about the "garbage in and garbage out" system of computing facilities. Thus, there is a need for coordination and understanding at the highest level between the researchers and analytical experts.

2.1.12 Interpretation of Results

In modern science having so many fields of expertise and study, it is not always possible for a researcher to become an expert in all his/her relevant fields of research. As a result, a multidisciplinary approach of research is gaining momentum day by day. An understanding of a research program among team members of research groups complements and supplements each other in interpreting the results of a research program.

2.1.13 Verification of Results

Once the results of a research program are interpreted, these should be verified/authenticated. Nowadays, taking advantage of Internet facility, various discussion boards are emerging out to discuss on a specific topic. Researchers can very well discuss the findings with the experts of the relevant field; they may post their findings in the web and invite discussions. They may also search the literature to gather knowledge about their findings. It may be noted that a research finding may be an expected or unexpected one. A good researcher should not bother about the expected one; he/she should have an open mind to accept both the expected and unexpected. Rather on getting the unexpected result, a researcher might be interested to verify the entire research program, starting from the formulation of the problem to the outcome of a research program.

2.1.14 Conclusion

Conclusion is the most important part of a research activity; it summarizes in a systematic manner about the whole research program, starting from the conceptualization of the program to the ultimate findings and their interpretations. In doing so, it must be kept/borne in mind that it should be neither too long nor too brief. The main objective of a conclusion, sometimes called as executive summary, is to provide an overall idea about the research problem concerned, its execution, and its salient findings. A conclusion provides a bird's eye view of the entire research program and its importance in brief. Conclusion may sometimes include recommendations with respect to the problem area undertaken in a research program.

2.1.15 Future Scope of the Research

It is said that a good research program leads to a number of future research programs. In every research report, a section highlighting how the present research program would be extended or continued to throw more light to the knowledge bank and to throw lights on how other related programs would be solved in the same line of the present research work. A good research worker should also report in this section the sort of falls or lacunae of the present research program and how these would be overcome in a future research program.

2.1.16 Bibliography

During the whole research process, a researcher is required to consult the existing body of literature. An alphabetic arrangement of this literature is needed not only to acknowledge the past works but also to help the future workers on the relevant aspects.

2.1.17 Appendix

Appendix is constituted of all technical details like questioner/schedule, specific methods used during data collection or analysis and laboratory methods requiring elaboration. This means all support information which are not presented in the main part but may help the readers in understanding the research program.

3 Research Problems

3.1 Research Problems

Research is an endeavor to know the facts from the unknown. All inventions are the results of *inquiry*. Doubts are often preferred over confidence, because doubts lead to *inquiry* and *inquiry* leads to invention. Research may be carried out being motivated by the desire to know or to gather knowledge or by the desire to solve problems on hand. In the latter case, one need not identify the problems, but in the former case, a researcher needs to identify the problems. One thing must be clear that when we talk about a research problem, it is generally having a reference to a descriptive or hypothesis testing type of research. An exploratory or formulative type of research generally does not start with identifications of the problems—its objective is to find out the problems or the hypothesis. Generally all the problems are not researchable; to be researchable, the problem must satisfy certain criteria:

1. It must refer to some difficulty experienced in the context of either theoretical or practical situations and should obtain the solution for the same.
2. It must be related to the group of individual/society. The study must be useful in solving burning problems towards enhancing the quality of life.
3. It must draw the attention of the experts, policy makers, and academicians, who are familiar with the subject.
4. A subject should not be taken as a research problem in which lots of works have already been done and very difficult to throw new lights.
5. Controversial subject should be avoided by an average researcher unless and otherwise there is a compelling situation or a researcher is extraordinary in nature.
6. Too narrow or too wide problems should be avoided.
7. A research program should have alternative outcomes for obtaining the objectives a researcher has in his/her mind.

A research problem is a type of problem which requires solution. A research problem is a problem for which a course of action is required to be set optimally to reach the objective. There should be alternative courses of actions to counter the unforeseen difficulties during the conductions of research. A research problem should be formulated taking all of the above into account.

A research problem is the brainchild of researchers. The hunger to know the unknowns and the desire to solve the compelling problems to provide some sorts of relief to the society are the motivational sources of research problems. While selecting a research problem, a researcher should not be too ambitious; he/she should be well aware of the resources and facilities available at his/her disposal and expertise required for the research along with other factors. While selecting research problems, the following guidelines may be followed; however, these guidelines are neither exclusive nor exhaustive or ordered.

P.K. Sahu, *Research Methodology: A Guide for Researchers in Agricultural Science, Social Science and Other Related Fields*, DOI 10.1007/978-81-322-1020-7_3, © Springer India 2013

3.1.1 Researcher Interest

Research is re-search, to know the unknowns. A research problem is the brainchild of a researcher. As such, a researcher should select a problem area in which he/she is interested in it and familiar with and has reasonable level of understanding. Though research organizations both in private and government have guidelines for the areas of research to be undertaken depending upon the objective of the organizations, a researcher should try to select such areas in which his/her interest and the organizational interest coincide. All these are required because to accomplish a fruitful research work, the probing attitude, tenacity of spirit, and dedications of researchers are most warranted.

3.1.2 Usefulness of the Topic

Most of the research problems arise out of the desire to obtain a fruitful solution to the pressing problems of the society. So a research problem should have directions such that its results are useful in solving the problem of the society. On the other hand, pure and basic research, though guided by the "knowledge for the sake of knowledge," must give rise to a number of application-oriented research problems. The novelty theoretical research lies in stimulating or motivating the applied or empirical researches. A research problem should not be chosen such that it will be very difficult to throw new lights on the respective area. Moreover, an average researcher should try to avoid selecting such problems which are controversial in nature. Only the established and renowned researchers in the respective field shall take up such researches. A research problem with too narrow or too vague objectives is criticized for its usefulness for the betterment of the society. Thus, whatever may be the type of research, a research problem should enrich the knowledge bank either on theoretical aspects or application aspects.

3.1.3 Resources Availability

Any research program is guided by three questions: "what to do," "where to do," and "what is available?" A researcher should have a clear-cut knowledge about the availability of resources at his/her disposal. With the experience and expertise available to the researchers, is he/she confident in completing a research program? What are the facilities in the form of clerical and technical assistance, instrumental facilities, computational facilities, and of course monitory and time available for conduction of the research work? A well-equipped researcher in the form of laboratory, library, and technical manpower can think of attaining maximum objectives of a research program and, thus, can formulate a research problem accordingly. Most of the applied research programs are time bound in nature, so while framing a research program, one must be careful about the time limit allowed for the purpose.

3.1.4 Availability of Data

Availability of data plays a vital role during the formations of a research problem, in a research based on secondary information. A researcher is required to examine whether the data for the program are available in plenty or not. For experimental research or survey type of research, these problems pose little problem, because a researcher generates the information for a specific purpose. A research problem formulated and to be tested with the help of secondary data is very susceptible to availability of the data.

3.2 Steps in the Formulation of Research Problem

Defining a research problem is a daunting task. Defining a problem involves a task of laying down boundaries within which a researcher should study a problem with a predetermined objective in view. Defining a research problem properly, clearly, and carefully is the most important part of any research study and should not be in any case accomplished hurriedly. A research problem should be defined in a systematic manner providing due considerations to all related points. Ideally a research problem is formulated following the paths described below.

(a) **Statement of the Problem**
A research problem is the brainchild of a researcher. In the first instance, a problem should be stated in a broad and general way. Ample scope should be there in defining and redefining research problems. In an organizational research generally, a researcher has little to do with this. But in case of an individual research, a statement of research problems plays a vital role; before stating the problem, a researcher must have a prima facie knowledge about the problem and its possible solutions. In the absence of such knowledge, the best way is to conduct a pilot study or pilot survey. In a conventional research program of academic institutes, the guide or the supervisor generally provides the problem in general terms. The researcher then takes up the problem and concretizes the areas and puts them in operational terms.

(b) **Understanding the Nature of the Problems**
The best way of understanding the nature of a problem is to discuss the problem with the experts in the relevant fields. With the advancement of computer and Internet facilities, discussion groups are being formed with the view to exchange ideas on different aspects. Researchers in modern days can also take help of such facilities. A researcher can enrich himself/herself using these techniques to find out how the problem originally came about and with what objective in view. Understanding the nature of the problem is also related to where a research is to be carried out. This is because of the two facts in which with the change of situations, the nature of the problem changes and also the type of unforeseen situations during the conduction of the study.

(c) **Surveying Available Literature**
Before defining a research problem along with a clear-cut understanding about the nature of the problem, a researcher should be thorough with relevant theories, reports, and documents in the field of study. Reviewing of the available literature, thus, forms an inseparable part of defining research problems. The review of literature enhances the knowledge bank, the confidence of a researcher, and directions of a research program to be taken up. It also helps researchers in getting an idea about the past works and their merits and demerits. It can also help in improving the methodology to be taken up in a research program. Contextual findings are the added information generated through a review of literature. Contextual findings can not only help in fine-tuning the research problems but also help in explaining the findings of the research during the course of the study. There are different sources of literatures like printed books and journals in the libraries, e-libraries, Internet, discussions, and exchange of documents from the experts in relevant fields. While reviewing the literature, one should be careful whether the knowledge relevant to theoretical framework is clarified and confirmed beyond doubt. Different theories, their criticisms, methodologies adopted, and their basis should be noted carefully. A careful examination is required with respect to what extent the findings of the literature can be generalized and used in the present perspective of the research.

(d) **Developing Ideas Through Discussions**
Having passed through the stages of the general statement of the problem, understanding the nature of the problem, and surveying the literature, a researcher is now comfortably placed to discuss research problems with the experts in the relevant field to give a proper shape to the problem. In the process, a researcher must discuss who has enough experience in the same area or working on similar problems. This is sometimes known as experience survey. This will help a

researcher in sharpening a research program and specifying the problem at hand.

(e) **Rephrasing/Refining the Research Problem**
The last step of the formation of a research problem is rephrasing or refining a research problem. The nature of the problem has been clearly understood, environment under which the research is to be carried out is clearly exposed to the researcher, experience from the experts has made it possible for the researcher to intersect and dissect the problems at hand, and the researcher is now in a better position to reframe or refine his/her research problem so as to make the research problem operationally and analytically viable to reach the objective of the program. While refining a research problem, a researcher should be careful in defining and using the technical terminologies, phrases, and words. He/she should clearly state the assumptions or the postulates relating to the research problem. A straightforward statement of the criteria for the selections of the problem should be provided. While refining the research problem, the time frame and the sources of data should be taken into consideration. Researcher should have clear idea about the scope of the investigation, limitations of the investigation, and prospects of furthering the study.

Thus, though defining a research problem is a herculean task, if properly followed, sequentially arranged, and meticulously executed, it may lead to the formulations of an ideal research problem. Starting from the general statement of the problem to the rephrasing of the problem, in each and every step, a researcher should be very careful and objective in mind.

4 Research Design

A research design is the blueprint of the different steps to be undertaken starting with the formulation of the hypothesis to drawing inference during a research process. The research design clearly explains the different steps to be taken during a research program to reach the objective of a particular research. This is nothing but advance planning of the methods to be adopted at various steps in the research, keeping in view the objective of the research, the availability of the resources, time, etc. As such, the research design raises various questions before it is meticulously formed. The following questions need to be clarified before the formulation of the research design:

1. What is the study?
2. Why the study is being undertaken?
3. Where should the study be carried out?
4. What kind of data are required?
5. What are the resources available, including time?
6. What should be the sampling design (if required)?
7. What would be the methods of data collection?
8. What are the possible analytical techniques that can be used?
9. What would be the mode of presentation of the findings?

Without having a clear-cut idea about the nature of the study, pure or applied, experimental or survey, exploratory or descriptive, etc., appropriate steps cannot be identified. Unless it is clearly known why a study is being undertaken, the objective of the study won't be clear. The study area where the research program is to be undertaken must be known, because the steps to be followed differ for different types of studies. The steps to be followed differ from survey design to experimental design, descriptive to exploratory design, and diagnostic to experimental design. If it is a laboratory experiment, the type of precision required is different from that of a field experiment. Depending upon the type of experiment, the type of research data required and the methods of data collection differ. In a survey type of research, one may have to adopt a specific sampling design or technique, which may be different from sampling techniques adopted for the selection of samples for experimental research studies. The research design is required to be framed in a way so as to utilize the available manpower, money power, and instrumental facilities within a given time frame in the best possible manner, and so as to reach to the objective of the study optimally. A research design then is a detailed outline of a plan of work to be undertaken during the process of conducting a research program, which may or may not require modifications as the research progresses.

P.K. Sahu, *Research Methodology: A Guide for Researchers in Agricultural Science, Social Science and Other Related Fields*, DOI 10.1007/978-81-322-1020-7_4,

4.1 Characteristics of a Good Research Design

A good research design is one that helps in reaching the objective of a research program in the best possible way. *Objectivity*, *reliability*, *validity*, and *generalization* are the main characteristic features of a good research design.

4.1.1 Objectivity

By objectivity, we mean observations free from bias, from the observer's point of view. When we say "blood is red," it is an objective statement, but when we say "blood is the most useful element in the human body," then the statement may not be a purely objective one, because of the fact that the human body cannot survive only with blood; for an effective functioning of the human body, other useful components are needed along with blood, meaning blood can be a useful component, but most probably one cannot assign the word "most" before useful. In many research programs, various types of instruments and measures are being used. Good research designs use only those instruments and measures which are precise in nature. Generally it is felt that objectivity can be achieved easily, but in actual situations, it is very difficult to attain the objectivity from a host of competitive available sources.

4.1.2 Reliability

Reliability refers to consistency and authenticity in responses. Consistency means the respondent should not provide a different answer every time the same question is asked by the investigators in different forms for a particular problem. An instrument is consistent if its response remains same under repeated observations, essentially under the same condition. Authentic information is obtained from a source which has the authority and credential to report about the problem. To verify the reliability of the responses, particularly in the social science researches, there should be provision for cross-checking the responses in the methods of collection of information.

4.1.3 Validity

While using any measuring instruments, a researcher should be sure that the instrument selected for use is a valid one. An originality test constructed for measuring originality should measure none other than originality. In literature, there are a number of procedures for establishing the validity of a test.

4.1.4 Generalization

Generalization of the outcome of a research program is one of the important points to be noted during the formulation of the research design. In most of the research programs, though the researcher starts with a particular problem for a particular situation, at the end of the day, the researcher would have the intent to explore the possibility of using such outcomes in other relevant fields and would precisely want to generalize the research outcome. That is, how best the information or solution generated by the present research work would be applied to a large group, from which the samples have been drawn. This is possible only when effective steps are taken during the formulation of the hypothesis, data collection, and of course analysis and interpretation of information. For example, if a sample size is large (>30 generally), then one can use the information applying the central limit theorem or normal approximation. But if a sample size is low, then specific probability distributions are required to be ascertained and followed; as such generalization of the result becomes restricted.

4.2 Types of Research Design

Research designs vary depending upon the type of research. By and large, research designs are framed in three types: (a) exploratory type of research, (b) descriptive type of research, and (c) hypothesis-testing type of research.

4.2.1 Research Design for Exploratory Research

The purpose of an exploratory research is the formulation of problem or the development of hypothesis. These can also be knowledge enhancing in nature. The major consideration of such design is the discovery of ideas and theories that have not been completely explored. As such, its research design must be quite flexible in nature, for the researcher explores with an open mind. Nobody knows what may happen in the next stage of exploration; everybody should remain alert with an expectation of unexpected outcome. Generally, there are three different methods of research designs in the context of exploratory research: (1) the survey of literature, (2) the experience survey, and (3) the analysis of insight-stimulating examples.

The survey of the concerned literature is the most fruitful and simple method of developing a research problem or hypothesis. This is nothing but a process of gathering knowledge from the past works of the concerned field of study. Hypotheses stated or formulated by previous workers may be reviewed and evaluated for their usefulness on the basis for future research. Sometimes, the creative work of previous workers may provide ample ground for the formulation of theories and hypotheses. In the process, a lot of effort, time, money, and energy could be saved if the researcher is well aware of the facts of what has been done in the past with respect to the proposed area under study.

An experience survey gives an opportunity to the researcher to get insight into the subject and its various facts and facets. Formal or informal discussions with experts, researchers, administrators, and professionals may provide effective clues toward the formulation of the problems. During the modern era of the Internet, different discussion groups may be formed and can play a vital role in sharing the knowledge about any specific field around the world. Previously, questionnaires were sent to respondents (experts, administrators, group of affected people, researchers, professionals), sitting at different places to express their views on a particular subject. Thus, an experience survey not only helps the researcher in developing the problem but also helps in studying the feasibility of such type of research.

Analysis of insight-stimulating examples is a very useful method for suggesting hypothesis and formulating problems. It is more useful in areas where there is little experience which serves as a guide. The essence of this method lies in the intensive study of the selected phenomenon in which the researcher is interested. Examination of records and informal interviews may help in the process of selecting examples. In general, the problems of such contrast and having striking futures in the relevant fields are considered to be relatively useful for the purpose.

Thus, the exploratory studies provide excellent opportunities to get an insight into the subject concerned, vis-a-vis formulation of hypothesis and problems. Whatever may be the methods or research designs, they must be flexible to accommodate the need of the hour, so as to achieve the objective of the research program.

4.2.2 Research Design for Descriptive Type of Research

Descriptive research studies deal with the description of the characteristics of a group or a particular situation. In this type of study, a researcher should clearly mention what should be measured and how it should be measured, for a particular population. The research design must be well equipped to protect the bias and to maximize the reliability, and at the same time, it must be rigid and should not be too flexible in placing attention on the following points: (1) what the study is about and why it is being undertaken, (2) designing the methods for data collection in the most efficient manner, (3) selection of the samples and processing and analysis of data after examinations, and (4) interpreting the results and drawing conclusions.

In a descriptive study, the first step is to specify the objective with sufficient clarity. The objective of a study must be based on its usefulness along with its background information. Once the objectives are fixed, the collection of information comes into play. The quality of data in sufficient amount is a need of any good research program. Data collection should be safeguarded against bias and reliability. Whatever method of data collection is selected, questions must be well examined, they must be unambiguous, and there should not be any influence on the part of the enumerator on the responses. It is better to use pretested data collection instruments. The sampling design to be followed must be an appropriate one; it must take care of the nature of populations from which the samples are to be drawn. If the samples are not drawn properly, the findings are bound to be affected and, subsequently, the generalization of the results suffers. Data collected from the samples must pass through the following criteria: completeness, comprehensibility, consistency, and reliability. Only the scrutinized data are required to be put under processing and analysis. The processing and tabulation of information give a firsthand idea about the nature of the information generated. A host of statistical tools are available for analyzing the data. Definitely, all of them may not be suitable in all kinds of research. Appropriate statistical tools, befitting the objective of the research program and the data collected for the purpose, must be used. Coding of data, if required, must be done carefully to avoid any error and without compromising the reliability factor.

Research is the process of finding new things, which remained hidden otherwise. In the stage of research design, people come to know about the new findings from a particular research program. The art and skill of interpreting results is crucial in any research program. How elaborative, elucidative, comprehensive, and understandable a report is depends on the expertise and effectiveness of a researcher. This is the most vital part of any research program, and it clearly indicates how the objectives of a program have been fulfilled or, otherwise, how it is going to improve upon the quality of life of the people, how it is effective to policy makers or decision makers, etc. Thus, the important task of interpretation of the results is to communicate the findings to the society so that the fruit of research changes the lifestyles of human beings.

4.2.3 Research Design for Hypothesis-Testing Research

A hypothesis-testing type of research is generally concerned with studies where a researcher wants to verify the hypothesis with respect to one or more objectives of the study. There is an argument whether the diagnostic type of research will be considered under this type of research design or the descriptive research design discussed in the previous section. In diagnostic type of research, there is a scope for testing the hypothesis, diagnosing the causal relationship among the parameters/variables in a research design. As such, we shall include a diagnostic type of research design under the broad category of research design for hypothesis testing. Thus, hypothesis-testing type of design can broadly be categorized under:

1. Research design for diagnostic studies
2. Research design for experimental studies

4.2.3.1 Research Design for Diagnostic Studies

Hypothesis refers to assertion about the population parameters. One or many hypotheses relating to the concerned population are tested in this type of research design. In the diagnostic type of research design, generally the observations are recorded from the existing population; an association between the variables and the related hypotheses is tested. A diagnostic type of study deals with problems as well as solutions. The main objective is to diagnose the problem accurately to work out for some solutions and, in the process, to find out the relationship among the variables associated. The success of a diagnostic study depends on (a) how the questions or questionnaires are framed and defined for the purpose, (b) how the concept used for the study is clearly defined, (c) how the area of the study is clearly stated and restricted, (d) how the methods and collection of information specific to the objective of the study are clearly understood or executed, (e) how best the efforts are taken to remove the bias from the information collected, (f) how best the statistical tools required for the purpose are adopted, and (g) how best the interpretation has been made only in the light of the observations and objective of the study. Too much generalization of the results of the study must be avoided.

4.2.3.2 Research Designs for Experimental Studies

In an experimental type of studies, a researcher wants to test the hypothesis with respect to the populations through experimentations. Experiments are designed and are conducted to get proper solutions to the problems at hand. The design for experimental studies has its origin in agricultural research. Prof. R.A. Fisher is the pioneer in this field of study. Experimental design may be of two types: (a) *informal* and (b) *formal*.

Informal Research Designs

Informal research designs are based on or associated with a less sophisticated form of analysis. Informal experimental design can be of three categories: (1) before and after without a control design, (2) after only with a control design, and (3) before and after with a control design.

1. **Before and After Without Control Design**
   In this type of design, variables are measured before and after the introduction of treatments. The impact of the treatment is measured by taking the differences in the measurements of a particular phenomenon at two different points of time. Hypothesis is tested whether the treatment has any effect on the level of phenomenon or not. Thus, $\overline{D} = \frac{1}{n}\sum_{i=1}^{n}(Y_i - X_i)$, where Y_{ij} and X_i are values of the phenomenon after and before the treatment. The experimenter measures the Y_is and X_is for the targeted population and tests the hypothesis H_0: $\overline{D} = 0$ using an appropriate test statistic.

Table 4.1 Before and after research design

Group	Before	After
Treatment	Y_i	Y_i'
Control	X_i	X_i'

2. **After Only with Control Design**
   In this type of experimentation, in addition to a treatment group, there will be one control group, which will not be treated. The level of phenomenon will be measured on both the treatment group and control group before and after treatment. The impact of the treatment will be studied against the control group. Let X_{ij}' be the observations of the *i*th individuals on the particular phenomenon for the control group after treatment. Similarly Y_i' is the value of the phenomenon of the *i*th individual of the treatment group after the treatments. Thus, we have $\overline{\delta} = \frac{1}{n}\sum_{i=1}^{n}(Y_i' - X_i')$.

   The researcher would be interested in testing H_o: $\overline{\delta} = 0$ using an appropriate test statistic.
3. **Before and After with Control Design**
   In this type of research design, the researcher is interested in comparing the *change* in the level of phenomenon for two groups of people, that is, treatment group and control group (Table 4.1).

   $$\overline{\delta} = \frac{1}{n}\frac{\sum_{i=1}^{n}(Y_i' - Y_i)}{\sum_{i=1}^{n}(X_i' - X_i)}$$;observations are recorded in both control and treatment groups before and after treatment. The researcher tests the H_o: $\overline{\delta} = 0$ using the appropriate test statistic.

Formal Experimental Design

Compared to informal experimental design, in the formal experimental design, the researcher has more control over the experiments. As such more sophisticated statistical tools would be used for the analysis of the data from the experiments.

Formal experiments are of two types: *laboratory experiments* and *field experiments*. In these types of designs, mostly the techniques of analysis of variance are applied. Depending upon the situation and the objective of the study, the analysis of data arising out of experimental designs is taken up as per one-way analysis of variance, two-way analysis of variance, *analysis of covariance*, and so on. It should be clearly noted that analysis of variance is based on certain assumptions. So the validity of the result is not independent of these assumptions. The experimental designs may be one-factor, two-factor, or more-factor experiments, depending upon the number of factors put under the experimentations with different levels of each factor. Whatever may be the number of factors, these designs are based on certain basic principles and follow three basic designs. The principles of designs of experiments are replication, randomization, and local control. The three basic designs are completely randomized designs, randomized block designs (randomized complete block design), and Latin square design.

Replication is the repeated application of the treatment among the experimental units. It helps in the estimation of the effects of different components of the experiments. Along with randomization, it leads to valid estimation. Along with local control, it leads to reductions in experimental error.

Randomization is the technique of unbiased allocation of treatment among the experimental units. Randomization along with replication provides valid estimations of the component experiment. It reduces bias on the part of the experimenter and others.

Local control is the technique by virtue of which experimental error is reduced.

The above three basic principles form the backbone of the experimental designs. Let us now discuss in brief the three basic designs.

Completely Randomized Design

A completely randomized design (CRD) is the simplest of all designs where only two principles of design of the experiments, that is, replication and randomization, are used. The principle of local control is not used in this design. As such a one-way analysis of variance is adopted in analyzing the data in this type of experimental design. The basic characteristic of this design is that the whole experimental area or all experimental units should be homogeneous in nature. One needs as many experimental units as the sum of the number of replications of all the treatments. If we have five treatments A, B, C, D, and E replicated 5, 4, 3, 3, and 5 times, respectively, then according to this design, we require the whole experimental area to be divided into 20 experimental units of equal size or 20 homogeneous experimental units. Thus, a completely randomized design is applicable only when the experimental area is homogeneous in nature. Under laboratory conditions, where other conditions including the environmental condition are controlled and all experimental units are homogeneous in nature, completely randomized design is the most accepted and widely used design.

Advantages

1. This is the simplest of all experimental design.
2. This is the only design with the flexibility in adopting different numbers of replications for different treatments. In a practical situation, it is very useful because sometimes researchers come across the problem of differential availability of experimental materials. Sometimes, response from particular experimental unit(s) may not be available; even then, the data can be analyzed with the help of the CRD design.

Disadvantages

1. The basic assumption of homogeneity of experimental units, particularly under field condition, is rare. That is why this design is suitable mostly in laboratory condition or greenhouse condition.
2. The principle of "local control" is not used in this design which is very efficient in reducing the experimental error. Thus, the experimental error in CRD is comparatively higher.
3. With the increase in the number of treatments especially under field condition, it becomes very difficult to use this design, because of the need to obtain more number of homogeneous experimental units.

Randomized Block Design or Randomized Complete Block Design (RBD/RCBD)

A randomized complete block design (RCBD) or simply randomized block design (RBD) is the most widely used design which takes into account the variability among the experimental units. A randomized block design uses all the three basic principles of experimental design, namely, (1) replication, (2) randomization, and (3) local control. In RBD, the whole variation is partitioned into mainly three components: (a) due to factors that are varied by an experimenter at his/her own wish, (b) due to variability in experimental units, and (c) due to extraneous factors—the experimental error. As such, a two-way analysis of variance is being followed. In a randomized block design, the whole experimental area is divided into a number of homogeneous block/groups, and each block/group consists of as many experimental units as the number of treatments. Blocking/grouping is done in such a way that the variation among the experimental units within the block/group is minimum (homogeneous) and the variation among the experimental units of different blocks/groups is maximum (heterogeneous).

Advantages

1. RBD is the simplest of all block designs.
2. Its layout is very simple.
3. It uses all the three principles of design of experiments.
4. It is more efficient compared to CRD.

Disadvantages

1. The number of treatments cannot be very large. If the number of treatments is very large, then it is very difficult to have a greater homogeneous block/group to accommodate all the treatments. In practice, the number of treatments in RBD should not exceed 12.
2. Like CRD, flexibility of using variable replication for different treatments is not possible.
3. Missing observation, if any, is to be estimated first and then analysis of data is to be undertaken.
4. It takes care of the heterogeneity of an experimental area in one direction only.

Latin Square Design

In some experimental areas, it is found that soil heterogeneity varies in two perpendicular directions, that is, north to south and east to west or south to north and west to east, or individual units can be grouped based on the two variable characteristics. BD cannot take care of such methodology; LSD has the capacity to handle two sources of variations independently from the blocking criterion as in the case of RBD. A Latin square is an arrangement of treatments in such a way that each treatment occurs once and only once in each row and each column. Because of this type of allocation of treatments, the total variations among the experimental units are partitioned into different sources, namely, row, column, treatments, and errors. If the number of treatments is also equal to the number of replications r for each treatment, then the total number of experimental units needed for this design is t x t. These t^2 units are arranged in t rows and t columns and the resulting LSD is an incomplete three-way design.

This type of experiment is mostly conducted or useful in field conditions or greenhouse conditions. The condition for the appearance of a treatment once and only once in each row and in each column can be achieved only if the number of replications is equal to the number of treatments, thereby is equal to the number of rows and columns treatments to be taken under LSD. This limits the LSD design's applicability in field experimentations. The number of treatments in LSD design should not be too many or too few. All these limitations have resulted in the limited use of Latin square design, in spite of its high potential in controlling experimental errors.

Advantages and Disadvantages

Advantages

Latin square design is an improvement over the other two basic designs, CRD and RBD, as it takes care of the heterogeneity or the two-way variations. Sometimes, when an experimenter does not have an idea about the soil heterogeneity among the experimental units or does not have the time to check the heterogeneity pattern, he/she can opt for the LSD design.

Disadvantages

1. The number of treatments equals the number of rows and columns.
2. The layout of the design is not as simple as in the case of CRD or RBD.
3. The analysis assumes that there are no interactions.
4. It requires *mainly* a square number of experimental units.

The LSD is suitable and more advantageous over CRD and RBD under specific field conditions.

Factorial Experiment

To save time and other resources and also to know the interaction effects of different factors in single factor experiments factorial experiments are set up. These are also suitable for agricultural field, laboratory, social science, and other studies in which the impact of more than one factor is required to be compared. It is our common experience that a particular variety of wheat or a crop is responding differentially under different irrigation schedule, different rates of fertilizers, different weed management practices, different crop protection practices etc. and an experimenter wants to know all the above. Moreover, in this type of experiments, the experimenter wants to know not only the level or the doses of individual factor giving better result but also wants the combination of the levels of different factors which is producing the better result. In a study related to the role of health drinks and physical exercise, one may be interested in knowing (1) which health drink is comparatively better, (2) what type of exercise is better, and (3) what combination of health drinks and exercise provides better health to the growing children. In socioeconomic studies, one may be interested in relating the economic status and educational status of the people in adopting modern agricultural practices. Factorial experiments are the methods for answering the questions related to more than one factor at a time for their individual effects as well as interaction effects in a single experiment.

Advantages

1. In factorial experiments, the effects of more than one factor at a time can be estimated.
2. In factorial experiments the interaction effects can be estimated which is not possible in single-factor experiments.
3. Factorial experiments are resource saving compared to single-factor experiments; factorial experiments could be set up with lesser number of replications.

Disadvantages

1. When the number of factors or the levels of factors or both increase, then the number of treatment combinations will also increase, resulting in a requirement of bigger experimental area/units and bigger block/group size. As the block/group size increases, it is very difficult to maintain homogeneity among the plots/units within the block/group. Thus, there is a possibility of increase in the experimental error vis-à-vis decrease in precision of experiment.
2. Factorial experiments are more complicated than single-factor experiments.
3. Failure in one experiment may result in a loss of huge amount of information compared to a single-factor experiment.

Types of Factorial Experiment

(a) *Based on the number of factors*: Depending upon the number of factors used in the experiment, a factorial experiment is a *two-factor, three-factor . . . p-factor* factorial experiment when the number of factors put under experimentation is 2, 3 . . . p, respectively.

(b) *Based on the level of factors*: Depending upon the equality or inequality in the levels of factors put under experimentation, a factorial experiment is either *symmetrical* or *asymmetrical*. If the numbers of levels for all factors are same, it is symmetrical, otherwise asymmetrical. For example, a two-factor factorial experiment with five varieties and five different doses of nitrogen is a symmetrical factorial experiment, but a factorial experiment with five varieties and any number of doses of nitrogen (not equal to five) is an asymmetrical factorial experiment. Symmetrical factorial experiments are denoted by p^q form, that is, q factors each at p levels. When asymmetrical factorial experiments are denoted by $p \times q \times r \times s$, where the first, 2nd, 3rd, and 4th factors have p, q, r, and s levels, respectively.

There are several competitive research designs; a researcher must decide upon the proper type of research design, which is befitting to the need of the research toward fulfilling the objective of the study. In doing so, a researcher must keep in mind the type of data required and its collection and analysis, so as to extract the hidden information from the subject area of the research. The planning of research design and its execution with utmost sincerity and accuracy lead to valid inference and generalization about the subject of the research undertaken.

5 Variables, Measurement, and Scaling Technique

The objective of a research is to search the hidden truth of the universe for the betterment of humanity. In doing, so a population is studied by its characteristics. The different characteristics of any population are also different in nature. If one wants to study the students (population), then the characteristics of these students are required to be studied. The different characteristics of students like age, intelligence, height, weight, economic conditions, race, gender, etc., are also required to be studied. All these characteristics mentioned above vary from student to student, and while studying the student population, one must take into account the above characteristics along with other varied characteristics. Similarly if one wants to study the production behavior of food grains of any country to meet the challenges of food and nutritional requirements, a researcher is required to study not only the time series information of each and every component of the food crops but also their variations over the different component states/provinces of the country concerned. Thus, studying a population is always related with studying "variabilities" of the characteristics/entities under consideration.

5.1 Variable

In this universe, there are two types of entities—the first one varies over different situations and the second one remains fixed. An entity which varies over different situations (e.g., time, individuals, places) is known as the *variable*. On the other hand, an entity which doesn't vary is known as the *constant*. Thus, if a variable is a thesis, then the constant is the antithesis. The universal gravitational constant (G), acceleration due to gravity (g), Avogadro number, etc., are some of the examples of constant. One can find that the nature of constant G and g are not same; G remains constant at any situation, but "g" varies slightly over situations like at the top of Mt Everest and at a ground level which is even being constant itself. Thus, we come across two types of constant, (1) *restricted constant* and (2) unrestricted constant. A constant like "g" is known as a restricted constant, whereas constants like G are known as an *unrestricted constant*.

The characteristics/variables which one generally studies during a research process can be of two different types: *measurable* and *nonmeasurable*. Height, weight, length, distance, marks obtained by a student in different subjects, income, number of family members, number of pods per plant, number of grains per panicle, number of insects per plant, area under disease, number of employees in a particular industry, turnover of a particular industry over the year, prices of different commodities at different markets at a particular point of time, prices of different commodities at different points of time at a particular market, etc., are few examples of characteristics/variables which could be measured. The above measurable characteristics/variables are known as quantitative characters/variables. There are certain characteristics like races (e.g., Aryans, Dravidians, Mongoloids), gender (e.g., male or female), complexion (e.g., good, fair, better),

P.K. Sahu, *Research Methodology: A Guide for Researchers in Agricultural Science, Social Science and Other Related Fields*, DOI 10.1007/978-81-322-1020-7_5, © Springer India 2013

color (e.g., red, green), and taste (e.g., sweet, sour, bitter) which cannot be measured as such but can be grouped, ranked, and categorized; these are known as *qualitative characters/variables*. The qualitative variables are also known as *attributes*. Each variable or characteristic is generally associated with the chance factor. In statistical sciences, a researcher is interested in studying not only the variable but also their probability distributions. As such, the variable with chance/probability factor—the variate—is of much interest rather than simply a variable. Thus, the variate is defined as variable with chance or probability factor. Rainfall is a variable which varies over days (say), but when it is associated with probability distributions, then it becomes a variate.

In literature, variables have been categorized into different types in accordance with its nature, purpose, use, etc. One can find the following types of variables in different literatures, though the types are neither exhaustive nor exclusive rather sometimes they are overlapping:

1. Continuous variable
2. Discrete variable
3. Dependent variable
4. Independent variable
5. Explanatory variables
6. Extraneous variable
7. Stimulus variable
8. Control variable
9. Dummy variable
10. Preference variable
11. Multiple response variable
12. Target variable
13. Weight variable
14. Operationally defined variable

5.1.1 Continuous Variable

A continuous variable can take any value within a given range. Plant height, length of panicle, length and breadth of leafs, heights of different groups of people, etc., are the examples of a continuous character/variable. These variables can take any value for the respective range. If one says that the plant height of a particular variety of paddy varies from 60 to 90 cm, which means taking out any plant from the field of a particular variety, its height will be within the range of 60–90 cm.

5.1.2 Discrete Variable

A discrete variable is one which takes only an integer value within a given range. For example, the number of grains per panicle of a particular variety of paddy varies between 40 and 60 grains. This means if one takes out any panicle of that particular variety, the number of grains in it will take any value within this range. But one cannot expect that the panicle taken at random should have a number of grains 50.6 or like that; it will be 50 or 51.

5.1.3 Dependent Variable

A dependent variable is a type of variable whose values are dependent on the values taken by the other variables and their relationship. Generally in relational studies, a variable is influenced/affected by other related variables. In a production function analysis, there exists a functional relationship between the output and the factors of production. Here the output is considered as dependent variable which depends on the factors of production like land, labor, capital, and management. In socioeconomic studies, the adoption index (dependent variable) with respect to the adoption of a particular technology may depend on a number of socioeconomic factors like age (x_1), caste (x_2), education (x_3), family type (x_4), social status (x_5), economic conditions (x_6), area under cultivation (x_7), and the size of holding (x_8). Thus, one can write $y = f(x_1, x_2, x_3, x_4, x_5, x_6, x_7, x_8)$. In this example, y is the dependent variable and $x_1 \ldots x_8$ are the independent variables. One can use a functional relationship to predict the values of a dependent variable for a given set of values of the variables $x_1 \ldots x_8$. As such, y is also known as predicted variable and $x_1 \ldots x_8$ are known as predictor variables.

5.1.4 Independent Variable

In any relational analysis, variables which help to predict the dependent variable using the functional relationship are known as independent variables. In the above example, in Sect. 5.1.3, $x_1 \ldots x_8$ are said to have no association among themselves and are termed as independent variables. These variables independently help in predicting the dependent variable. Though there is no silver lining between the predictor variable and the independent variables, generally in regression analysis, predictor variables are synonymous to independent variables. But it is not necessary that the predictor variable must be an independent once.

5.1.5 Explanatory Variables

Independent variables are sometimes known as explanatory variables. Any variable which explains the response/dependent/predicted variable is known as explanatory variable. In a simple regression analysis, there are only one predictor and one response variable. In a multiple regression analysis generally, there is one response or predicted variable with more than one predictor/explanatory/independent variables. In the case of system of simultaneous equations model, there may be more than one response variable and more than one predictor/independent/explanatory variable. Moreover, the response variable(s) in one equation may be the explanatory variable in the other equation.

5.1.6 Extraneous Variable

In a relational analysis, independent and dependent variables are not only the variables present in the system; some other variables known as extraneous variables may have an impact on the relationship between the independent and dependent variables. For example, the relationship between the performance in driving and salary may be influenced by factors like age, gender, academic background, conditions of the road, and conditions of the car. Thus, factors like age, gender, academic background, conditions of the road, conditions of the car, etc., are the extraneous factors/variables in a relational study of driving performance and salary. Extraneous variables are those variables whose information is obtained from the outside a periphery of study. Generally the values of these variables are not directly obtained from the system under study, but these variables may affect the dependent/response/predicted variables. Extraneous variables may again be of two types: (a) *participant variables* and (b) *situational variables*. Participant variables are extraneous variables which are related to the individual characteristics of each participant. Thus, in the above example, age, gender, and academic background are the participant variables. On the other hand, situational variables or extraneous variables are related to environmental factors like conditions of the road and climatic conditions etc. Generally under experimental research conditions, extraneous variables are controlled by researchers. These are more pertinent in the case of a system of simultaneous equations model; the information about these variables is taken from the outside of the system to solve the model.

5.1.7 Stimulus Variable

The idea of stimulus and response variables is familiar in agriculture, socioeconomic, and clinical studies. A stimulus is a type of treatment applied to the respondents to record their response. In clinical studies generally, the doses, concentrations, different chemicals, etc., form a stimulus, whereas the response may be in the form of *quantal response* or quantitative response. When a stimulus is applied to a record response, the response may be either-or type, or it may be measurable. In an either-or type of response, a respondent will either respond or not respond after being applied with the stimulus. The different concentrations of a particular chemical in controlling a particular pest of a particular crop may kill the pest or may not; the response is either to kill or not to kill. On the other hand, applications of insulin at a particular dose can help in reducing the blood sugar level, a measurable response. In socioeconomic studies, stimulus variables may be in the form of action variable like documentary film, field demonstration, and method demonstration, an effect which would be measured in the form of adoptions or non-adoptions of a particular technology in which documentary/field demonstration/method demonstrations were used.

5.1.8 Control Variable

Control variables are nothing but the independent variables in relational studies, which can affect the relationship between the dependent variables and the independent variables and which could be controlled by making them constant or by eliminating them from the model. Thus, control variables are the type of independent variables which could be controlled by a researcher to study effectively the effects of other independent variables. Thus, independent variable can be categorized into two groups, that is, the *control variable group* and *the moderator variable group*. In comparison to the control variable groups, the effect of moderator variables is studied, keeping the control variables at constant or eliminating or minimizing them. Depending upon the objective of a research, it is up to a researcher to determine the moderator variables and the control variables.

5.1.9 Dummy Variable

In many research studies, particularly concerned with the qualitative characters, it is very difficult to guide/put a study under mathematical treatment. To overcome this problem, one of the techniques is to assign numbers against the quality parameters. For example, in a study concerned with gender-related issues, the male may be assigned number 1, while 0 for the female or vice versa. In a study of plant type, a bushy type of plants may be assigned number 1, erect type number 2, and tree type number 3, and so on. Thus, in each of the above cases, quality characters are designated by different numbers. As such, quality characters take the values 0, 1, 2, etc., and are known as dummy variables. Dummy variables have special implications, particularly in relational analysis. For example, instead of an ordinary simple regression analysis in case of numeric variables, one should go for *probit*, *logit*, and *nomit* regression analysis when encountered with dummy variables.

5.1.10 Preference Variable

These are generally discrete type of variables whose values are either in decreasing or increasing order. For example, in a survey of acceptability of a film by the audience, respondents were asked to grade the film by using five different codes which are as follows: (1) excellent, 1; (2) very good, 2; (3) good, 3; (4) poor, 4; and (5) bad, 5. It may be noted here that there is no relationship between the difference in grades 1 and 2 with that of the difference between any two consecutive grades and vice versa. Similarly, in a study of constraint analysis, farmers were asked to indicate the importance of the following constraints in accordance with their preferences by using codes from 1 to 8: (1) weather, (2) finance, (3) irrigation, (4) marketing, (5) price of input, (6) price of output, (7) nonavailability of good quality seed, and (8) nonavailability of appropriate technology. The farmers are to arrange the above eight constraints from 1 to 8 with 1 being allotted to the most important constraint and 8 to the least important constraint.

5.1.11 Multiple Response Variable

In multiple response variables, a variable can assume more than one value. In a social, economic, market research, etc., it becomes very difficult for respondents to select a particular option against the other alternatives; rather, they opt for combinations of absence; a typical example is the use of modern-day high-tech mobile phones. If a respondent is asked to indicate the purposes of using mobile phones in his/her daily life, the respondent could score more than one category out of the following options: (1) talking to people at distant places, (2) using GPS for daily life activities, (3) using camera for getting records, (4) listening to music, (5) used for simple calculations, (6) using as a miniature form of computers, etc. To a user talking to people at different places along with listening to music may be of equal and top priority. To others, it may be the talking to people at distant places along with the use of miniature form of computers be of equal or top priority. Thus, in both cases, the user has no preference to select one at topmost priority rather than other ones. As such, there are multiple responses to the query: why a user likes to have a mobile?

5.1.12 Target Variable

A target variable is almost synonymous to that of a dependent variable in a classical regression analysis. The main objective is to target a variable (predicted variable) whose value (s) is required to be predicted taking help of the values of the other variables and also the relational form.

5.1.13 Weight Variable

A weight variable specifies the weightage to be given to different data sets/subsets. A weight variable may be continuous or discrete in nature. In a given data set, if "0" is assigned to any row of data, then that particular row could be ignored. On the other hand, a weightage given to rows 1, 2, 3, etc., means the rows concerned have different importance.

5.1.14 Operationally Defined Variable

Before conducting any experiment, it is imperative to have a firm operational definition for all the variables under consideration. An operational definition of a certain variable describes how variables are defined and measured within the framework of a research study. If one wants to test the performance of driving, then before taking the variables to score the performance of a driver, it is required to have clear-cut a definition of "a driver"—what do we mean by driver?

5.1.14.1 Criteria for the Selection of Variables

Having formulated the problem of a research work, the objective and the specific objective of the studies and the hypothesis to be tested and a guideline or an abstract idea about the type of variables to be studied become visible to a researcher. At the first instance, a researcher investigates what is the information required and what is the information available from the study—experimental or nonexperimental. In fact a researcher at this point of time will be among the host of variables. But the most pertinent thing is that how many variables are available and how many variables are required to be studied under a framework of a study (under the given time, money, technical experts, instrumental availability, etc.). Generally while selecting a variable, a researcher should keep in mind the following points: (1) objective of the study, (2) specific objective of the study, (3) hypothesis to be tested, (4) variables which should be mutually exclusive (nonoverlapping), (5) variables which should be clearly understood by a researcher, (6) techniques that are available or to be developed to measure a variable, (7) the number of variables that should not be too many or too few, and (8) availability of time and resources to researchers. The variables to be selected should be guided or commensurated with the objective and specific objective of a research program. The variable to be included must be clearly defined; any ambiguity in defining the variables may jeopardize the findings of a research program. The variables to be selected must be related with the hypothesis to be tested in a study, and must be nonoverlapping in nature. The variable which could not be measured with sufficient level of accuracy or which measurement availability of technical staff is in dearth may be avoided.

5.2 Measurement

Human instinct is to quantify the things in our daily life. As a result, we measure height, weight, age, length, etc. We also try to measure how good a painting is, how good a story is, how good a drama is, how best a new technology is adopted by a farmer is, etc. Measurement is relatively complex and must be a "to do" type of activity. By and large, by measurement, we mean the process of assigning numbers to objects or observations. The level of measurement is a function which dictates numbers to be assigned to the objects or observations. Measurement is a process of mapping aspects of a domain onto other aspects of a range according to some rule of correspondence. In assigning numbers to objects or observations, there are two aspects to be considered: (a) assigning numbers in respect of the property of some object and (b) assigning numbers relative to others. The second aspect is relatively difficult. Measuring social integrity, intelligence, varietal adjustments, etc., requires close observations and attentions compared to measuring physical quantities like height, weight, and age. In the measuring process, one needs to have some scales. In fact, in literature, one can find different scales of measurement; sometimes, these are also known as levels of measurement. Mostly one can find a (a) *nominal*, (b) *ordinal*, (c) *interval*, and (d) *ratio scale/levels of measurement.*

(a) **Nominal Scale**

In this type of scaling system, numbers are assigned to objects just to level them. The numbers provided on the jersey of a player are simply to recognize a player with that particular number. It has nothing to do with the capability of the player or his/her order or grade. Nominal scales just provide an easy way of tracking the objects or people or events. Nominal scale has no order or distant relationship or no arithmetic origin. Nominal scales are generally useful in social and economical research studies and also in ex post facto research. The variables measured under a nominal scale can be put to get a frequency percentage, mode, median, chi-square test, etc.

(b) **Ordinal Scale**

An ordinal scale is a one-step upgradation of a nominal scale. Ordinal scales place events in order, but the intervals between two consecutive orders may not be equal. Ranking and gradation are examples of ordinal scales. In an ordinal scale of measurement, there is a sense of greater than or less than, and *events are categorized not only to provide numbers but also to indicate the order of events*. This scale has no absolute zero point. Though the intervals are unequal, the ordinal scale provides more information than the nominal scales by virtue of their ordering properties of the events. Cricket players may be ranked or graded according to their performance popularity. Thus, the top 10 players will be having numbers assigned to them starting from 1 to 10, with the most popular performer being assigned as number 1. In economic surveys, families may be categorized into poor, lower middle class, upper middle class, higher income group, etc. Thus, this grouping not only assigns particular group to a family concerned, but also it describes its position in relative to others. The variables measured with ordinal scales can be put under statistical calculations like median, percentile, rank correlation coefficient, and chi-square test.

(c) **Interval Scale**

Interval scales of measurement are further improved over an ordinal scale of measurement. In this type of measurement, scale numbers are assigned to objects or events which can be ordered like those of ordinal scale *with an added feature of equal distance between the scale values*. An interval scale can have an arbitrary zero, but it may not be possible to determine what could be called an absolute zero. In this type of scaling, for example, one can say that the increase in temperature from 40 to 50°C is the same as that of 80–90°C, but one cannot be sure that 80°C temperature is twice as warm as 40°C temperature. Incorporations of concepts of equality of intervals make the interval scale a more powerful measurement than the ordinal scale. The mean and standard deviation are the appropriate measures of central tendency and dispersion, respectively, using the interval scale of measurement.

(d) **Ratio Scale**

Demerit of an interval scale of having no absolute zero point of measurement is being overcome in a ratio scale. A ratio scale is also a type of an interval scale with an equal interval between the consecutive scales along with the added feature of having the true zero point on the scale. Thus, the presence of zero point on the scale benefits the scale in comparing two events, with their respective positions. The ratio scales have wider acceptability and use. Generally, almost all statistical tools are usable with the variables measured in a ratio scale.

5.2.1 Causes for Error in Measurement

For any research project to have meaningful research findings, the measurements taken during the process should be precise and accurate. Any failure or shortfall on this aspect may seriously affect the research findings. As such, every researcher should have an attempt to minimize the error of measurement. Mainly there are four sources of errors in measurement: (a) the respondent, (b) the situation, (c) the measurer, and (d) the instrument.

(a) **The Respondent**

Reluctance on the part a respondent may be due to the fact that either he/she has tremendous negative feelings or may have very little knowledge about the subject concerned, but very much reluctant to express his/her ignorance. In addition to this, fatigue, apathy, anxiety, etc., may cause in responding accurately.

(b) **The Situation**

The situation during the time of collection of information plays a great role in manifesting an effective research output. A situation (strained/easy) during the time of data collection, particularly for socioeconomic researches, may assist or distort information needed for a research project. Sometimes, a respondent may feel that anonymity is not assured, and then he/she may be very reluctant to provide authentic information. The presence of somebody may hinder or ease out an information extraction process.

(c) **The Measurer**

The behavior and attitude of a measurer or a surveyor, and interviewer, and the instrument used is the most important thing in getting reliable and valid information. Particularly in social studies, the style of interactions of interviewers with the respondents may encourage or discourage the information gathering process. Faulty recording or tabulation of information may also hamper the authenticity and creditability of the information.

(d) **The Instrument**

Faulty or an unstandardized instrument may give rise to inaccurate information. Similarly, in social and market studies, complex languages beyond the comprehension of a respondent must be avoided. A researcher must know how to take a measurement correctly under the given resource and time situations. Whatever may be the planning and hypothesis, if information is collected not up to the standard, it has some adverse effects on the output of a research program.

5.2.2 Criteria for Good Measurement Scale

Validity, reliability, and practicality are the three major criteria one should consider while evaluating any measurement tool. Validity refers to the degree to which the instrument measures what is supposed to be measured. Validity specifies the utility of a measurement scale; it actually measures the true differences among the values measured. Depending upon the nature of a research problem and the judgment of a researcher, validity can take a great role.

In literature one can find three types of validity: (a) *content validity*, (b) *criteria validity*, and (c) *construct validity*. Content validity is related to a measure of adequate coverage of a topic under study. The coverage of a study is one of the most important criteria in evaluating a research project. On the other hand, criteria validity relates to an ability to predict or estimate the existence of some current conditions. It is mainly concerned with the fact of the power of measuring of some other empirical studies. However, a research work is able to predict the correlation, and other theoretical propositions are measured through construct validity. Construct validity compares the result of a particular research work with other works. Test for reliability refers to consistency when measurements are taken repeatedly. Reliability doesn't contribute to validity; similarly, a reliable instrument may not be a valid instrument. Reliability is measured in terms of stability and equivalence aspects. Stability refers to the consistency of results under repeated measurements, while equivalence aspects consider the amount of error that may be introduced by the different investigators or different samples of the items being studied. The test of practicability is measured in terms of economy, convenience, and interpretability. Economy refers to a trade-off between the budget allocated for a particular project and the budget for an ideal project. While convenience suggests the ease of handling an instrument or measurement, the interpretability consideration is supported or supplemented by detailed interaction for administering the test. The key scoring points and evidence about the reliability are the guide for interpreting results.

5.2.3 Stages of Techniques for Developing Measurement Tools

The measurement tools can be developed by following systematic stages of development. By and large, four stages of development in measurement tools can be identified:

(a) Stage one: development of concept
(b) Stage two: specifying the dimensions of the concept developed
(c) Stage three: indicator selection
(d) Stage four: indexing

The concept of development relates to understanding the conceptual framework of a study to be taken up by researchers. In applied research, the conceptual frameworks are mostly obtained as a result of previous theoretical works. As such conceptualization is a concern of mainly theoretical research study. Having a fixed conceptual part of a study, it is now the work of a researcher to find out the periphery of a research work, precisely to find out the dimensions of a research study. For example, in a market study, the dimensions may be the reputation of the manufacturers, the quality of the product, the acceptance of the product to the customers, the leadership of the manufacturing organizations, their social responsibilities, etc. Fixing the concepts and dimensions of a research study leads in finding out the measurement criteria. Indicators for measuring the concepts and dimensions identified in the previous stages must be worked out. Indicators are the means of measuring the knowledge, opinion, expectations, satisfaction, etc., of the respondents. In doing so, a researcher must have alternative options for judging the stability and verification of the responses received. Generally, the studies are related to multiple dimensions; as such, it is now essential to have a particular numerical figure taking the various indicators into consideration. The simplest way of getting an index is to sum up the scores of each and every response. Thus, the formation of index plays a vital role in bringing the responses of the different characteristics of a particular study from a particular respondent into numerical values by which respondents can be compared. It is better to have indicators measured in terms of unit-less measures so that the ultimate index is also a unit-less quantity.

5.3 Scaling and Its Meaning

Quantitative or measurable indicators have the privilege of being considered as mathematical manipulations or treatments. But most of the socioeconomic, psychometric studies are being measured through subjective or abstract indicators. So, it is very difficult to aggregate the measurements taken on abstract concepts to bring to one into unison. Moreover, while measuring the attitudes or opinions, the problem of valid measurement comes into play. Scaling technique has come to rescue such problem. Scaling is a technique by which one attempts to determine quantitative measures for subjective or abstract concepts. Thus, by scaling, we mean assigning numbers to various subjective/abstract concepts like degree of opinion and attitude. The whole process of scaling is based on one or more of the following bases: (a) *subject orientations*, (b) *response form*, (c) *degree of subjectivity*, (d) *scale properties*, (e) *dimensionality*, and (f) *scale construction technique*. In a subject orientation, a scale is designed to measure the characteristics of respondents on a particular stimulus presented to the respondents. This can be done by allowing the respondent to go through the stimulus thoroughly and grade it, or it can also be done by asking the respondent to judge the stimulus on various dimensions. In a response form, responses are categorized into different scales: they may be a rating scale or a ranking scale. In a rating scale, the respondent rates a stimulus without having any direct reference to other stimuli; comparison among stimuli is absent. On the other hand, in a ranking scale, different stimuli are ranked 1, 2, 3. . ., etc., by the respondent. Scales may be based on subjectivity (preferential or non-preferential) of the respondent concerned. As discussed in the scales of measurement, scales may be developed on the basis of the nominal, ordinal, interval, or ratio nature. *A ratio scale is the most important scale which possesses all the properties of the other three scales, along with a unique origin*. Scales may be developed as unidimensional or multidimensional. In a unidimensional scale, only one attribute could be measured, but in multidimensional scales, objects maybe categorized/indexed based on multiple attributes—multivariate analysis.

5.3.1 Scaling Technique

As it has already been mentioned, rating and ranking scaling techniques are mostly used in social and business studies. In a rating scale technique, responses about the characteristics of a particular object are judged without any reference to other similar objects. The objects may be judged good, fair, better, best, etc., or average, below average, above average, and so on. The problem with this type of scaling is that nobody knows how far the two objects differ with a particular characteristic rated as fair and better or better and best. Another interesting feature of this type of scaling is the questions, how many points are to be provided in a scaling technique? Is it a two-point, three-point, four-point,. . ., scaling? As we go on increasing the number of points in a scaling technique, sensitivity increases but more and more difficulty arises in differentiating the object of two consecutive rates; compromising three- to seven-point scales is generally used for all practical purposes. Rating scales are again subdivided into graphic rating scale or itemized rating score (numeric scale). In a graphic rating scale, the respondents are allowed to select one out of the different alternatives. On the other hand, in an itemized rating scale, respondents come across a with series of statements relating to the feature of the objects and selects one which they find suitable or best reflect the characteristics of the objects. Thus, in the previous case, there might be a difference in the level of understanding of the respondents and the researcher who have developed different points in graphic rating scales. On the other hand, the merit of an itemized rating scale is that more and more information is provided to the respondent, thereby increasing his/her confidence on the features of the objects and ultimately judging these objects efficiently.

Ranking scales are mostly based on comparative basis. In this type of scaling technique, respondents are required to compare either pairwise comparison taking two objects at a time, ranking several objects in order, or placing the objects into different groups developed following interval scales. Based on the methods used, the ranking scales are used in three approaches: (a) *method of pair comparison*, (b) *method of rank order*, and (c) *method of successive intervals*. In method of pair comparison, a pairwise comparison is taken up. As a result, if there are N numbers of objects to be compared, then one needs to compare ${}^{N}C_2$ pairs. If N is very large, then the number of combinations will also be very large. On the other hand, in method of rank order, a respondent orders or ranks all objects according to the specified criteria into a, b, c ... or 1, 2, 3 .... In many situations, respondents are asked to rank a particular object or to rank the first three and so on. Thus, the need of ranking all the objects under consideration doesn't arise at all. The problem with this type of method is that the carelessness or too much sensitivity on the part of the respondent may lead to bad information or time-consuming affairs, respectively. When there are a large number of objects for which the selections are to be made, it becomes very difficult to use the previous two methods. Instead, one can use the method of successive intervals. The large number of objects is sorted to different groups formed by taking interval scales.

5.3.2 Errors in Scaling Techniques

If a respondent is not very careful while rating or ranking, errors may occur. Mainly there could be three types of error: *error of leniency, error of central tendency*, and *error of halo effect*. The error of leniency takes place when respondents are not so serious in taking the job of rating/ranking on their own hands. On the other hand, the error of central tendency occurs when respondents are very much reluctant to provide the highest/lowest rate/rank. Thus, in grading students, a teacher may be reluctant in judging the answer script thoroughly but put the grade leniently. Contrarily, if a teacher never wishes to provide the highest rank or the lowest rank but rather puts the average rank, then errors of central tendency occur. Errors of systematic bias or halo effects occur when a respondent carries or has preconceived ideas about the objects under rating. In most of the cases, the raters carry a generalized impression from one subject to another.

6 Sampling Design

Collecting quality information/data is a prerequisite of any research program. And there are mainly two different methods of collecting data: (a) *census method* and (b) *sample survey method*. In census method, data are collected from individual members of the population. On the other hand, in sample survey method, information is collected from the units of sample drawn for the purpose following definite methods of the sampling techniques. A comparative account of two methods of data collection is given in Table 6.1.

Research is pursuit towards knowing the unknowns in this universe. In doing so, it is not possible to take care of all the aspects; rather the researchers quite often select only a few aspects from this universe as the purpose of study. In most of the research studies, the usual approach is to generalize the findings, that is, to draw inference about the population characteristics based on sample characteristics. The whole process depends heavily on the success of drawing sample from the population. If the sample(s) drawn is not a proper representative part of the population, then it may lead to wrong decisions with regard to generalization of the findings. To avoid this problem, we may think of checking the whole population, that is, each and every member of the population. This is simply not possible, mainly because of time, labor, cost involvement, and other difficulties. Sometimes, it is not possible also to identify each and every member of the population (infinite population). Thus, we need to have a proper sample following a statistical technique so as to obtain valid inference about population characteristics based on sample observations and avoid taking any wrong decision.

Sampling theory can be visualized as consisting of three major components: (a) *selection of proper sample*, (b) *collection of information from the sample*, and (c) *analysis of information to draw inferences about the population as a whole*. Before discussing the above three components, let us have a look into the definitions and characteristics of some of the terminologies related to this aspect.

Population: A population is a collection or totality of well-defined objects (entities). The observations or entities could be anything like persons, plants, animals, and objects (like nut bolts, books, pens, pencils, medicines, and engines). An individual member of the population is known as *element or unit* of the population.

Population size (N) is generally referred to as the number of observations in the population. Depending upon the size of the population, a population may be finite or infinite. *A finite population* is a population that contains a finite number of observations, for example, the population of students in a particular university and population of books/journals in a library.

An *infinite population* is a population that contains an infinite number of observations or units or elements, for example, the number of hairs in a particular head and the number of plants in a forest.

P.K. Sahu, *Research Methodology: A Guide for Researchers in Agricultural Science, Social Science and Other Related Fields*, DOI 10.1007/978-81-322-1020-7_6, © Springer India 2013

Table 6.1 Comparison of sample survey method and census method of data collection

Sl no.	Sample survey method	Census method
1	Only a representative part of the population (sample) comes under investigation	Every element of the population comes under investigation
2	Comparatively less accurate, if not done properly	Accurate
3	Economical	Costly
4	Less time- and labor-consuming	Time- and labor-consuming
5	Possible in case of infinite population	Not possible for infinite population
6	Possible for large population	Rarely possible and feasible for large population
7	With both sampling and non-sampling errors	With absent sampling errors
8	Nonresponse errors can be solved	Nonresponse problem cannot be worked out
9	Parameters are to be estimated and tested	Parameters are directly worked out
10	Used frequently	Rarely used (e.g., population census)

Parameter is a real valued function of the population values. For example,

$$\text{Population mean} = \bar{Y} = \frac{1}{N}\sum_{i=1}^{N} Y_i,$$

$$\text{Population variance} = \sigma_Y^2 = \frac{1}{N}\sum_{i=1}^{N} (Y_i - \bar{Y})^2;$$

$$\text{Population coefficient of variation} = C_Y = \frac{\sigma_Y}{Y}.$$

Sample: A sample is a representative part of a population. If the sample fails to represent the population adequately, then there is every chance of drawing wrong inference about the population based on such sample because of the fact that it will overestimate or underestimate some population characteristics. Let us suppose that we want to know the average economic background of the students of an agriculture college. If the college is a coeducation college and consists of students from different parts of the country and one draws a sample (1) of either boys or girls only, or (2) from a particular class, or (3) from the students of a particular state only, then the average economic condition obtained from the sample may fail to infer about the true average economic condition of the students of the college (the population). This type of sample is called a *biased sample*. On the other hand, an *unbiased sample* is statistically almost similar to its parent population, and thus, inference about the population based on this type of sample is more reliable and acceptable than from a biased sample.

The *Sample size* (n) of a sample is the number of elements/units with which the sample is constituted of. Sample size is determined by a number of factors, namely, (1) objective and scope of the study, (2) nature of population and sampling unit, (3) sampling technique and estimation procedure to be used, (4) structure of variability in the population, (5) structure of time and cost component, and (6) size of the population. *An efficient and optimum sample size either minimizes the mean squared error of the estimator for a fixed cost or minimizes the cost for a fixed value of mean squared error*. Fixing of optimum sample size becomes complicated when more than one parameter is to be estimated or more than one variable is under study. In fact, it is very difficult to have a fixed rule for getting the sample size. However, based on past information or information gathered through the pilot study conducted before the main study and giving due consideration to the above decisive factors, sample sizes are fixed for specific studies.

Generally a sample is regarded as *large sample* if the sample size $n \geq 30$; otherwise, *small sample*.

Krejcie and Morgan in their 1970 article "Determining Sample Size for Research Activities" (*Educational and Psychological Measurement*, #30, pp. 607–610) have used the following formula to determine sampling size provided the population is definite.

Table 6.2 Sample size corresponding to different population sizes

Population size (*N*)	Sample size (*n*)	Population size (*N*)	Sample size (*n*)	Population size (*N*)	Sample size (*n*)	Population size (*N*)	Sample size (*n*)
10	10	150	108	460	210	2,200	327
15	14	160	113	480	214	2,400	331
20	19	170	118	500	217	2,600	335
25	24	180	123	550	226	2,800	338
30	28	190	127	600	234	3,000	341
35	32	200	132	650	242	3,500	346
40	36	210	136	700	248	4,000	351
45	40	220	140	750	254	4,500	354
50	44	230	144	800	260	5,000	357
55	48	240	148	850	265	6,000	361
60	52	250	152	900	269	7,000	364
65	56	260	155	950	274	8,000	367
70	59	270	159	1,000	278	9,000	368
75	63	280	162	1,100	285	10,000	370
80	66	290	165	1,200	291	15,000	375
85	70	300	169	1,300	297	20,000	377
90	73	320	175	1,400	302	30,000	379
95	76	340	181	1,500	306	40,000	380
100	80	360	186	1,600	310	50,000	381
110	86	380	191	1,700	313	75,000	382
120	92	400	196	1,800	317	100,000	384
130	97	420	201	1,900	320		
140	103	440	205	2,000	322		

Estimation of sample size in research using Krejcie and Morgan's table is a commonly employed method (Krejcie and Morgan 1970).

$$S = \chi^2 NP(1-P)/d^2(N-1) + \chi^2 P(1-P)$$

S = the required sample size
χ^2 = the table value of chi-square for one degree of freedom at the desired confidence level
N = the population size
P = the population proportion (assumed to be .50 since this would provide the maximum sample size)
d = the degree of accuracy expressed as a proportion (.05)

Table 6.2 is given in DK Lal Das' Design of Social Research for the purpose. Based on Krejcie and Morgan's (1970) table for determining sample size, for a given population of 500, a sample size of 217 would be needed to represent a cross section of the population. However, it is important for a researcher to consider whether the sample size is adequate to provide enough accuracy to base decisions on the findings with confidence.

A *statistic* is a real valued function of the sample values. For example, the sample mean $= \bar{y} = \frac{1}{n}\sum_{i=1}^{n} y_i$, sample variance $= s'^2_y = \frac{1}{n}\sum_{i=1}^{n}(y_i - \bar{y})^2$; sample coefficient of variation $= c_y = \frac{s'_y}{\bar{y}}$.

An *estimator* is a statistic used to estimate the population parameter and is a random variable as its value differs from sample to sample, and the samples are selected with specified probability laws. The particular value, which the estimator takes for a given sample, is known as an *estimate*.

An estimator t is said to be an *unbiased estimator* of parameter θ if $E(t) = \theta$, where $E(t) = \sum_{i=1}^{M_0} t_i p_i$ and the probability of getting the ith sample is p_i, and t_i ($i = 1, 2, 3 \ldots\ldots M_0$) is the estimate, that is, the value of estimator "t" based on this sample for the parameter θ, M_0 being the total number of possible samples for the specified probability scheme. On the other hand, if $E(t) \neq \theta$, the estimator is said to be a *biased* estimator of parameter θ and the bias of t is given as $B(t) = E(t) - \theta$. The difference between estimate t_i based on the ith sample and parameter θ, that is, $(t_i - \theta)$, may be called the error of the estimate.

The mean squared error (MSE). The MSE of an estimator t of θ is $M(t) = E(t-\theta)^2 = \sum_{i=1}^{M_o} p_i(t_i - \theta)^2 = V(t) + [B(t)]^2$, where $V(t)$ (the variance of t) is defined as $E[t - E(t)]^2 = \sum_{i=1}^{M_o} p_i[t_i - E(t)]^2$.

Given two estimators t_1 and t_2 for the population parameter θ, the estimator t_1 is said to be more efficient than t_2 if $\text{MSE}(t_1) < \text{MSE}(t_2)$.

The selection of proper method to obtain a representative sample is of utmost importance. The selection of sample is oriented to answer the following major queries:

(a) To select a representative part (sample) of the population under interest
(b) To fix the size of the sample
(c) To ensure good quality of information and subsequent inference
(d) To estimate parameters, minimizing errors towards valid inferences about the population

Once the purpose and objective of the study is fixed, one has to prepare a suitable sampling plan to fulfill the objective of the study. The *basic components of sampling plans* are:

(a) Definition of population and sampling units
(b) Preparation of sampling frame
(c) Scope of study area or domain
(d) Time period allowed
(e) Amount of cost permissible
(f) Coverage, that is, type of information (qualitative or quantitative) to be collected
(g) Type of parameters to be estimated
(h) Sampling design
(i) Selection of sample and collection of data through trained investigators

All the above steps aim at reducing the sampling error at a given cost within limited resources.

6.1 Errors in Sample Survey

There are mainly two types of errors associated with estimates worked out from the sample: (a) sampling error and (b) non-sampling error.

(a) *Sampling Error*: If we use different samples, drawn following exactly the same way from the same population, it will be found that the estimates from each sample differ from the other even if the same questionnaires, instructions, and facilities are provided for the selection of all the samples. This difference is termed as sampling error.
(b) *Non-sampling Error*: Non-sampling errors are attributed mainly to differential behavior of respondents as well as interviewers. Thus, difference in response, difficulties in defining, difference in interpretations, and inability in recalling information and so on are the major sources of non-sampling errors.

6.2 Selection of Sample (Sampling Technique)

Depending upon the nature and scope of the investigation and the situations under which the study is being carried out, appropriate sampling technique is being chosen. Available sampling techniques can broadly be categorized into two categories: (a) *probability sampling* and (b) *non-probability sampling*. Probability sampling helps researchers to estimate the extent to which the sample statistics are likely to adhere to the population parameters. Different types of sampling techniques available are:

Probability sampling

1) Simple random sampling
2) Varying probability sampling
3) Stratified sampling
4) Systematic sampling
5) Cluster sampling
6) Multistage sampling
7) Multiphase and double sampling
8) Sampling on two occasions
9) Inverse sampling
10) Sampling technique for rapid assessment (STRA)

Non-probability sampling

1) Quota sampling
2) Judgment sampling
3) Purposive sampling
4) Snowball sampling
5) Accidental sampling

Besides the above, some complex and mixed sampling techniques like (a) two-stage or three-stage sampling with stratification, (b) double sampling for stratification, and (c) sampling on successive occasions are useful in studies related to socioeconomic, agronomic, and animal husbandry aspects.

6.2.1 Probability Sampling

6.2.1.1 Simple Random Sampling

Simple random sampling is the most widely used, simplest method of drawing sample from a population such that each and every unit in the population has an equal probability of being included in the sample. Simple random sampling is of two different types: (a) *simple random sampling with replacement* (*SRSWR*) and (b) *simple random sampling without replacement* (*SRSWOR*).

From a population of N units, we select one unit by giving equal probability $1/N$ to all units with the help of random numbers. A unit is selected, noted, and returned to the population before drawing the second unit, and the process is repeated "n" times to get a simple random sample of "n" units. This procedure of selecting a sample is known as "simple random sampling with replacement (SRSWR)." On the other hand, if the above procedure is continued till "n" *distinct* units are selected ignoring all repetitions, a "simple random sampling without replacement (SRSWOR)" is obtained.

How to use random number from a random number table for drawing a random sample from a finite population is illustrated in the following example.

Example 6.1. Let 48 different lots of bulbs, each containing 100 bulbs, be studied for defective bulbs, and the following figures give the number of defective bulbs in each lot. We are to draw two random samples of 10 lots: (1) with replacement and (2) without replacement.

Lot no.	1	2	3	4	5	6	7	8	9	10	11	12
No. of defective bulbs/lot	2	12	10	8	9	11	3	8	10	12	7	9
Lot no.	13	14	15	16	17	18	19	20	21	22	23	24
No. of defective bulbs/lot	4	5	7	12	14	8	10	0	5	6	10	12
Lot no.	25	26	27	28	29	30	31	32	33	34	35	36
No. of defective bulbs/lot	14	15	2	5	15	17	8	9	12	13	14	5
Lot no.	37	38	39	40	41	42	43	44	45	46	47	48
No. of defective bulbs/lot	6	7	8	12	11	9	12	15	18	16	9	10

The sampling units, that is, lots, are already serially numbered from 1 to 48. Now $N = 48$ is a two-digit number.

Method 1 (Direct Approach)

Since $N = 48$ is a two-digit number, we shall use a two-digit random number table, consider random numbers from 01 to 48, and reject numbers greater than 48 and 00. One can start from any point (number) in the random number table, which is arranged in rows and columns. We can move in any random way we like; it can be vertically downward or upward, to the right or to the left. Let us start at random from a number in the 1st row and 12th column and move vertically downward. The numbers selected from the random number table 6.7 are given in Table 6.3:

The random samples of size 10 with replacement and without replacement consist of the unit numbers 12, 4, 36, 36, 32, 18, 11, 45, 15, and 32 and 12, 4, 36, 32, 18, 11, 45, 15, 27, and 33, respectively. While selecting the random numbers according to the SRSWR, we have kept some random numbers, namely, 36 and 32, more than once because these units after selection were returned to the population. But no repetition of random number is found in SRSWOR method.

Demerit of the direct approach is that a large number of random numbers are rejected simply because these are more than the population size.

Table 6.3 Selection of random numbers for SRSWR and SRSWOR scheme using direct method

Random numbers taken from the table 6.7	Selected random numbers	
	SRSWR	SRSWOR
12	12	12
4	04	4
36	36	36
80	–	–
36	36	–
32	32	32
95	—	—
63	–	–
78	—	—
18	18	18
94	—	—
11	11	11
87	—	—
45	45	45
15	15	15
32	32	–
71	—	—
77	–	–
55	–	–
95	–	—
27	–	27
33	–	33

— means random number drawn from random number table is more than 48.

Method 2 (Remainder Approach)

To reduce time and labor, the commonly used "remainder approach" is employed to avoid the rejection of random numbers. The greatest two-digit number, which is a multiple of 48, is 96, and we consider two-digit numbers from 01 to 96, rejecting the numbers greater than 96 and 00. By using two-digit random numbers as above, we prepare Table 6.4.

The random samples of size 10 with replacement and without replacement consist of the variety numbers 12, 4, 36, 32, 36, 32, 47, 15, 30, and 18 and 12, 4, 36, 32, 47, 15, 30, 18, 46, and 11, respectively.

In the direct approach, as many as 22 random numbers were selected to obtain a random sample of ten only. On the contrary, only 12 random numbers were sufficient in the second "remainder approach" to draw the sample of 10. Thus, with the help of the remainder approach, one can save time as well as labor in drawing a definite simple random sample.

Table 6.4 Selection of random numbers for SRSWR and SRSWOR scheme using remainder approach

Random numbers	Remainder when divided by 48	Selected random numbers	
		SRSWR	SRSWOR
12	12	12	12
4	04	04	4
36	36	36	36
80	32	32	32
36	36	36	–
32	32	32	–
95	47	47	47
63	15	15	15
78	30	30	30
18	18	18	18
94	46	–	46
11	11	–	11

If a sample of n units is drawn from a population of N units with SRSWR, then there are N^n ordered possible samples and the probability of getting any sample is $P(s) = 1/N^n$. If a sample is drawn with SRSWOR, then there are $({}^NC_n)$ unordered possible samples and $P(s) = 1/({}^NC_n)$. It is further to be noted that the probability of getting i^{th} unit at r^{th} draw is $P(\text{U}_{i_r}) = 1/N,\ i = 1,\ 2,\ 3,\ \ldots,\ N$ and $r = 1, 2, 3, \ldots, n$ for both SRSWR and SRSWOR.

In case a sample of n units is drawn from a population of N units with SRSWR and the sample values are $(y_1, y_2, y_3, \ldots, y_n)$, then the sample mean $\bar{y} = \frac{1}{n}\sum_{i=1}^{n} y_i$ is an unbiased estimator of the population mean $\bar{Y}$, that is, $E(\bar{y}) = \bar{Y}$, and a sampling variance of the sample mean, that is, $V(\bar{y})$ is equal to $\frac{\sigma^2}{n}$ where σ^2 is the population variance. It is also observed that the sample mean square (s^2) is an unbiased estimator of the population variance, that is $E(s^2) = \sigma^2$ where $s^2 = \frac{1}{n-1}\sum_{i=1}^{n}(y_i - \bar{y})^2$. Thus, the standard error of the sample mean is given by $\text{SE}(\bar{y}) = \frac{\sigma}{\sqrt{n}}$ and the estimated $\text{SE}(\bar{y}) = \frac{s}{\sqrt{n}}$.

$$\text{For SRSWOR } E(\bar{y}) = \bar{Y},\ V(\bar{y}) = \frac{N-n}{N-1}\frac{\sigma^2}{n}$$

$$= \frac{N-n}{N}\frac{S^2}{n}$$

$$= (1-f)\frac{S^2}{n}, E\left(s^2\right) = S^2$$

where $S^2 = \frac{1}{N-1}\sum_{i=1}^{n}(Y_i - \bar{Y})^2$ is the population mean square and $f = \frac{n}{N}$ = sampling fraction.

The factor $\frac{N-n}{N} = (1-f)$ is a correction factor for the finite size of the population and is called the finite population correction factor. It is noted that $V_{WOR}(\bar{y}) = \frac{N-n}{N-1}\frac{\sigma^2}{n} = \left(1 - \frac{n-1}{N-1}\right) \times \frac{\sigma^2}{n} \to \frac{\sigma^2}{n}$ which is equal to $V(\bar{y})$, when $N \to \infty$.

6.2.1.2 Varying Probability Sampling (Probability Proportional to Size Sampling)

In many practical situations, the units vary in size and the variable under study is directly related to the size of the unit. In this case, probabilities may be assigned to all the units proportional to their sizes and be employed probability proportional to size sampling technique. Let there be "N" units in a population having sizes $X_1, X_2, X_3, \ldots, X_N$, the total of the sizes of all the units in the population being $\sum_{i=1}^{n} X_i = X$. In this method, the probability of selecting ith unit is X_i/X. The selection of sample according to probability proportional to size sampling with replacement can be made in two different ways: (a) *cumulative total method* and (b) *Lahiri's method*.

Cumulative Total Method

Let the size of the ith unit in the population be X_i ($i = 1, 2, 3, \ldots, N$), and suppose that these are integers. Then the procedure is given in the following steps:

1. Assign 1 to X_1 numbers to the first unit, X_1+1 to (X_1+X_2) numbers to the second unit, X_1 X_2+1 to X_1+X_2+X_3 numbers to the third unit, and so on.
2. Select a random number from 1 to X from the random number table using the method for selection of sample according to simple random sampling technique.
3. The unit in whose range the random number falls is taken in the sample.
4. Steps in (2) and (3) are repeated "n" times to get a sample of size "n" with probability proportional to size with replacement.

Table 6.5 Selection of sample using cumulative total method

Farm no.	Farm area (cents) (X_i)	Cumulative total	Numbers associated	PPSWR
1	20	20	1–20	
2	15	35	21–35	
3	10	45	36–45	
4	30	75	46–75	
5	60	135	76–135	√
6	100	235	136–235	
7	40	275	236–275	
8	25	300	276–300	
9	35	335	301–335	
10	45	380	336–380	
11	50	430	381–430	
12	55	485	431–485	
13	20	505	486–505	
14	10	515	506–515	
15	70	585	516–585	√
16	75	660	586–660	
17	90	750	661–750	√
18	65	815	751–815	
19	40	855	816–855	√
20	45	900	856–900	

Example 6.2. Suppose there are 20 farms. We are to select 5 farms with probability proportional to size with replacement.

Solution. With size being the area of farms by cumulative total method, we have Table 6.5 prepared:

Select five random numbers from 1 to 900 from the random number table. Random numbers selected from the random number table at the 1st row and 9th column and moving vertically downward are 843, 122, 735, 82, and 559. The units associated with these numbers are 19th, 5th, 17th, 5th, and 15th, respectively. Thus, according to PPS, with replacement the sample contains farms with serial numbers 5, 5, 15, 17, and 19.

Lahiri's Method

According to Lahiri's (1951) method of PPS selection, we do not require accumulation of the sizes at all. It consists in selecting a number at random from 1 to N and noting down the unit with the corresponding serial number provisionally.

Table 6.6 Selection of sample using PPSWR (Lahiri's method)

Sl. no. of farms	Farm area (cent) (X_i)	Farm selected
1	76	
2	50	√
3	30	
4	26	
5	40	
6	50	√
7	34	
8	27	√
9	60	
10	90	√

Another random number is then chosen from 1 to M, where M is the maximum of the sizes of the N units or some convenient number greater than the maximum size. If the second random number is smaller or equal to the size of the unit provisionally selected, the unit is selected into the sample. If not, the entire procedure is repeated until a unit is finally selected. For selecting a sample of n units with PPSWR, the above procedure is repeated till "n" units are selected.

Example 6.3. Let there be ten farms having farm sizes of X_1 to X_{10} as given below. We are to select four farms with PPSWR according to Lahiri's method (Table 6.6).

In this example, $N = 10$, $M = 90$. First, we have to select a random number from 1 to 10 and a second random number from 1 to 90. Referring to the random number table, the pair is (8, 23). Here, $23 < X_8 = 27$. Hence, the 8th unit is selected in the sample. We choose another pair, which is (2, 48). Here, $48 < X_2 = 50$. Thus, the 2nd unit is selected in the sample. We choose another pair, which is (7, 55). Here, $55 > X_7 = 34$. So the pair (7, 55) is rejected. Choosing two pairs, we can have (10, 70) and (6, 38). Here, the 10th and 6th units are selected in the sample. Hence, the sample consists of farms with serial numbers 8, 2, 10, and 6.

PPS sampling takes care of the heterogeneity in the population as well as the varying sizes of the population units/elements. Thus, it uses ancillary information in the form of size of units. This will help us in getting more efficient estimator of the population.

A random number can also be generated from the function menu of MS Excel following the method given below. Assume that we want to have 20 random numbers between 1 and 900.

It may be noted that in this process, the random number changes as we change the

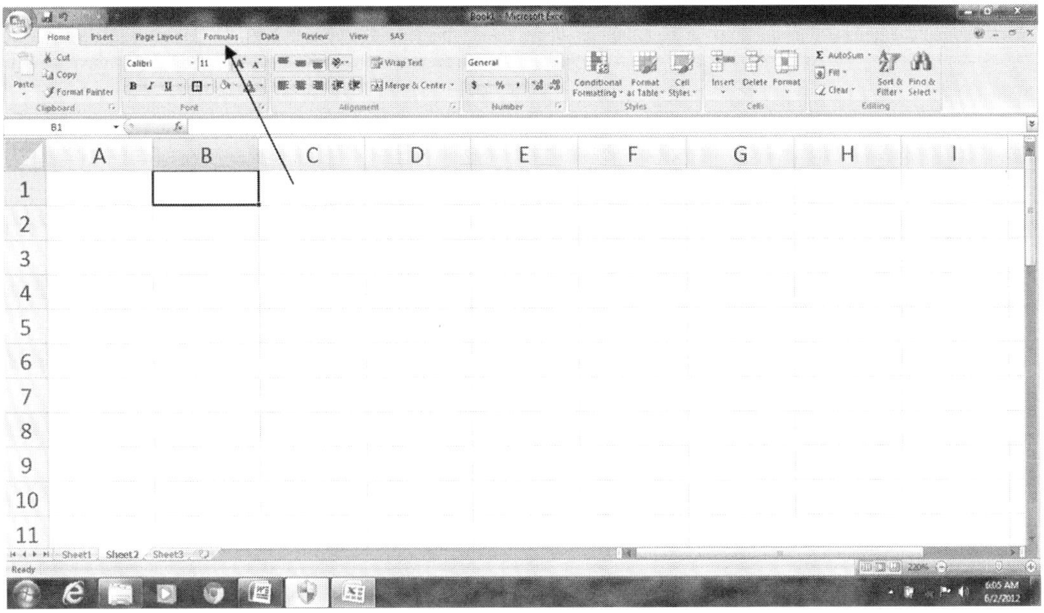

Step 1: Click on Formulas as shown above in any Excel spreadsheet to get the following screen.

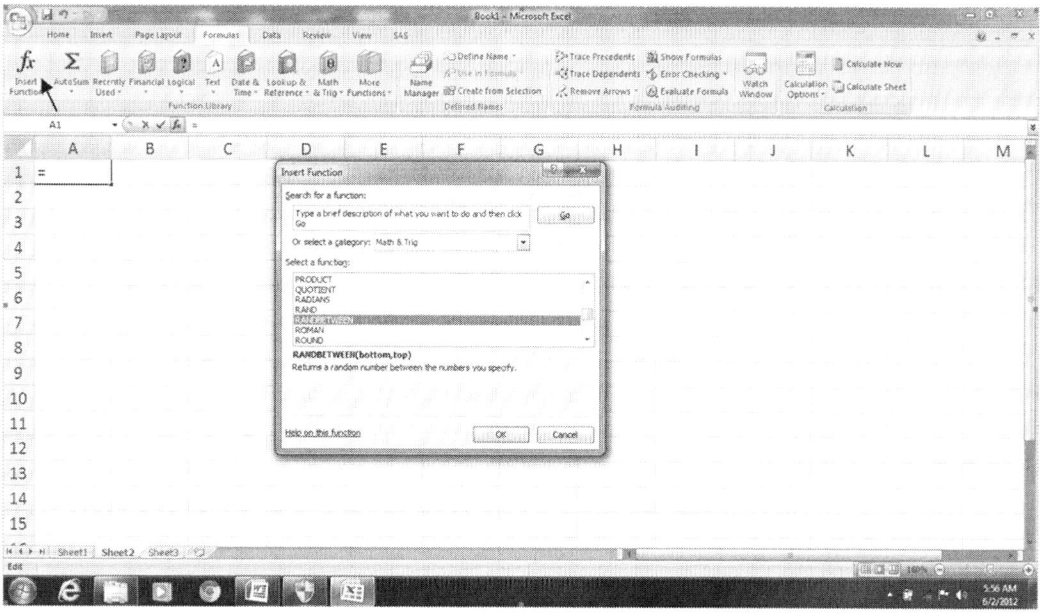

Step 2: Click on Insert Function menu as shown above. The small window will appear, and from the different options, scroll to select RANDBETWEEN and click OK to get the following window.

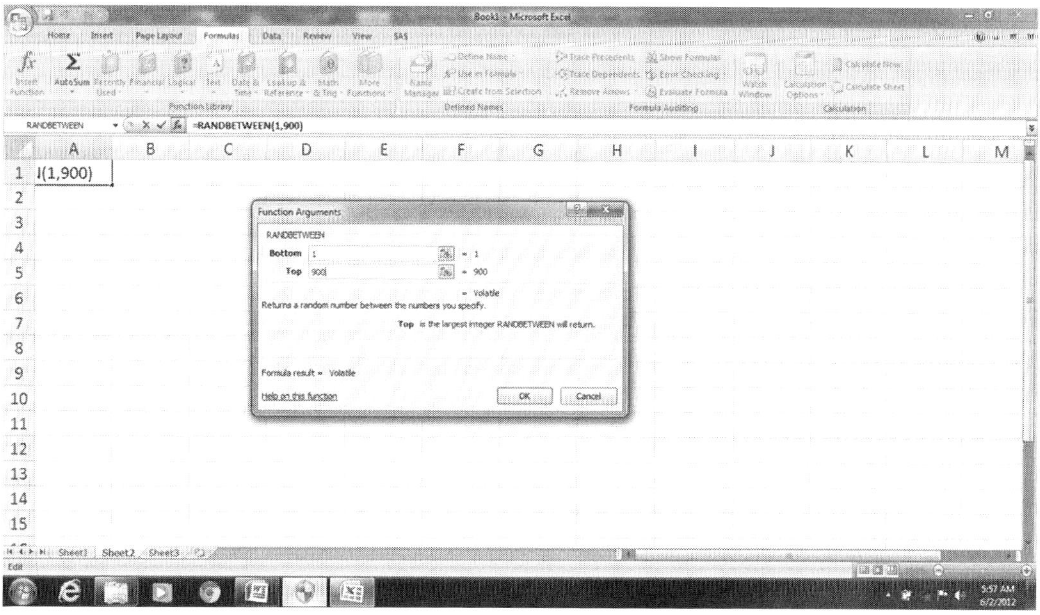

Step 3: Put the lowest and the highest values in bottom and top spaces, respectively, of the above window and click OK to get the following screen.

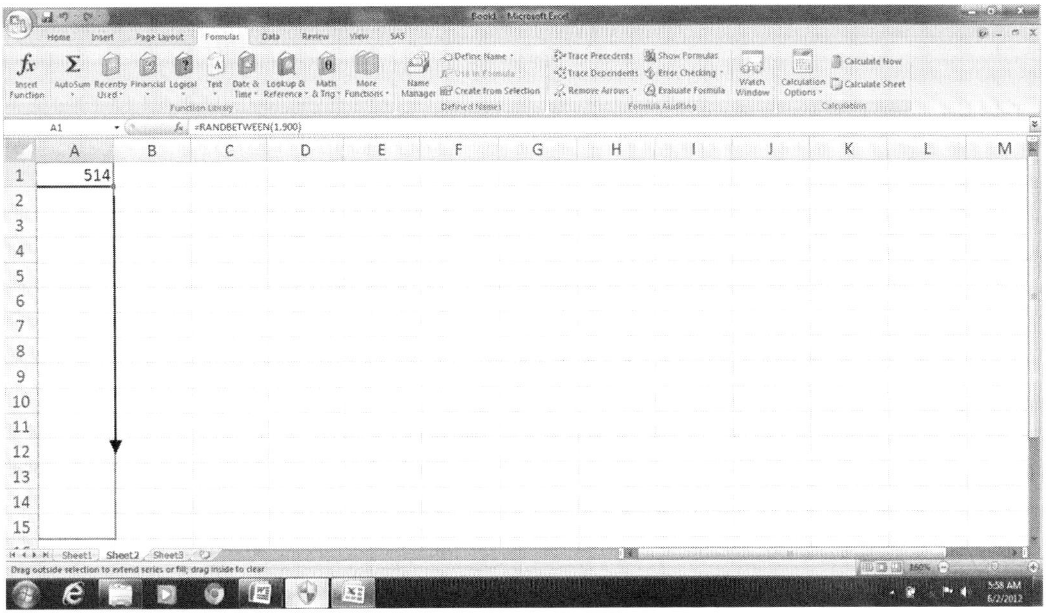

Step 4: The first random number is generated. Click on the right lower corner of the first cell, hold it, and drag it to more than 20 cells to get random numbers as given below.

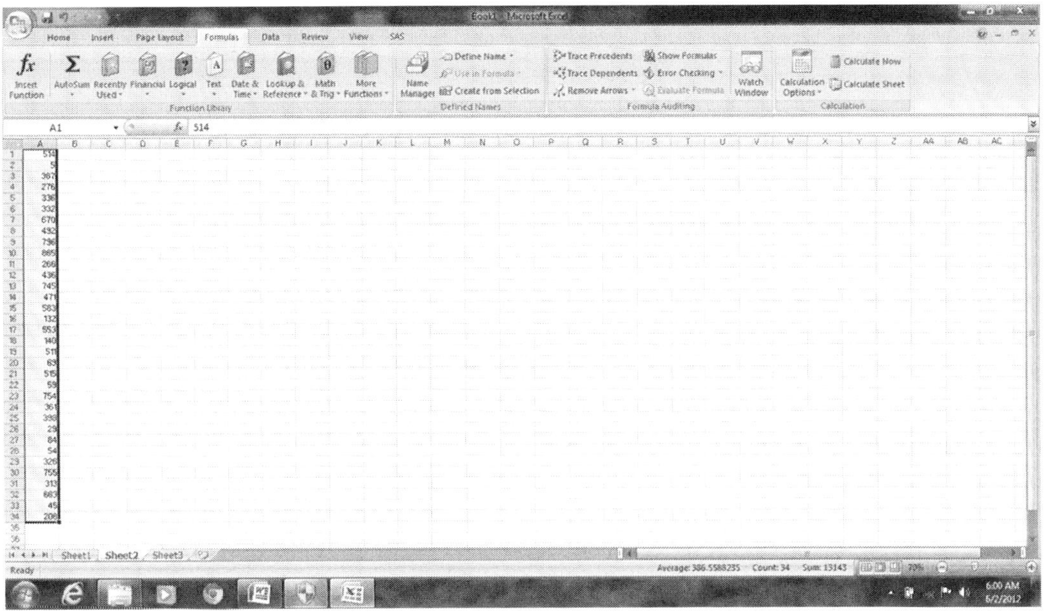

Step 5: The above random number generated may contain repetitions; repetitions are allowed in SRSWR, but to get distinct random numbers for SRSWOR, discard the repetitive ones and take only the distinct random numbers.

selection or repeat the process. So it is better to generate more random numbers than what is actually required and transform these into values using Copy, Paste Special, Values command immediately.

6.2.1.3 Stratified Sampling

For a heterogeneous population, stratified random sampling is found to be better compared to SRS and simple random sampling with varying probability.

In stratified random sampling method, a population of size "N" is divided into subpopulations (called strata), which are homogeneous within and heterogeneous among themselves. Random samples are drawn from each stratum separately. Age, gender, educational or income status, geographical location, soil fertility pattern, stress level, species of fish, etc., are generally used as stratifying factors. The efficiency of the stratified random sampling design, relative to the simple random sampling design, will be high only if an appropriate stratification technique is used.

Number of strata: Stratification can be done to the extent such that the strata are not too many in number because too many strata do not necessarily bring down the variance proportionately.

Let there be P strata in a population with mean $\bar{Y}$ and variance σ^2, and let N_h be the sizes of hth stratum with mean $\bar{Y}_h$ and variance σ_h^2 ($h = 1, 2, 3, \ldots, P$) so that $N = N_1,+N_2+\ldots\ldots+N_L$. One can write

$$\bar{Y} = \sum_{h=1}^{P} W_h \bar{Y}_h,$$

$$\sigma^2 = \sum_{h=1}^{P} W_h \sigma_h^2 + \sum_{h=1}^{P} W_h (\bar{Y}_h - \bar{Y})^2,$$

where $W_h = \frac{N_h}{N}$.

Let us take a random sample of size n by selecting n_h individuals from hth stratum such that $\sum_{h=1}^{P} n_h = n$. Let $\bar{y}_h$ and s_h^2 be the sample mean and sample mean square for the nth stratum where

$$\bar{y}_h = \frac{1}{n_h}\sum_{j=1}^{n_h} y_{hj} \text{ and } s_h^2 = \frac{1}{n_h - 1}\sum_{j=1}^{n_h}\left(y_{hj} - \bar{y}_h\right)^2.$$

Unbiased estimator for the population mean $\bar{Y}$ and the population total Y are given by

$$\hat{\bar{Y}} = \bar{y}_{st} = \sum_{h=1}^{L} W_h \bar{y}_h \text{ and } \hat{Y} = N\bar{y}_{st},$$

and their estimated variances are given by

$$\hat{V}(\bar{y}_{st}) = \sum_{h=1}^{L} W_h{}^2(1 - f_h)\frac{s_n^2}{n_h} \quad \hat{V}(N\bar{y}_{st}) = N^2\hat{V}(\bar{y}_{st}),$$

where $f_h = \frac{n_h}{N_h}$.

Allocation of Sample Size to Different Strata

Three methods of allocation of sample size to various strata in the population are given below:

1. *Equal allocation*: Total sample size is divided equally among the strata, that is, sample n_h to be selected from hth stratum such that $n_h = n/L$.
2. *Proportional allocation*: In proportional allocation, $n_h \propto$ (proportional to) N_h, that is, $N_h = nW_h$; $h = 1, 2, 3, \ldots, P$.
3. *Optimum allocation*: This is based on minimization of $V(\bar{y}_{st})$ under a given cost C'. The simplest cost function is of the form $C = C_0 + \sum_{n=1}^{P} C_h n_h$, where C_0 is the overhead cost, C_h is the cost of sampling a unit from the hth stratum, and C is the total cost. To solve this problem, we have

$$n_h = (C' - C_0)\frac{W_h S_h/\sqrt{C_h}}{\sum_{h=1}^{L} W_h S_h \sqrt{C_h}},$$

$h = 1, 2, 3, \ldots, L$, where S_h^2 is the population mean square for the hth stratum.

A particular case arises when $C_h = C''$, that is, if the cost per unit is the same in all the strata. In this case,

$$n_h = n\frac{W_h S_h}{\sum_{h=1}^{L} W_h S_h} = n\frac{N_h S_h}{\sum_{h=1}^{L} N_h S_h}.$$

This allocation is also known as *Neyman allocation*.

Stratified random sampling gives more representative sample, that is, more accuracy in sampling and efficient estimation of population parameters. However, it is more costly than simple random sampling and involves complicated analysis and estimation procedure.

6.2.1.4 Systematic Sampling

In this method of sampling, only the first unit is selected with the help of a random number, and the rest of the units of the sample get selected automatically according to some predesigned pattern. Suppose there are 500 population units numbered from 1 to 500 in some order and we want to draw a sample size of 50, We have $500 = 50 \times 10$, where 50 is the sample size and 10 is an integer; let a random number less than or equal to "10" be selected and every 10th unit thereafter; such a procedure is termed *linear systematic sampling*. If $N \neq nk$, where N = population size, n = sample size, and k = an integer, and every kth unit is included in a circular manner till the whole list is exhausted, it will be called *circular systematic sampling*. Systematic sampling is used in various surveys like in census work, forest surveys, milk yield surveys, and fisheries.

6.2.1.5 Cluster Sampling

When the population is very wide or big, say for countrywide survey, it may not be feasible to take sample units directly from the population itself. Moreover, this type of sampling is used when the population size is very large, and stratification is also not feasible to the best possible way because of nonavailability of full information on each and every element of the population. Resource constraint is also a major factor. In such cases, auxiliary/secondary information like block list, village list, and subdivision lists is used in probability sampling. In cluster sampling, the whole population is divided into a number of clusters, each consisting of several sampling units. Cluster size may definitely vary from cluster to cluster. Then some clusters are selected at random out of all the clusters. Clusters need not necessarily be natural aggregates; these may be virtual or hypothetical or artificial also, like grids in the map. The best size of cluster depends on the cost of collecting information from the clusters and the resulting variance. The objective is to reduce both the cost and variance, and for that we can have a pilot survey also, if felt necessary.

Cluster sampling is useful where listing of population units is not available; for example, in a crop survey, the list of plots may not be available, but the list of villages may be available. Here, villages will be treated as clusters similarly. In animal husbandry, the list of cattle may not be available, but the list of rearers may be available. In such case, rearers will be considered as clusters.

A population of NM units is divided into N clusters having M units each. Let Y_{ij} be the value of the character y under study for jth observation corresponding to ith cluster ($i = 1, 2, 3, \ldots, N$ and $j = 1, 2, 3, \ldots, M$). The population mean $\bar{Y}$ is defined as

$$\bar{Y} = \frac{1}{NM}\sum_{i}^{N}\sum_{j}^{M} Y_{ij} = \frac{1}{N}\sum_{i=1}^{N}\bar{Y}_i,$$

where $\bar{Y}_i$ is the ith cluster mean. A sample of n clusters is drawn with SRSWOR, and all the units in the selected clusters are surveyed. An unbiased estimator of the population mean $\bar{Y}$ is given by

$$\hat{\bar{Y}}_c = \frac{1}{n}\sum_{i=1}^{n}\bar{y}_i,$$

and its estimated variance is

$$\hat{V}\left(\hat{\bar{Y}}_c\right) = \frac{N-n}{N}\frac{s_b^2}{n},$$

where $\bar{y}_i = \frac{1}{M}\sum_{j=1}^{M} y_{ij}$ = mean for the ith selected cluster and $s_b^2 = \frac{1}{n-1}\sum_{i=1}^{n}\left(\bar{y}_i - \hat{\bar{Y}}_c\right)^2$

6.2.1.6 Multistage Sampling

Multistage sampling is the extension of clustering in more than one stage. Instead of taking blocks/villages directly as clusters, one may take districts at the first stage, then blocks/villages from the selected districts and farmers from the villages, etc. Thus, districts are the first stage units (fsu) (or primary stage units (psu)), blocks/villages are the second stage units, and farmers are the third stage units or the respondents. A multistage sampling is a two-stage/three-stage, etc., sampling depending upon the number of stages or clusters. If one is interested in counting the number of grains per panicle in a wheat field experiment, a three-stage sampling with individual hills as the "fsu" and individual panicles as the "ssu" and grains in panicle as "tsu" for sampling may be used.

Multistage sampling is very flexible. The whole process depends on the expertise of the supervisor. For two-stage sampling with a population divided into N first stage units and having M second stage units in each; the population mean $\bar{Y} = \frac{1}{NM}\sum_{i}^{N}\sum_{j}^{M} Y_{ij} = \frac{1}{N}\sum_{i=1}^{N}\bar{Y}_i$. If a sample of n fsu is selected from N fsu with SRSWOR and a sample of m ssu is selected from each selected fsu with SRSWOR, then an unbiased estimator for $\bar{Y}$ is given by

$$\hat{\bar{Y}}_t = \frac{1}{n}\sum_{i=1}^{n}\bar{y}_i \text{ with its estimated variance}$$

$$\hat{V}\left(\hat{\bar{Y}}_t\right) = (1 - f_1)\frac{s_b^2}{n} + \frac{f_1(1 - f_2)}{nm}s_2^2,$$

where

$$\bar{y}_i = \frac{1}{m}\sum_{j=1}^{m} y_{ij}, \quad s_b^2 = \frac{1}{n-1}\sum_{i=1}^{n}\left(\bar{y}_i - \hat{\bar{Y}}_t\right)^2 \text{ and}$$

$$s_2^2 = \frac{1}{n(m-1)}\sum_{i=1}^{n}\sum_{j=1}^{m}\left(y_{ij} - \bar{y}_i\right)^2,\ f_1 = \frac{n}{N}, f_2 = \frac{m}{M}$$

6.2.1.7 Multiphase and Double (Two-Phase) Sampling

Sampling in two or more phases is known as two-phase or multiphase sampling. This is another way of tackling the large population. The usual procedure is to take a large sample of size "n'" from the population of N units to observe the x-values of the auxiliary character and to estimate the population parameter (say mean), while a subsample of size "n" is drawn from "n'" to study the character under consideration.

Let the problem be the selection of a sample of families with probability proportional to income. But the problem is that we do not have any information on the income of the families. We can take an initial random sample of families having varied income and collect information on their incomes; then a subsample is taken from the initial sample with probability proportional to the size of income. This will serve as the test sample for the character under study from a selected sample on the basis of family income.

Thus, in multiphase sampling, every sample is obtained from previous sample.

Multiphase sampling should not be confused with multistage sampling. In multistage sampling, the sampling units at each stage are the clusters of units of the next stage, and the ultimate observable units are selected in stages, sampling at each stage being done from each of the sampling units or clusters selected in the previous stage.

Multiphase or double sampling is very flexible and valuable for survey purpose, particularly where the auxiliary information may not be available always. But it is tedious, monotonous, and less accurate.

In many situations, the population mean $\bar{X}$ of an auxiliary character x is known. Then the ratio estimator for $\bar{Y}$ is given by $\bar{y}_R = \frac{\bar{y}}{\bar{x}}\bar{X}$, where $\bar{y}$ and $\bar{x}$ are sample means of the characters y and x, respectively, based on a sample of size n drawn with SRSWOR. It may happen that $\bar{X}$ is not known and in this case, the ratio of estimation cannot be used to estimate the population mean $\bar{Y}$. The usual procedure in such a situation is to use the technique known as two-phase or double sampling. A preliminary random sample without replacement (SRSWOR) of size n' is taken as the information on x is collected. A subsample of size n' is drawn (SRSWOR) from n' and information on y is measured. The ratio estimator for $\bar{Y}$ based on double sampling is given by

$$\bar{y}_{\text{Rd}} = \frac{\bar{y}}{\bar{x}}\bar{x}',$$

where $\bar{x}$ and $\bar{y}$ are subsample means of x and y, respectively, and $\bar{x}'$ is the sample mean in the first sample. It is noted that $\bar{y}_{\text{Rd}}$ is biased and the estimator of $V(\bar{y}_{\text{Rd}})$ to the first-order approximation is given by

$$\hat{V}(\bar{y}_{\text{Rd}}) = \left(\frac{1}{n'} - \frac{1}{N}\right)s_y^2 + \left(\frac{1}{n} - \frac{1}{n'}\right)\left(s_y^2 + \hat{R}^2 s_x^2 - 2\hat{R}s_{yx}\right)$$

where $\hat{R} = \frac{\bar{y}}{\bar{x}}, s_{yx} = \frac{1}{n-1}\sum_{i=1}^{n}(y_i - \bar{y})(x_i - \bar{x})$

6.2.1.8 Sampling in Two Occasions

In agriculture, plant characters are commonly measured at different growth stages of the crop. For example, tiller number in rice may be measured at 30, 60, 90, and 120 days after transplanting or at the tillering, flowering, and harvesting stages. If such measurements are made on the same plants at all stages of observation, the resulting data may be biased because plants that are subjected to frequent handling may behave differently from others. In this situation, if sampling on successive occasions is done according to a specific rule, with partial replacement of sampling units, it is known as *rotation sampling*.

On the first occasion, a SRSWOR sample "s" of size "n" is selected from a population of "N" units; on the second occasion, a SRSWOR sample "s_1" of "m" units from "s" and a SRSWOR sample "s_2" of "u" ($= n - m$) units from the $(N - n)$ units in the population not included in "s" are selected. The estimate for the population mean $\bar{Y}_2$ on the second occasion is given by $\hat{\bar{Y}}_2 = W\bar{y}_{2n} + (1 - W)\bar{y}'_{2m}$

where $\bar{y}_{2n} = \sum_{s_2}\frac{y_{2i}}{u}$ and $\bar{y}'_{2m} = \bar{y}_{2m} - b(\bar{y}_{1m} - \bar{y}_{1n})$ are the simple mean for $\bar{Y}_2$ based on the replaced (unmatched) portion and the regression estimator for $\bar{Y}_2$ based on the matched portion, respectively, b being the sample regression coefficient of y_2 on y_1 based on matched part of s_1, W being a constant, and $\bar{y}_{1h} = \sum_{s}\frac{y_{1i}}{n}$, $\bar{y}_{hm} = \sum_{s_1} yhi/m$, $(h = 1, 2)$

The alternatives to the above scheme are the schemes based on (1) *complete matching* (i.e., retaining the sample of the first occasion to the second occasion completely) or (2) *complete replacement* (i.e., drawing a single independent sample at the second occasion and taking the sample mean as the estimator of $\overline{Y_2}$ in both cases). The partial matching procedure is better than the other two procedures; for all practical purposes, one should retain 25–30% units at the second occasion for the gain in the efficiency.

6.2.1.9 Inverse Sampling

Inverse sampling is generally used for the estimation of a rare population and is a method of sampling in which the drawing of a sample unit continues until certain specified conditions dependent on the results of those drawings have been fulfilled, for example, until a given number of individuals of specified type have emerged. For example, if we want to draw a sample of population affected by pox disease in which at least "k" has been affected by "chicken pox," while doing so, we do not know about the exact size of the sample because unless and otherwise we have "k" number of persons affected by chicken pox, the drawing will continue and the sample size will go on increasing. Thus, though costly and time and labor- consuming, this sampling gives due weightage to rare elements in the population.

Let P denote the proportion of units in the population possessing the rare attribute under study. Therefore, NP number of units in the population will possess the rare attributes. To estimate P, units are drawn one by one with SRSWOR. Sampling is discontinued as soon as the number of units in the sample possessing the rare attribute (a predetermined number, m) is reached. Let us denote by n the number of units required to be drawn in the sample to obtain m units possessing the rare attribute. An unbiased estimator of P is given by

$$\hat{P} = \frac{m-1}{n-1}$$

and an unbiased estimator of the variance of $\hat{P}$ is

$$\hat{V}(\hat{P}) = \frac{\hat{P}(1-\hat{P})}{(n-2)}\left[1 - \frac{n-1}{N}\right].$$

6.2.1.10 Sampling Technique for Rapid Assessment (STRA)

In exigencies like flood, diseases, tsunami, and Aila, we may not get enough time, money, or manpower to undertake a usual sampling technique, but in any way, we need a reliable estimate on some population parameters like damage of crops, houses, cattle, and human lives in a very short time so that the next suitable steps may be taken. Thus, to obtain an estimate of the total area damaged due to the outbreak of a particular pest in block, one may select "n" points ("n" being small, say, $n = 15$) haphazardly at different locations and calculate the mean intensity of outbreak from a rectangular area of, say, A = 5×5 m^2 at these points. Let $I_1, \ldots, I_n$ be the intensities of pest infestation in these "n" pseudorandomly selected rectangles and $\bar{I} = \frac{1}{n} \sum_{i=1}^{n} I_i$ be the sample mean. Then the estimated infestation would be

$$\hat{M} = \left(\frac{A}{a}\right)\bar{I} \qquad (7.1)$$

where A is the area (in m^2) under the block.

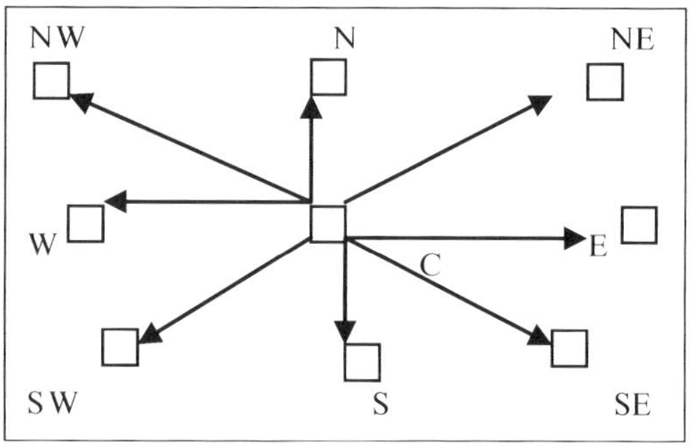

To obtain the rapid estimate of the intensity of a disease in a crop field, one may locate "n" subplots (say $n = 9$ one each in every direction and one at center) in a pseudorandom manner, each plot being a square of the same area, say, "a." The square may be of side 1 or 2 m. Let p_i be the damaged proportion of the crop in the Ith plot, $I = 1, \ldots, n$. Then $\bar{p} = \left(\frac{1}{n}\right)\sum_{i=1}^{n} p_i$ would be the required estimate. The total area damaged in the field would be estimated as $\hat{A}_p = \left(\frac{A}{a}\right)\bar{p}$

where A is the area of the field. The estimated variance would be

$$v(\hat{A}_p) = \left(\frac{A}{a}\right)^2 v(\bar{p}).$$

The sampling techniques for rapid assessment (STRA) may also be used for obtaining estimates of the yields of various crops.

6.2.2 Non-probability Sampling

1. *Quota Sampling*: Each interviewer is assigned to interview a fixed number of respondents (quota) that are taken as representation of the whole sample.
   If these are different groups in a population and a sample is drawn in such a way that a fixed number of units from different groups required to be included in that sample, say, $n_1, n_2, \ldots, n_i$ from the ith group to provide random sample of size $n = n_1+n_2+\ldots+n_i$. But these $n_1, n_2, \ldots, n_i$ number of units are selected by the interviewer in a nonrandom fashion.
   This procedure of sampling is less costly, is convenient, does not require any sampling frame, and gives quick reply. But as has been mentioned earlier, it is criticized heavily because of its nonrandom selection of elements. The success of the entire procedure depends on the skill and personal efficiency of the interviewer.
2. *Judgment Sampling*: Basic to this type of sampling is to select a sample of desired size,

giving priority to the purpose of the study, and the elements of the sample are selected in such a way so as to fulfill the objective of the study. This method is the simplest but may lead to biased and inefficient sample depending upon the efficiency of the investigator.

3. *Purposive Sampling*: The selection of elements to be included in the sample is entirely made on the basis of the choice of the researcher. Though easy to operate, it gives hardly a representation of the parent population and as such the results are biased and unsatisfactory.
4. *Snowball Sampling*: In researches, particularly in the field of social studies, related to sensitive issues like drug abuse and HIV victims in which the respondents are very much hesitant because of social taboos, reasons, or otherwise, this type of sampling is very useful. In this type of study, the researcher finds out initially one or two or a few respondents for the study purpose and then takes the help of these respondents to find more and more respondents on specific aspects. This is because of the fact that this type of respondents may be hesitant to present themselves when approached by a stranger rather than a fellow friend. Thus, in this type of studies, size of sample goes on increasing as one gets more and more number of respondents.
5. *Accidental Sampling*: This type of sampling is mostly used in evaluative types of research studies. Suppose one wants to know the extent of adoption of a particular technology, say, birth control program/hybrid rice technology in a certain community. In this method of selecting the respondents, the researchers/interviewers take information from the respondents who happen to meet the researchers first, and in the process, the researchers may take information from the first 100 respondents whom the researchers/interviewers have met and who are willing to share information. Samples collected in this process are subject to biasness, and the only way to overcome the bias is to conduct parallel studies.

6.3 Execution of the Sampling Plan

A good sampling plan is an "essential" condition for the study to be undertaken, but it is not sufficient to have a good sample leading to efficient and accurate estimation of the population parameter. It is necessary to have a "good" sampling plan suitable for the study followed by the "efficient execution" of the sampling plan. A "good" sampling plan, if not executed properly, may give "bad" (unreliable, inaccurate) results, leading to wastage of entire time, energy, and money used for the investigation. For the efficient execution of the sampling plan, the investigator responsible for the data collection must possess the necessary qualifications. They must be properly trained before the data collection. They must be taught how to handle the equipment and to make correct observations and measurements and note them down carefully. A proper supervision of the fieldwork is a must, and scrutiny and editing of the collected data is essential. Precautions must be taken to identify the sample units, in specifying the units(s) of measurements at various stages, and to minimize the error in recording the data. Pilot survey may be undertaken to select the suitable sampling plan among the alternative plans. An efficient execution of sampling plan cannot only reduce both the sampling and non-sampling errors but also helps in reducing the cost of the study.

Table 6.7 Random number table

	00–04	05–09	10–14	15–19	20–24	25–29	30–34	35–39	40–44	45–49
00	54463	22662	65905	70639	79365	67382	29085	69831	47058	08186
01	15389	85205	18850	39226	42249	90669	96325	23248	60933	26927
02	85941	40756	82414	02015	13858	78030	16269	65978	01385	15345
03	61149	69440	11286	88218	58925	03638	52862	62733	33451	77455
04	05219	81619	10651	67079	92511	59888	84502	72095	83463	75577
05	41417	98326	87719	92294	46614	50948	64886	20002	97365	30976
06	28357	94070	20652	35774	16249	75019	21145	05217	47286	76305
07	17783	00015	10806	83091	91530	36466	39981	62481	49177	75779
08	40950	84820	29881	85966	62800	70326	84740	62660	77379	90279
09	82995	64157	66164	41180	10089	41757	78258	96488	88629	37231
10	96754	17675	55659	44105	47361	34833	86679	23930	53249	27083
11	34357	88040	53364	71726	45690	66334	60332	22554	90600	71113
12	06318	37403	49927	57715	50423	67372	63116	48888	21505	80182
13	62111	52820	07243	79931	89292	84767	85693	73947	22278	11551
14	47534	09243	67879	00544	23410	12740	02540	54440	32949	13491
15	98614	75993	84460	62846	59844	14922	48730	73443	48167	34770
16	24856	03648	44898	09351	98795	18644	39765	71058	90368	44104
17	96887	12479	80621	66223	86085	78285	02432	53342	42846	94771
18	90801	21472	42815	77408	37390	76766	52615	32141	30268	18106
19	55165	77312	83666	36028	28420	70219	81369	41943	47366	41067
20	75884	12952	84318	95108	72305	64620	91318	89872	45375	85436
21	16777	37116	58550	42958	21460	43910	01175	87894	81378	10620
22	46230	43877	80207	88877	89380	32992	91380	03164	98656	59337
23	42902	66892	46134	01432	94710	23474	20423	60137	60609	13119
24	81007	00333	39693	28039	10154	95425	39220	19774	31782	49037
25	68089	01122	51111	72373	06902	74373	96199	97017	41273	21546
26	20411	67081	89950	16944	93054	87687	96693	87236	77054	33848
27	58212	13160	06468	15718	82627	76999	05999	58680	96739	63700
28	70577	42866	24969	61210	76046	67699	42054	12696	93758	03283
29	94522	74358	71659	62038	79643	79169	44741	05437	39038	13163
30	42626	86819	85651	88678	17401	03252	99547	32404	17918	62880
31	16051	33763	57194	16752	54450	19031	58580	47629	54132	60631
32	08244	27647	33851	44705	94211	46716	11738	55784	95374	72655
33	59497	04392	09419	89964	51211	04894	72882	17805	21896	83864
34	97155	13428	40293	09985	58434	01412	69124	82171	59058	82859
35	98409	66162	95763	47420	20792	61527	20441	39435	11859	41567
36	45476	84882	65109	96597	25930	66790	65706	61203	53634	22557
37	89300	69700	50741	30329	11658	23166	05400	66669	48708	03887
38	50051	95137	91631	66315	91428	12275	24816	68091	71710	33258
39	31753	85178	31310	89642	98364	02306	24617	09609	83942	22716
40	79152	53829	77250	20190	56535	18760	69942	77448	33278	48805
41	44560	38750	83635	56540	64900	42915	13953	79149	18710	48805
42	68328	83378	63369	71381	39564	05615	42451	64559	97501	65747
43	46939	38689	58625	08342	30459	85863	20781	09284	26333	91777
44	83544	86141	15707	96256	23068	13782	08467	89469	93842	55349
45	91621	00881	04900	54224	46177	55309	17852	27491	89415	23466
46	91621	00881	04900	54224	46177	55309	17852	27491	89415	23466
47	55751	62515	21108	80830	02263	29303	37204	96926	30506	09808
48	85156	87689	954963	88842	00664	55017	55539	17771	69448	87530
49	07521	56898	12236	60277	39102	62315	12239	07105	11844	01117

7 Collection of Data

Research is a process of knowing the unknown things for the betterment of humanity. In the process, explorations of the different sources of information are carried out by a researcher to expose the so long unexposed truth. Thus, information is the prerequisite for achieving the objectives of any research program. Information may be qualitative or quantitative/numerical. Data refers to different kinds of numerical information. In any research program, a researcher is always in search of a suitable mechanism/process for the collection of information. Thus, data collection plays a vital role in any research program. Depending upon the research design, in particular, the objective of a research program and the types of information required are fixed. The next task is data collection. Depending upon the sources of information, data may be (1) primary or (2) secondary.

1. Primary Data
   Primary data are those data which are collected by a researcher afresh and for the first time with specific research objectives in mind. Thus, primary data are original in nature.
2. Secondary Data
   Secondary data are those data which are collected by someone, an agency, an organization, etc., but are being used by some other users. So secondary data are not collected by a user himself/herself; rather the user is using the information generated by some other users.
   Weather data is the primary data to the department of meteorology, as a part of their mandatory duties, but when this meteorological information is used by any researcher as an auxiliary or supportive information to facilitate the findings of a research program, then this data becomes secondary. Secondary information might have undergone a thorough statistical process to some extent. But primary data are raw in nature; that's why sometimes these are known as raw data, which definitely require processing. As a result, the methods of collection of these two types of data differ significantly. In case of primary data, the method for collection is required, whereas in secondary data, it requires a mere compilation.

7.1 Methods of Collection of Primary Data

As it has already been mentioned, the primary data are original in nature and collected with specific objectives in mind. Thus, the methods of the collection of primary data play a vital role in a research process. It requires an efficient planning and execution. Depending upon the type of a research design, primary data may be obtained from experimental fields or through a survey type of study. Experiment is an investigation to explore the hidden fact under the objectives. On the other hand, in a descriptive type of research, the information is collected through a sample survey technique or a census survey method from the existing area under study. Among the several methods of collection of primary data, the following methods are mostly used: (a) *observation*

P.K. Sahu, *Research Methodology: A Guide for Researchers in Agricultural Science, Social Science and Other Related Fields*, DOI 10.1007/978-81-322-1020-7_7, © Springer India 2013

method, (b) *interview method*, (c) *questionnaire method*, (d) *schedule method*, and (e) *other methods*.

7.1.1 Observation Method

In an observation method, data can be collected from structured experiments as well as from descriptive research studies. Though it is commonly used in behavioral/social sciences, it has also an ample application in experimental sciences. The power of observation in a scientific research program may lead to an icebreaking discovery. Observation is a planned, carefully and thoughtfully selected method of data collection. A scientific method of observation can result accurate findings and conclusions. In observation, a researcher can observe the elements under study even without asking anything. If observation is taken accurately, then subjective bias may also get reduced. While applying the observation method of data collection, a researcher should keep in mind (a) *what to observe*, (b) *how observations are required to be noted*, (c) *how to ensure the accuracy of the information*, and so on. Observations can be of two kinds: (a) structured *observation* and (b) *unstructured observation*. In a structured observation, the definition of units to be observed, the style of observation, the method of recording, the standardization of condition of data collection and selection of the pertinent data of observation, etc., are all settled well ahead of data collection. On the other hand, when data collection takes place without the above characteristics thought/settled in advance, then the process is known as unstructured observations. Generally the structured observation process is used in a descriptive type of studies, whereas the unstructured observations are mostly used in an exploratory type of research.

Depending upon the participation or nonparticipation of a researcher/investigator/enumerator, the observation methods of data collection can again be divided into *participant* and *nonparticipant* methods. In a participant method of observation, the observer (i.e., the researcher/the investigator/enumerator) acts as a member of a group in which he/she shares his/her experiences. Thus, in the participant method, the added advantage is that the observer can extract information which were not thought earlier but may be found suitable while discussing. Nonparticipant observation method of data collection leads the observer to record the feelings/experiences of others maybe without disclosing the identity. As such, this method is also known as *disguised observation*. In a disguised method of observation, participants are free to exchange their experiences without being cautious by the presence of an investigator. But in the process, the investigator may not be in a position to ask for any additional explanation or experience arising out of discussion. Both participant and nonparticipant observation methods are applicable in social as well as experimental studies.

Observation methods of data collection may again be grouped into *controlled* and *uncontrolled* observations. When observations are recorded in accordance with the prearranged plans of the observer, which involves experimental procedures, the observation is known as controlled observation. On the other hand, in an uncontrolled observation, the observations take place in normal natural orders, without requiring any preplan for the recording of the observations. In social sciences, particularly in the studies of human behavior, uncontrolled observations provide natural and complete behavior of the human being/society. These are mostly applicable in an exploratory type of a research study. Control observation mostly takes place in laboratory or field experiments.

7.1.2 Interview Method

This is one of the most common methods of data collection, particularly in social and behavioral sciences. Oral communication is the main theme behind such method. Different questions, sometimes called stimuli are presented, to the respondents to record their responses. Interviews can broadly be classified into three categories: *personal interview, telephonic interview, and chatting*.

7.1.2.1 Personal Interview

In a personal interview method of data collection, a researcher or an interviewer generally asks some questions and notes down the respondent's responses, while in a direct personal interview method, an interviewer collects the information directly from the respondent. This is comparatively easy, and data collection takes place then and there. But, in some cases, it may not be possible to contact directly the person concerned due to various reasons; in such cases, indirect oral examinations can be conducted and of course followed by cross-examinations from other persons who have sufficient knowledge about the problem under investigation. An indirect method of interview is mostly useful if respondents are high profile in nature or are members of the commission and committee appointed by the government for specific investigations.

A personal interview can again be *structured* and *unstructured*. As usual, a structured personal interview is concerned with the use of a set of predetermined questions with standardized technique of data recording. These are mostly useful in descriptive and experimental type of research studies. An unstructured personal interview method is characterized by a freestyle approach of an investigator for getting the answers to questions as per situations the investigator finds them suitable. In this process, the investigator has a greater freedom to include and exclude some of the answers and also to inquire some relevant supplementary questions if he/she found them suitable. The unstructured personal interview method depends greatly on the capability of an interviewer for its success. In an exploratory type of research or a formulative type of studies, an unstructured personal interview is mostly useful.

Interview is a method of exposing the hidden factors at the heart of the respondent. The skill and capability of an interviewer greatly influence the outcome of an interview. In many of the situations, a specific and skilled interviewer is required to conduct the process. As such interviews can further be classified into three categories: (a) *focus interview*, (b) *clinical interview*, and (c) *nondirective interviews*. In a focus interview methods, an interviewer is free to design and prepare a sequence of questions. Mostly this is a conversation-based interview in which an interviewer concentrates in getting information from the respondent about the subject in which the respondent has enough experience. Clinical interview is concerned with the recording of information about the feelings and opinions of individuals about their experiences in their own lives, whereas in a nondirective interview method, information on a particular aspect is recorded from the respondents. In this method, the work of an interviewer is to stimulate the respondent to go on talking about his/her feelings, beliefs, and experiences on a particular aspect.

Merits and Demerits of Personal Interview

Though a personal interview is an excellent method of data collection, particularly in the field of social studies and behavioral sciences, it has certain advantages and weaknesses. There is a saying that "if there is a challenge, then face it." A personal interview method of data collection is such an activity which inspires researchers not only face the challenges of solving burning problem of the society but also collect the information for the purpose. Though a personal interview method is a challenging task to an interviewer or a researcher, it has still the following advantages:

1. More and more information at a greater depth could be obtained.
2. By overcoming the resistance, if any, during the recording of information, an interviewer or researcher can lead to yield a perfect information bank.
3. Flexibility is the beauty of a personal interview method.
4. The factors of nonresponse, nonreturned, and ill response can be minimized.
5. Because of greater interactions, an interviewer develops a greater confidence on human behavior.
6. With the help of charisma, greater acceptability, and capability, newer dimension of research may also be worked out.

7. Both the interviewer and respondent become further educated.
8. Misinterpretation could be avoided.

In spite of the above strengths, there are also certain weaknesses of a personal interview method of data collection:

1. It is very expensive and time consuming, particularly when the area of study is big.
2. The whole method of a personal interview depends on the quality of an interviewer. As such for a big study, one needs to employ a number of interviewers, wherein they differ among themselves and thereby causing differences in the quality of information.
3. Nonresponse from potential respondents like higher officials in the government/private machineries is one of the major problems with this type of interview.
4. The presence of an interviewer may overstimulate or may cause shyness to the respondent, thereby hampering the quality of the data.
5. Selecting, providing training, and supervising an interviewer are the challenging tasks of a researcher.

Criteria for Better Interviewing

1. An interviewer should be selected carefully; only those persons who are well acquainted and honest and have the intelligence to capture the essence of the interview should be selected for this purpose.
2. A selected interviewer should be trained adequately so that there would be no ambiguity arises in understanding the questions and the expected responses.
3. An interviewer should be well behaved, honest, sincere, hard working, and impartial. Interviewing is an art of extracting the inner heart of the respondent.
4. An interviewer must enjoy the confidence and faith of the respondent.
5. An interviewer's approach must be friendly, courteous, conversational, and unbiased.
6. An interviewer should refrain from asking undesirable and unwarranted questions.
7. An interviewer should always try to create an atmosphere of mutual understanding, belief, and faith.

7.1.2.2 Telephonic Interview

With the advancement of communication technology, telephonic interview is gaining momentum. An interviewer should contact a respondent over a telephone. Telephonic methods of interviews include phone calls, SMS, and emails. Respondents are asked to provide a suitable time slot during which they could be talked over the telephone. During the interview, questions are asked and responses may be noted or recorded; the process facilitates the interview of the respondents at different places at different times. Short message service (SMS) and multimedia message service (MMS) are also used for interviewing over telephones, and an interview may also be conducted through emails. Questionnaires can be sent as an email attachment to which the respondent replied. Like the other methods, a telephonic interviewing method has merits and demerits.

Merits

1. It is faster than any conventional method of personal interviewing or data collections through questionnaires or schedules.
2. It is cost-efficient.
3. It is easily manageable, no questions of training or supervising the field staff, etc.
4. People located at distant places over the world may be contacted for the purpose of the interview.
5. It is lesser time consuming than the other methods.
6. Telephonic interview can be recorded.

Demerits

1. All intended respondents may not have a telephone connection.
2. Sometimes it is difficult to collect the telephone numbers of all potential respondents.
3. Respondents may refuse to response without facing the interviewers.

4. If the survey is comprehensive in nature and requires much time, the method may not be a suitable one.
5. For a telephonic interview, questions must be short and to the point; this may prove sometimes difficult to handle.
6. Time provided to answer the question is comparatively limited.

7.1.2.3 Chatting

With the increasing use of Internet facilities, chatting has become the most popular way of communicating among people. Various common interest groups have come into existence, taking advantage of the facilities. Simply chatting or video chatting helps more than one person, sitting at distant places all over the world, to communicate their ideas, thoughts, and beliefs instantly. This type of interviewing, particularly the video chatting, is as good as personal interviewing method of data collection. As such, the competence of an interviewer not only over the subject under study but also the communication technology plays a great role on the successful completion of a process. This type of interviewing is very fast and has almost all the merits of telephonic methods of data collection. One of the most demerits of this method is that each and every respondent is required to be a member of the group for which an interview is to be conducted. The disturbances during the conversations or chatting may cause additional problems in this method. However, with the advancement of technology, this type of interviewing is bound to gain momentum day by day. In business and corporate sectors, this type of interviewing is highly appreciated. In many corporate sectors, employers recruit their employees through telephonic or chatting interviewing methods.

7.1.3 Questionnaire Method

One of the most conventional methods of data collection, particularly in wider areas having big inquiries, is the questionnaire method of primary data collection. In this method, a questionnaire is prepared befitting to the objective of the study and sent generally by post to the respondents with a request to answer the questionnaires. Nowadays, questions are also sent via an email attachment using Internet facilities. The respondents are expected to read the questions and answer them. During the framing of the questionnaire, the questions may be set in a sequential order. The researcher sets the questions in such a way that the respondent can have an idea about the logical sequences and chronology of the questions. This is known as structured questionnaire. The questionnaire can also be nonstructured, that is, following no sequence. This may lead to problems in understanding the questions by the respondent. Questionnaires may be disguised or nondisguised in nature. In a disguised questionnaire method, the objective of the questionnaire is not clearly spelt out, whereas in a nondisguised questionnaire method, the objective of the study is made clear to the respondent. Like other methods of collecting information, a questionnaire method has also certain advantages and disadvantages. The success of this method, though, depends on the ability of a researcher to frame the questionnaires and the cooperation from the respondent. The following are some of the merits and demerits of questionnaire method:

Merits

1. Most useful in comprehensive study.
2. Interviewer's biases are reduced.
3. In the absence of an interviewer, the answers of the respondents are free in nature.
4. Sufficient times are provided to the respondent.
5. Questionnaire can be sent to any phone number of respondents in any part of the world.

Demerits

1. This method of data collection does not apply to uneducated or illiterate respondents.
2. It is time consuming.
3. Lengthy questionnaires may bore the respondent and thereby reducing the number of responses.

4. Postal delay may lead to ill response.
5. The response of the respondent may not be accurate.
6. The possibility of getting more information like that of an interview method is reduced.

For a data collection method to be successful, researchers dealing with the questionnaire should pay attention to the following features: (1) introducing letter, (2) the size of the questionnaires, (3) sequence of questions, (4) simplicity, (5) clarity, (6) nonpersonal, and (7) pretesting.

Each and every questionnaire should reach to the respondent with a request from the researcher in a very humble approach clearly mentioning the need for the study and the help needed from the respondent. The size of the questionnaire is a matter of great concern. In the present and fast-moving world, time is precious to each and every one. If one gets very lengthy questionnaire, the respondent gets bored. The question should be kept as minimal as possible. Repetitions and unnecessary questions should be avoided. Questions should be framed in a logical sequence so that both the respondent and the researcher become very much clear about each other with respect to the objective of the study vis-a-vis the questions framed. Questions should be framed in a free-flowing sequence. If the sequencing of the questions is such that the respondent may get confused, then their immediate reflection will be in how to answer the questions. Simplicity in language and communication is the art of framing a good questionnaire. A researcher should communicate to the respondent in a very simple and sober manner using simple language. The simpler the language, the easier is the way to reach the respondent. Any ambiguous question should be avoided. Question should be very short, simple, and informative. Clarity in the nature of questioning is the most important factor in getting an actual response. The degree of clarity of the questions asked and their wordings are the most important things. Questions should be easily understood; it should convey only one message at a time. The researcher should try to avoid asking any personal questions which may make the respondent hesitant in answering those questions. Questions related to personal income, income tax paid, sale tax paid, and extramarital relationship should be avoided. Asking personal information greatly influences the overall response of the respondent. As such, personal questions if at all required to be asked should be placed at the end of the questionnaire. Instructions in filling out the questionnaire will provide an additional advantage to the respondent and may encourage him/her to respond positively to the request of the researcher. The researcher should clearly clarify the code, groups, or other special notations used in the questionnaire. A declaration on the part of the researcher stating that the information provided by the respondent should be kept confidential and should not be used for other purposes other than the purpose mentioned in the study without prior permission from the respondent may encourage the respondents in providing feedback to the questionnaires. For comprehensive and big studies, pretesting of questionnaire is essential. Pretesting of questionnaire means application of the questionnaire to a small group of respondents before it is being used for the main study. This is also known as pilot survey. There are several advantages of a pilot survey. The researcher can modify or adjust, incorporate some new questions, and delete some unwanted or nonresponsive questions based on the feedback from the pilot survey.

7.1.4 Schedule Method

The method of data collection through a schedule is almost similar to that method of data collection through questionnaires with a slight difference, that is, an enumerator or a researcher prepares a list of questions and takes it to the respondents to fill them in during the process/time of interviewing. While the questionnaires are being filled in by the respondents themselves, the schedule is most likely to be filled in by an enumerator or investigator. The added advantage of data collection through schedule is that the enumerator or investigator clarifies or explains the questions which seem to be difficult to answer by the respondent. The success of this

Table 7.1 Differences between the questionnaire and schedule

Sl no.	Questionnaire	Schedule
1	Sent through mail	Carried by an enumerator or investigator
2	Filled in by the respondent	Filled in by an enumerator or the enumerator helps the respondent in filling in
3	No manpower is required	Trained enumerator or investigators are required
4	Cheaper or cost-effective	Costly
5	No questions of supervision is required	Enumerator or investigators are required to be supervised
6	Has more nonresponse	Has comparatively low nonresponse
7	Respondent's identify is unknown	Identity of respondent is fixed
8	May be time-consuming	Can be completed within a given time frame
9	No personal contact (in general)	Personal contact is established
10	Risk of misinformation or more wrong information is seen	Quality of information depends on the quality of the enumerator
11	Not applicable for illiterate	May be applicable for illiterate
12	Worldwide collection of information is possible	Worldwide collection of information is not possible

method of data collection depends on the capability of the enumerator. The enumerator should be trained in such a way so that he/she becomes competent enough about the objective of the project and his/her duties. Though this method is time and resource consuming, a good, qualified, and capable enumerator can deliver wonderful information even beyond the scope of the research studies (Table 7.1).

7.1.5 Other Methods of Data Collection

Besides the above methods, there are other methods also for data collection particularly in the field of social, economic, psychometric, business, etc., studies. Among the different methods, (1) *projective method of data collection*, (2) *warranty card method*, (3) *audit method*, (4) *consumer panels*, (5) *mechanical device method*, (6) *depth interview method*, (7) *content analysis method*, and (8) *PRA method are important.*

7.1.5.1 Projective Method of Data Collection

It is mostly used in psychological studies to project the feelings, behaviors, and understanding of the respondents on a particular aspect. It is an indirect method of interview in which a respondent is put in a situation wherein information about his/her personality and inner mind is extracted consciously by an interviewer but unconsciously by him. This method is mostly used in situations where the respondents are unwilling to respond to direct questions—this might be because of the fact that they have no clear cut idea, opinion, and thought of feelings or situations where direct questioning is undesirable. There are different methods of projective techniques: (a) *word association test*, (b) *sentence completion test*, (c) *story completion test*, (d) *verbal projection test*, (e) *pictorial technique*, (f) *play technique*, (g) *quiz*, (h) *sociometry*, etc.

In a word association test, the number of disjoint words is provided to the respondent, and a response is sought about the first thought or word that the individual associates with each word. The interviewer notes the response, promotional expressions, etc., to judge the attitude of the respondents.

Instead of providing a list of words, a list of incomplete sentences is provided to the respondents with the request to complete this. Analysis of the replies serves as an indicator for the conception of the respondent about the subject and also the attitude towards it. Both the word testing and sentence completion are quick and easy to use but may not be easy to analyze numerically. Moreover, the entire success of word association and sentence completion test depends on the championship of the interviewer.

A story completion test is one step ahead of a word association and sentence completion test. A part of the story is provided to the respondent and asked to complete the same. In the process, the intention, thought, feelings, and the attitude of the respondent towards the subject of the story are recorded and analyzed.

In a verbal projection method, respondents are requested to comment on the subject about the feelings of others. For example, the question may be asked why people get married. From the answer to the question, the feelings of the people about marriage are being noted through the respondents.

In a pictorial technique, pictures of different forms, types, and subjects in series are presented to respondents, and the responses are recorded and analyzed to draw the personality structure and attitude of the respondents. There are different methods of pictorial techniques; among these, the important are (a) *thematic appreciation test (TAT)*, (b) *Rosenzweig test*, (c) *Rorschach test*, (d) *Holtzman inkblot test (HIT)*, and (e) *Tomkins–Horn picture arrangement test*.

A Play technique is the simplest form of understanding the inner feelings and the attitudes of the younger. In this method, different types of dolls (different races, different genders, different dresses, different religions, etc.) depending upon the subject of study are provided to the children or younger persons with the request to arrange these. The arrangements made by the respondents are the reflection of the state of mind, feelings, and attitude towards the society.

The different types of short questions are framed to test memorization and analytical ability of the respondents, supposing that the memorization capability and analytical ability are related to the likings, dislikings, and inner feelings of the respondents. It relates to answering the research problem: what are the subject areas in which people are interested in?

The motives of the respondents are studied to trace the flow of information and examine the ways in which new ideas are diffused. Sociograms are constructed to categorize the respondent into innovators, leaders, early adopters, laggards, etc.

7.1.5.2 Warranty Card Method

In market and business research, the warranty card method is mostly used. It is our common experience that when the consumers purchase any durable items, the post card (or in different form) size of information sheet is provided to the consumer to collect information not only about the product but also about certain other points of business interest. Analyzing the feedback from the consumer, the efficiency of the business house could be enhanced. Analyzing the feedback forms, the feelings, the attitude of the consumers, their expectations or discontent, etc., can be obtained.

7.1.5.3 Audit Method

There are different audit methods to know the attitudes of the consumers. Distributor or store audit method and pantry audit methods are mostly used in market and business research. Distributors or manufacturers get the retail stores audited through a salesman at a regular time interval. The information collected is used to assess the market size, market share, seasonality and cyclical behavior, trend, and so on of a particular item or a group of items. Thus, the process of auditing the retail stores by a distributor or a manufacturer is known as auditing method of data collection. Auditing is generally followed at an interval of time for a long period. In the process, one can have an idea about the impact of promotional program (if any) taken during the period of investigations. Auditing can also be made to estimate the consumption pattern of the consumers—known as *pantry audit method*. In a pantry audit method, an investigator collects the information on the types and quantities consumed of different commodities at different points of time and at different places. The same thing can also be obtained from the respondents. This type of pantry audit method of data collection can also be panel as well as one time. These methods give the investigators huge amount of information maybe through questionnaires, personal interviews, or other methods. The only demerit is that personal preferences of the consumers may not be identified in this method or the effect of sale promotional activities may not be measured accurately.

7.1.5.4 Consumer Panels

A consumer panel method is a specialized type of pantry audit method. In this data collection method, the consumers or a set of consumers are studied on a regular basis. The assumption or the understanding of this method is that a selected consumer will maintain a daily detail record of his/her consumption, and the same could be made available to researchers as, and when, necessary. The consumer panel method of data collection may be *transitory or continuing*. In a transitory consumer panel method, information on the consumers' consumption details is taken before and after basis. That is, the information is taken from the selected consumer before introducing a particular phenomenon (sales promotional scheme like 20 % extra in biscuit packets of particular brand, etc.) and again after the occurrence of a particular phenomenon. Generally, consumers are not aware about the promotional scheme so long these are not introduced in the market. Thus, the effect of such phenomenon would be measured with such method of data collection. On the other hand, a continuing consumer panel is generally set up to study the consumer behaviors on particular aspects over a period of time. Like the market survey of consumers' good, in studies like television/radio viewership/listeners, care must be taken so that consumer panels selected for that purpose must represent the whole consumer community.

7.1.5.5 Mechanical Device Method

As the name suggests, this is a method of collecting data through mechanical devices. Mechanical devices like camera, psychogalvanometer, motion picture camera, and audiometer are the principal devices used for general purposes. In agriculture, the instruments like leaf area meter and seed counter are in use to get information on particular aspects of interest.

7.1.5.6 Depth Interview Method

A special type of interviewing to get data is the depth interview method. Generally, this type of method is used to unearth the motives and desires of the respondents. Thus, this technique is more on psychological than mechanical one. The success of this method depends on the skills of an interviewer and the endurance of the respondents to reply to the queries, as most of depth interviews involve considerable time. The depth interviews may be projective or non-projective in nature depending upon the objective and the type of questions.

7.1.5.7 Content Analysis Method

It has already been mentioned that during the formulation of the subject matter of research studies, related documents are essential in the whole process. A researcher is to study and analyze the content of the documents (books, journals, magazines, Internet resources, etc.) to facilitate the whole research process. The content analysis may be a qualitative as well as quantitative one.

7.1.5.8 PRA and RRA Method

This method is mostly used under two situations, particularly in social sciences, when a researcher does not have a complete idea about the problems faced by the people and/or when information is needed quickly. The purpose of this method is to get an understanding and complexity of the problem rather than of getting absolutely accurate data. In this method, people participate in sharing their views, action–interaction takes place, and information is recorded.

7.2 Collection of Secondary Data

Secondary data are available only if required to be collated or compiled. Generally, secondary data are available from different sources like the government/private/industrial/business/research/voluntary organizations. These are generally found in journals, booklets, books, monograms, technical reports, etc. In India, each and every state government and central government along with their different establishments publishes different data on various aspects. Before utilizing any secondary data, a researcher should be very much careful about the originality or authenticity of the data. In many of the cases, it is found that the data on the same aspect in various organizations has been reported in different ways. So the reliability, accuracy, and adoptability of the data are the most important features. The sources of secondary data may be under published or unpublished. The published data are generally available from various publications of the central/provincial/local bodies, publications of different international bodies or organizations, technical/scientific/trade journal, and books/monographs/magazines/newspaper published by various government/nongovernment/business/voluntary establishments. Reports and documents prepared by different research scholars, universities, and establishments also serve as good sources of secondary data. Among the unpublished sources of information are diaries, letters, biographies, autobiographies, and other relevant unpublished documents.

Before using a secondary data, a researcher must be careful whether to use or not to use the available secondary data. This can be judged on the basis of parameters like reliability, suitability, and adequacy. Reliability refers to the characteristic features of the sources of information. A researcher should be clear about the following essential points: (1) *who collected the data*, (2) *where the data was collected*, (3) *what were the methods of data collection*, (4) *were the required methods followed properly*, (5) *what is the time of data collection*, and (6) *were the data collected at the desired level of accuracy?* A researcher must be aware about the fact that whether the secondary data available are suitable for the present research purpose. That means the adoptability of the data must be verified under the given situations. The original objective, scope, and nature of data collection must be scrutinized thoroughly.

7.3 Case Study

Case study refers to an in-depth study on various qualitative and quantitative aspects of a particular community, social unit, institution, etc. The main emphasis of this type of study is to study the units under considerations vertically rather than horizontally. For intensive investigation, this type of method of data collection is generally used. Minute details of the selected unit are studied intensively. Being exhaustive and intensive in nature, a case study method enables a researcher to understand fully the behavioral pattern of the concerned unit. Understanding the relationship of a social unit with the social factors and the forces involved in the surrounding environment is the major aim of case studies.

7.4 Criteria for Selections of Appropriate Method of Data Collection

So far, we have discussed the different methods of collections of primary and secondary data. Among the competitive methods of data collection, a researcher must optimally select the method for a specific research program. In doing so, a researcher is required to follow the following aspects of a research program: (1) *the nature, scope, and objective of the study*, (2) *availability of fund*, (3) *availability of time*, (4) *availability of technical person*, and (5) *precision required*.

7.4.1 Nature, Scope, and Objective of the Study

The nature, scope, and objective of a study decide the type of information required for the purpose. Whether a research program is exploratory, descriptive, etc., in nature decides the type of data collection method. The objective and scope of the study guide whether to use the primary data or secondary data. If secondary data are to be used, then what are the sources, what is the reliability of, and at what accuracy they have been collected, all these things should be critically examined before taking any decision in this regard. If primary data are to be recorded, then what type of data is needed, qualitative or quantitative; does it require assistance from experts or sophisticated instrument, etc.?

7.4.2 Availability of Fund

Financial resource is a pivotal point or lifeline of any good or bad research program. Depending upon the availability of fund, research projects are exhaustive/intensive. In developing countries, where financial resources are one of the major constraints in conducting good research work, the guardians generally advise "try to do, whatever you can do within the limited resources, don't hanker for that which is not available at your disposal." When funds are available, a researcher is at his/her liberty to frame a research process; otherwise, he/she has to keep in mind the available resources at his/her disposal. If resources are available, one can make the program exhaustive/intensive, make use of sophisticated instrument, can take help from technical experts, and also can minimize the time required or otherwise.

7.4.3 Availability of Time

As discussed already, the different methods of data collection require different time periods for a successful completion. A researcher must be aware about the time period allowed to fulfill the objective of the study. A researcher must always make an endeavor to complete the data collection within the shortest possible time period at his/her disposal, keeping in mind the financial resources available for the purpose.

7.4.4 Availability of Technical Person

While discussing the different methods of data collection, it is noted that some methods of data collection require the help from the experts. In a schedule/observation/interview and other methods of data collection, the quality of the data depends greatly on the understanding of the subject/questions of the enumerator/investigator, behavior, intellectuality, and training taken by the enumerator. One cannot use highly sophisticated instrument for data collection in the absence of a highly trained expert in recording and retrieving the data from the sophisticated instruments.

7.4.5 Precision Required

Methods of data collection are greatly influenced by the precision required for the study. For example, length data can be taken with the help of screw gauge, slide calipers, centimeter scales, or tape. All these are coming under an experimental and observational data collection method. Now, the question is, to which one should one apply? Clearly, the precision with which the above instruments work is different, and a researcher needs to desire which technique or which method he/she should apply. Similarly, the different methods of data collection, may it be primary or secondary, have different levels of precision or accuracy. A researcher is to desire the best methods of data collection befitting to the objective of a research program commensurating with the available resources and the given time frame.

8 Processing and Analysis of Data

8.1 Processing of Information

The information/data collected/collated either from primary or secondary sources at the initial stage is known as *raw data*. Raw data is nothing but the observation recorded from individual units. Raw data, particularly the primary data, can hardly speak anything unless and otherwise arranged in order or processed. Data are required to be processed and analyzed as per the requirement of a research problem outlined. Working with data starts with the scrutiny of data; sometimes it is also known as *editing* of data. There are several steps to follow before a set of data is put under analysis befitting with the objectives of a particular research program.

Though the order of the steps are not unique and may change according to the need and objective of a study, the following steps are generally followed: (1) *scrutiny/editing of data*, (2) *arrangement of data*, (3) *coding of data*, (4) *classification of data*, and (5) *presentation of data*. The first three steps, that is, scrutiny, arrangement, and coding of data may interchange the order depending upon the situation. If the number of observations is few, one can go for scrutiny at the first stage; otherwise, it is better to arrange the data in ascending or descending order. We shall demonstrate the whole procedure by taking the following example.

Example 8.1. The following table gives data on the yield (q/ha) of paddy of 130 varieties.

15.50	25.50	26.50	24.50	28.50	29.50	26.50	53.50	46.50	43.50	77.50	63.50	51.50
24.80	12.80	21.80	23.80	25.80	27.80	58.80	80.80	24.80	33.80	77.80	74.80	70.80
18.00	19.00	22.00	21.00	23.00	26.00	40.00	68.00	20.00	30.00	83.00	55.00	80.00
20.00	21.00	22.00	27.00	28.00	26.00	41.00	37.00	61.00	68.00	78.00	39.00	51.00
19.00	28.00	27.00	26.00	29.00	30.00	68.00	48.00	33.00	38.00	23.00	66.00	37.00
23.50	32.50	15.50	17.50	18.50	19.50	87.50	29.50	18.50	65.50	45.50	64.50	44.50
17.50	16.50	19.50	22.50	23.50	21.50	57.50	66.50	52.50	60.50	51.50	56.50	80.50
19.00	21.00	23.00	25.00	27.00	24.00	36.00	74.00	32.00	23.00	80.00	16.00	45.00
20.00	20.00	21.00	22.00	23.00	24.00	59.00	39.00	71.00	20.00	46.00	30.00	19.00
20.00	24.00	25.00	26.00	27.00	28.00	41.00	30.00	49.00	70.00	85.00	27.00	24.00

P.K. Sahu, *Research Methodology: A Guide for Researchers in Agricultural Science, Social Science and Other Related Fields*, DOI 10.1007/978-81-322-1020-7_8, © Springer India 2013

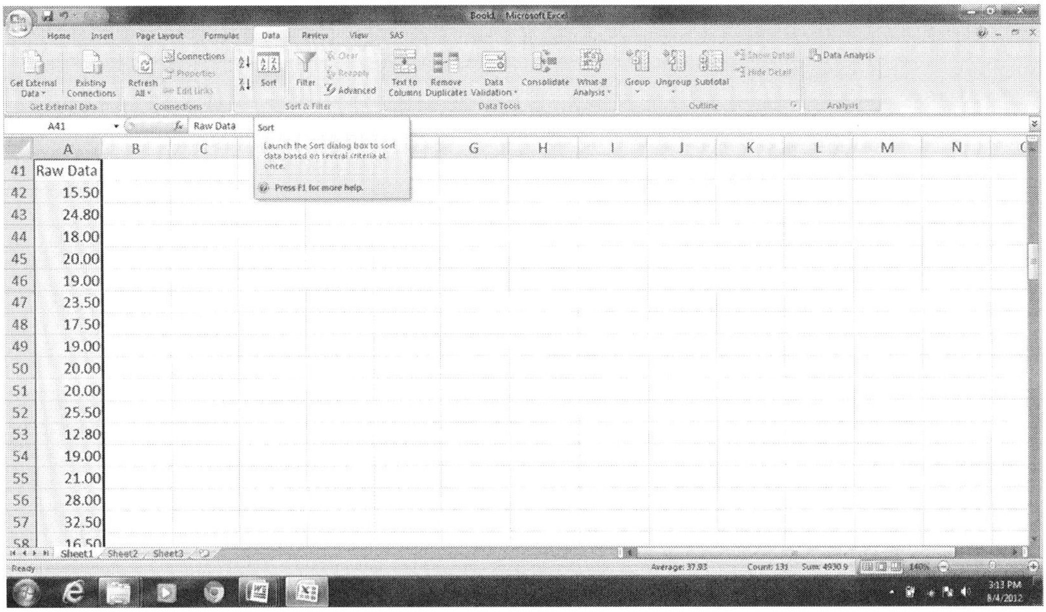

Slide 8.1: Step 1 showing the entry or the transformation of data to Excel data sheet

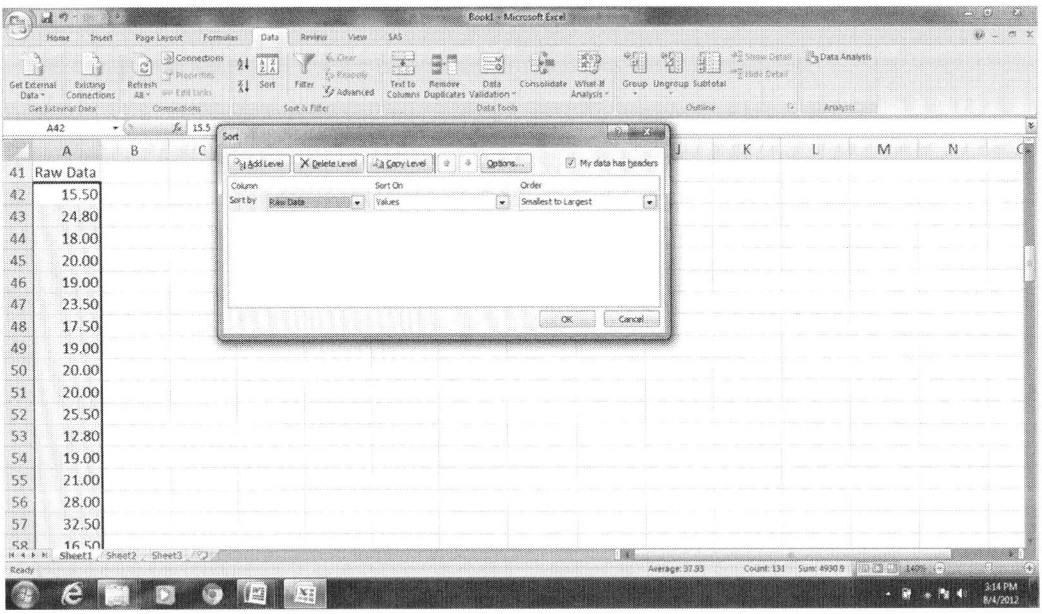

Slide 8.2: Step 2 showing the sorting of data in Excel data sheet

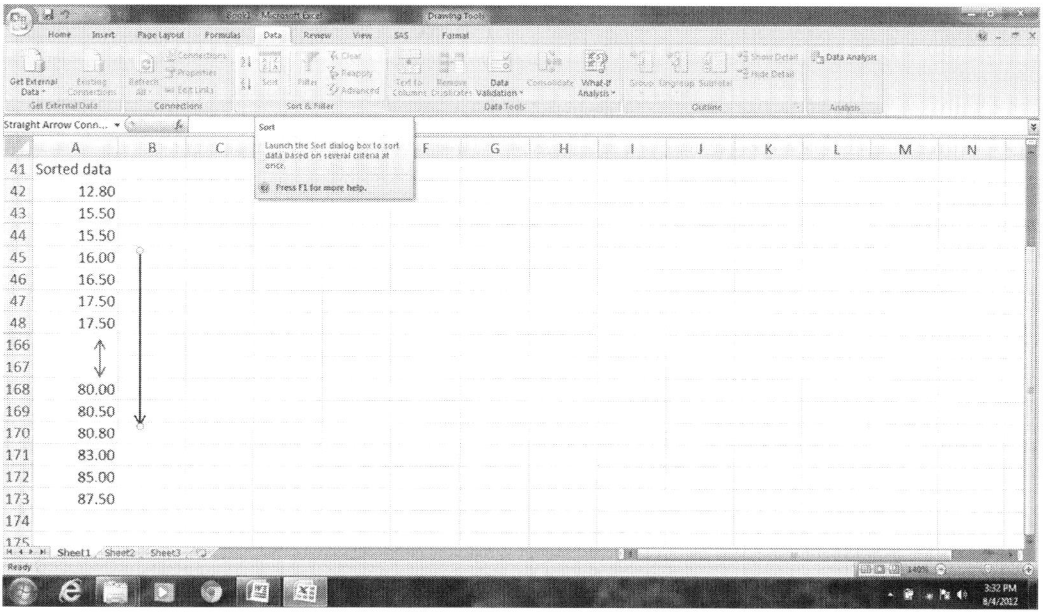

Slide 8.3: Step 3 showing the output of sorted data in Excel data sheet

From the arrangement of the above data set, it is clear that the yield of paddy has a maximum value 87.50 q/ha and a minimum value 12.80 q/ha, thereby having a yield difference of 74.8 q/ha between the highest yielder and the lowest yielder.

8.1.1 Scrutiny and Arrangement of Data

Raw data set is put under careful examination to find out the existence of any abnormal/doubtful observation, to detect errors and omissions, if any, and to rectify these. Editing/scrutiny of data ensures the accuracy, uniformity, and consistency of data. If the observations are few in number, during scrutiny, one can have an overall idea about the information collected or collated. If the number of observations is large, then one may go for arrangement of observations in order, that is, either ascending or descending order and then go for scrutiny. Scrutiny and arrangement of data help to have preliminary knowledge about the nature of the data set which may not be possible (particularly when a number of observations are large) otherwise. Either scrutiny followed by arrangement (for small number of observations) or arrangement followed by scrutiny would help.

1. To know the maximum and minimum values of the observations.
2. Whether the values of the observations are consistent with the area of interest under consideration or not. That means whether there is any possibility of the data set containing outlier or not.
3. Whether the data set could be used for further analysis towards fulfilling the objective of the study or not.

   Arrangements of data can be made using SORT command in MS Excel or similar command in other similar software.

Example 8.2. For discrete data, one may think of the formation of array. The formation of array taking the number of effective tiller (ET) data for 100 varieties of paddy is demonstrated as follows:

Raw data:

Variety	ETL	Variety	ETL	Variety	ETL	Variety	ETL
1	16	26	17	51	18	76	16
2	15	27	22	52	20	77	26
3	10	28	6	53	23	78	20

(continued)

(continued)

Variety	ETL	Variety	ETL	Variety	ETL	Variety	ETL
4	11	29	8	54	9	79	21
5	12	30	16	55	13	80	19
6	13	31	12	56	14	81	7
7	14	32	13	57	8	82	9
8	7	33	14	58	7	83	14
9	8	34	7	59	9	84	22
10	9	35	8	60	13	85	23
11	10	36	11	61	7	86	14
12	11	37	17	62	8	87	26
13	14	38	9	63	11	88	18
14	16	39	10	64	24	89	20
15	18	40	12	65	16	90	19
16	20	41	18	66	4	91	17
17	7	42	17	67	6	92	9
18	8	43	16	68	5	93	10
19	4	44	8	69	6	94	12
20	6	45	6	70	7	95	23
21	15	46	4	71	10	96	20
22	8	47	16	72	11	97	21
23	12	48	22	73	4	98	12
24	15	49	26	74	8	99	10
25	16	50	21	75	12	100	18

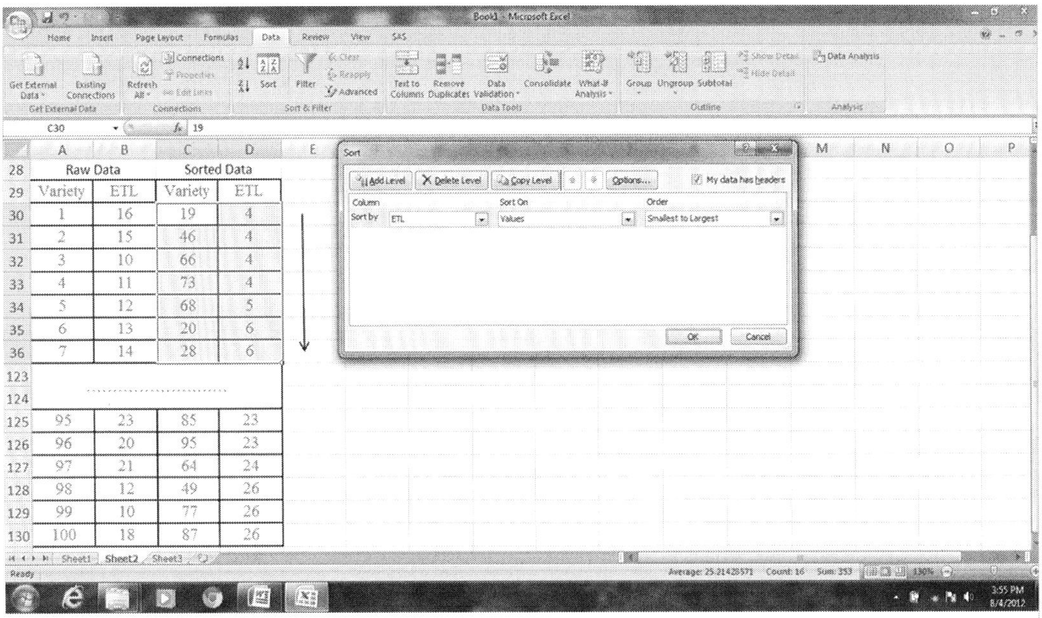

Slide 8.4: Step 1 showing the entry or the transformation of data to Excel data sheet

From the arrangement of the above data set on the number of effective tier per hill (ETL) of 100 varieties of paddy, it is clear that the ETL has a minimum value 4 and a maximum value 26; thereby having a range of $26 - 4 = 22$ number of ETL among the varieties of paddy under consideration. After the formation of data, it takes the following shape (Tables 8.1 and 8.2).

It is found from the array that there are four varieties having ETL 4 numbers per hill; one

Table 8.1 ETL data arranged in ascending order

ETL									
4	7	8	9	11	13	15	16	19	22
4	7	8	9	11	13	15	17	19	22
4	7	8	10	11	13	15	17	20	22
4	7	8	10	12	13	16	17	20	23
5	7	8	10	12	14	16	17	20	23
6	7	8	10	12	14	16	18	20	23
6	7	9	10	12	14	16	18	20	24
6	8	9	10	12	14	16	18	21	26
6	8	9	11	12	14	16	18	21	26
6	8	9	11	12	14	16	18	21	26

Table 8.2 Summary of ETL data arranged in ascending order

From the above array, one can have the following summary:

ETL	4	5	6	7	8	9	10	11	12	13	14	15	16	17	18	19	20	21	22	23	24	26
Frequency	4	1	5	7	9	6	6	5	7	4	6	3	8	4	5	2	5	3	3	3	1	3

Table 8.3 Tally marking and formation of frequency distribution of ETL data

ETL	4	5	6	7	8	9	10	11	12	13	14	15	16	17	18	19	20	21	22	23	24	26
Tally marks	////	/	卌	卌 //	卌 ////	卌 /	卌 /	卌	卌 //	////	卌 /	///	卌 ///	////	卌	//	卌	///	///	///	/	///
Frequency	4	1	5	7	9	6	6	5	7	4	6	3	8	4	5	2	5	3	3	3	1	3

variety is having 5 ETL per hill, five varieties is having ETL six, and so on. This formation of array is done with the help of SORT key in Microsoft Excel computer package as shown above in case of example with the yield of paddy. Otherwise, conventionally we may take help of the tally marking. Process for tally marking is presented below (Table 8.3).

8.1.2 Coding of Data

Sometimes the information collected may be qualitative in nature like male/female, black/yellow/white/green, determinate/indeterminate, and educated/illiterate. Coding refers to the process of assigning numerals or other symbols to the responses so that these could be categorized. Coding should be made in such a way that these are nonoverlapping and all the observations are categorized in one of the categories framed for the purpose. That means the coding should be made in such a way that categories are exclusive and exhaustive in nature. Generally, the numerical information does not require coding. Coding helps researchers in understanding the data in a more meaningful way.

8.1.3 Classification/Grouping

While dealing with a huge number of observations, it is sometimes very difficult to have a concise idea about the information collected. So the first idea comes to mind, that is, to have a logical classification (formation of groups) in accordance with some common characteristic(s)/classification or grouping, may be one of the solutions.

The first question in classification comes to mind is, *how many classes one should make*? There is no hard-and-first rule as to fix the number of classes. However, a general guideline as given below is followed while making the classes:

(a) Classes should be well defined and exhaustive.
(b) Classes should not be overlapping.
(c) Classes should be of equal width as far as possible.
(d) The number of classes should not be too few or too many.
(e) Classes should be devoid of an open-ended limit.
(f) Classes should be framed in such a way that each and every class should have some observation.

Table 8.4 Determination of the no. of classes according to the two methods of classification for example data 8.1 and 8.2

	Example 8.1		Example 8.2	
Formula	*N*	Class	*N*	Class
Yule	130	8.44 = 8	100	7.91 = 8
Struge	130	8.02 = 8	100	7.64 = 8

While deciding the number of classes or groups, the general idea is to have minimum variations among the observations of a particular class/group and maximum variation among the groups/classes.

Following the above guidelines and the formulae given below, classes may be formed.

1. *Yule formula*: $K = 2.5 \times N^{1/4}$
2. *Sturge formula*: $K = 1 + 3.322 \log_{10}N$,

   where N is the number of observations and K is the number of classes.

Thus, according to the above two formulae, the number of classes for Example 8.1 and Example 8.2 is given in Table 8.4.

Generally, the range of date can be extended in both sides (i.e., at the lower end as well as at the upper end) so as to (1) make the no. of classes as a whole number and (2) avoid the class width as a fraction. For example, in the paddy yield, the maximum value is 87.5 and the minimum is 12.8. So to have 8 classes, the class width becomes 9.33, a fraction not advisable for the convenience of further mathematical calculation. To avoid this, one can increase the data range in both sides to 10 and 90, respectively, for the lower and upper sides, thus making the $(90 - 10)/8 = 10$ class width a whole number. It should emphatically be noted that making the class width whole number is not compulsory; one can very well use a fractional class width also.

8.1.3.1 Method of Classification

When both the upper limit and lower limit of a particular class are included in the class, it is known as *inclusive* method of classification. While in other methods one of these limits is not included in the respective class, it is known as *exclusive* method of classification (Table 8.5).

The above two classifications, that is, the inclusive method of classification and the exclusive method of classification, are for Examples 8.2 and 8.1, respectively. For Example 8.2 of the effective tiller per hill, that is, for discrete variable, both the class limits are included in the respective class. But for Example 8.1, that is, for yield variable (a continuous character), exclusive method of classification has been used.

Table 8.5 Inclusive and exclusive method of classification of data

Inclusive method		Exclusive method[a]	
ETL Classes	Frequency	Yield classes	Frequency
4–6	10	10–20	16
7–9	22	20–30	54
10–12	18	30–40	14
13–15	13	40–50	11
16–18	17	50–60	10
19–21	10	60–70	10
22–24	7	70–80	8
25–27	3	80–90	7

[a]Upper limits are not included in a particular class

Table 8.6 Making continuous classes from discrete classes

ETL classes		
Discrete	Continuous	Frequency
4–6	3.5–6.5	10
7–9	6.5–9.5	22
10–12	9.5–12.5	18
13–15	12.5–15.5	13
16–18	15.5–18.5	17
19–21	18.5–21.5	10
22–24	21.5–24.5	7
25–27	24.5–27.5	3

May it be discrete or continuous, different statistical measures in subsequent analysis of the data may result in a fractional form. As such, generally discrete classes are made continuous by subtracting "$d/2$" from the lower class limit and adding "$d/2$" to the upper class limit, where "d" is the difference between the upper limit of a class and the lower limit of the following class. The above classification wrt ETL may be presented in the form given in Table 8.6.

Thus, the constructed class limits are known as *lower class boundary* (*3.5, 6.5, . . .,24.5*) and *upper class boundary* (*6.5, 9.5, . . ., 27.5*), respectively, in case of continuous distribution. The class *width or the class interval* is equal to the difference between the upper class boundary and the lower class boundary of the class interval.

Table 8.7 Frequency distribution table of yield (q/ha) in 130 paddy varieties

Yield classes	Frequency (f_i)	Mid value (x_i)	Cumulative frequency (CF<)	Cumulative frequency (CF≥)	Relative frequency	Frequency density
10–20	16	15	16	130	0.123077	5.33
20–30	54	25	70	114	0.415385	18.00
30–40	14	35	84	60	0.107692	4.67
40–50	11	45	95	46	0.084615	3.67
50–60	10	55	105	35	0.076923	3.33
60–70	10	65	115	25	0.076923	3.33
70–80	8	75	123	15	0.061538	2.67
80–90	7	85	130	7	0.053846	2.33

Table 8.8 Frequency distribution table of effective tillers per hill (ETL) in 100 paddy varieties

ETL discrete classes	ETL continuous classes	Frequency (f_i)	Mid value (x_i)	Cumulative frequency (CF<)	Cumulative frequency (CF≥)	Relative frequency	Frequency density
4–6	3.5–6.5	10	5	10	100	0.10	3.33
7–9	6.5–9.5	22	8	32	90	0.22	7.33
10–12	9.5–12.5	18	11	50	68	0.18	6.00
13–15	12.5–15.5	13	14	63	50	0.13	4.33
16–18	15.5–18.5	17	17	80	37	0.17	5.67
19–21	18.5–21.5	10	20	90	20	0.10	3.33
22–24	21.5–24.5	7	23	97	10	0.07	2.33
25–27	24.5–27.5	3	26	100	3	0.03	1.00

Mid Value: Mid value of a class is the average of the lower limit/boundary and the upper limit/ boundary. Mid values are generally taken as representatives of different classes. In Table 8.1, the mid values for different classes are 15, 25, 35,. . ., etc.

Frequency Density: As we know it, "density" is mass per unit volume, that is, $d = m/v$ gm/cc where m in gram is the mass for a "v" (cc) of a matter. Similarly, frequency density is the frequency of a particular class per unit of class width. In Table 8.2, the frequency density of the class 3.5–6.5 is 10/3 = 3.333. Similarly, for class 6.5–9.5 is 22/3 = 7.33. Frequency density indicates the concentration of observations in different classes of a frequency distribution table per unit of class width.

Relative Frequency: Relative frequency is defined as the proportion of observation in a particular class to a total number of observations. Thus, in Table 8.2, the relative frequency of the class 3.5–6.5 is 10/100 = 0.10. Sometimes, the relative frequency is expressed in percentage also.

Cumulative Frequency: Cumulative frequency of a class is defined as the number of observations up to a particular class (*less than type*) or above a particular class (*greater than type*). In our previous example, if we take the class 12.5–15.5, then the cumulative frequency (lesser than type) for the class is 63. This means that there are 63 varieties of rice which have less than 15.5 number of ETL per hill. Similarly, the cumulative frequency (greater than type) for the same class is 50. That means the number of varieties having 12.5 or more than 12.5 number of ETL per hill is 39. Cumulative frequency gives an instant idea about the distribution of frequencies among the classes and the cutoff points (Tables 8.7 and 8.8).

8.1.4 Presentation of Information

Edited/scrutinized data can either be used for the application of statistical methodologies and/or presented in a suitable form to present and concise the information from the recorded data. In the following sections, discussion has been made on different forms of data presentation so that some of the critical observations (without going for in-depth analysis) can be extracted meaningfully from the data as such.

Different forms of presentation of data are:

(1) Textual form, (2) Tabular form, (3) Diagrammatic form

1. *Textual Form*: In a textual form of data presentation, information is presented in a form of a paragraph. In many of the research papers or articles, while discussing the findings of the research outcome, this method is adopted for explanation.
2. *Tabular Form*: It is the most widely used form of data presentation. A large number of data can be presented in a very efficient manner in a table. At the same time, it can bring out some of the essential features of the data. Frequency distribution tables presented in previous section are the example of efficient use of tables to present data. A table consists of the following parts: (1) title, (2) stub, (3) caption, (4) body, and (5) footnote.

   Title: The title of a table gives a brief description of the content or the subject matter presented in a table. Generally, the title is written in short and concise form such that it becomes easily visible and eye-catching at a glance and through light to the content of the table.

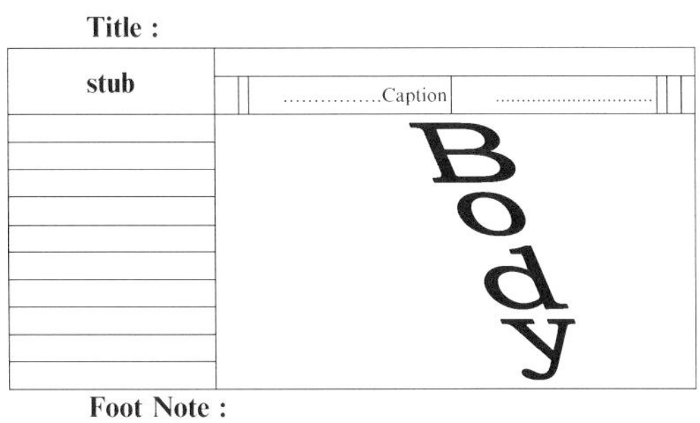

Fig. : Different parts of a table

   Stub: A table is divided into a number of rows and columns. Stub is used to describe the contents of the rows of a table. Different classes represent the rows of the table, and the heading "classes" at the top left corner of the table is the stub. With the help of this stub, one can extract the features of the rows in a table.

   Caption: Caption describes the content of each and every column. Thus, "mid value," "frequency," etc., are the captions for the different columns in Tables 8.1 and 8.2. With the help of the "mid value" or "frequency," one can understand how the mid values or the frequencies are changing over different classes (stub).

   Body: Relevant information is given in the body of a table.

   Footnote: Footnotes are not compulsory but may be used to indicate the source of information or a special notation (if) used in the table. Though a tabular form is more appealing than a textual form of presentation, it is only applicable to literate and educated persons.
3. *Diagrammatic Form*: Keeping in mind the variety of users, this form of representation is more convincing and appealing than the other forms of data presentation. This form of presentation is easily understood by any person, layman as well as an educated person. Different diagrammatic forms of presentation are (a) *line diagram*, (b) *bar diagram*, (c) *histogram*, (d) *frequency polygon*, (e) *cumulative frequency curve or Ogive*, (f) *pie charts*, (g) *pictorial diagrams*, (h) *maps*, etc.; within each type, there may be variant types.

In the following section, taking the yield data, different forms of diagrammatic presentation as obtained by using MS Excel are presented below.

(a) A frequency line for discrete as well as for continuous distributions can be represented graphically by drawing ordinates equal to the frequency on a convenient scale at different values of the variable, X. For the example of yield, we shall have different yield classes on the horizontal X-axis and frequencies on the vertical Y-axis as shown in Fig. 8.1.

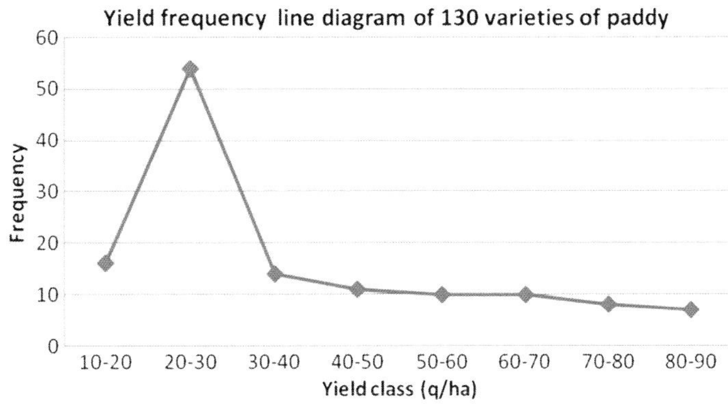

Fig. 8.1 Line diagram

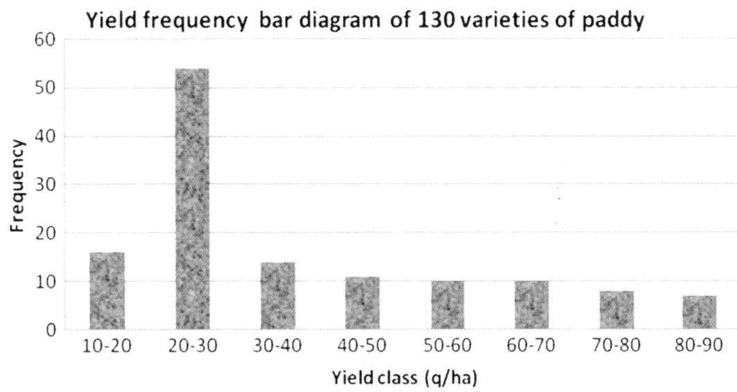

Fig. 8.2 Rice yield data and its corresponding bar diagram

(b) *Bar diagram*: Instead of drawing a line joining the class frequencies, one represents the frequencies in the form of bars. In bar diagrams, equal bases on a horizontal (or vertical) line are selected, and rectangles are constructed with length proportional to the given frequencies on a suitably chosen scale. The bars should be drawn at equal distances from one another (Fig. 8.2).

A more complicated form of bar diagrams is the *clustered* column/bar, *stacked* column and bar, and *100% stacked* column/bar diagram. In clustered bar diagrams, values of the same item for different categories are compared. While in a stacked column, the proportions of the values across the categories are shown. In 100% stacked bar, a comparison of each category is made in such a way to make the following example. With the help of educational status data of five different blocks, let us demonstrate these graphs in Table 8.9, Figs. 8.3 and 8.4.

Table 8.9 Educational status data (%) of five different blocks

Block	Illiterate	School	Graduation	Post graduation
Block1	23.7	34.6	18.0	23.7
Block2	13.2	42.2	23.8	20.8
Block3	18.0	33.0	30.0	19.0
Block4	19.3	14.0	34.0	32.7
Block5	26.5	29.5	24.0	20.0

(c) *Histogram*: Histogram is almost similar to that of a bar diagram for discrete data; the only thing is that the reflection of nonexistence of any gap between two consecutive classes is also reflected by leaving no gap between two consecutive bars. Continuous grouped data are usually represented graphically by a histogram. The rectangles are drawn with bases corresponding to the true class intervals and with heights proportional to the frequencies. With all the class intervals equal, the areas of a rectangle also represent the corresponding frequencies. If the class intervals are not all equal, then the heights are to be suitably adjusted to make the area proportional to the frequencies (Fig. 8.5).

(d) *Frequency Polygon*: If the midpoints of the top of the bars in histogram are joined by straight lines, then a frequency polygon is

Fig. 8.3 Stacked bar diagram of educational status of five blocks

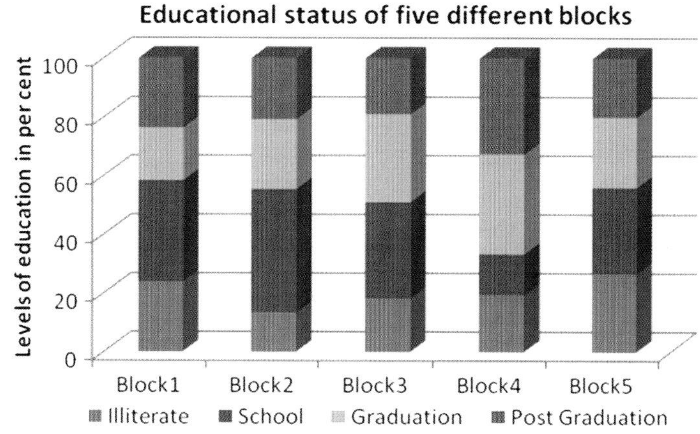

Fig. 8.4 Clustered bar diagram of educational status of five blocks

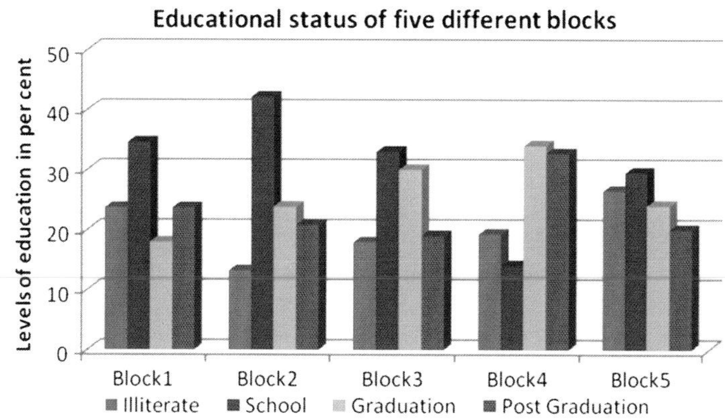

Fig. 8.5 Yield frequency histogram of 130 varieties of paddy

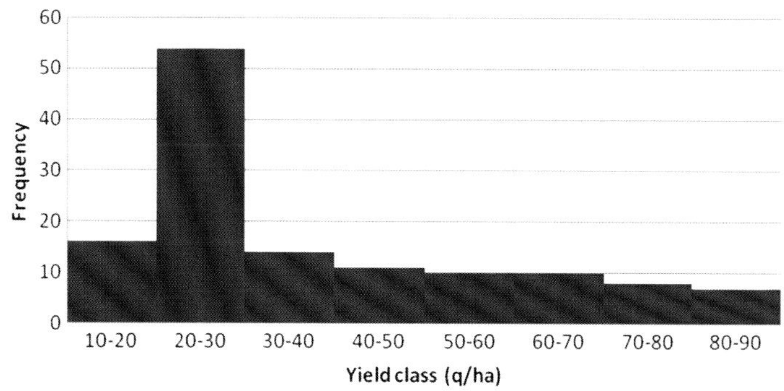

obtained. To complete the polygon, it is customary to join the extreme points at each end of the frequency polygon to the midpoints of the next higher and lower class intervals on a horizontal line (class axis here). For the purpose, generally two hypothetical classes, one before the lowest actual class and another at the last of the highest actual class, are added with zero observation in both cases so that the frequency line diagram completes a bounded area with horizontal *X*-axis as shown below (Fig. 8.6):

(e) *Pie Chart*: The basic idea behind the formation of a pie diagram is to take the whole frequencies in 100% and present it in a circle with 360° angle at the center. In the frequency distribution table, ordinary frequency or relative frequency can effectively be used

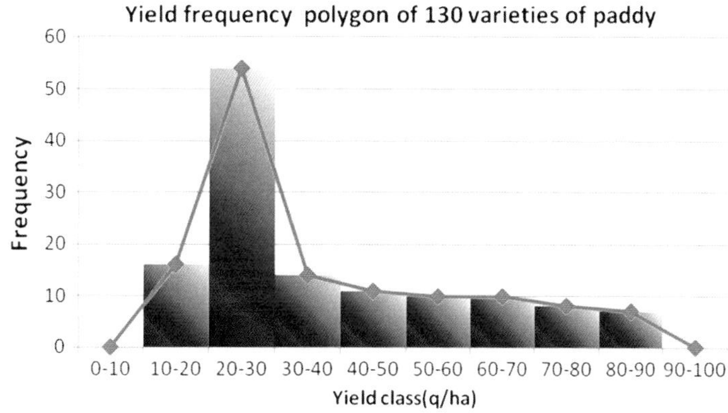

Fig. 8.6 Yield frequency polygon and histogram of 130 varieties of paddy

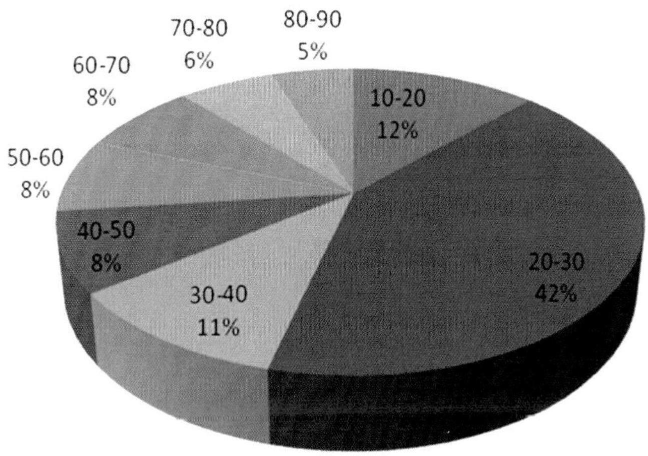

Fig. 8.7 Pie diagram of yield frequency for 130 varieties of paddy

in the form of a pie diagram. Thus, for example, the yield data following a pie chart is prepared with class frequencies (Fig. 8.7). The advantage of pie diagram is that different characteristics measured in different units and or under different situations can be compared with the help of this diagram. Moreover, along with other diagrammatic data presentation, this method is also appealing to both illiterate and educated persons. Taking the example of an expenditure pattern of two groups of people, pie diagram could be used for comparison (Table 8.10 and Fig. 8.8).

Table 8.10 Expenditure pattern of two groups of people

Expenditure (Rs)	Group-1	Group-2
Food	1,800	1,600
Housing	6,500	4,500
Clothing	4,000	3,400
Education	1,250	1,650
Medical	800	500
Others	650	350
Total	15,000	12,000

(f) *Cumulative Frequency Curve (Ogive): Partitioning the whole data set can very well be made with the help of a cumulative frequency graph, also known as OGIVE.* It is of two different types, that is, "less than type" and "more than equal to type." For "less than type," one plots the points with the upper boundaries of the classes as abscissa and the corresponding cumulative frequency as ordinates. The points are joined by a freehand smooth curve. For "more than equal to type" one plots the points with the lower boundaries of the classes as abscissas and the corresponding cumulative frequencies as ordinates. Then, the points are joined by a freehand smooth curve (Fig. 8.9).

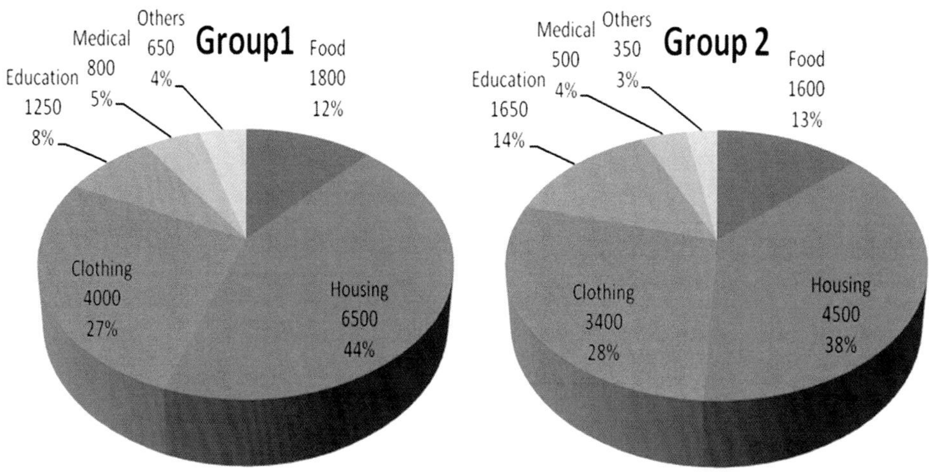

Fig. 8.8 Pie diagram showing the expenditure pattern of two groups of people

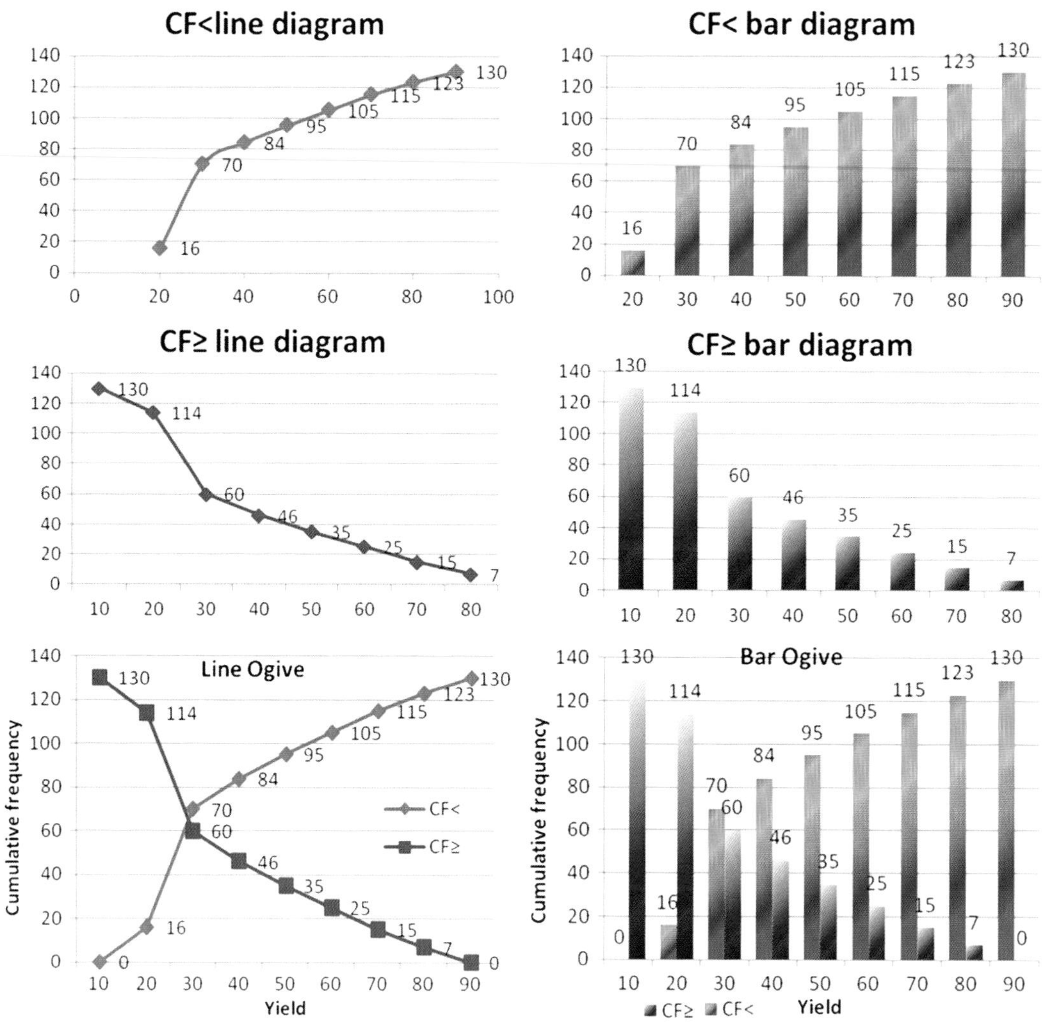

Fig. 8.9 Different forms of cumulative frequency graphs (Ogives)

Fig. 8.10 Pictorial presentation of a frequency distribution of the no. of insect per hill

Class	Frequency	Pictures
3.5 - 6.5	12	
6.5 - 9.5	21	
9.5 -12.5	18	
12.5-15.5	13	
15.5-18.5	16	
18.5-21.5	11	
21.5-24.5	9	

(g) *Pictorial Diagram*: To make the information lively and easy to understand by any user, sometimes information is presented in pictorial forms. Instead of a bar diagram or line diagram or pie chart, one can use pictures in the diagrams. Let us take the example of the no. of particular insects per hill of 100 varieties of paddy; the frequencies can be represented with suitably scaled figure. If we assume that each insect represents four similar insects, then the frequency distribution can be represented in the following forms (Fig. 8.10): This type of pictorial representation is easy to understand even by the layman and is more eye-catching. But the problem with this type of representation is that if the number of figures does not exactly match with the frequency, then some of the drawings are to be kept incomplete like classes 4–7 and 10–13.

(h) *Maps*: Statistical maps are generally used to represent the distribution of particular parameters like a forest area in a country, paddy-producing zone, different mines located at different places in a country, rainfall pattern, population density, etc. The map shown in Fig. 8.11 gives the rainfall distribution during 1-6-12 to 25-7-12 for the whole of India. The essence of a pictorial diagram or map lies in their acceptability to a wide range of users including the illiterate people. This type of data representation is easily conceived by any person, but utmost care should be taken to make the statistical map true to the sense and scale, etc.

It should clearly be noted that all types of presentation are not suitable for all types of data, at all situations, and to all users. The appropriate type of presentation is to be decided on the basis of the type of information, the objective of the presentation, and the person concerned for whom the presentation is basically meant.

8.2 Analysis of Data

Once after the processing and data presentation, it is now imperative for a researcher to explain and describe the nature of the research in a deeper sense. In the first attempt, the researcher tries to explain/describe the nature of the information through measures of central tendency, measures of dispersion, measures of asymmetry, etc., taking one variable at a time—known as univariate analysis. In his/her next endeavor, he/she tries to find out the association among the variables—which are coming under either bivariate (taking two variables at a time) or multivariate analysis (taking more than two variables at a time). Once after completion of the description (through univariate/bivariate/

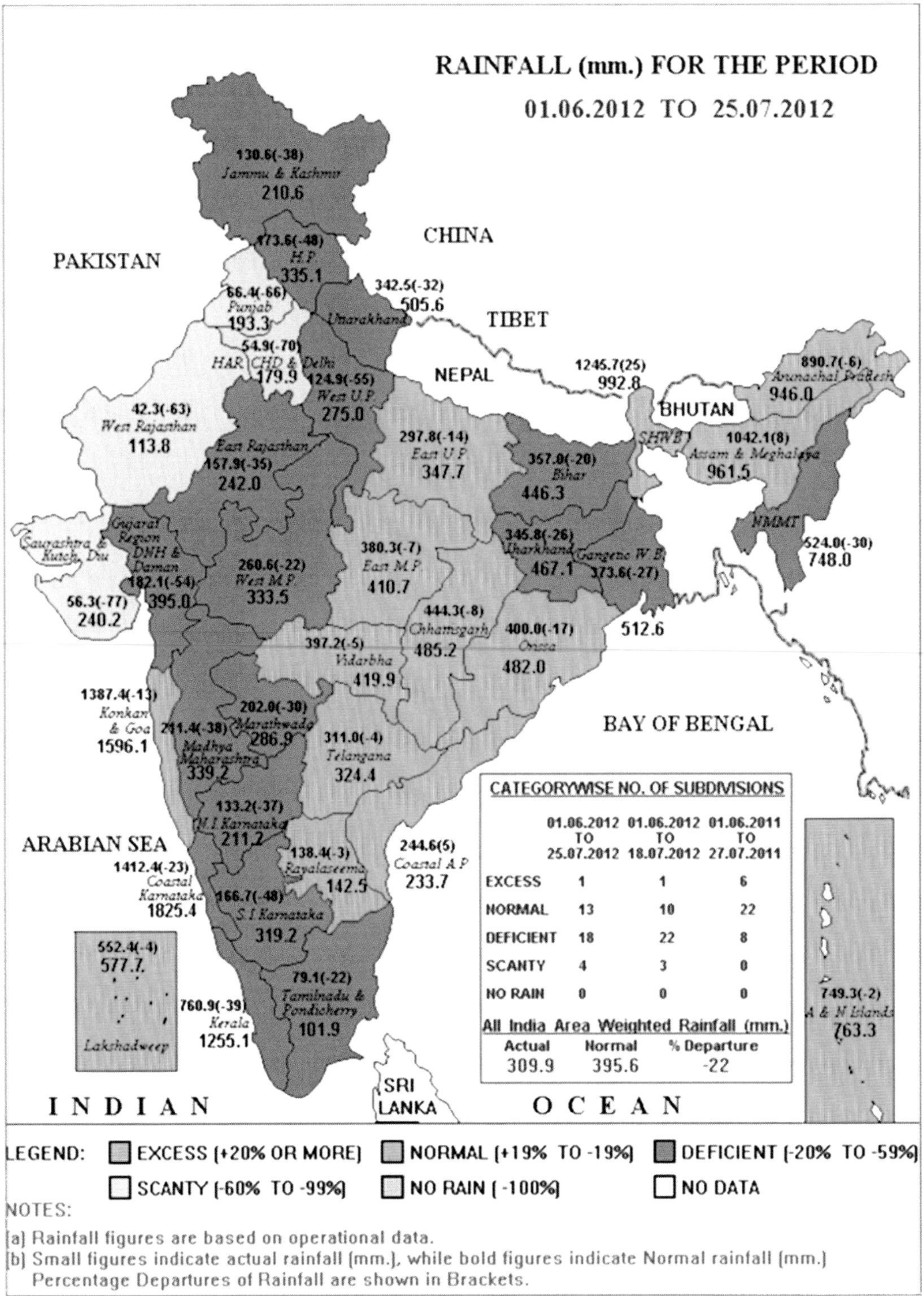

Fig. 8.11 Pictorial presentation of rainfall distribution in India (Source: India meteorological department; http://www.google.com.in, accessed on 18-8-12)

multivariate analysis), the researcher tries to infer or draw conclusion about the population characteristics from the sample behaviors studied so far through inferential statistics. In the following few sections, we shall discuss the descriptive statistics up to bivariate levels. Multivariate and inferential statistics will be taken up in the subsequent chapters.

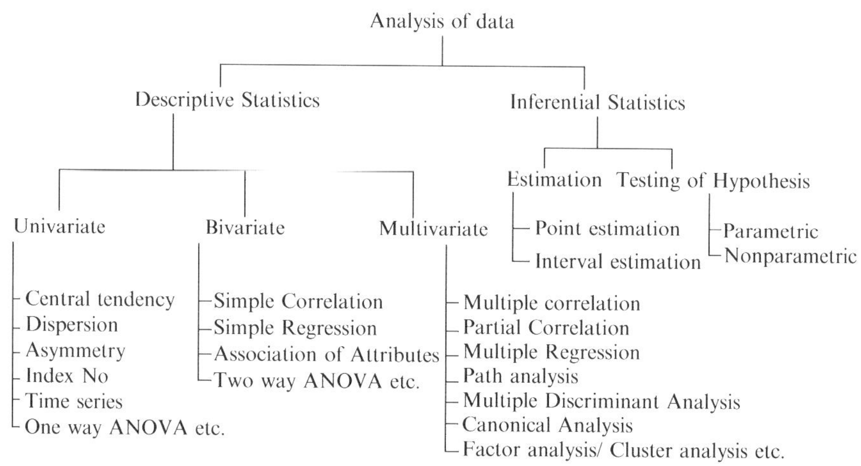

In an attempt to summarize the information/ data, a researcher is always in search of certain value(s) that can represent a set of information given in a big table or otherwise. Thus, we are always in search of such a measure, which can describe the inherent characteristics of a given set of information. Generally, a researcher is in search of the two types of measures, one to find out the central value around which the observations are supposed to lie and another measure for the spread/dispersion/scatteredness of the observations.

8.2.1 Measures of Central Tendency and Locations

Given a set of data, we are in search of a typical value below and above which the observations tend to cluster around. Thus, the tendency of the observations to cluster around a central value is known as central tendency. Sometimes, researchers want to have an idea about specific locations below and above which there is a certain percentage of population. For example, one may be interested in knowing the 50th percentile value, that is, the value below and above which there are half the population, or the 3rd quartile value, that is, the value below which there are 75% of the population and so on. That means we are in search of certain locational values.

According to Yule, a good measure should have the following characteristics:

(a) It should be defined rigidly without any ambiguity
(b) It should be based on all observations
(c) It should be easy to calculate
(d) It should be easy to understand
(e) It should be readily acceptable to mathematical treatments
(f) It should be the least influenced by sampling fluctuations

The Different Measures of Central Tendency Are:

(a) Mean
(b) Median
(c) Mode
(d) Midpoint average, etc.

8.2.1.1 Mean

Means are of three different types: *arithmetic mean, geometric mean, and harmonic* mean.

Arithmetic Mean

The arithmetic mean of a set of observations is their sum divided by a number of observations and is denoted by $\bar{x}$ or μ, that is,

$$\bar{x} = \mu = \frac{\sum_{i=1}^{n} x_i}{n}$$

where $x_1, x_2, \ldots, x_n$ are the values of the first, second, third,. . ., nth observation.

The arithmetic mean for grouped/classified data is given by $\bar{x} = \frac{\sum_{i=1}^{n} f_i x_i}{\sum_{i=1}^{n} f}$, $\left[\sum_{i=1}^{n} f_i = N\right]$, where x_i's and f_i's are the mid values and the frequencies of the ith (i = 1, 2, 3, . . ., n) class.

Properties of Arithmetic Mean

1. It is readily defined.
2. It is easy to calculate.
3. It is easy to understand.
4. It is based on all observations.
5. It is readily acceptable for mathematical treatments.
6. The arithmetic mean of the "n" number of constants (say $x = A$) is also the constant.
7. Arithmetic mean depends on the transformation of data. If $Y = a + b\ X$ be the transformation of X data to Y; where a and b are constants and $\bar{x}$ is the arithmetic mean of the variable X, then $\bar{Y} = a + b\bar{x}$.

   Let us suppose that variable X has an arithmetic mean of 57 and that this variable is related with another variable Y as $Y = 15 + 2.5X$. The arithmetic mean of Y then would be $(15 + 2.5 \times 57) = 157.5$
8. The overall mean of "k" number of arithmetic means from "k" number of samples with observations $n_1, n_2, n_3, \ldots, n_k$ and the arithmetic means $\bar{x}_1, \bar{x}_2, \bar{x}_3, \ldots, \bar{x}_k$ are the weighted average of the arithmetic means. Thus, for

Samples	1	2	3	4	5	. . .	k
No. of observations	n_1	n_2	n_3	n_4	n_5	. . .	n_k
AM	$\bar{x}_1$	$\bar{x}_2$	$\bar{x}_3$	$\bar{x}_4$	$\bar{x}_5$	. . .	$\bar{x}_n$

$$\bar{x} = \frac{\sum_{i=1}^{k} n_i \bar{x}_i}{\sum_{i=1}^{k} n_i}.$$

Example 8.3. The following table gives the number of students and their average height (cm.) of the four different classes of undergraduate courses along with their respective means. Find the overall average age of the undergraduate students.

Class	1st year	2nd year	3rd year	4th year
No. of students	135	132	140	135
Average height in cm.	166	175	188	190

The overall average age of the undergraduate students are given by

$$\bar{x} = \frac{\sum_{i=1}^{k} n_i \bar{x}_i}{\sum_{i=1}^{k} n_i}, \text{ here } k = 4,$$

$$\text{so } \bar{x} = [135 \times 166 + 132 \times 175 + 140 \times 188 + 135 \times 190]/542 = 179.85 \text{ cm.}$$

9. Arithmetic mean cannot be worked out simply by an inspection of data.
10. Arithmetic mean is highly affected by the missing observations or a few large or small observations.
11. Arithmetic mean sometimes seems to be misleading, particularly in the presence of an outlier.
12. AM cannot be calculated if any of the observation is missing.
13. AM cannot be used for open-ended classes because in that case, it is difficult to get the mid value of the class.

Geometric Mean

The geometric mean of a set of "n" observations is defined as the nth root of the product of all observations.

Let $x_1, x_2, \ldots, x_n$ be the "n" number of observations of variable "X"; the geometric mean then is given by

$$X_g = (x_1.x_2 \ldots x_n)^{1/n} = \left(\prod x_i\right)^{1/n}$$

Taking logarithm of both the sides, we have

$$\log(X_g) = \frac{1}{n}\log\left(\prod x_i\right)$$
$$= \frac{1}{n}.[\log x_1 + \log x_2 + \cdots + \log x_n]$$
$$= \frac{1}{n}\sum_{i=1}^{n}\log x_i = A(\text{say})$$

so, $X_g = \text{Antilog}(\log x_g) = \text{Antilog}(A)$.

Thus, the logarithm of a geometric mean is the arithmetic mean of logarithm of the observations.

For grouped/classified data on variable X having $x_1, x_2, \ldots, x_n$ as class mid values with the respective frequencies of $f_1, f_2, f_3, \ldots, f_n$, the geometric mean is given by

$$X_g = \left(x_1^{f_1}.x_2^{f_2}\ldots x_n^{f_n}\right)^{1/\sum_{i=1}^{n} f_i} = \left(\prod x_i^{f_i}\right)^{1/\sum_{i=1}^{n} f_i}.$$

For grouped frequency data, x_i is taken as the mid value of the ith class.

With the help of a log conversion or scientific calculator, one can easily find out the geometric mean.

Example 8.4. Find the geometric mean of the following observations: 20,24,28,32,36,40,44, and 48.

Solution. There are eight numbers of observations. If we denote the geometric mean by G, then $G = (20 \times 24 \times 28 \times 32 \times 36 \times 40 \times 44 \times 48)^{1/8} => \text{Log}(G) = 1/8[\text{Log}(20) + \text{Log}(24) + \text{Log}(28) + \text{Log}(32) + \text{Log}(36) + \text{Log}(40) + \text{Log}(44) + \text{Log}(48)]$

x	Log(x)
20	1.3010
24	1.3802
28	1.4472
32	1.5051
36	1.5563
40	1.6021
44	1.6435
48	1.6812

Thus, $\log(G) = 1/8\ (12.1166) = 1.5146$

So, the antilog of $\log(G) = G = \text{antilog}\ (1.5146) = 32.7039$. So the geometric mean of the given numbers is 32.7039.

Example 8.5. Find the geometric mean from the following frequency distribution.

Variable values	15	20	20	25	30
Frequency	3	4	3	4	6

Solution. This is a simple frequency distribution and we have the geometric mean $X_g = \left(x_1^{f_1}.x_2^{f_2}\ldots x_n^{f_n}\right)^{1/\sum_{i=1}^{n} f_i}$. Taking the logarithm of both sides, we have Log(x_g)

$$\frac{1}{\sum_{i=1}^{n} f_i}\log\left(\prod_{i=1}^{n} x_1^{fi}\right) = \frac{1}{20}[3\log(15) + 4\log(20) + 3\log(20) + 4\log(25) + 6\log(30)]$$
$$= \frac{1}{20}[3 \times 1.1760 + 4 \times 1.3010 + 3 \times 1.3010 + 4 \times 1.3979 + 6 \times 1.4771)]$$
$$= \frac{1}{20}[27.0892] = 1.35446.$$
$$X_g = \text{Alog}(1.35446) = 22.618.$$

Thus, the geometric mean of the above simple frequency distribution is 22.618.

Example 8.6. Find the geometric mean from the following frequency distribution of the stem borer.

No of insect	4.5–5.5	5.5–6.5	6.5–7.5	7.5–8.5	9.5–10.5	10.5–11.5	11.5–12.5
Frequency	14	18	23	16	11	19	4

Solution. From the given information, we make the following table:

X_i	Frequency(f_i)	Log (X_i)	f_i. Log (x_i)
5	14	0.69897	9.78558
6	18	0.77815	14.0067
7	23	0.84509	19.4370
8	16	0.90309	14.4494
9	11	0.95424	10.4966
10	19	1	19.0000
11	4	1.04139	4.16556

$$\text{Thus, } \log(G) = \frac{1}{\sum_{i=1}^{n} f_i} f_1 \sum_{i=1}^{n} \log(x_i) = \frac{91.340}{105}$$

$$= 0.869$$

$$G = \text{Alog}(0.869) = 7.396.$$

Properties

1. It is rigidly defined.
2. It is not so easy to calculate or understand the physical significance.
3. It takes care of all observations.
4. Mathematical treatments are not as easy as in the case of arithmetic mean.
   If $G_1, G_2, \ldots, G_k$ are the geometric means of the samples having $n_1, n_2, \ldots, n_k$ numbers of the observations, then the combined geometric mean is given by

$$G = \left(G_1^{n_1}.G_2^{n_2}\ldots G_k^{n_k}\right)^{1/\sum_{i=1}^{k} n_i} = \prod_{i=1}^{k} G_i^{1/\sum_{i=1}^{k} n_i}$$

$$\text{or } \log G = \frac{1}{\sum_{i=1}^{1} n_i} \sum_{i=1}^{k} n_i \log[G_i].$$

5. If all observations are equal to a constant, say A, then the geometric mean is also equal to A.
6. Geometric mean is not influenced very much by the inclusions of a few large or small observations unlike the arithmetic mean.
7. Geometric mean can't be calculated if any of the observation is zero.
8. GM is useful in the construction of index numbers.
9. As GM gives greater weights to smaller items, it is useful in economic and socioeconomic data.

Harmonic Mean

The harmonic mean of a set of "n" observations is the reciprocal of the arithmetic mean of the reciprocals of the observations.

Let $x_1, x_2, \ldots, x_n$ be the n number of observations. So the arithmetic mean of the reciprocals of the observations is $\frac{\sum_{i=1}^{n} \frac{1}{x_i}}{n}$. Hence, the harmonic mean, HM $= \frac{n}{\sum_{i=1}^{n} \frac{1}{x_i}}$.

For grouped data:

Let $x_1, x_2, \ldots, x_n$ be the mid values of the "n" number of classes with the respective frequencies of $f_1, f_2, f_3, \ldots, f_n$; then, the harmonic mean is given by $\frac{\sum_{i=1}^{n} f_i}{\sum_{i=1}^{n} f_i/x_i}$.

Example 8.7. Find the harmonic mean of the following observations: 15,20,25,30,35,40,45, and 50.

Solution. Here, the number of observations is 8. So their harmonic mean is given by

$$\text{HM} = \frac{n}{\sum_{i=1}^{n} \frac{1}{x_i}} = \frac{8}{\sum_{i=1}^{8} \frac{1}{x_i}}$$

$$= \frac{8}{\frac{1}{15} + \frac{1}{20} + \frac{1}{25} + \frac{1}{30} + \frac{1}{35} + \frac{1}{40} + \frac{1}{45} + \frac{1}{50}}$$

$$= \frac{8}{0.2857} = 28.00.$$

Example 8.8. Find the harmonic mean from the following frequency distribution.

$$\text{HM} = \frac{\sum_{i=1}^{n} f_i}{\sum_{i=1}^{n} f_i/x_i} = \frac{\sum_{i=1}^{5} f_i}{\sum_{i=1}^{5} f_i/x_i}$$

$$= \frac{5+4+3+4+4}{\frac{5}{15}+\frac{4}{20}+\frac{3}{18}+\frac{4}{12}+\frac{4}{16}} = \frac{20}{1.149} = 17.406.$$

Properties

1. It is rigidly defined.
2. It is easier to calculate than geometric mean.
3. It is easy to understand and calculate on the basis of all observations.
4. The harmonic mean of the "n" number of constants is the constant.
5. If any of the observation is zero, then the harmonic mean cannot be defined.

Use of the Different Types of Means:
One has to be selective while choosing the type of means to be used in different situations.

Arithmetic mean is widely used in most of the situations where the data generally do not follow any definite pattern. *Geometric mean* is generally used in a series of observations, both discrete and continuous data, where the values are changing in geometric progression (observation changes in a definite ratio). The average rate of depreciation, compound rate of interest, etc. are some examples where geometric mean can effectively be used. GM is useful in the construction of index numbers. As GM gives greater weights to smaller items, it is useful in economic and socioeconomic data. The *harmonic mean* use is very restricted though it has ample uses in practical fields particularly under changing scenario.

Example 8.9. The price of petrol/diesel/kerosene oil changes frequently, particularly over the growing seasons, and let us assume that a farmer has a fixed amount of money on fuel expenses for running pumps, etc., from his/her monthly farm budget. So, the use of fuel is to be organized in such a way that the above two conditions are satisfied (the monthly expenditure on fuel remains constant and the prices of fuel changes over the months); that is, the objective is to get an average price of fuel per unit which suggests the amount of average consumption of fuel.

Solution. Let the monthly expenditure on fuel be Rs "F" and the prices of fuel of "K" for the consecutive months be $p_1, p_2, \ldots, p_k$, respectively. Then, average monthly consumption of fuel is given by

$$= \frac{KF}{\frac{F}{p_1}+\frac{F}{p_2}+\cdots+\frac{F}{p_K}}$$

$$= \frac{KF}{F\left(\frac{1}{p_1}+\frac{1}{p_2}+\cdots+\frac{1}{p_K}\right)}$$

$$= \frac{K}{\sum_{i=1}^{K} 1/p_i} = \text{Harmonic mean of price of fuel.}$$

The relationship of Arithmetic Mean, Geometric Mean, and Harmonic Mean:

Let $x_1, x_2, x_3, \ldots, x_n$ be the n positive values of variable "X"; then, for arithmetic mean, $A = \frac{1}{n}\sum_{i=1}^{n} x_i$, geometric mean, $G = \left(\prod_{i=1}^{n} x_i\right)^{1/n}$, and harmonic mean, $H = \frac{n}{\sum_{i=1}^{n}\frac{1}{x_i}}$; $A \geq G \geq H$.

Thus, for a given set of data which is meant to be used must be decided based on the nature of the data and its purpose of use.

Example 8.10. Four different blends of teas at Rs 200, Rs 250, Rs 400, and Rs 150 per kilogram. He/she then mixed these four types of teas that are to be sold. What could be the minimum selling price (per kilogram) of tea?

Solution. Let a shopkeeper spend Rs(X) for each quality of tea; then, he has spent altogether $4x$ the amount of his/her money. In the process, he/she has bought $X/200$, $X/250$, $X/400$, and $X/150$ kg of tea, respectively, for four different qualities of tea. So the average purchasing price of tea is $4X/(X/200 + X/250 + X/400 + X/150)$; this is nothing but the harmonic mean of the individual prices. Thus, the average purchasing price is

$$= 4/(1/200 + 1/250 + 1/400 + 1/150)$$
$$= 4/(0.005 + 0.004 + 0.0025 + 0.006)$$
$$= 4\ /0.0175 = 228.57 \text{ per kilogram of tea.}$$

To meet the other charges and to have some profit by selling the same product, he/she should sell the tea mixture at more than Rs 228.57 per kilogram of tea.

8.2.1.2 Median

The median divides the whole set of data into two equal halves; below and above the median, there are equal numbers of observations. The median of a set of observations is defined as the value of the middle-most observation when the observations are arranged either in ascending or descending order.

Median

$$= \begin{cases} \frac{n+1}{2}\text{th observation when } n \text{ is odd} \\ \frac{1}{2}\left[\frac{n}{2}\text{th observation} + \left(\frac{n}{2}+1\right)\text{th observation}\right] \\ \text{when } n \text{ is even.} \end{cases}$$

Of course, at first, the observations should be arranged in increasing or decreasing order. Then, the median value is to be worked out.

For grouped data, the median is calculated as

$$\text{Me} = x_l + \frac{N/2 - F\text{me} - 1}{f\text{me}}.\text{CI},$$

where x_l is the lower class boundary of the median class,

N is the total frequency, $F_{\text{me}-1}$ is the cumulative frequency (less than type) of the class preceding the median class, f_{me} is the frequency of the median class, and CI is the width of the median class.

Steps in Calculating the Median:

1. Identify the median class, that is, the class having the N/2th observation from the cumulative frequency (less than the type) column.
2. Identify the lower class boundary (x_l), the class width (CI), and the frequency (f_{me}) values of the median class.
3. Identify the cumulative frequency (less than the type) of the class preceding the median class ($F_{\text{me}-1}$).
4. Use the above values in the formula for median.
5. Median will be having the same unit as that of the variable.

Example 8.11. The number of insects per plant for the 10 plants of a particular variety of rose is given as follows: 17, 22, 21, 13, 29, 23, 15, 16, 25, 27. The objective is to find out the median value of the variable number of insect per plant.

Solution. Now if we arrange the data in ascending order, then it will become 13, 15, 16, 17, 21, 22, 23, 25, 27, and 29. The middle-most observations are the 5th and 6th observations; hence, the median value for the above data set is 21.5, that is, (21 + 22)/2 = 21.5.

Example 8.12. Find the median value of yield (q/ha) from the following frequency table.

Yield classes	Frequency (f_i)	Mid value (x_i)	Cumulative frequency (CF<)
10–20	16	15	16
20–30	54	25	70
30–40	14	35	84
40–50	11	45	95
50–60	10	55	105
60–70	10	65	115
70–80	8	75	123
80–90	7	85	130

1. The total number of observation is 130; N/2 = 65.
2. The median class is 20–30.
3. The lower boundary (x_l) is 20, the class width (CI) is 10, and the frequency (f_{me}) is 54 of the median class.
4. The cumulative frequency (less than the type) of the class preceding the median class ($F_{\text{me}-1}$) is 16.

So the median of the above frequency distribution is

$$\text{Me} = 20 + \frac{65 - 16}{54} \times 10 = 29.074 \text{ q/ha}.$$

Alternatively, the approximate median value can also be worked out from the intersection point of the two cumulative frequency (less than and more than that type) curves.

Properties of Median:
1. Median is easy to calculate and understand.
2. For its calculation, it does not require to have all observations.
3. For qualitative data also, median can be worked out.
4. Median cannot be put under mathematical treatments like AM and GM.
5. Arrangements of data are necessary for the calculation of median.

Uses of Median: Median is a useful measure for both quantitative and qualitative characters. Using the measure in agriculture, socioeconomic and other field researchers divide the whole population into parts for any subsequent action-oriented research program.

Percentiles, Deciles, and Quartiles (Partition Values):
By modifying the formula for median, different percentiles/deciles/quartiles can very well be worked out to divide the whole population into as many groups as one would like to have. Just by substituting $N/2$ in the median formula by "$Np/100$," "$Nd/10$," or "$Nq/4$" and the corresponding cumulative frequencies (less than the type), where "p," "d," and "q" denote the pth percentile, dth decile, and qth quartile, respectively, one can get different percentile/decile/quartile values.

Thus, the formula for some percentiles, deciles, or quartiles is as follows:

50th percentile or P_{50}

$$= x_l + \frac{50N/100 - F_{p_{50} - 1}}{f_{p_{50}}}.\mathrm{CI} = \text{Median}$$

where x_l is the lower class boundary of the 40th percentile class,

N is the total frequency, $F_{p_{50}-1}$ is the cumulative frequency (less than the type) of the class preceding the 50th percentile class, $f_{p_{50}}$ is the frequency of the 50th percentile class, and CI is the width of the 40th percentile class.

$$\text{8th decile or } D_8 = x_l + \frac{8N/10 - F_{d_8 - 1}}{f_{d_8}}.\mathrm{CI}$$

where x_l is the lower class boundary of the 8th decile class,

N is the total frequency, $F_{d_8 - 1}$ is the cumulative frequency (less than the type) of the class preceding the 8th decile class, f_{d_8} is the frequency of the 8th decile class, and CI is the width of the 8th decile class.

$$\text{3rd quartile or } Q_3 = x_l + \frac{3N/4 - F_{q_3 - 1}}{f_{q_3}}.\mathrm{CI}$$

where x_l is the lower boundary of the 3rd quartile class,

n is the total frequency, Fq_{3-1} is the cumulative frequency, (less than the type) of the class preceding the 3rd quartile class, $f_{q_3}$is the frequency of the 3rd quartile class, and CI is the width of the 3rd quartile class.

Example 8.13. Find the P_{50}, D_8, and Q_3 values of yield (q/ha) from the following frequency table.

Yield classes	Frequency (f_i)	Mid value (x_i)	Cumulative frequency (CF<)
10–20	16	15	16
20–30	54	25	70
30–40	14	35	84
40–50	11	45	95
50–60	10	55	105
60–70	10	65	115
70–80	8	75	123
80–90	7	85	130

$$P_{50} = x_l + \frac{50N/100 - F_{p_{50} - 1}}{f_{p_{50}}}.\mathrm{CI}$$

$$= 20 + \frac{50.130/100 - 16}{54} \times 10 = 29.074$$

$$D_8 = x_l + \frac{8N/10 - F_{d_8 - 1}}{f_{d_8}}.\mathrm{CI}$$

$$= 50 + \frac{8 \times 130/10 - 95}{10} \times 10 = 59$$

$$Q_3 = x_l + \frac{3N/4 - F_{q_3 - 1}}{f_{q_3}}.\mathrm{CI}$$

$$= 50 + \frac{3 \times 130/4 - 95}{10} \times 10 = 52.5$$

Thus, from the above results, one can find that 50% (P_{50}) plants have 29.074 q/ha or less yield, 75% (Q_3) plants have 52.5 q/ha or less yield, and 80% (D_8) plants 59 q/ha or less yield.

8.2.1.3 Mode

The mode of a variable is defined as the value of the observation having a maximum frequency.

Example 8.14. The number of accidents per day on a specific road is given below:

15, 8, 7, 15, 10, 12, 13, 9, 10, 15, 8, 9, 10, 13,9,12, 10, 8, 10, and 7. Find the mode of the number of accidents per day.

No. accidents per day	7	8	9	10	12	13	15
Frequency	2	3	3	5	2	2	3

From the above arrangement, it is found that the maximum frequency is for the 10 accidents per day. Hence, the mode is 10.

From a *grouped/classified frequency distribution*, the mode cannot be worked out as above. The mode from a grouped frequency distribution is calculated with the help of the following formula:

$$\mathrm{Mo} = x_l + \frac{f_{mo} - f_{mo-1}}{(f_{mo} - f_{mo-1}) + (f_{mo} - f_{mo+1})}.\mathrm{CI},$$

where x_l is the lower class boundary of the modal class,

f_{me-1} is the frequency of the class preceding the modal class,

f_{mo} is frequency of the modal class,

f_{me+1} is the frequency of the class following the modal class, and

CI is the width of the modal class.

Steps in Calculating the Mode:

1. Identify the modal class, that is, the class having the maximum observation from the frequency column.
2. Identify the lower class boundary (x_l), the class width (CI), and the frequency (f_{mo}) values of the modal class.
3. Identify the frequency of the class preceding the modal class (f_{m-1}) and also the frequency of the class following the modal class (f_{m+1}).
4. Use the above values in the formula for mode.
5. Mode will be having the same unit as that of the variable.

Example 8.15. Find the P_{50}, D_8, and Q_3 values of yield (q/ha) from the following frequency table.

Yield classes	Frequency (f_i)	Mid value (x_i)	Cumulative frequency (CF<)
10–20	16	15	16
20–30	54	25	70
30–40	14	35	84
40–50	11	45	95
50–60	10	55	105
60–70	10	65	115
70–80	8	75	123
80–90	7	85	130

1. The total number of observation is 130.
2. The modal class is 20–30.
3. The lower class boundary (x_l) is 20, the class width (CI) is 10, and the frequency (f_m) is 54 of the modal class.
4. The frequency of the class preceding the modal class (f_{m-1}) is 16 and the frequency of the class following the modal class (f_{m+1}) is 14.

So the mode of the above frequency distribution is

$$\mathrm{Mo} = 20 + \frac{54 - 16}{(54 - 16) + (54 - 14)} \times 10$$

$$= 24.87\,\mathrm{q/ha}.$$

Properties of Mode:

(a) Mode is easy to calculate and understand.

(b) For its calculation, it does not require to have all observations.

(c) Mode cannot be put under mathematical treatments like AM and GM.

(d) It is least affected by the presence of extreme values.

(e) A distribution may have one or more (when two or more values have same frequency) modes. If a distribution has more than two modes, it is said to be a multimodal distribution.

Uses of Mode: Mode is of little use unless the number of observations is very high. Mode can best be used in case of qualitative characters like a race of people, awareness pattern of people of certain locality, and types of crop or cropping pattern grown in a particular locality.

8.2.1.4 Midpoint Range

Midpoint range is simply the arithmetic mean of the lowest and highest value of a given set of data. If L and U are the lowest and highest values of a given set of data, respectively, the midpoint range (MDr) is $(L + U)/2$. As such, it is devoid of the many good properties of average mentioned in this section for different averages. It takes care of only the two extreme values and as such affected by these values. Even then it can provide some sort of information about the data.

Selection of Proper Measure of Central Tendency: All the measures of central tendency cannot be used everywhere. The selection of appropriate measure should be based on merits and demerits of the measures of central tendency. Qualitative data can be measured through median and mode, but these two measures are not based on all observations. Among the three means, the AM$\geq$ GM $\geq$ HM, the AM suffers from extreme value while the GM is suitable for data which changes in definite ratio or rate. Thus, the nature of data and the objective of the study along with the characteristics of the measures are the major point of consideration during the selection of appropriate measure of central tendency.

8.2.2 Measures of Dispersion, Skewness, and Kurtosis

The essence of analysis of research data is to unearth the otherwise hidden truth from a set of data. In this direction, if the measures of central tendency be the search for a value around which the observations have the tendency to center around, then dispersion is a search for a spread of the observations within a given data set. Thus, if central tendency is the thesis, then dispersion is the antithesis. The tendency of the observations of any variable to remain scattered/dispersed from a central value or any other value is known as dispersion of the variable. A researcher must have good knowledge about the central tendency and the dispersion of the research data he/she is handling to discover the truth that had remained hidden so long. This is more essential because neither the measure of central tendency nor the measure of dispersion in isolation can reveal the nature of the information.

Let us take the following example:

Example 8.16. To measure the innovation index, ten persons from each of the society were studied, and the following table gives the indices individually.

	Innovation index									
Society 1	15.00	16.50	16.00	18.00	17.00	18.50	19.00	18.50	17.50	14.00
Society 2	10.50	21.50	22.50	9.00	24.50	11.50	23.00	20.00	17.00	10.50

From the above, one can have the arithmetic means, $\bar{X}_1 = (15 + 16.5 + \cdots + 14)/10 = 7$ and $\bar{X}_2 = (10.50 + 21.50 + \cdots + 10.50)/10 = 17$. Thus, the central tendency measured in terms of arithmetic mean for innovation index for the above two societies is same. But a critical examination of the data reveals that in society 1, the innovation remains in between 14 and 19; on the other hand, in society 2 it remains between 9 and 24.5. Thus, the innovation indices are more dispersed in society 2 than in society 1, in spite of having the same average performance for both the societies. Thus, from the above results, two conclusions can be drawn: (a) the maximum innovation potentiality and innovation variability of society 1 is less than in society 2 and (b) given a better attention, people in society 2 can be better innovative by attaining its full potentiality.

Measures of Dispersion: Like the measures of central tendency, there is a need to have measures for dispersion also. In fact, different measures of dispersions are available in the theory of statistics. Before going into the details on the discussion of the different measures of dispersion, let us try to examine the characteristics of a good measure of dispersion. There should not be any ambiguity in defining a measure; it should be clear and rigid in definition. Unless a measure is convincing, that is, easily understood and applicable by the user, it is of least importance. For further application of a measure, it should be put easily under mathematical treatments. In order to reflect the true nature of the data, a good measure should try to take care of all the observations, and it should lay equal importance to each and every observation without being affected by the extreme values.

Measures of dispersion

Absolute measure	**Relative measures**
a) Range	a) Coefficient of quartile deviation
b) Mean deviation	b) Coefficient mean deviation
c) Quartile deviation	c) Coefficient of variation
d) Standard deviation	
e) Moments	

8.2.2.1 Absolute Measures of Dispersion

The absolute measures of dispersion have the units according to those of variables, but relative measures are pure number; as such are *unit-free* measures. Thus, unit-free measures can be used to compare distributions of different variables measured in different units.

(a) *Range*: A range of a set of observations is the difference between the maximum value and the minimum value of a set of data. Thus, $Rx = X_{max} – X_{min}$ is the range of variable "X" for a given set of observations. In the above example, the ranges for two societies are $19 - 14 = 5$ and $24.5 - 9 = 15.5$, respectively, for society 1 and society 2.

Uses of Range: Range can be used in any type of continuous or discrete variables. Range has its uses in the field of stock market (daily variation), in meteorological forecasting, in statistical quality control, etc.

Advantages and Disadvantages of Range:

1. Range is rigidly defined and can be calculated easily.
2. Though range is not based on all the observations, it cannot be worked out if there are missing values.
3. Range is very much affected by sampling fluctuations.

In spite of all these drawbacks, range is being used in many occasions only because of its simplicity and of having a first-hand information on the variation of the data.

(b) *Mean Deviation*: The mean deviation of variable "X" (x_1, x_2, $x_3, \cdots$, x_n) about any arbitrary point "A" is defined as the mean of the absolute deviation of the different values of the variable from the arbitrary point "A" and is denoted as $\mathrm{MD}_A = \frac{1}{n}\sum_{i=1}^{n} X_i - A$, $\frac{1}{\sum_{i=1}^{n} f_i} \times \sum_{i=1}^{n} f_i |X_i - A|$ for grouped data.

Similarly, instead of taking the deviation from the arbitrary point "A," one can take the deviation from the arithmetic mean, median, or mode also. Actually, the mean deviation of variable "X" is defined as the mean of the absolute deviation of different values of the variable from the arithmetic mean $\bar{X}$ of the variable and is denoted as MD $= \frac{1}{n}\sum_{i=1}^{n} X_i - \bar{X}$, $\frac{1}{\sum_{i=1}^{n} f_i}\sum_{i=1}^{n} f_i|X_i - \bar{X}|$ for grouped data.

Similarly, the mean deviation from median is denoted as $\mathrm{MD_{Me}} = \frac{1}{n}\sum_{i=1}^{n} X_i - \mathrm{Me}$, $\frac{1}{\sum_{i=1}^{n} f_i}\sum_{i=1}^{n} f_i|X_i - \mathrm{Me}|$ for grouped data, and mean deviation from mode is $\mathrm{MD_{Mo}} = \frac{1}{n}\sum_{i=1}^{n} X_i - \mathrm{Mo}$, $\frac{1}{\sum_{i=1}^{n} f_i}\sum_{i=1}^{n} f_i|X_i - \mathrm{Mo}|$ for grouped data

For a grouped frequency distribution, x_i is taken as the mid value of the ith class.

(c) *Quartile Deviation*: Quartile deviation is defined as half of the difference between the 3rd and 1st quartile values QD $= \frac{Q_3 - Q_1}{2}$; that is why, it is also called as semi–inter-quartile range.

(d) *Standard Deviation*: Standard deviation is defined as the positive square root of the arithmetic mean of the square of the deviations of the observations from arithmetic mean and is written as $\sigma_X = +\sqrt{\frac{1}{n}\sum_{i=1}^{n}(Xi - \bar{X})^2}$ for raw data and $\sigma_X = +\sqrt{\frac{1}{\sum_{i=1}^{n} fi}\sum_{i=1}^{n} fi(X_i - \bar{X})^2}$ for grouped data, where x_i ($i = 1, 2, 3, \ldots, n$) are the mid values of the respective classes with the frequency of f_i ($i = 1, 2, 3, \ldots, n$) and $\bar{X}$ is the mean of the variable X. The squared quantity of the standard deviation is known as the *variance* of the variable.

Thus, *variance* is the mean squared deviation from the mean for a given set of data and is written as

$$\sigma_X^2 = \frac{1}{n}\sum_{i=1}^{n}(Xi - \bar{X})^2$$

$$= \frac{1}{n}\sum_{i=1}^{n} Xi^2 - \bar{X}^2 \text{ for raw data and}$$

$$\sigma_X^2 = \frac{1}{\sum_{i=1}^{n} f_i}\sum_{i=1}^{n} f_i(X_i - \bar{X})^2$$

$$= \frac{1}{\sum_{i=1}^{n} f_i}\sum_{i=1}^{n} f_i X_i^2 - \bar{X}^2$$

for grouped/classified data.

Properties of Variance:

1. Variance is rigidly and clearly defined.
2. Variance has the range between 0 to ∞; when all the observations are equal (i.e., the variable remains no longer a variable), then variance/standard deviation is 0.
3. Variance is based on all observations.
4. Variance cannot be worked out for frequency distribution with open-ended classes.
5. Variance is amenable to mathematical treatments.
6. Variance or standard deviation is least affected by the sampling fluctuations.
7. Variance does not depend on the change of origin but depends on the change of scale. That means, two variables "X" and "Y" are related in the form of $Y = a + bX$, where a and b are two constants known as the change of origin and scale, respectively; if the variance of "X" be σ_x^2 then the variance of "Y" would be $b^2\sigma_x^2$

$$\sigma_y^2 = \frac{1}{\sum_{i=1}^{n} f_i}\sum_{i=1}^{n} f_i(Y_i - \bar{Y})^2$$

$$= \frac{1}{\sum_{i=1}^{n} f_i}\sum_{i=1}^{n} f_i\ (a + bX_i - a - b\bar{X})^2$$

$$= \frac{1}{\sum_{i=1}^{n} f_i}\sum_{i=1}^{n} b^2 f_i\ (X_i - \bar{X})^2 = b^2\sigma_X^2.$$

Thus, the variance of Y depends only on the change of scale, not on the change of origin. Similarly, $\sigma_y = \mathrm{s.d.}(y) = |b|\sigma_X = |b|\mathrm{s.d.}(x)$.

8. The combined variance of "k" number of samples with n_1, n_2, n_3, …, n_k and $\bar{X}1$, $\bar{X}2$, $\bar{X}3$,…,$\bar{X}k$ as number of observations and means, respectively, is given by $\sigma^2 = \frac{1}{\sum_{i=1}^{k} n_i}\left[\sum_{i=1}^{k} n_i\sigma_i^2 + \sum_{i=1}^{k} n_i d_i^2\right]$, where σ_i^2 is the variance of the ith sample with "n_i" observations, and by $d_i = x_i - \bar{X}$, where $\bar{X}$ is the combined mean of all the samples.

Example 8.17. The following data gives the general features of plant height (ft) of maize plants for two different samples. Find the composite variance of the maize plants.

Characteristics	Sample 1	Sample 2
Sample size	65	35
Mean height (ft)	5.34	5.49
Sample variance (ft^2)	3.89	2.25

Solution. Combined mean height is given by $\bar{X} = [n_1\bar{X}_1 + n_2\bar{X}_2]/[n_1 + n_2] = 5.4$ and combined variance is given by $\sigma^2 = \frac{1}{\sum_{i=1}^{2} n_i}\left[\sum_{i=1}^{2} n_i{\sigma_i}^2 + \sum_{i=1}^{2} n_i{d_i}^2\right] = 2.21$. So one can conclude that the average height and the combined variance of the samples are 6.4 ft and 2.21 ft^2, respectively.

Example 8.18. If the relation between two variables is given as $Y = 25 + 4.5X$ and the $\sigma_X^2 = 4.2$, then find the standard deviation of Y.

Solution. We know that $\sigma_Y^2 = b^2\ \sigma_X^2$, here $b = 4.5$, so $\sigma_Y^2 = (4.5)^2 \times 4.2 = 85.05$.

Standard deviation $\sigma_Y = +\sqrt{(85.05)} = 9.22$.

Thus, from the discussion of the above measures of dispersion, one can find that the standard deviation/variance follows almost all the qualities of a good measure. As a result of which, this is being extensively used as measure of dispersion.

(e) Moments

The rth raw moment about an arbitrary point "A" is the mean of the rth power of the deviation of the observations from the point "A" and is denoted as

$$\mu'_r(A) = \frac{1}{N}\sum_{i=1}^{N}(x_i - A)^r$$

for raw data and for grouped data

$$\mu'_r(A) = \frac{1}{\sum_{i=1}^{n} f_i}\sum_{i=1}^{n} f_i(x_i - A)^r$$

where variate X takes the values x_i (i =1,2, 3,4,…,N) for raw data and the mid value x_i (i = 1,2,3,4,…,n) with respective frequency of f_i (i = 1,2,3,…,n) for n classes/groups and $\sum_{i=1}^{n} f_i = N$.

If we take $A = 0$, then we get the moment about the origin

$$\upsilon_r = \frac{1}{\sum_{i=1}^{n} f_i}\sum_{i=1}^{n} f_i x_i^r \text{ and putting } A = \bar{x}, \text{ and}$$

$$m_r = \frac{1}{\sum_{i=1}^{n} f_i}\sum_{i=1}^{n} f_i(x_i - \bar{x})^r,$$

the rth central moment.

It can be noted that

$$\mu'_0(A) = \upsilon_0 = m_0 = 1 \text{ and}$$
$$\upsilon_1 = \bar{x},\ m_1 = 0,\ m_2 = \sigma^2$$

If $y = (x-c)/d$, then $m_r(x) = d^r\ m_r(y)$, where "c" and "d" are constants. Since $x_i = c + dy_i$, i =1,2,3,4,…,n, and $\bar{x} = c + d\bar{y}$, we have

$$m_r(x) = \frac{1}{\sum_{i=1}^{n} f_i}\sum_{i=1}^{n} f_i(c + dy_i - c - d\bar{Y})^r$$
$$= \frac{1}{\sum_{i=1}^{n} f_i}\sum_{i=1}^{n} f_i d^r(y_i - \bar{Y})^r = d^r m_r(y)$$

8.2.2.2 Relative Measures of Dispersion

Relative measures of dispersions are mainly coefficients based on the absolute measures, also known as coefficients of dispersion, are unit-free measures, and are mostly ratios or percentages.

1. *Coefficient of dispersion based on range*: It is defined as $\frac{X_{max} - X_{min}}{X_{max} + X_{min}}$; X_{max} and X_{min} are

the maximum and minimum values of variable "*X*."

2. *Coefficient of dispersion based on quartile deviation*: It is defined as $\frac{\frac{Q_3-Q_1}{2}}{\frac{Q_3+Q_1}{2}} = \frac{Q_3-Q_1}{Q_3+Q_1}$
3. *Coefficient of dispersion based on mean deviation from mean / median / mode / arbitrary point*: This measure is given by $\frac{\text{MD mean/median/mode etc.}}{\text{Mean/Median/Mode}}$
4. *Coefficient of dispersion based on standard deviation*: It is defined as $\frac{\sigma_X}{\bar{X}}$, where σ_X and $\bar{X}$ are the standard deviation and arithmetic mean of variable "*X*," respectively. The more widely used and customary coefficient of dispersion based on standard deviation is the "*coefficient of variation (CV)*" $\frac{\sigma_X}{\bar{X}} \times 100$.

8.2.2.3 Skewness and Kurtosis

Neither the measure of central tendency nor the measure of dispersion alone is sufficient to extract the inherent characteristics of a given set of data. We need to combine both these measures together. We can come across with a situation where two frequency distributions have the same measures of central tendency as well as measure of dispersion but they differ widely in their nature.

Example 8.20. Given below are the two frequency distributions for a number of nuts per bunch with respective means as 12.62 and 12.11 and standard deviations as 3.77 and 3.484, respectively. Thus, both the distributions have almost the same measure of central tendency as well as measure of dispersion. But a close look at the graphs drawn for the two frequency distributions shows that they differ widely in nature (Fig. 8.12).

Along with the measures of dispersion and central tendency, there should be certain measures which can provide the exact picture of the given data set. *Skewness* and *kurtosis* talk about the nature of the frequency distribution.

The *skewness* of a frequency distribution is the departure of the frequency distribution from symmetricity.

A frequency distribution is either *symmetric* or *asymmetric/skewed*. Again an asymmetric/ skewed distribution may be *positively skewed* or *negatively skewed*.

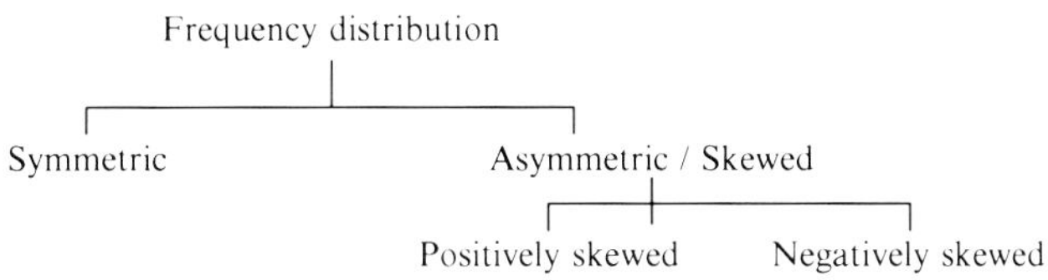

Example 8.21. Let us take the example of the panicle length in three different varieties of paddy. The frequency distributions and the corresponding graphical representations are given in Figs. 8.13, 8.14, and 8.15.

Among the different measures, the measures based on moments are mostly used and are given as $\sqrt{\beta_1} = \gamma_1 = \frac{m_3}{\sqrt{m_2^3}}$, where m_3 and m_2 are the 3rd and 2nd central moments, respectively. It may be noted that all the measures of skewness have no units; these are pure numbers and equal to zero when the distribution is symmetric. Further, it is to be noted that $\beta_1 = \frac{m_3^2}{m_2^3} \geq 0$. The sign of $\sqrt{\beta_1}$ depends on the sign of m_3.

Kurtosis refers to the peakedness of a frequency distribution. While skewness refers to the horizontal property of the frequency distribution, kurtosis refers to the vertical nature of the frequency distribution. According to the nature of peak, a distribution is *leptokurtic*, *mesokurtic*, or *platykurtic* in nature. Kurtosis is measured in terms of $\beta_2 = \frac{m_4}{m_2^2}$, where m_4 and m_2 are the 4th and 2nd central moments, respectively. If $\beta_2 > 3, \beta_2 = 3$, the distribution is mesokurtic, and if $\beta_2 < 3$, the distribution is platykurtic. In the case of a normal distribution mentioned in Chap. 6, if this is taken as standard, the quantity $\beta_2 - 3$ measures the what is known as excess of kurtosis. If $\beta_2 > 3$, the distribution is leptokurtic, and $\beta_2 = 3$, the distribution is

a

Distribution 1

Class	Frequency
3.5-6.5	11
6.5-9.5	27
9.5-12.5	45
12.5-15.5	54
15.5-18.5	34
18.5-21.5	8
Mean	12.62
SD	3.77

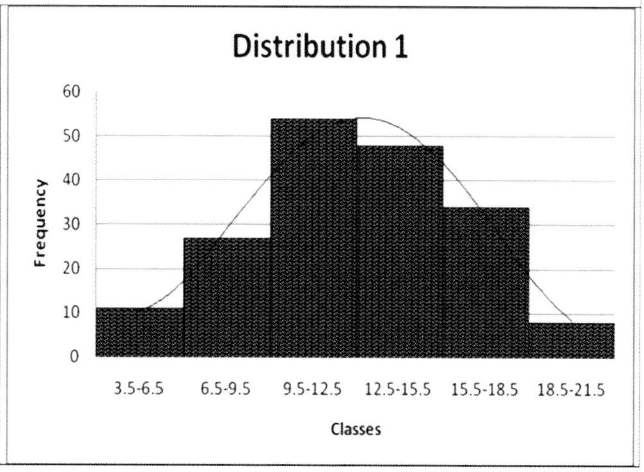

b

Distribution 2

Class	Frequency
3.5-6.5	9
6.5-9.5	22
9.5-12.5	64
12.5-15.5	33
15.5-18.5	27
18.5-21.5	4
Mean	12.11
SD	3.48

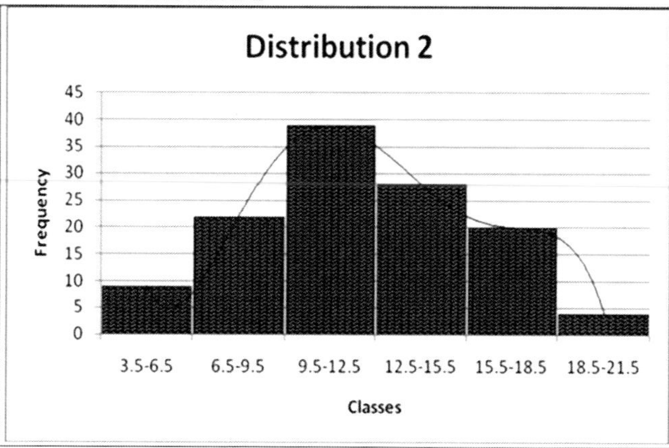

Fig. 8.12 (**a**) Graphical presentation of a no. of nuts per plant in 179 plants of coconut, (**b**) Graphical presentation of a no. of nuts per plant in 159 plants of coconut

Symmetric Distribution

Class	Frequency
2.5-7.5	14
7.5-12.5	18
12.5-17.5	22
17.5-22.5	30
22.5-27.5	22
27.5-32.5	18
32.5-37.5	14
Mean	20
Median	20
Mode	20

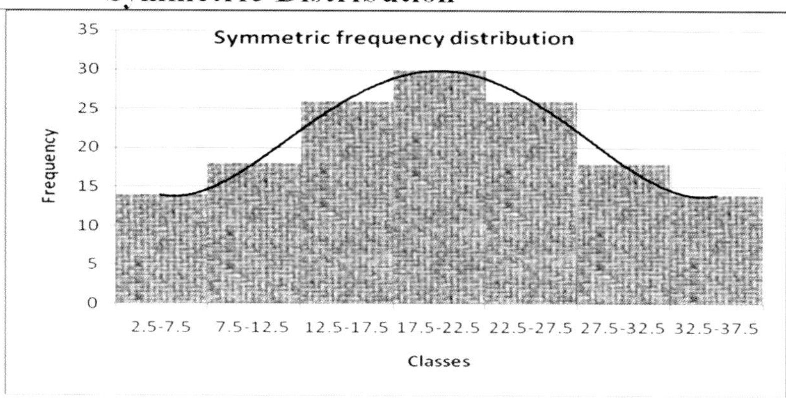

Fig. 8.13 Graphical presentation of a symmetric distribution

Positively Skewed Distribution

Class	Frequency
2.5-7.5	14
7.5-12.5	24
12.5-17.5	34
17.5-22.5	28
22.5-27.5	12
27.5-32.5	9
32.5-37.5	7
Mean	17.148
Median	16.32
Mode	15.6

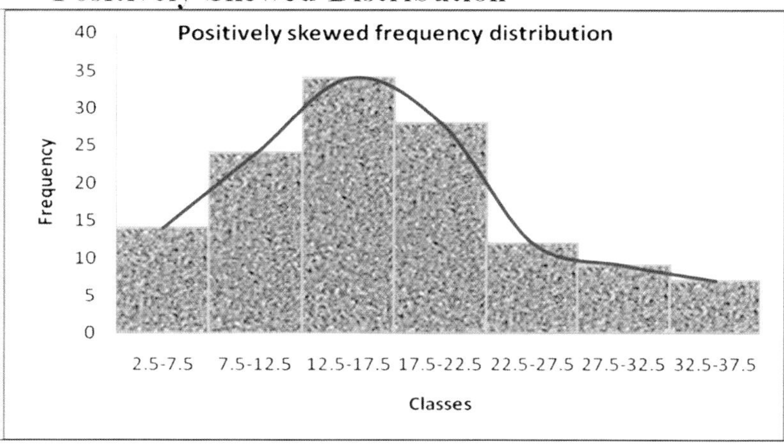

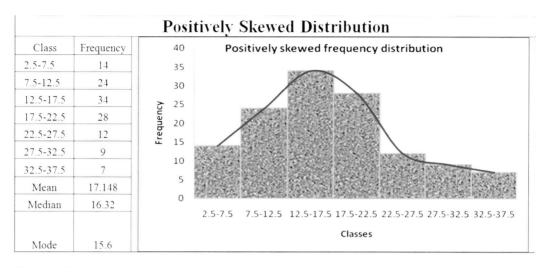

Fig. 8.14 Graphical presentation of a positively skewed distribution

Negatively Skewed Distribution

Class	Frequency
2.5-7.5	10
7.5-12.5	26
12.5-17.5	34
17.5-22.5	41
22.5-27.5	7
27.5-32.5	6
32.5-37.5	4
Mean	16.67
Median	18.32
Mode	21.28

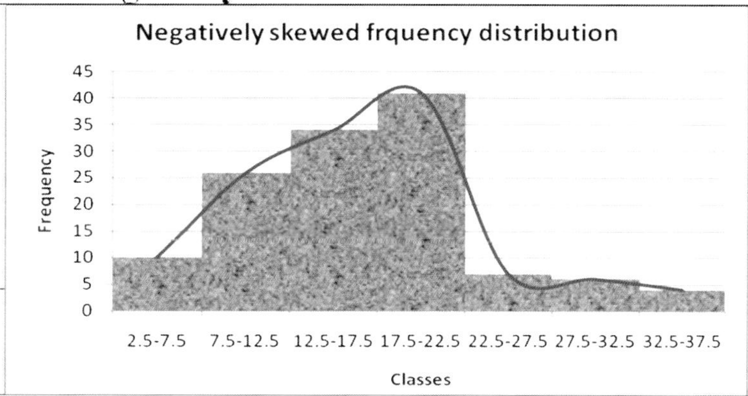

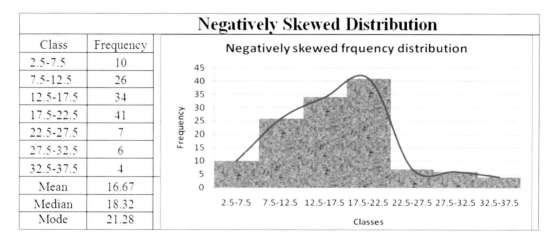

Fig. 8.15 Graphical presentation of a negatively skewed distribution

mesokurtic, and the distribution is platykurtic in nature. It can be shown that $\beta_2 \geq 1 + \beta_1$.

Example 8.22. The frequency distribution of the length of the leaflet of the three varieties of tube rose flower is given in Figs. 8.16, 8.17 and 8.18. It is clear from the corresponding figures that the three distributions vary widely with respect to the peakedness of the frequency graph.

A leptokurtic distribution means an aggregation of more observations (frequency) in a particular class or a couple of classes. Mesokurtic distribution follows the general norm that at the lower classes, there are a fewer number of frequencies, and as the class value increases, the number of observations (frequency) also increases then reaches to the peak, and as the class value increases further, the number of observations (frequency) decreases. On the other hand, in a platykurtic distribution, a good number of classes have almost the same higher observations and thereby, forming a platelike structure at the top of the frequency distribution.

Thus, to know the nature of the data, the measures of central tendency and measures of dispersion along with skewness and kurtosis of the frequency distribution are essential.

With the help of the following example, we shall demonstrate how these analyses could be taken up using SAS 9.3.

Example 8.23. The following table gives the yield attributing characters along with the yield for 37 varieties. We are to summarize the data in terms of the different measures of dispersion, central tendency, skewness, and kurtosis.

Class	Frequency
2.5-7.5	6
7.5-12.5	8
12.5-17.5	13
17.5-22.5	72
22.5-27.5	16
27.5-32.5	7
32.5-37.5	6
Mean	20.04
μ2	38.48
μ4	6060.72
β2	4.0942

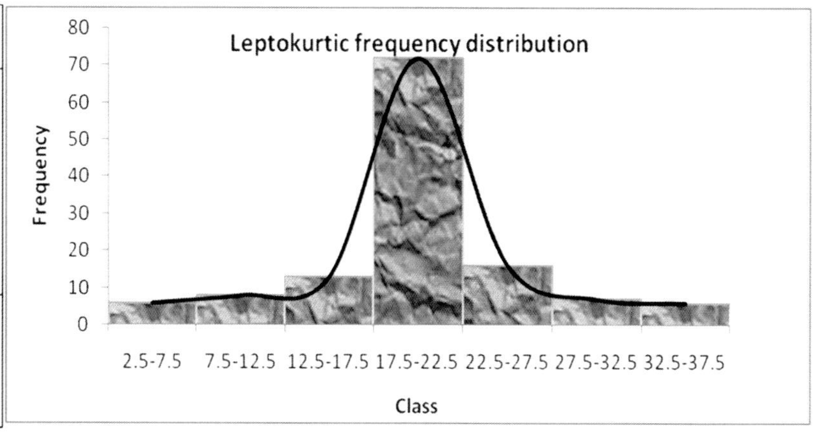

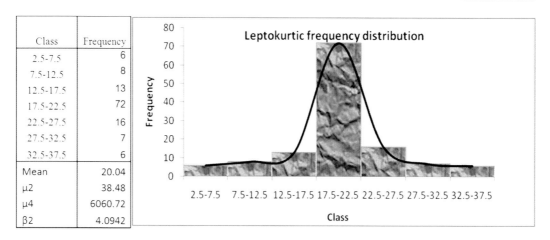

Fig. 8.16 Graphical presentation of a leptokurtic distribution

Class	Frequency
2.5-7.5	8
7.5-12.5	12
12.5-17.5	25
17.5-22.5	42
22.5-27.5	25
27.5-32.5	10
32.5-37.5	6
Mean	19.61
μ2	51.41
μ4	7440.35
β2	2.82

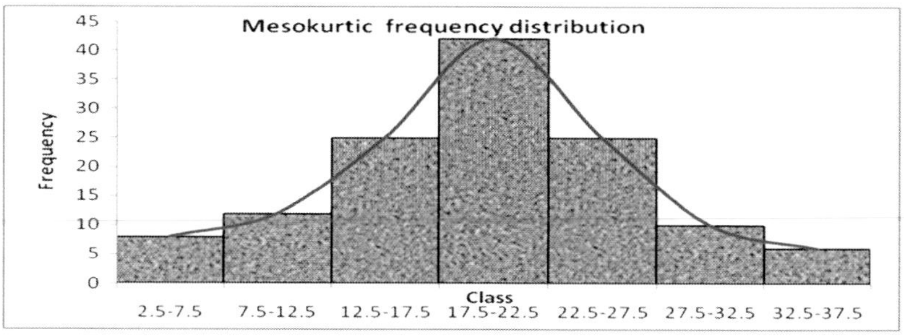

Fig. 8.17 Graphical presentation of a mesokurtic distribution

Class	Frequency
2.5-7.5	6
7.5-12.5	16
12.5-17.5	24
17.5-22.5	28
22.5-27.5	27
27.5-32.5	18
32.5-37.5	9
Mean	20.63
μ2	62.76
μ4	8796.45
β2	2.23

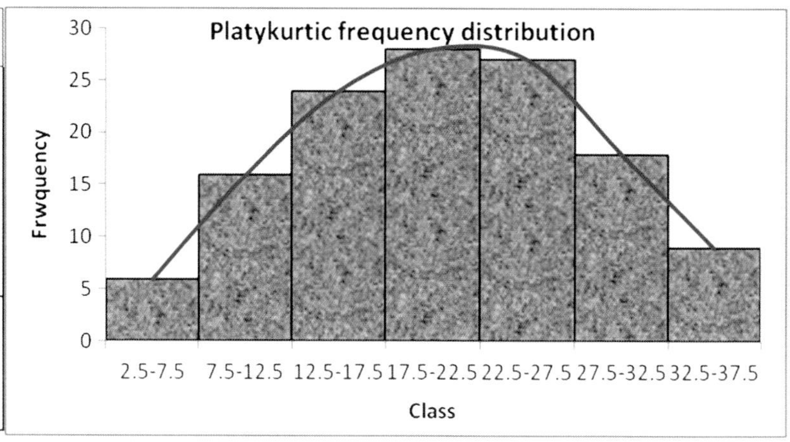

Fig. 8.18 Graphical presentation of a platykurtic distribution

SL No	X1	X2	X3	X4	X5	X6	X7	X8	X9	X10	X11	X12	X13	X14	X15	X16	X17	X18	X19	Y
1	32.34	151.81	140.47	21.23	162.67	107.88	23.20	31.33	9.47	149.37	7.55	2.01	195.77	1663.67	313.38	1.93	1.93	9.96	9.49	345.85
2	29.85	140.71	140.95	20.33	150.00	104.49	23.20	33.17	8.87	144.04	7.82	2.11	222.68	1675.33	298.72	1.93	2.45	14.50	9.45	354.41
3	32.38	138.77	144.30	18.63	146.00	93.17	22.23	33.33	9.00	145.20	7.90	2.14	219.74	1672.33	302.74	1.90	2.44	15.45	9.33	358.12
4	38.33	138.31	140.06	17.67	156.33	92.60	22.93	34.50	8.73	141.57	7.67	2.15	228.32	1587.33	263.92	1.57	2.59	15.63	7.93	341.39
5	37.81	140.23	135.62	18.70	153.67	103.54	23.07	33.50	8.77	133.96	7.72	2.15	207.93	1621.00	277.86	1.53	2.40	13.75	7.73	323.82
6	35.02	132.90	128.79	18.47	143.00	108.60	23.33	29.00	9.03	145.32	7.97	2.27	189.59	1629.00	267.74	1.34	2.41	13.37	6.91	369.82
7	32.25	136.10	123.41	19.93	130.00	108.25	23.13	30.33	9.70	149.08	7.80	2.12	161.48	1657.33	264.72	1.72	2.45	13.76	8.44	373.14
8	32.56	137.11	116.18	18.87	143.67	100.95	23.73	28.17	9.47	133.45	7.83	2.08	146.38	1621.67	241.44	1.74	2.25	12.26	7.98	353.28
9	34.10	147.68	125.74	19.67	153.00	98.60	24.13	30.33	9.67	132.12	7.88	2.08	162.86	1634.33	253.30	1.95	2.74	15.22	8.94	350.18
10	37.05	153.77	123.49	19.10	160.00	100.17	24.07	30.00	9.23	137.06	8.58	2.27	184.37	1702.67	263.88	1.77	2.74	15.05	8.18	394.78
11	43.70	161.64	124.89	19.07	166.33	112.08	27.57	25.93	10.30	147.31	10.19	2.73	239.65	1785.67	286.02	1.77	2.73	14.98	8.67	489.41
12	41.96	160.23	114.79	19.60	143.67	107.01	26.27	24.83	10.23	135.48	9.17	2.40	237.48	1693.00	284.65	1.86	2.20	11.29	8.54	416.26
13	38.29	144.75	115.58	20.03	143.00	107.58	26.63	26.17	9.80	129.08	8.97	2.52	201.06	1609.00	276.24	1.74	2.47	12.21	8.01	384.56
14	33.84	155.17	117.81	19.33	157.00	102.07	25.27	24.83	9.17	130.27	7.72	2.16	178.28	1496.00	294.98	1.67	2.29	11.29	7.76	327.73
15	34.56	148.67	119.07	16.67	164.33	116.45	25.87	24.50	9.33	130.48	8.55	2.43	188.79	1511.67	288.41	1.32	2.33	11.52	6.68	363.85
16	32.46	156.00	121.01	20.10	162.67	115.48	25.50	22.50	9.07	138.49	8.38	2.25	202.79	1547.33	307.51	1.44	2.05	10.52	7.22	371.94
17	34.01	156.99	138.86	21.83	144.00	117.92	25.07	27.40	9.10	151.50	10.06	2.54	216.33	1662.67	316.73	1.42	2.66	14.93	7.46	505.86
18	34.58	162.46	145.61	23.90	149.33	104.79	25.17	29.67	9.20	144.95	10.46	2.60	220.73	1685.33	313.12	1.49	2.58	14.54	7.77	502.73
19	34.22	166.22	153.17	22.73	155.67	108.22	24.67	33.83	9.07	150.81	10.01	2.56	208.66	1738.67	288.29	1.85	3.05	17.96	9.29	495.12
20	33.13	162.80	143.02	23.10	144.00	109.56	24.03	30.60	8.93	131.19	8.64	2.34	168.35	1668.33	247.84	1.55	2.40	13.12	7.58	359.93
21	34.27	168.36	135.34	21.20	163.33	115.47	23.13	26.50	11.93	132.39	7.50	1.98	131.32	1711.33	239.02	2.07	2.19	12.33	8.40	325.47
22	31.85	163.56	129.03	20.53	154.67	102.78	24.60	29.00	12.87	117.91	7.89	2.02	117.85	1653.00	234.97	1.81	2.02	10.43	7.41	298.67
23	29.36	160.86	129.24	21.17	193.67	99.50	20.80	37.60	12.50	117.78	7.99	2.11	158.37	1701.67	211.97	2.13	2.38	12.65	8.86	299.57
24	35.01	154.62	140.06	21.53	171.00	90.33	22.53	42.50	9.50	129.01	8.64	2.37	218.03	1692.00	244.12	1.66	2.60	13.66	8.31	356.87
25	35.71	161.44	145.13	22.40	181.00	98.85	21.87	36.73	9.67	131.64	8.67	2.62	263.96	1764.00	289.06	1.87	2.88	15.42	9.13	372.53
26	38.24	159.29	144.58	19.73	161.67	93.00	24.57	37.50	10.57	140.35	9.84	2.84	223.82	1744.67	312.47	1.96	2.65	14.10	9.53	453.42
27	35.36	147.69	144.01	21.93	182.33	104.51	24.37	36.50	10.87	135.28	10.06	2.70	216.55	1747.33	289.87	1.71	2.53	13.53	8.67	450.74
28	33.81	139.24	145.32	22.97	178.00	105.69	23.20	34.50	9.90	137.24	10.00	2.45	197.75	1698.67	286.04	1.49	2.02	10.29	7.83	449.22
29	34.44	139.44	146.65	22.63	193.00	112.40	23.90	29.50	8.50	155.76	8.94	2.08	238.45	1624.00	333.34	1.39	2.01	10.28	7.65	457.62
30	36.05	151.53	154.25	23.53	194.00	103.77	24.10	31.17	8.73	149.60	8.58	2.25	236.20	1542.33	306.56	1.79	2.04	10.45	8.28	420.00
31	37.34	157.67	151.84	21.63	195.33	102.57	24.90	33.50	10.90	161.70	8.42	2.20	232.85	1582.33	291.33	1.76	2.05	10.76	8.12	442.93
32	33.13	151.16	143.56	24.53	191.00	98.26	25.00	29.50	11.03	135.25	7.69	2.20	203.45	1588.33	250.95	2.15	1.80	9.63	7.10	331.91
33	30.20	144.68	133.70	21.73	176.00	95.76	22.50	35.83	10.50	140.91	6.96	1.94	154.35	1611.00	277.85	1.89	2.00	10.19	7.25	321.00
34	30.26	140.64	119.93	21.93	178.67	94.23	20.77	37.67	9.30	128.65	7.12	2.13	159.38	1601.00	286.54	1.89	1.99	10.35	7.28	283.70
35	31.27	141.59	123.96	19.37	180.67	105.51	18.00	35.33	8.93	133.87	6.65	1.91	205.68	1648.00	302.31	1.74	2.23	12.42	8.15	276.98
36	33.17	150.82	131.18	18.80	178.67	107.36	17.37	31.17	9.23	143.59	6.68	1.92	206.11	1634.00	306.61	2.08	2.12	11.87	9.09	300.94
37	32.15	151.23	137.13	22.37	179.67	112.17	20.13	30.83	8.90	143.36	6.83	1.81	203.41	1650.33	304.86	2.12	2.15	11.52	9.27	295.61

1. Analysis Through MS Excel
   If one wants to perform the above analyses using MS Excel software, then one should follow the following steps:

Step 1: Go to Data Analysis submenu of Data in MS Excel. Select Descriptive Statistics as shown below.

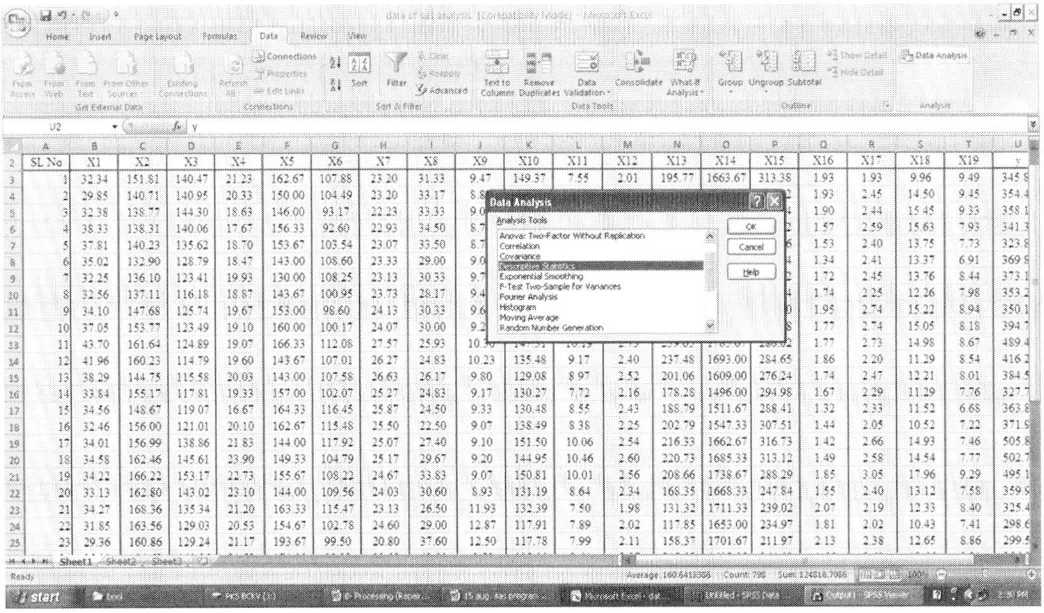

Slide 8.5: Step 1 showing the entry or the transformation of data to Excel data sheet and Data Analysis menu selection

Step 2: Provide the input range and the output range (where the output is to be placed up on analysis), and tick on to Summary Statistics as shown below.

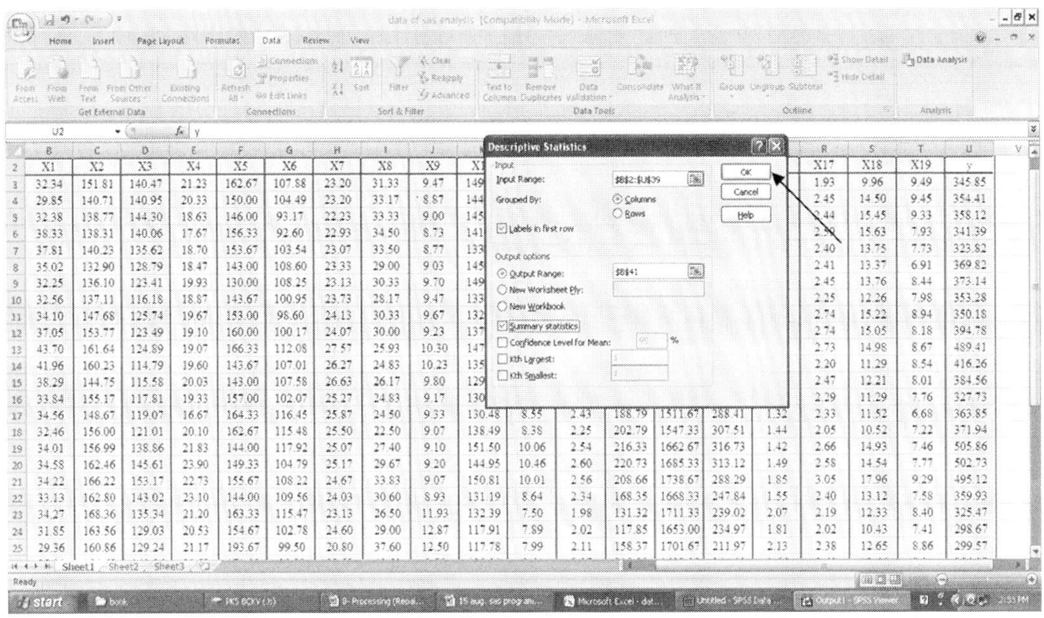

Slide 8.6: Step 2 showing the selection of appropriate menus in Descriptive Data Analysis menu of MS Excel

Step 3: Click OK to get the window as given below containing the output for each and every character.

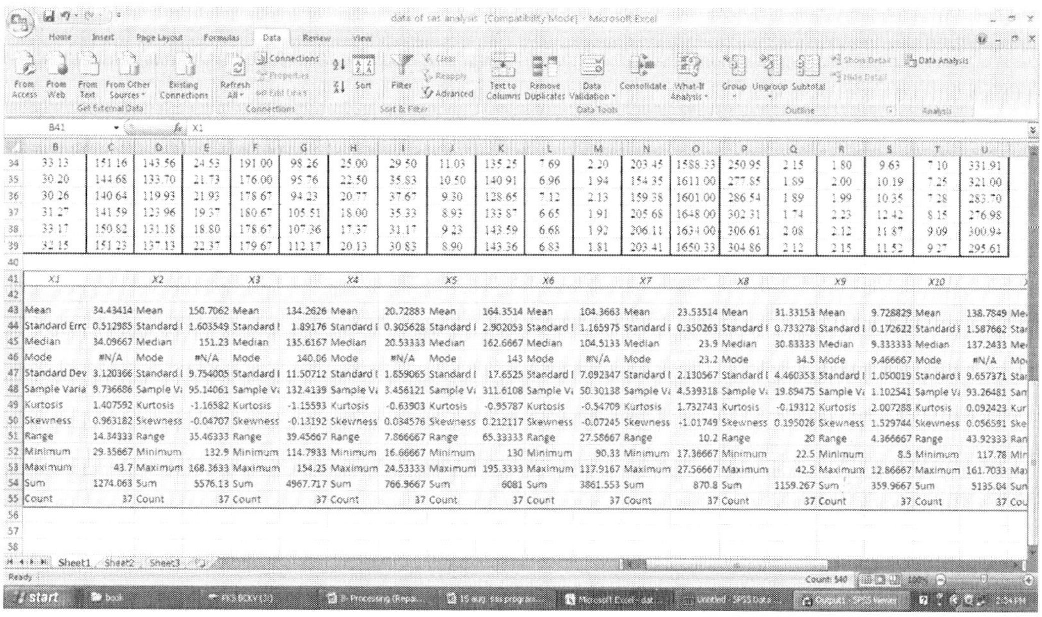

Slide 8.7: Step 3 showing the output in Descriptive data Analysis menu of MS Excel

2. Analysis Through SPSS

Step 1: When these data are transferred to SPSS 11.5 Data Editor either copying from the sources or importing, it looks like the following slide.

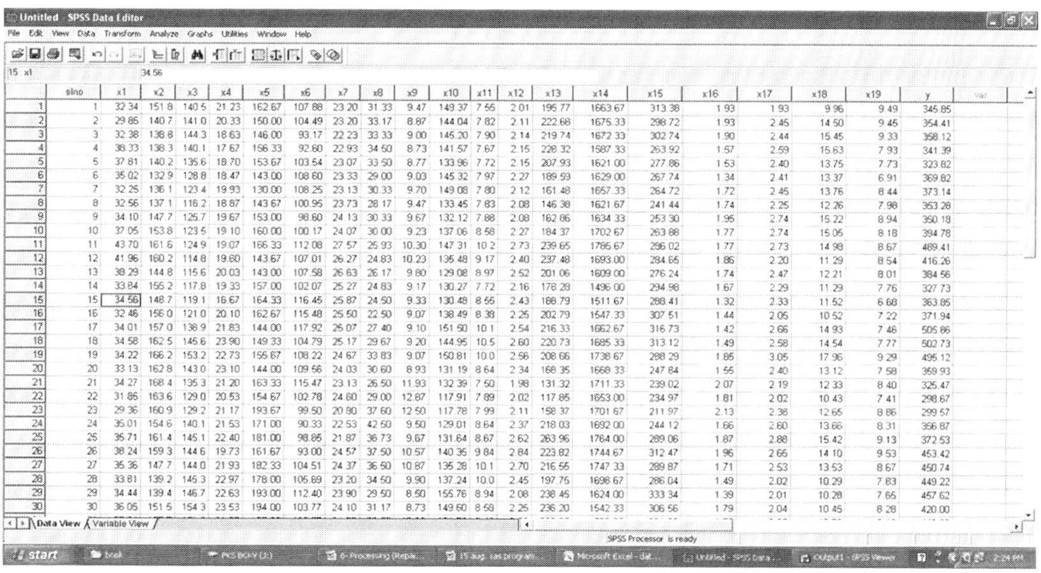

Slide 8.8: Step 1 showing the entry or transferred data in SPSS Data Editor

Step 2: Go to Analysis menu followed by Descriptive Statistics and Descriptive as shown below.

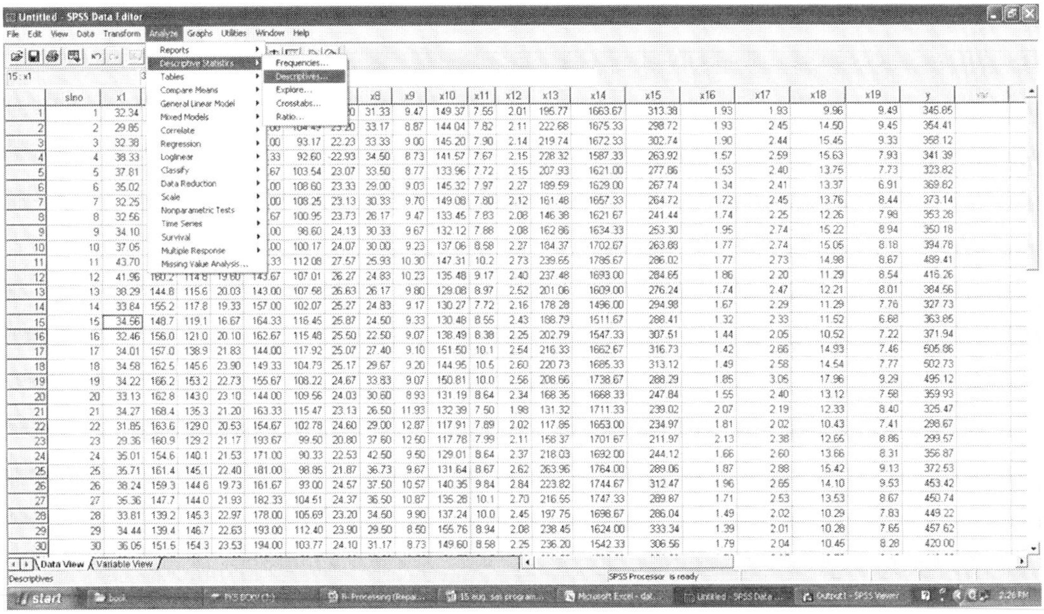

Slide 8.9: Step 2 showing the selection of appropriate Data Analysis menu in SPSS analysis

Step 3: Select the variable for which one wants the Descriptive Statistics and then click on to Option menu as shown below.

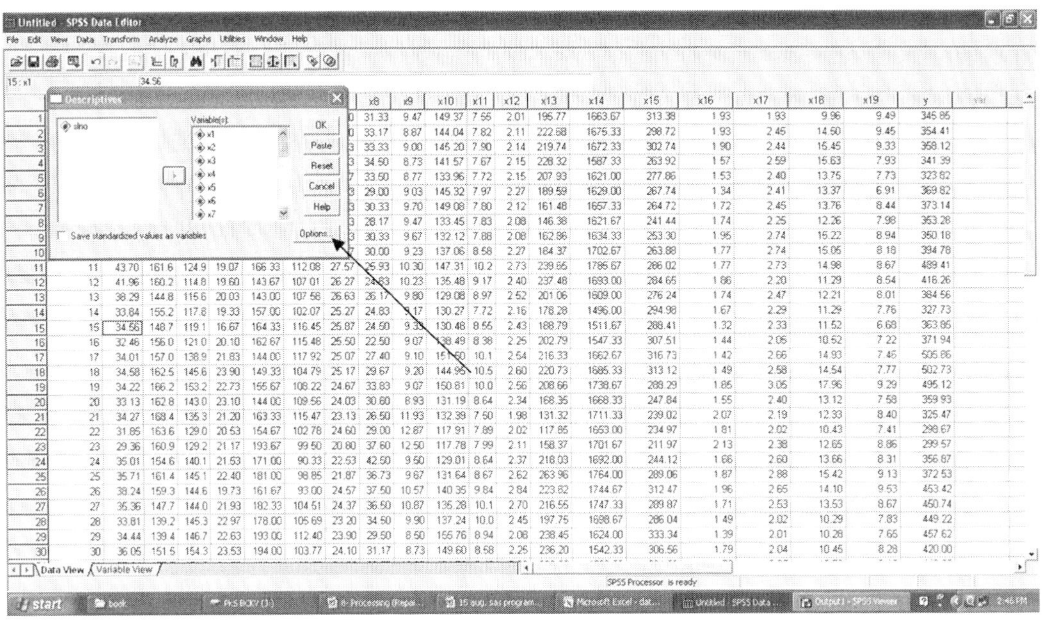

Slide 8.10: Step 3 showing the selection of appropriate variables for Analysis menu in SPSS

Step 4: From Options, select the required measures as shown below and click on to Continue.

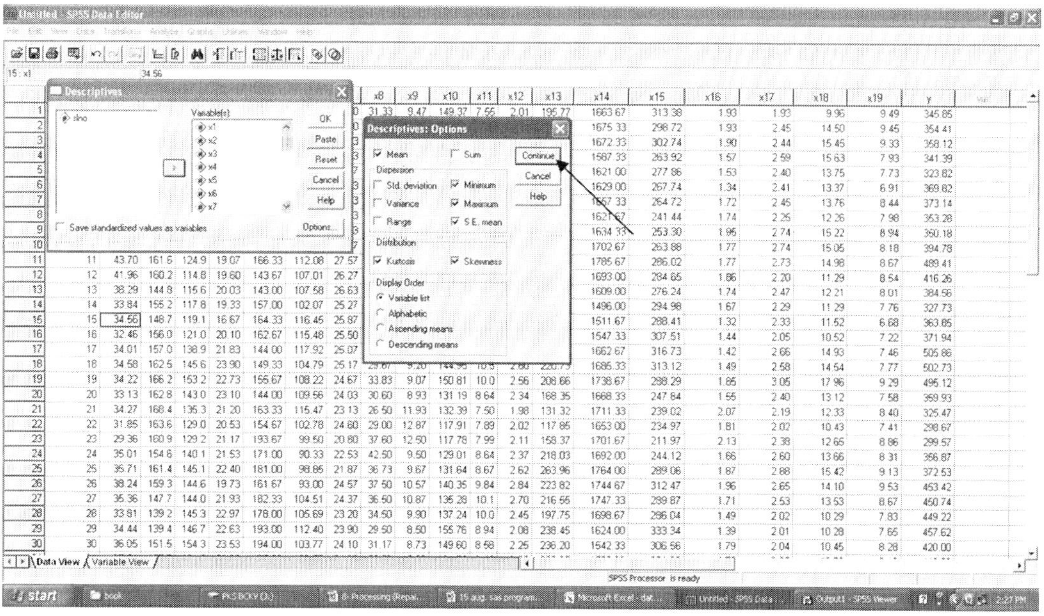

Slide 8.11: Step 4 showing the selection of appropriate Analysis submenu of SPSS

Step 5: Click on to OK as given below to get the output.

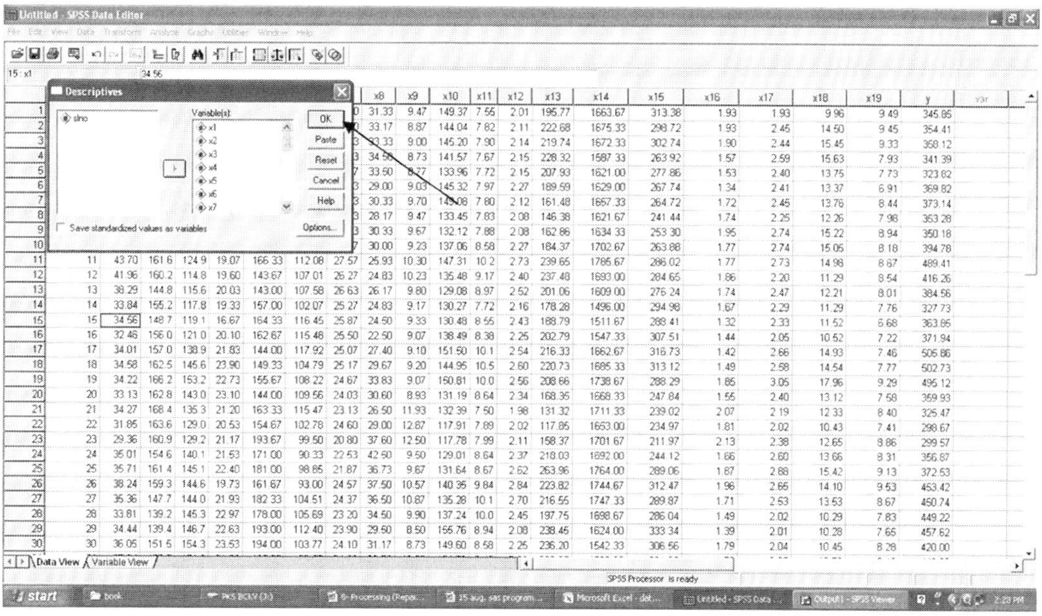

Slide 8.12: Step 4 showing the appropriate command to get the output in Analysis submenu of SPSS

The following output slide would be obtained.

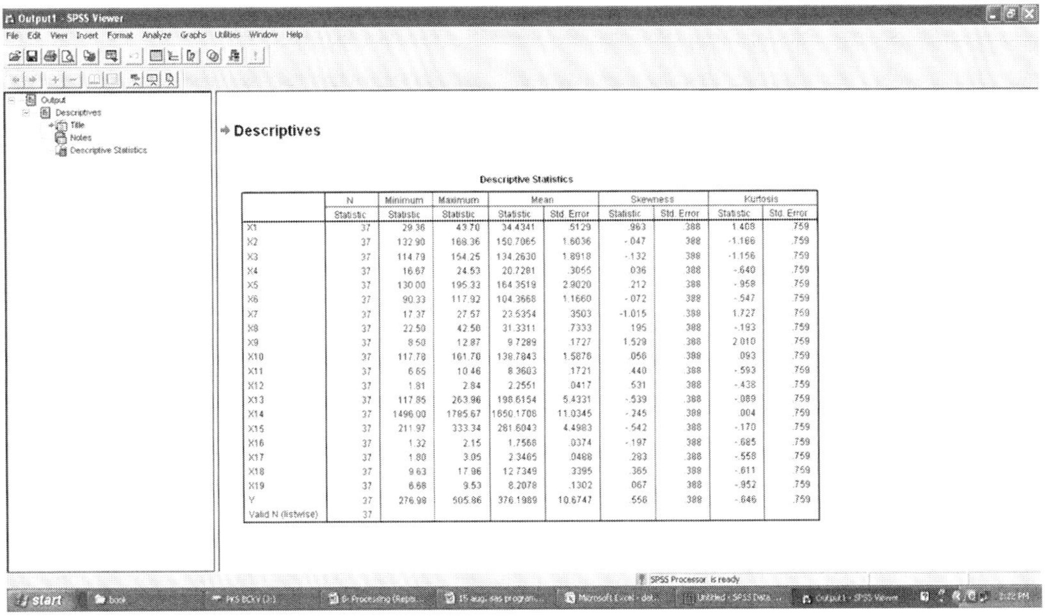

Slide 8.13: Figure showing the output for selective descriptive statistics using SPSS

3. Analysis Using SAS 9.3

The above data, if entered/copied in SAS editor, will look like the one given below in two continued slides:

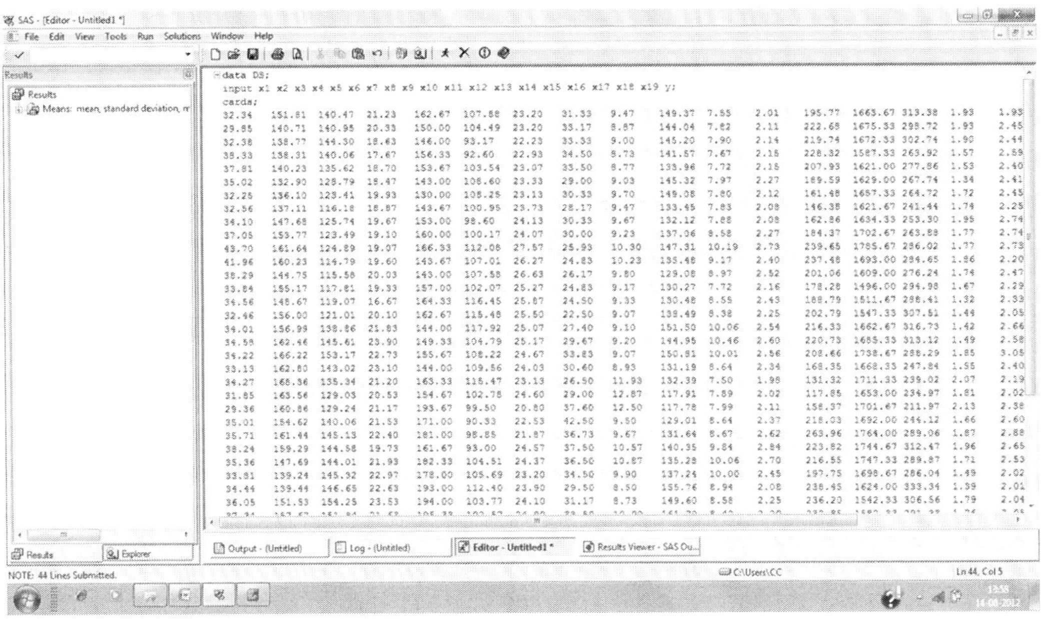

Slide 8.14: Step 1 showing the entered or transferred data in SAS Data Editor

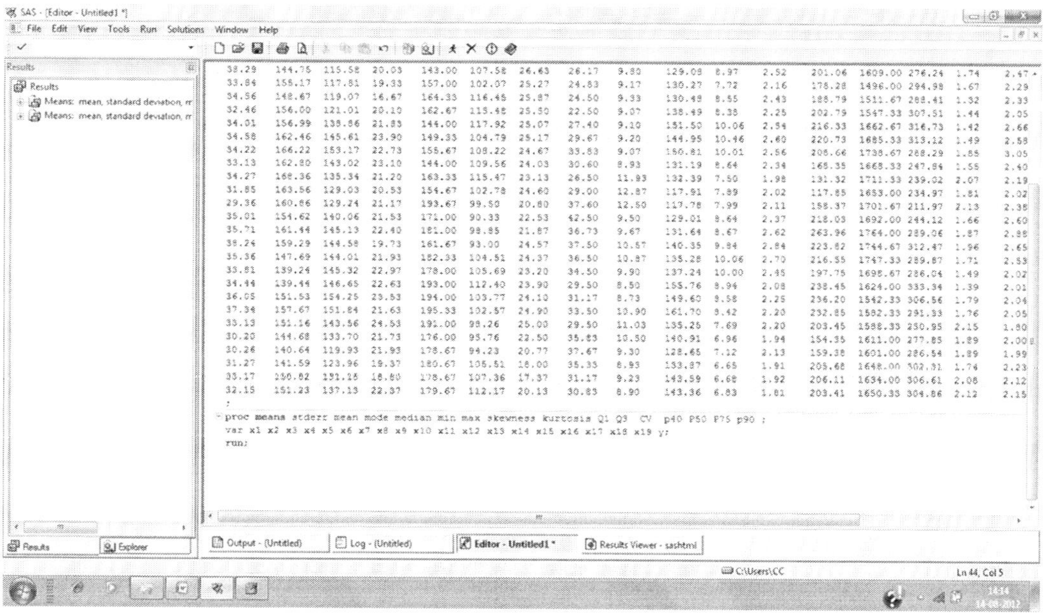

38.29	144.75	115.58	20.03	143.00	107.58	26.63	26.17	9.80	129.08	8.97	2.52	201.06	1609.00	276.24	1.74	2.47
33.84	155.17	117.81	19.33	157.00	102.07	25.27	24.83	9.17	130.27	7.72	2.16	178.28	1496.00	294.98	1.67	2.29
34.56	148.67	119.07	16.67	164.33	116.45	25.87	24.50	9.33	130.49	8.55	2.43	188.79	1511.67	288.41	1.32	2.33
32.46	156.00	121.01	20.10	162.67	115.48	25.50	22.50	9.07	138.49	8.38	2.25	202.79	1547.33	307.51	1.44	2.05
34.01	156.99	138.86	21.83	144.00	117.92	25.07	27.40	9.10	151.50	10.06	2.54	216.33	1662.67	316.73	1.42	2.66
34.58	162.46	145.61	23.90	149.33	104.79	25.17	29.67	9.20	144.95	10.46	2.60	220.73	1685.33	313.12	1.49	2.58
34.22	166.22	153.17	22.73	155.67	108.22	24.67	33.83	9.07	180.81	10.01	2.56	208.66	1738.67	288.29	1.85	3.05
33.13	162.80	143.02	23.10	144.00	109.56	24.03	30.60	8.93	131.19	8.64	2.34	165.35	1668.33	247.84	1.55	2.40
34.27	168.36	135.34	21.20	163.33	115.47	23.13	26.50	11.83	132.39	7.50	1.98	131.32	1711.33	239.02	2.07	2.19
31.85	163.56	129.03	20.53	154.67	102.78	24.60	29.00	12.87	117.91	7.89	2.02	117.85	1653.00	234.97	1.81	2.02
29.36	160.86	129.24	21.17	193.67	99.50	20.80	37.60	12.50	117.78	7.99	2.11	158.37	1701.67	211.97	2.13	2.38
35.01	154.62	140.06	21.53	171.00	90.33	22.53	42.50	9.50	129.01	8.64	2.37	218.03	1692.00	244.12	1.66	2.60
35.71	161.44	145.13	22.40	181.00	98.85	21.87	36.73	9.67	131.64	8.67	2.62	263.96	1764.00	289.06	1.87	2.88
38.24	159.29	144.58	19.73	161.67	93.00	24.57	37.50	10.57	140.35	9.84	2.84	223.82	1744.67	312.47	1.96	2.65
35.36	147.69	144.01	21.93	182.33	104.51	24.37	36.50	10.87	135.28	10.06	2.70	216.55	1747.33	289.87	1.71	2.53
33.81	139.24	145.32	22.97	178.00	105.69	23.20	34.50	9.90	137.24	10.00	2.45	197.75	1698.67	286.04	1.49	2.07
34.44	139.44	146.65	22.63	193.00	112.40	23.90	29.50	8.50	155.76	8.94	2.08	238.45	1624.00	333.34	1.39	2.01
36.05	151.53	154.25	23.53	194.00	103.77	24.10	31.17	8.73	149.60	9.58	2.25	236.20	1542.33	306.56	1.79	2.04
37.34	157.67	151.84	21.63	195.33	102.57	24.90	33.50	10.90	161.70	9.42	2.20	232.85	1582.33	291.33	1.76	2.05
33.13	151.16	143.56	24.53	191.00	98.26	25.00	29.50	11.03	135.25	7.69	2.20	203.45	1588.33	250.95	2.15	1.80
30.20	144.68	133.70	21.73	176.00	95.76	22.50	35.83	10.50	140.91	6.96	1.94	154.35	1611.00	277.85	1.89	2.00
30.26	140.64	119.93	21.93	178.67	94.23	20.77	37.67	9.30	128.65	7.12	2.13	159.38	1601.00	286.54	1.89	1.99
31.27	141.59	123.96	19.37	180.67	105.51	18.00	35.33	8.93	133.87	6.65	1.91	205.68	1648.00	302.31	1.74	2.23
33.17	190.82	131.18	18.80	178.67	107.36	17.37	31.17	9.23	143.59	6.68	1.92	206.11	1634.00	306.61	2.08	2.12
32.15	151.23	137.13	22.37	179.67	112.17	20.13	30.83	8.90	143.36	6.83	1.81	203.41	1650.33	304.86	2.12	2.15

```
;
proc means stderr mean mode median min max skewness kurtosis Q1 Q3  CV  p40 P50 P75 p90 ;
var x1 x2 x3 x4 x5 x6 x7 x8 x9 x10 x11 x12 x13 x14 x15 x16 x17 x18 x19 y;
run;
```

Slide 8.15: Step 2 showing the entered or transferred data and command for getting the required descriptive statistics using SAS Data Editor

We want to have the mean, standard error, median, mode, maximum–minimum values, skewness, kurtosis, 1st and 3rd quartiles, 40th, 50th, 75th, and 90th percentile values for each and every character. The commands for execution are provided in the last three lines of the above slide. Upon the execution of the said command, the following output as given below in two consecutive slides is obtained.

a

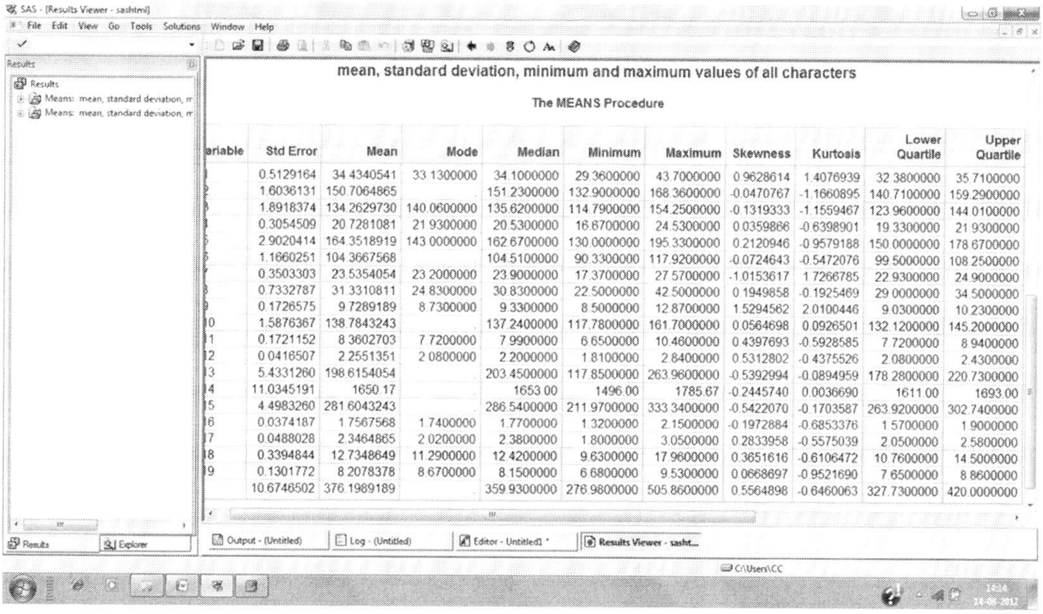

mean, standard deviation, minimum and maximum values of all characters

The MEANS Procedure

ariable	Std Error	Mean	Mode	Median	Minimum	Maximum	Skewness	Kurtosis	Lower Quartile	Upper Quartile
	0.5129164	34.4340541	33.1300000	34.1000000	29.3600000	43.7000000	0.9628614	1.4076939	32.3800000	35.7100000
	1.6036131	150.7064865	.	151.2300000	132.9000000	168.3600000	-0.0470767	-1.1660895	140.7100000	159.2900000
	1.8918374	134.2629730	140.0600000	135.6200000	114.7900000	154.2500000	-0.1319333	-1.1559467	123.9600000	144.0100000
	0.3054509	20.7281081	21.9300000	20.5300000	16.6700000	24.5300000	0.0359866	-0.6398901	19.3300000	21.9300000
	2.9020414	164.3518919	143.0000000	162.6700000	130.0000000	195.3300000	0.2120946	-0.9579188	150.0000000	178.6700000
	1.1660251	104.3667568	.	104.5100000	90.3300000	117.9200000	-0.0724643	-0.5472076	99.5000000	108.2500000
	0.3503303	23.5354054	23.2000000	23.9000000	17.3700000	27.5700000	-1.0153617	1.7266785	22.9300000	24.9000000
	0.7332787	31.3310811	24.8300000	30.8300000	22.5000000	42.5000000	0.1949858	-0.1925469	29.0000000	34.5000000
	0.1726575	9.7289189	8.7300000	9.3300000	8.5000000	12.8700000	1.5294562	2.0100446	9.0300000	10.2300000
0	1.5876367	138.7843243	.	137.2400000	117.7800000	161.7000000	0.0564698	0.0926501	132.1200000	145.2000000
1	0.1721152	8.3602703	7.7200000	7.9900000	6.6500000	10.4600000	0.4397693	-0.5928585	7.7200000	8.9400000
2	0.0416507	2.2551351	2.0800000	2.2000000	1.8100000	2.8400000	0.5312802	-0.4375526	2.0800000	2.4300000
3	5.4331260	198.6154054	.	203.4500000	117.8500000	263.9600000	-0.5392994	-0.0894959	178.2800000	220.7300000
4	11.0345191	1650.17	.	1653.00	1496.00	1785.67	-0.2445740	0.0036690	1611.00	1693.00
5	4.4983260	281.6043243	.	286.5400000	211.9700000	333.3400000	-0.5422070	-0.1703587	263.9200000	302.7400000
6	0.0374187	1.7567568	1.7400000	1.7700000	1.3200000	2.1500000	-0.1972884	-0.6853376	1.5700000	1.9000000
7	0.0488028	2.3464865	2.0200000	2.3800000	1.8000000	3.0500000	0.2833958	-0.5575039	2.0500000	2.5800000
8	0.3394844	12.7348649	11.2900000	12.4200000	9.6300000	17.9600000	0.3651616	-0.6106472	10.7600000	14.5000000
9	0.1301772	8.2078378	8.6700000	8.1500000	6.6800000	9.5300000	0.0668697	-0.9521690	7.6500000	8.8600000
	10.6746502	376.1989189	.	359.9300000	276.9800000	505.8600000	0.5564898	-0.6460063	327.7300000	420.0000000

Slide 8.16a: Step 3 showing the output for required descriptive statistics using SAS

b

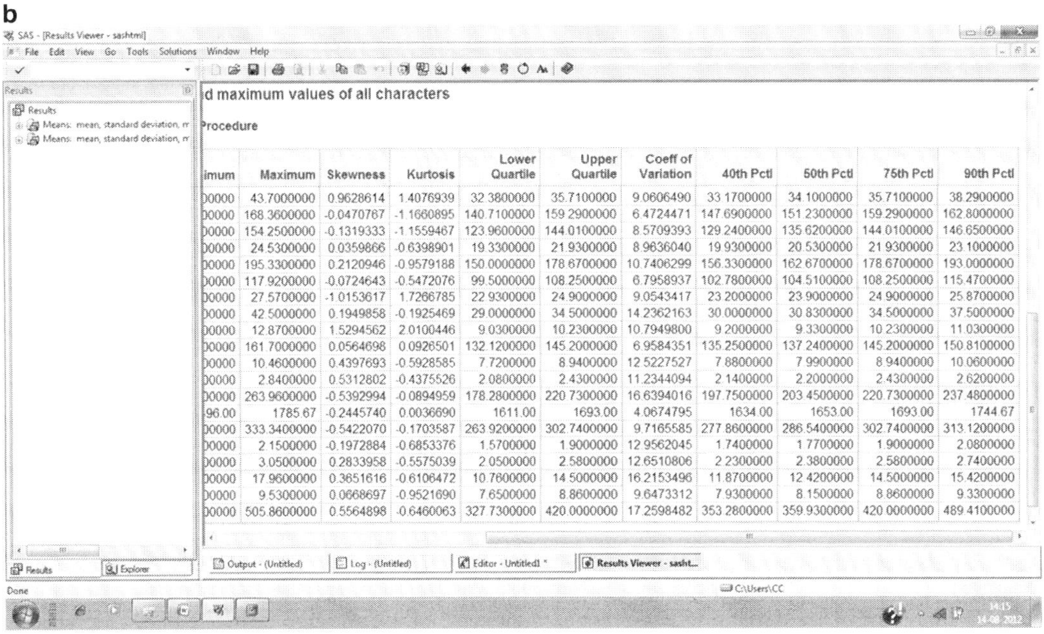

d maximum values of all characters

Procedure

imum	Maximum	Skewness	Kurtosis	Lower Quartile	Upper Quartile	Coeff of Variation	40th Pctl	50th Pctl	75th Pctl	90th Pctl
00000	43.7000000	0.9628614	1.4076939	32.3800000	35.7100000	9.0606490	33.1700000	34.1000000	35.7100000	38.2900000
00000	168.3600000	-0.0470767	-1.1660895	140.7100000	159.2900000	6.4724471	147.6900000	151.2300000	159.2900000	162.8000000
00000	154.2500000	-0.1319333	-1.1559467	123.9600000	144.0100000	8.5709393	129.2400000	135.6200000	144.0100000	146.6500000
00000	24.5300000	0.0359866	-0.6398901	19.3300000	21.9300000	8.9636040	19.9300000	20.5300000	21.9300000	23.1000000
00000	195.3300000	0.2120946	-0.9579188	150.0000000	178.6700000	10.7406299	156.3300000	162.6700000	178.6700000	193.0000000
00000	117.9200000	-0.0724643	-0.5472076	99.5000000	108.2500000	6.7958937	102.7800000	104.5100000	108.2500000	115.4700000
00000	27.5700000	-1.0153617	1.7266785	22.9300000	24.9000000	9.0543417	23.2000000	23.9000000	24.9000000	25.8700000
00000	42.5000000	0.1949858	-0.1925469	29.0000000	34.5000000	14.2362163	30.0000000	30.8300000	34.5000000	37.5000000
00000	12.8700000	1.5294562	2.0100446	9.0300000	10.2300000	10.7949800	9.2000000	9.3300000	10.2300000	11.0300000
00000	161.7000000	0.0564698	0.0926501	132.1200000	145.2000000	6.9584351	135.2500000	137.2400000	145.2000000	150.8100000
00000	10.4600000	0.4397693	-0.5928585	7.7200000	8.9400000	12.5227527	7.8800000	7.9900000	8.9400000	10.0600000
00000	2.8400000	0.5312802	-0.4375526	2.0800000	2.4300000	11.2344094	2.1400000	2.2000000	2.4300000	2.6200000
00000	263.9600000	-0.5392994	-0.0894959	178.2800000	220.7300000	16.6394016	197.7500000	203.4500000	220.7300000	237.4800000
96.00	1785.67	-0.2445740	0.0036690	1611.00	1693.00	4.0674795	1634.00	1653.00	1693.00	1744.67
00000	333.3400000	-0.5422070	-0.1703587	263.9200000	302.7400000	9.7165585	277.8600000	286.5400000	302.7400000	313.1200000
00000	2.1500000	-0.1972884	-0.6853376	1.5700000	1.9000000	12.9562045	1.7400000	1.7700000	1.9000000	2.0800000
00000	3.0500000	0.2833958	-0.5575039	2.0500000	2.5800000	12.6510806	2.2300000	2.3800000	2.5800000	2.7400000
00000	17.9600000	0.3651616	-0.6106472	10.7600000	14.5000000	16.2153496	11.8700000	12.4200000	14.5000000	15.4200000
00000	9.5300000	0.0668697	-0.9521690	7.6500000	8.8600000	9.6473312	7.9300000	8.1500000	8.8600000	9.3300000
00000	505.8600000	0.5564898	-0.6460063	327.7300000	420.0000000	17.2598482	353.2800000	359.9300000	420.0000000	489.4100000

Slide 8.16b: Step 3 showing the output for the required descriptive statistics using SAS

Along with the values for different measures of different component characters, graphical representations can also be made.

From the above-described analysis, using three different software clearly reveals that there are limitations, specifications, and methods in order to get the desired output. But the user must have a clear idea about the information being analyzed, requirement, and knowledge about the statistical method to verify the results obtained and about the method of handling the appropriate software. Otherwise, there are few chances of mis-conclusion, vagueness, superfluousness, or unwarranted/incorrect conclusion.

8.3 Correlation Analysis

While conducting research works, a researcher needs to deal with a number of factors/variables at a time, instead of a single variable/factor. And all these variables may not be independent of each other; rather they tend to vary side by side. Most of the growth/social/economic and other variables are found to follow the above characteristics. For example, while dealing with yield component analysis of any crop, it is found that yield is an ultimate variable contributed/influenced/affected by a number of other factors. If we consider the yield of paddy, then one can find that the factors like the number of hills per square meter, number of tillers per hill, number of effective/panicle-bearing tillers per hill, length of the panicle, number of grains per panicle, and test (1,000 grain) weight of grains are influencing the yield. Variation in one or more of the above-mentioned factors results in variations of the yield. Thus, yield may vary because of variation in the number of hills per square meter or variations in the number of tillers per hill or so on. Again, yield may vary because of variation in the number of hill per square meter and/or in the number of tiller per hill and other factors. When we consider variations in one variable due to variations in any other variable, then it becomes a bivariate case. On the other hand, when the variations of more than two variables are considered at a time, it becomes a multivariate case. The problem of measuring the degree of association among the variables is considered through a correlation and regression analysis. In this section, we shall first study the correlation and then go for the study of the form of the association, that is, the regression analysis.

The simplest and widely used measure of correlation is the measure of Karl Pearson's *correlation coefficients. Correlation coefficient measures the degree of closeness of the linear association between any two variables* and is given as

$r_{xy} = \frac{\mathrm{Cov}(x,y)}{S_x.S_y}$, where (x_1,y_1), (x_2,y_2), (x_3,y_3),..., (x_n,y_n) are n pairs of observations and

$$\text{(i) } \mathrm{Cov}(x,y) = \frac{1}{n}\sum_{i=1}^{n}(x_i-\bar{x})(y_i-\bar{y}) = S_{xy}$$

$$\text{(ii) } s_x^2 = \frac{1}{n}\sum_{i=1}^{n}(x_i-\bar{x})^2 \text{ and}$$

$$\text{(iii) } s_y^2 = \frac{1}{n}\sum_{i=1}^{n}(y_i-\bar{y})^2$$

$$\text{Thus, } r_{xy} = \frac{\frac{1}{n}\sum_{i=1}^{n}x_iy_i-\bar{x}\bar{y}}{\sqrt{\left(\frac{1}{n}\sum_{i=1}^{n}x_i^2-\bar{x}^2\right)\left(\frac{1}{n}\sum_{i=1}^{n}y_i^2-\bar{y}^2\right)}}.$$

It may be noted that we have considered two variables x and y irrespective of their dependency.

Properties of Correlation Coefficient:

1. The correlation coefficient between any two variables is independent of change of origin and scale in value but depends on the signs of scales.

   Let us consider (x_1,y_1), (x_2,y_2), (x_3,y_3),..., (x_n,y_n), the n pairs of observations for the two characters x and y having the means $\bar{x}$ and $\bar{y}$ and the variances S_x^2 and S_y^2. We take another two variables such that, $x_i = a + bu_i$ and $y_i = c + dv_i$ $\Rightarrow \bar{x} = a + b\bar{u}$ and $\bar{y} = c + d\bar{v}$, and $S_x^2 = b^2S_u^2$ and $S_y^2 = d^2S_v^2$; $i = 1,2,3,\ldots,n$ and a, b, c, and d are constants, and a and c are changes of origin and b and d are changes of scale.

   So,

$$\mathrm{Cov}(x,y) = \frac{1}{n}\sum_i(x_i-\bar{x})(y_i-\bar{y})$$
$$= bd\mathrm{Cov}(u,v).$$

   Thus, the correlation coefficient between x and y becomes

$$r_{xy} = \frac{\mathrm{Cov}(x,y)}{\sqrt{S_x^2.S_y^2}} = \frac{b.d}{|b|.|d|}.r_{uv}.$$

   Thus, the numerical value of r_{xy} and r_{uv} are the same. But, the sign of r_{uv} depends on the sign of b and d. If both b and d are of the same sign, then $r_{xy} = r_{uv}$; on the other hand, if b and d are of the opposite signs, then $r_{xy} = -r_{uv}$.
2. The correlation coefficient r_{xy} lies between -1 and $+1$, that is, $-1 \le r_{xy} \le +1$.
3. The correlation coefficient between x and y is same as the correlation coefficient between y and x.
4. Being a ratio, a correlation coefficient is a unit-free measure. So, it can be used to compare the degree of the linear association between the different pairs of the variables.
5. The two independent variables are uncorrelated, but the converse may not be true. Let us consider the following two variables:

x	−4	−3	−2	−1	0	1	2	3	4	$\sum x = 0$
y	16	9	4	1	0	1	4	9	16	$\sum y = 60$
xy	−64	−27	−8	−1	0	1	8	27	64	$\sum xy = 0$

Therefore, $r_{xy} = \frac{\mathrm{Cov}(x,y)}{S_x.S_y} = \frac{\frac{1}{n}\sum_{i=1}^{n}(x_iy_i)-\bar{x}.\bar{y}}{S_x.S_y} = \frac{\frac{1}{9}.0-0.\frac{60}{9}}{S_x.S_y} = 0.$

Clearly, the relationship is $y = x^2$ between x and y. Thus, the zero correlation coefficient between the two variables does not necessarily mean that the variables are independent.

Examples 8.24. Given below are the plant height (cm) and the no. of fruits per plant in eight varieties of lady fingers. Find out whether the height and the weight of the boys are correlated or not.

Height (X)	205	206	208	207	209	207	210	212
No of fruit (Y)	70	71	75	68	75	71	72	74

At first, we calculate the following quantities:

$$\bar{x} = \frac{1}{8}\sum_{i=1}^{8} x_i = 208, \quad \bar{y} = \frac{1}{8}\sum_{i=1}^{8} y_i = 72.$$

We have

(i) $\text{Cov}(x, y) = \frac{1}{n}\sum_{i=1}^{n}(x_i - \bar{x})(y_i - \bar{y}) = \frac{1}{8}\sum_{i=1}^{n} x_i y_i - \bar{x}\bar{y} = 3$

(ii) $s_x^2 = \frac{1}{n}\sum_{i=1}^{n}(x_i - \bar{x})^2 = \left(\frac{1}{n}\sum_{i=1}^{n} x_i^2 - \bar{x}^2\right) = 5.143$

(iii) $s_y^2 = \frac{1}{n}\sum_{i=1}^{n}(y_i - \bar{y})^2 = \left(\frac{1}{n}\sum_{i=1}^{n} y_i^2 - \bar{y}^2\right) = 6.286$

$$r_{xy} = \frac{\frac{1}{n}\sum_{i=1}^{n} x_i y_i - \bar{x}\bar{y}}{\sqrt{\left(\frac{1}{n}\sum_{i=1}^{n} x_i^2 - \bar{x}^2\right)\left(\frac{1}{n}\sum_{i=1}^{n} y_i^2 - \bar{y}^2\right)}} = \frac{3}{\sqrt{(5.143)(6.286)}} = 0.527.$$

So the correlation coefficient between the height and weight of 8 boys of class nine is 0.603.

Example 8.25. Using the same problem given in Example 8.21, we shall try to get the correlation coefficients among the variables using MS Excel, SPSS, and SAS as shown in the following slides:

(a) Using MS Excel software

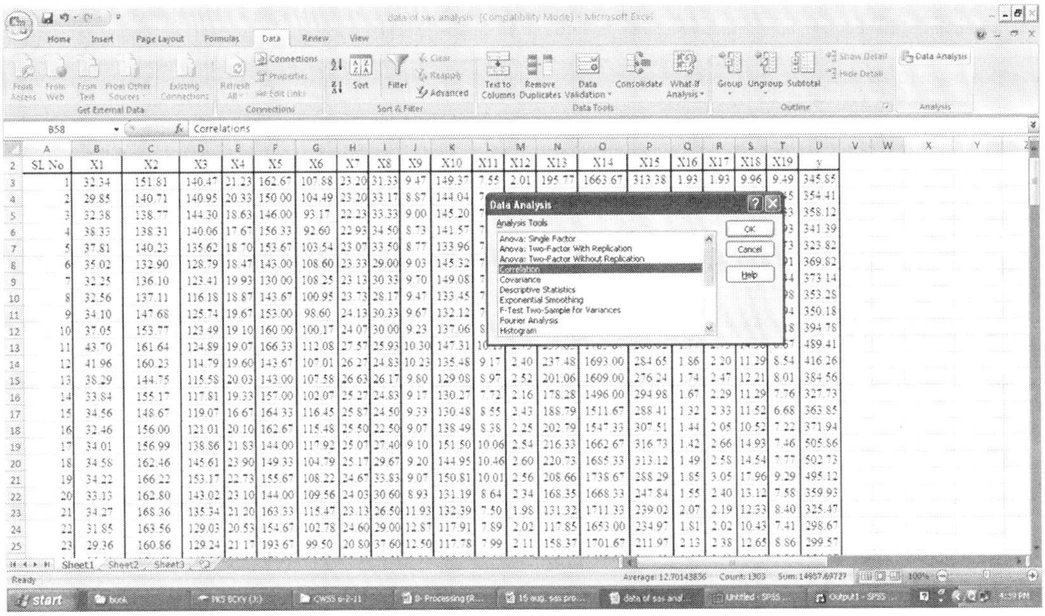

Slide 8.17: Step 1 showing the entered or transferred data and selection of Correlation Analysis menu in MS Excel work sheet

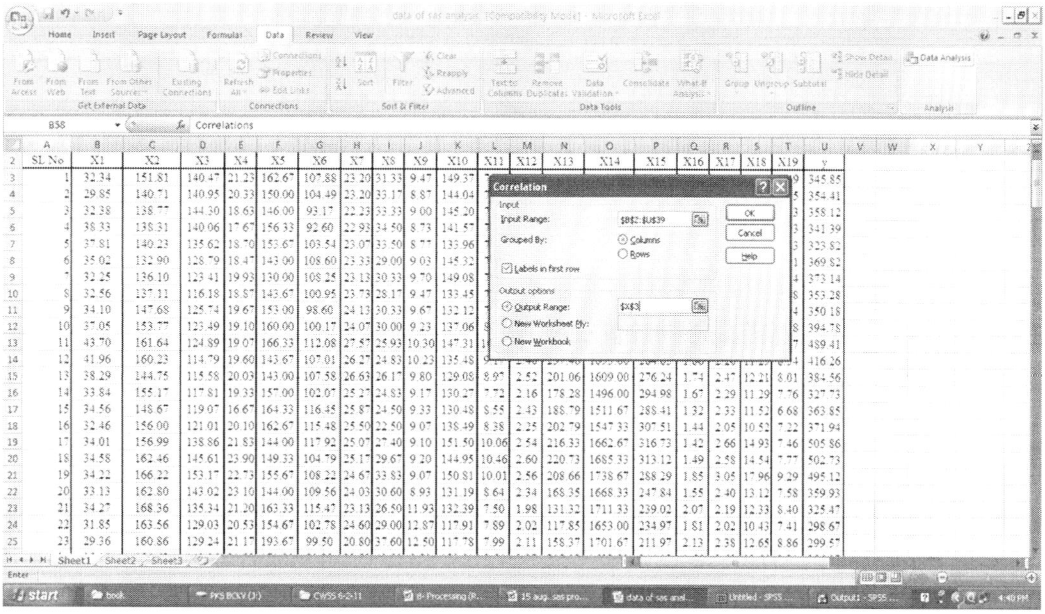

Slide 8.18: Step 2 showing the entered or transferred data and selection of data range and other commands in Correlation Analysis menu in MS Excel

	X1	*X2*	*X3*	*X4*	*X5*	*X6*	*X7*	*X8*	*X9*	*X10*	*X11*	*X12*	*X13*	*X14*	*X15*	*X16*	*X17*	*X18*	*X19*	*Y*
X1	1																			
X2	0.22	1																		
X3	-0.03	0.18	1																	
X4	-0.24	0.34	0.61	1																
X5	-0.13	0.11	0.38	0.43	1															
X6	0.07	0.19	-0.14	0.01	-0.14	1														
X7	0.55	0.28	-0.07	0.02	-0.33	0.24	1													
X8	-0.24	-0.14	0.47	0.25	0.39	-0.69	-0.54	1												
X9	-0.04	0.45	-0.06	0.13	0.22	-0.11	0.12	0.07	1											
X10	0.2	-0.15	0.49	0.15	0.04	0.26	0.11	-0.08	-0.43	1										
X11	0.52	0.36	0.3	0.27	-0.13	0.19	0.65	-0.08	0.03	0.21	1									
X12	0.58	0.33	0.2	0.12	-0.14	0.03	0.59	0.03	0.02	0.07	0.9	1								
X13	0.53	0.01	0.44	0.11	0.24	-0.01	0.12	0.14	-0.44	0.52	0.43	0.48	1							
X14	0.23	0.35	0.27	0.19	-0.07	-0.06	-0.05	0.37	0.28	0.00	0.45	0.44	0.18	1						
X15	0.12	-0.12	0.27	0.09	0.13	0.28	0.00	-0.11	-0.57	0.65	0.22	0.18	0.63	-0.09	1					
X16	-0.2	0.26	0.05	0.18	0.29	-0.34	-0.37	0.27	0.45	-0.17	-0.4	-0.31	-0.17	0.28	-0.24	1				
X17	0.36	0.22	0.07	-0.21	-0.4	-0.1	0.19	0.17	-0.19	0.01	0.48	0.58	0.27	0.53	-0.06	-0.12	1			
X18	0.26	0.11	0.17	-0.22	-0.44	-0.13	0.06	0.2	-0.24	0.12	0.36	0.43	0.26	0.53	-0.05	-0.05	0.95	1		
X19	0.07	0.2	0.29	0.01	0.02	-0.18	-0.26	0.34	0.02	0.19	0.04	0.05	0.29	0.58	0.16	0.62	0.37	0.42	1	
Y	0.52	0.25	0.4	0.27	-0.1	0.27	0.61	-0.13	-0.13	0.56	0.92	0.76	0.52	0.36	0.42	-0.4	0.42	0.35	0.07	1

Slide 8.19: Step 3 showing the output in Correlation Analysis menu in MS Excel

(b) Using SAS software, the same analysis could be done as follows:

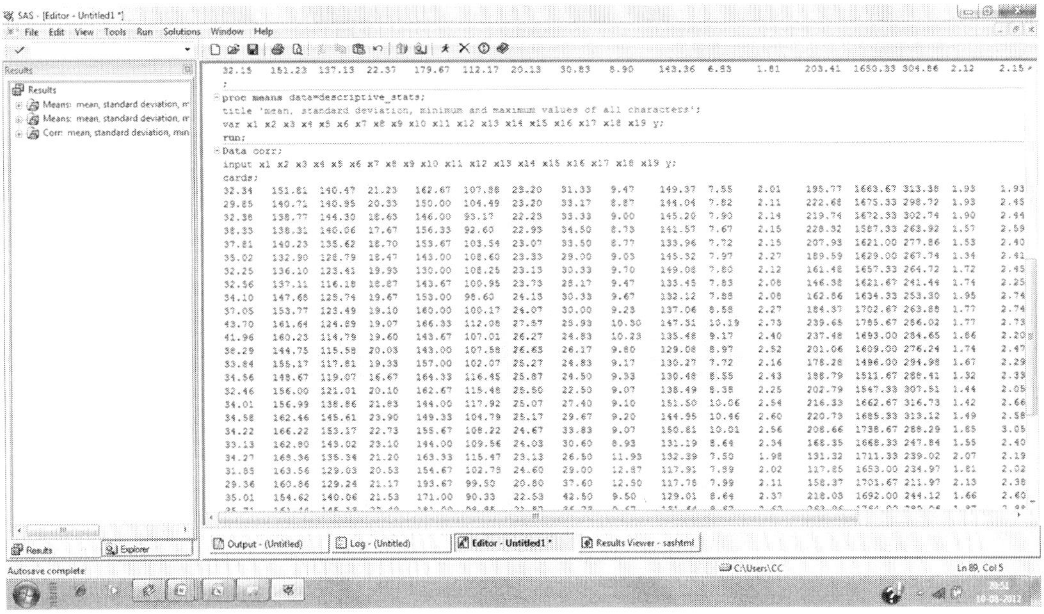

Slide 8.20: Step 1 showing the data input for correlation analysis using SAS

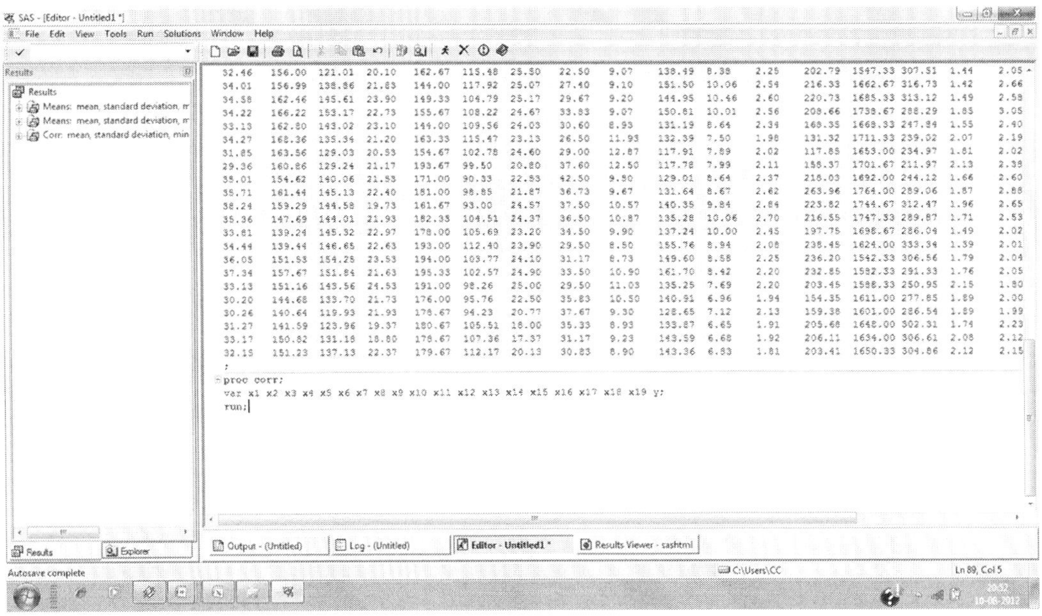

Slide 8.21: Step 2 showing the data and the command for correlation analysis using SAS

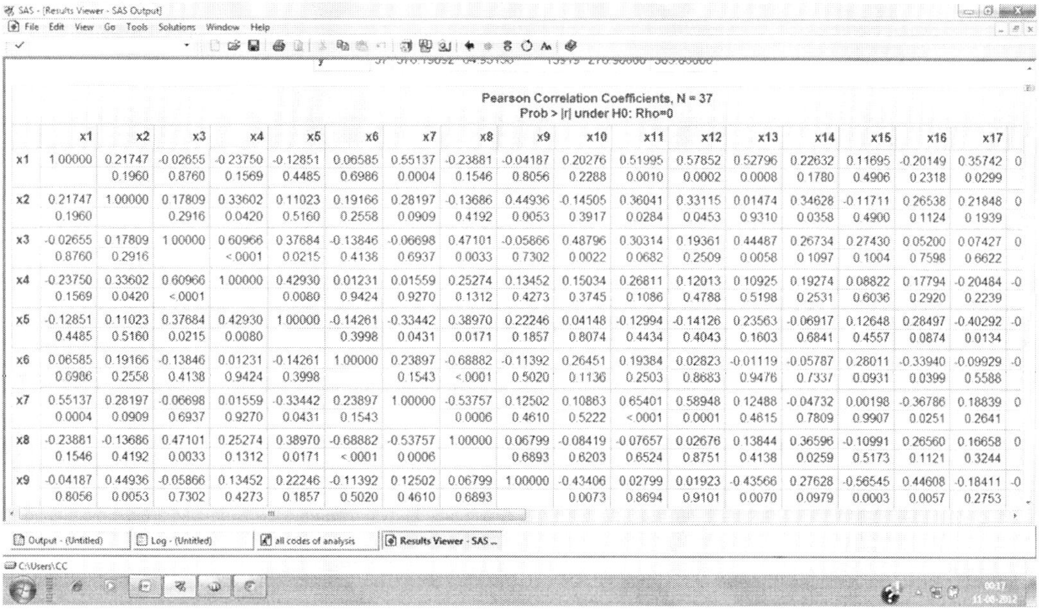

Pearson Correlation Coefficients, N = 37
Prob > |r| under H0: Rho=0

	x1	x2	x3	x4	x5	x6	x7	x8	x9	x10	x11	x12	x13	x14	x15	x16	x17	
x1	1.00000	0.21747	-0.02655	-0.23750	-0.12851	0.06585	0.55137	-0.23881	-0.04187	0.20276	0.51995	0.57852	0.52796	0.22632	0.11695	-0.20149	0.35742	0
		0.1960	0.8760	0.1569	0.4485	0.6986	0.0004	0.1546	0.8056	0.2288	0.0010	0.0002	0.0008	0.1780	0.4906	0.2318	0.0299	
x2	0.21747	1.00000	0.17809	0.33602	0.11023	0.19166	0.28197	-0.13686	0.44936	-0.14505	0.36041	0.33115	0.01474	0.34628	-0.11711	0.26538	0.21848	0
	0.1960		0.2916	0.0420	0.5160	0.2558	0.0909	0.4192	0.0053	0.3917	0.0284	0.0453	0.9310	0.0358	0.4900	0.1124	0.1939	
x3	-0.02655	0.17809	1.00000	0.60966	0.37684	-0.13846	-0.06698	0.47101	-0.05866	0.48796	0.30314	0.19361	0.44487	0.26734	0.27430	0.05200	0.07427	0
	0.8760	0.2916		<.0001	0.0215	0.4138	0.6937	0.0033	0.7302	0.0022	0.0682	0.2509	0.0058	0.1097	0.1004	0.7598	0.6622	
x4	-0.23750	0.33602	0.60966	1.00000	0.42930	0.01231	0.01559	0.25274	0.13452	0.15034	0.26811	0.12013	0.10925	0.19274	0.08822	0.17794	-0.20484	-0
	0.1569	0.0420	<.0001		0.0080	0.9424	0.9270	0.1312	0.4273	0.3745	0.1086	0.4788	0.5198	0.2531	0.6036	0.2920	0.2239	
x5	-0.12851	0.11023	0.37684	0.42930	1.00000	-0.14261	-0.33442	0.38970	0.22246	0.04148	-0.12994	-0.14126	0.23563	-0.06917	0.12648	0.28497	-0.40292	-0
	0.4485	0.5160	0.0215	0.0080		0.3998	0.0431	0.0171	0.1857	0.8074	0.4434	0.4043	0.1603	0.6841	0.4557	0.0874	0.0134	
x6	0.06585	0.19166	-0.13846	0.01231	-0.14261	1.00000	0.23897	-0.68882	-0.11392	0.26451	0.19384	0.02823	-0.01119	-0.05787	0.28011	-0.33940	-0.09929	-0
	0.6986	0.2558	0.4138	0.9424	0.3998		0.1543	<.0001	0.5020	0.1136	0.2503	0.8683	0.9476	0.7337	0.0931	0.0399	0.5588	
x7	0.55137	0.28197	-0.06698	0.01559	-0.33442	0.23897	1.00000	-0.53757	0.12502	0.10863	0.65401	0.58948	0.12488	-0.04732	0.00198	-0.36786	0.18839	0
	0.0004	0.0909	0.6937	0.9270	0.0431	0.1543		0.0006	0.4610	0.5222	<.0001	0.0001	0.4815	0.7809	0.9907	0.0251	0.2641	
x8	-0.23881	-0.13686	0.47101	0.25274	0.38970	-0.68882	-0.53757	1.00000	0.06799	-0.08419	-0.07657	0.02676	0.13844	0.36596	-0.10991	0.26560	0.16658	0
	0.1546	0.4192	0.0033	0.1312	0.0171	<.0001	0.0006		0.6893	0.6203	0.6524	0.8751	0.4138	0.0259	0.5173	0.1121	0.3244	
x9	-0.04187	0.44936	-0.05866	0.13452	0.22246	-0.11392	0.12502	0.06799	1.00000	-0.43406	0.02799	0.01923	-0.43566	0.27628	-0.56545	0.44608	-0.18411	-0
	0.8056	0.0053	0.7302	0.4273	0.1857	0.5020	0.4610	0.6893		0.0073	0.8694	0.9101	0.0070	0.0979	0.0003	0.0057	0.2753	

Slide 8.22: Step 3 showing the SAS output for correlation analysis

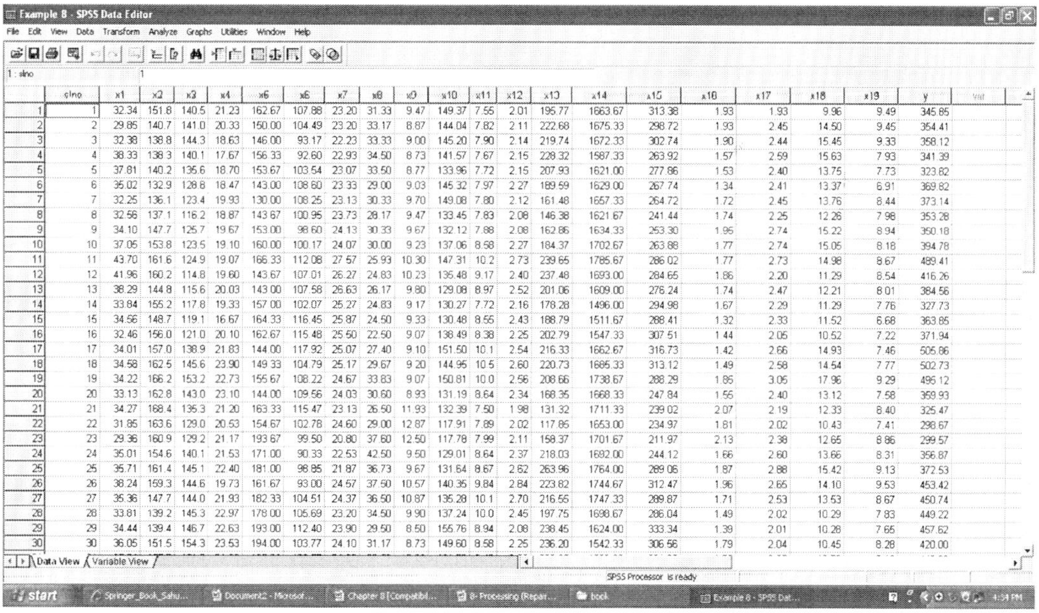

	slno	x1	x2	x3	x4	x5	x6	x7	x8	x9	x10	x11	x12	x13	x14	x15	x16	x17	x18	x19	y
1	1	32.34	151.8	140.5	21.23	162.67	107.88	23.20	31.33	9.47	149.37	7.55	2.01	195.77	1663.67	313.38	1.93	1.93	9.96	9.49	345.85
2	2	29.85	140.7	141.0	20.33	150.00	104.49	23.20	33.17	8.87	144.04	7.82	2.11	222.68	1675.33	298.72	1.93	2.45	14.50	9.45	354.41
3	3	32.38	138.8	144.3	18.63	146.00	93.17	22.23	33.33	9.00	145.20	7.90	2.14	219.74	1672.33	302.74	1.90	2.44	15.45	9.33	358.12
4	4	38.33	138.3	140.1	17.67	156.33	92.60	22.93	34.50	8.73	141.57	7.67	2.15	228.32	1587.33	263.92	1.57	2.59	15.63	7.93	341.39
5	5	37.81	140.2	135.6	18.70	153.67	103.54	23.07	33.50	8.77	133.96	7.72	2.15	207.93	1621.00	277.86	1.53	2.40	13.75	7.73	323.82
6	6	35.02	132.9	128.8	18.47	143.00	108.60	23.33	29.00	9.03	145.32	7.97	2.27	189.59	1629.00	267.74	1.34	2.41	13.37	6.91	369.82
7	7	32.25	136.1	123.4	19.93	130.00	108.25	23.13	30.33	9.70	149.08	7.80	2.12	161.48	1657.33	264.72	1.72	2.45	13.76	8.44	373.14
8	8	32.56	137.1	116.2	18.87	143.67	100.95	23.73	28.17	9.47	133.45	7.83	2.08	146.38	1621.67	241.44	1.74	2.25	12.26	7.98	353.28
9	9	34.10	147.7	125.7	19.67	153.00	98.60	24.13	30.33	9.67	132.12	7.88	2.08	162.86	1634.33	253.30	1.95	2.74	15.22	8.94	350.18
10	10	37.05	153.8	123.5	19.10	160.00	100.17	24.07	30.00	9.23	137.06	8.58	2.27	184.37	1702.67	263.88	1.77	2.74	15.05	8.18	394.78
11	11	43.70	161.6	124.9	19.07	166.33	112.08	27.57	25.93	10.30	147.31	10.2	2.73	239.65	1785.67	286.02	1.77	2.73	14.98	8.67	489.41
12	12	41.96	160.2	114.8	19.60	143.67	107.01	26.27	24.83	10.23	135.48	9.17	2.40	237.48	1693.00	284.65	1.86	2.20	11.29	8.54	416.26
13	13	38.29	144.8	115.6	20.03	143.00	107.58	26.63	26.17	9.80	129.08	8.97	2.52	201.06	1609.00	276.24	1.74	2.47	12.21	8.01	384.56
14	14	33.84	155.2	117.8	19.33	157.00	102.07	25.27	24.83	9.17	130.27	7.72	2.16	178.28	1496.00	294.98	1.67	2.29	11.29	7.76	327.73
15	15	34.56	148.7	119.1	16.67	164.33	116.45	25.87	24.50	9.33	130.48	8.55	2.43	188.79	1511.67	288.41	1.32	2.33	11.52	6.68	363.85
16	16	32.46	156.0	121.0	20.10	162.67	115.48	25.50	22.50	9.07	138.49	8.38	2.25	202.79	1547.33	307.51	1.44	2.05	10.52	7.22	371.94
17	17	34.01	157.0	138.9	21.83	144.00	117.92	25.07	27.40	9.10	151.50	10.1	2.54	216.33	1662.67	316.73	1.42	2.66	14.93	7.46	505.86
18	18	34.58	162.5	145.6	23.90	149.33	104.79	25.17	29.67	9.20	144.95	10.5	2.60	220.73	1685.33	313.12	1.49	2.58	14.54	7.77	502.73
19	19	34.22	166.2	153.2	22.73	155.67	108.22	24.67	33.83	9.07	150.81	10.0	2.56	208.66	1738.67	288.29	1.85	3.05	17.96	9.29	496.12
20	20	33.13	162.8	143.0	23.10	144.00	109.56	24.03	30.60	8.93	131.19	8.64	2.34	168.35	1668.33	247.84	1.55	2.40	13.12	7.58	359.93
21	21	34.27	168.4	135.3	21.20	163.33	115.47	23.13	26.50	11.93	132.39	7.50	1.98	131.32	1711.33	239.02	2.07	2.19	12.33	8.40	325.47
22	22	31.85	163.6	129.0	20.53	154.67	102.78	24.60	29.00	12.87	117.91	7.89	2.02	117.85	1653.00	234.97	1.81	2.02	10.43	7.41	298.67
23	23	29.36	160.9	129.2	21.17	193.67	99.50	20.80	37.60	12.50	117.78	7.99	2.11	158.37	1701.67	211.97	2.13	2.38	12.65	8.86	299.57
24	24	35.01	154.6	140.1	21.53	171.00	90.33	22.53	42.50	9.50	129.01	8.64	2.37	218.03	1692.00	244.12	1.66	2.60	13.66	8.31	356.87
25	25	35.71	161.4	145.1	22.40	181.00	98.85	21.87	36.73	9.67	131.64	8.67	2.62	263.96	1764.00	289.06	1.87	2.88	15.42	9.13	372.53
26	26	38.24	159.3	144.6	19.73	161.67	93.00	24.57	37.50	10.57	140.35	9.84	2.84	223.82	1744.67	312.47	1.96	2.65	14.10	9.53	453.42
27	27	35.36	147.7	144.0	21.93	182.33	104.51	24.37	36.50	10.87	135.28	10.1	2.70	216.55	1747.33	289.87	1.71	2.53	13.53	8.67	450.74
28	28	33.81	139.2	145.3	22.97	178.00	105.69	23.20	34.50	9.90	137.24	10.0	2.45	197.75	1698.67	286.04	1.49	2.02	10.29	7.83	449.22
29	29	34.44	139.4	146.7	22.63	193.00	112.40	23.90	29.50	8.50	155.76	8.94	2.08	238.45	1624.00	333.34	1.39	2.01	10.28	7.65	457.62
30	30	36.05	151.5	154.3	23.53	194.00	103.77	24.10	31.17	8.73	149.60	8.58	2.25	236.20	1542.33	306.56	1.79	2.04	10.45	8.28	420.00

Slide 8.23: Data imputed for correlation analysis through SPSS

The following is the output of the same analysis through SPSS.

SPSS output

Correlations

		X1	X2	X3	X4	X5	X6	X7	X8	X9	X10	X11	X12	X13	X14	X15	X16	X17	X18	X19	Y
X1	Pearson Correlation	1.000	0.217	-0.027	-0.238	-0.129	0.066	0.551	-0.239	-0.042	0.203	0.520	0.579	0.528	0.226	0.117	-0.201	0.357	0.258	0.066	0.522
	Sig. (2-tailed)	.	0.196	0.876	0.157	0.448	0.699	0.000	0.155	0.806	0.229	0.001	0.000	0.001	0.178	0.491	0.232	0.030	0.123	0.698	0.001
	N	37	37	37	37	37	37	37	37	37	37	37	37	37	37	37	37	37	37	37	37
X2	Pearson Correlation	0.217	1.000	0.178	0.336	0.110	0.192	0.282	-0.137	0.449	-0.145	0.360	0.331	0.015	0.346	-0.117	0.265	0.218	0.115	0.198	0.254
	Sig. (2-tailed)	0.196	.	0.292	0.042	0.516	0.256	0.091	0.419	0.005	0.392	0.028	0.045	0.931	0.036	0.490	0.112	0.194	0.498	0.240	0.129
	N	37	37	37	37	37	37	37	37	37	37	37	37	37	37	37	37	37	37	37	37
X3	Pearson Correlation	-0.027	0.178	1.000	0.610	0.377	-0.138	-0.067	0.471	-0.059	0.488	0.303	0.194	0.445	0.267	0.274	0.052	0.074	0.174	0.287	0.403
	Sig. (2-tailed)	0.876	0.292	.	0.000	0.022	0.414	0.694	0.003	0.730	0.002	0.068	0.251	0.006	0.110	0.100	0.760	0.662	0.303	0.085	0.013
	N	37	37	37	37	37	37	37	37	37	37	37	37	37	37	37	37	37	37	37	37
X4	Pearson Correlation	-0.238	0.336	0.610	1.000	0.429	0.012	0.016	0.253	0.135	0.150	0.268	0.120	0.109	0.193	0.088	0.178	-0.205	-0.223	0.008	0.274
	Sig. (2-tailed)	0.157	0.042	0.000	.	0.008	0.942	0.927	0.131	0.427	0.374	0.109	0.479	0.520	0.253	0.604	0.292	0.224	0.186	0.964	0.101
	N	37	37	37	37	37	37	37	37	37	37	37	37	37	37	37	37	37	37	37	37
X5	Pearson Correlation	-0.129	0.110	0.377	0.429	1.000	-0.143	-0.334	0.390	0.222	0.041	-0.130	-0.141	0.236	-0.069	0.126	0.285	-0.403	-0.442	0.022	-0.099
	Sig. (2-tailed)	0.448	0.516	0.022	0.008	.	0.400	0.043	0.017	0.186	0.807	0.443	0.404	0.160	0.684	0.456	0.087	0.013	0.006	0.896	0.560
	N	37	37	37	37	37	37	37	37	37	37	37	37	37	37	37	37	37	37	37	37
X6	Pearson Correlation	0.066	0.192	-0.138	0.012	-0.143	1.000	0.239	-0.689	-0.114	0.265	0.194	0.028	-0.011	-0.058	0.280	-0.339	-0.099	-0.135	-0.183	0.268
	Sig. (2-tailed)	0.699	0.256	0.414	0.942	0.400	.	0.154	0.000	0.502	0.114	0.250	0.868	0.948	0.734	0.093	0.040	0.559	0.426	0.278	0.109
	N	37	37	37	37	37	37	37	37	37	37	37	37	37	37	37	37	37	37	37	37
X7	Pearson Correlation	0.551	0.282	-0.067	0.016	-0.334	0.239	1.000	-0.538	0.125	0.109	0.654	0.589	0.125	-0.047	0.002	-0.368	0.188	0.059	-0.263	0.613
	Sig. (2-tailed)	0.000	0.091	0.694	0.927	0.043	0.154	.	0.001	0.461	0.522	0.000	0.000	0.461	0.781	0.991	0.025	0.264	0.729	0.115	0.000
	N	37	37	37	37	37	37	37	37	37	37	37	37	37	37	37	37	37	37	37	37
X8	Pearson Correlation	-0.239	-0.137	0.471	0.253	0.390	-0.689	-0.538	1.000	0.068	-0.084	-0.077	0.027	0.138	0.366	-0.110	0.266	0.167	0.203	0.338	-0.135
	Sig. (2-tailed)	0.155	0.419	0.003	0.131	0.017	0.000	0.001	.	0.689	0.620	0.652	0.875	0.414	0.026	0.517	0.112	0.324	0.227	0.041	0.427
	N	37	37	37	37	37	37	37	37	37	37	37	37	37	37	37	37	37	37	37	37
X9	Pearson Correlation	-0.042	0.449	-0.059	0.135	0.222	-0.114	0.125	0.068	1.000	-0.434	0.028	0.019	-0.436	0.276	-0.565	0.446	-0.184	-0.237	0.018	-0.126
	Sig. (2-tailed)	0.806	0.005	0.730	0.427	0.186	0.502	0.461	0.689	.	0.007	0.869	0.910	0.007	0.098	0.000	0.006	0.275	0.158	0.915	0.458
	N	37	37	37	37	37	37	37	37	37	37	37	37	37	37	37	37	37	37	37	37
X10	Pearson Correlation	0.203	-0.145	0.488	0.150	0.041	0.265	0.109	-0.084	-0.434	1.000	0.215	0.064	0.522	0.003	0.647	-0.174	0.013	0.119	0.187	0.562
	Sig. (2-tailed)	0.229	0.392	0.002	0.374	0.807	0.114	0.522	0.620	0.007	.	0.202	0.707	0.001	0.985	0.000	0.302	0.940	0.483	0.269	0.000
	N	37	37	37	37	37	37	37	37	37	37	37	37	37	37	37	37	37	37	37	37
X11	Pearson Correlation	0.520	0.360	0.303	0.268	-0.130	0.194	0.654	-0.077	0.028	0.215	1.000	0.895	0.429	0.452	0.223	-0.400	0.484	0.360	0.038	0.920
	Sig. (2-tailed)	0.001	0.028	0.068	0.109	0.443	0.250	0.000	0.652	0.869	0.202	.	0.000	0.008	0.005	0.185	0.014	0.002	0.029	0.826	0.000
	N	37	37	37	37	37	37	37	37	37	37	37	37	37	37	37	37	37	37	37	37
X12	Pearson Correlation	0.579	0.331	0.194	0.120	-0.141	0.028	0.589	0.027	0.019	0.064	0.895	1.000	0.479	0.439	0.179	-0.316	0.580	0.423	0.049	0.763
	Sig. (2-tailed)	0.000	0.045	0.251	0.479	0.404	0.868	0.000	0.875	0.910	0.707	0.000	.	0.003	0.007	0.288	0.057	0.000	0.009	0.775	0.000
	N	37	37	37	37	37	37	37	37	37	37	37	37	37	37	37	37	37	37	37	37
X13	Pearson Correlation	0.528	0.015	0.445	0.109	0.236	-0.011	0.125	0.138	-0.436	0.522	0.429	0.479	1.000	0.180	0.630	-0.171	0.271	0.259	0.293	0.525
	Sig. (2-tailed)	0.001	0.931	0.006	0.520	0.160	0.948	0.461	0.414	0.007	0.001	0.008	0.003	.	0.286	0.000	0.312	0.105	0.122	0.078	0.001
	N	37	37	37	37	37	37	37	37	37	37	37	37	37	37	37	37	37	37	37	37
X14	Pearson Correlation	0.226	0.346	0.267	0.193	-0.069	-0.058	-0.047	0.366	0.276	0.003	0.452	0.439	0.180	1.000	-0.095	0.278	0.530	0.532	0.579	0.358
	Sig. (2-tailed)	0.178	0.036	0.110	0.253	0.684	0.734	0.781	0.026	0.098	0.985	0.005	0.007	0.286	.	0.577	0.096	0.001	0.001	0.000	0.029
	N	37	37	37	37	37	37	37	37	37	37	37	37	37	37	37	37	37	37	37	37
X15	Pearson Correlation	0.117	-0.117	0.274	0.088	0.126	0.280	0.002	-0.110	-0.565	0.647	0.223	0.179	0.630	-0.095	1.000	-0.241	-0.063	-0.053	0.162	0.415
	Sig. (2-tailed)	0.491	0.490	0.100	0.604	0.456	0.093	0.991	0.517	0.000	0.000	0.185	0.288	0.000	0.577	.	0.151	0.712	0.756	0.339	0.011
	N	37.000	37.000	37.000	37.000	37.000	37.000	37.000	37.000	37.000	37.000	37.000	37.000	37.000	37.000	37.000	37.000	37.000	37.000	37.000	37.000
X16	Pearson Correlation	-0.201	0.265	0.052	0.178	0.285	-0.339	-0.368	0.266	0.446	-0.174	-0.400	-0.316	-0.171	0.278	-0.241	1.000	-0.116	-0.051	0.615	-0.400
	Sig. (2-tailed)	0.232	0.112	0.760	0.292	0.087	0.040	0.025	0.112	0.006	0.302	0.014	0.057	0.312	0.096	0.151	.	0.494	0.766	0.000	0.014
	N	37	37	37	37	37	37	37	37	37	37	37	37	37	37	37	37	37	37	37	37
X17	Pearson Correlation	0.357	0.218	0.074	-0.205	-0.403	-0.099	0.188	0.167	-0.184	0.013	0.484	0.580	0.271	0.530	-0.063	-0.116	1.000	0.945	0.369	0.416
	Sig. (2-tailed)	0.030	0.194	0.662	0.224	0.013	0.559	0.264	0.324	0.275	0.940	0.002	0.000	0.105	0.001	0.712	0.494	.	0.000	0.025	0.010
	N	37	37	37	37	37	37	37	37	37	37	37	37	37	37	37	37	37	37	37	37
X18	Pearson Correlation	0.258	0.115	0.174	-0.223	-0.442	-0.135	0.059	0.203	-0.237	0.119	0.360	0.423	0.259	0.532	-0.053	-0.051	0.945	1.000	0.421	0.345
	Sig. (2-tailed)	0.123	0.498	0.303	0.186	0.006	0.426	0.729	0.227	0.158	0.483	0.029	0.009	0.122	0.001	0.756	0.766	0.000		0.009	0.037
	N	37	37	37	37	37	37	37	37	37	37	37	37	37	37	37	37	37	37	37	37
X19	Pearson Correlation	0.066	0.198	0.287	0.008	0.022	-0.183	-0.263	0.338	0.018	0.187	0.038	0.049	0.293	0.579	0.162	0.615	0.369	0.421	1.000	0.074
	Sig. (2-tailed)	0.698	0.240	0.085	0.964	0.896	0.278	0.115	0.041	0.915	0.269	0.826	0.775	0.078	0.000	0.339	0.000	0.025	0.009	.	0.665
	N	37	37	37	37	37	37	37	37	37	37	37	37	37	37	37	37	37	37	37	37
Y	Pearson Correlation	0.522	0.254	0.403	0.274	-0.099	0.268	0.613	-0.135	-0.126	0.562	0.920	0.763	0.525	0.358	0.415	-0.400	0.416	0.345	0.074	1.000
	Sig. (2-tailed)	0.001	0.129	0.013	0.101	0.560	0.109	0.000	0.427	0.458	0.000	0.000	0.000	0.001	0.029	0.011	0.014	0.010	0.037	0.665	.
	N	37	37	37	37	37	37	37	37	37	37	37	37	37	37	37	37	37	37	37	37

Slide 8.24: Showing the SPSS output for correlation analysis

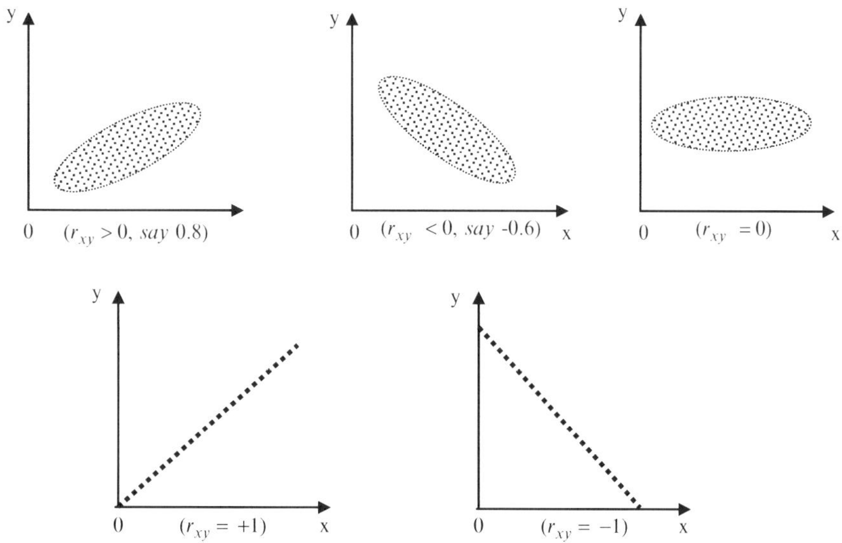

Fig. 8.19 Graphical presentation of different types of correlation coefficient

Table 8.11 Bivariate frequency distribution table

Mid value	Mid value (y)							
(x)	y_1	y_2	y_3	...	y_j	...	y_m	Total
x_1	f_{11}	f_{12}	f_{13}	...	...	...	f_{1m}	$f_{1.}$
x_2	f_{21}	f_{22}	f_{23}	...	...	...	f_{2m}	$f_{2.}$
x_3	f_{31}	f_{32}	f_{33}	...	...	...	f_{3m}	$f_{3.}$
...	...	...	...	...	...	...	...	...
x_i	...	...	...		f_{ij}	...	f_{1m}	$f_{1.}$
...	...	...	...	...	...	...	...	...
x_n	f_{n1}	f_{n2}	f_{n3}	...	...	...	f_{nm}	$f_{n.}$
Total	$f_{.1}$	$f_{.2}$	$f_{.3}$	...	$f_{.j}$		$f_{.m}$	$\sum_i \sum_j f_{ij} = N$

Significance of the Values of Correlation Coefficients:

1. A positive correlation coefficient between any two variables means both the variables move in the same direction.
2. A negative correlation between any two variables means if there is an increase or decrease in one variable, then there will be a decrease/increase, respectively, in the other variables.
3. As the numerical value of correlation coefficient approaches 1, the degree of linear association becomes more and more intense (Fig. 8.19).

Correlation Coefficient of Bivariate Frequency Distribution:

Let the two variables X and Y have a joint frequency distribution as follows (Table 8.11):where

$$\sum_{i=}^{n}\sum_{j=1}^{m} f_{ij} = \sum_{i=1}^{n} f_{i.} = \sum_{j=1}^{m} f_{.j} = N.$$

From the data given in Table 8.11, we shall calculate the means and variances using the marginal frequencies and the corresponding mid values of the classes. The covariance between the variables will be calculated by taking cell frequencies and the corresponding mid values of different class combinations of variables X and Y. Marginal frequency refers to the total frequencies of each and every class of the two variables under consideration. Thus, the marginal frequency corresponding to the mid value y_j is $f_{.j}$ and that for x_i is $f_{i.}$. So,

$$\bar{x} = \frac{1}{N}\sum_{i=1}^{n}(f_{i.}.x_i), \bar{y} = \frac{1}{N}\sum_{j=1}^{m}(f_{.j}.y_j).$$

Similarly,

$$S^2{}_x = \frac{1}{N}\sum_{i=1}^{n} f_{i.}.x^2{}_i - \bar{x}^2, S^2{}_y = \frac{1}{N}\sum_{j=1}^{m} f_{.j}.y^2{}_j - \bar{y}^2$$

and

$$\text{Cov}(x, y) = \frac{1}{N}\sum_{i,j} f_{ij}x_iy_j - \bar{x}.\bar{y} = S_{xy}.$$

Example 8.26. The following table gives the bivariate frequency distribution of the plant height and the no. of insects per plant. Using the information, find the correlation coefficient between the plant height and the no. of insects per plant.

	No. of insects per plant					
Height	1–3	3–5	5–7	7–9	9–11	11–13
10–20	0	1	0	0	0	0
20–30	1	2	0	0	0	0
30–40	0	3	2	1	0	0
40–50	1	3	2	1	1	0
50–60	2	2	3	1	0	0
60–70	0	1	2	2	2	0
70–80	0	2	3	3	1	0
80–90	0	1	4	3	2	1

Solution.

	No. of insects per plant										
		1–3	3–5	5–7	7–9	9–11	11–13				
		Mid value(x_2)									
Height	Mid value (x_1)	2	4	6	8	10	12	$f_{i.}$	$f_{i.}x_i$	$f_{i.}x_i^2$	$x_i\sum_{j=1}^{m} f_{ij}y_j$
10–20	15	0	1	0	0	0	0	1	15	225	60
20–30	25	1	2	0	0	0	0	3	75	1,875	250
30–40	35	0	3	2	1	0	0	6	210	7,350	1,120
40–50	45	1	3	2	1	1	0	8	360	16,200	1,980
50–60	55	2	2	3	1	0	0	8	440	24,200	2,090
60–70	65	0	1	2	2	2	0	7	455	29,575	3,380
70–80	75	0	2	3	3	1	0	9	675	50,625	4,500
80–90	85	0	1	4	3	2	1	11	935	79,475	7,140
Total ($f_{.j}$)		4	15	16	11	6	1	53	3165	209,525	20,520
$f_{.j}y_j$		8	60	96	88	60	12	324	Check		
$f_{.j}y_j^2$		16	240	576	704	600	144	2,280			
$y_j\sum_{i=1}^{n} f_{ij}x_i$		360	2,860	6,120	5,960	4,200	1,020	20,520			

$$\bar{x} = \frac{1}{N}\sum_{i=1}^{n}(f_{i.}.x_i) = [15 + 75 + 210 + \cdots + 935]/78 = 3165/53 = 59.716$$

Similarly,

$$\bar{y} = \frac{1}{N}\sum_{j=1}^{m}(f_{.j}.y_j) = [\ 8 + 60 + \cdots + 12]/78 = 324/53 = 6.113$$

$$S^2{}_x = \frac{1}{N}\sum_{i=1}^{n} f_{i.}.x^2{}_i - \bar{x}^2 = 209525/53 - (59.716)^2 = 387.301$$

Similarly,

$$S_y^2 = \frac{1}{\sum_{j=1}^{m} f_{.j}} \sum_{j=1}^{m} f_{.j} . y_{.j}^2 - \bar{y}^2$$

$$= 2280/53 - (6.113)^2 = 5.405$$

$$\text{Cov}(x, y) = S_{xy} = \frac{1}{N} \sum_{i,j} f_{ij} x_i y_j - \bar{x}.\bar{y}$$

$$= 20520/53 - 59.716 \times 6.113$$

$$= 22.126$$

$$\therefore r_{xy} = \frac{\text{Cov}(X,Y)}{S_X S_y} = \frac{22.126}{\sqrt{(387.301)(5.405)}}$$

$$= 0.484$$

Limitations:

1. Correlation coefficient assumes linear relationship between two variables.
2. High correlation coefficient does not mean high direct association between the variables under a multiple variables consideration problem. The high correlation coefficient between the two variables may be due to the influence of the third and/or fourth variable influencing both the variables. Unless and otherwise the effect of the third and/or fourth variable on the two variables is eliminated, the correlation between the variables may be misleading. To counter this problem, path coefficient analysis discussed in the 2nd volume of this book is useful.
3. If the data are not homogeneous to some extent, correlation coefficient may give rise to misleading conclusions.
4. For any two series of values, the correlation coefficient between the variables can be worked out, but there should be logical basis for any correlation coefficient; one should be sure about the significance of correlation coefficient; otherwise, the correlation is known as *nonsense correlation or spurious* correlation.

8.4 Correlation Ratio

Generally, in a real-life situation, we come across a nonlinear type of relationships among the variables. For example, when the varying doses of nitrogenous fertilizer to respond nonlinearly with the yield, it will be unwise to assume that as we go on increasing the dose of nitrogen, the yield will increase linearly. The yield will increase initially as we go on increasing the doses of fertilizer, reaches to maximum and then decreases. So a curvilinear relationship will be appropriate, and the use of correlation coefficient (r) to measure the degree of association will be misleading.

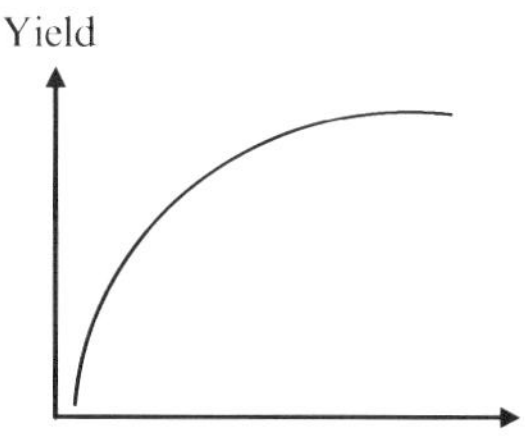

"Correlation ratio" "η" *is the appropriate measure of degree of nonlinear relationship between two variables.* Correlation ratio measures the concentration of points about the curve fitted to exhibit the nonlinear relationship between the two variables.

Correlation ratio (η) of Y on X is defined as $\eta_{yx}^2 = 1 - \frac{S_{ey}^2}{S_y^2}$

where

$$S_{ey}^2 = \frac{1}{N} \sum_i \sum_j f_{ij} \left(y_{ij} - \bar{y}_i \right)^2 \text{ and}$$

$$S_y^2 = \frac{1}{N} \sum_i \sum_j f_{ij} \left(y_{ij} - \bar{y} \right)^2$$

(Table 8.12).

Let the value y_{ij} occur with the frequency f_{ij}. Thus, the frequency distribution of two variables X and Y can be arranged as given in Table 8.11. Further, $\bar{y}_i = \frac{T_i}{n_i}$ and $\bar{y} = \frac{T}{N}$ are the means of ith array and the overall mean, respectively.

Table 8.12 Bivariate frequency distribution table for correlation ratio analysis

x_i					
y_{ij}	x_1	x_2	... x_i	... x_n	
y_{i1}	f_{11}	f_{21}	... f_{i1}	... f_{n1}	
y_{i2}	f_{12}	f_{22}	... f_{i2}	... f_{n2}	
y_{i3}	f_{13}	f_{23}	... f_{i3}	... f_{n3}	
...	....	...	... ...	... ...	
y_{ij}	f_{1j}	...	... f_{ij}	... f_{nj}	
...	...	...	... ...	... ...	
y_{im}	f_{1m}	f_{2m}	... f_{im}	... f_{nm}	
$n_i = \sum_{j=1}^{m} f_{ij}$	n_1	n_2	... n_i	... n_n	$\sum_i^n \sum_j^m f_{ij} = \sum_{i=1}^n n_i = N$
$T_i = \sum_{j=1}^{m} f_{ij} y_{ij}$	T_1	T_2	... T_i	... T_n	$\sum_i^n \sum_j^m f_{ij} y_{ij} = \sum_{i=1}^n T_i = T$
$\sum_{j=1}^{m} f_{ij} y_{ij}^2$	$\sum_{j=1}^{m} f_{1j} y_{1j}^2$	$\sum_{j=1}^{m} f_{2j} y_{2j}^2$	... $\sum_{j=1}^{m} f_{ij} y_{ij}^2$	... $\sum_{j=1}^{m} f_{nj} y_{nj}^2$	$\sum_i^n \sum_j^m f_{ij} y_{ij}^2$

We have

$$S_y^2 = \frac{1}{N} \sum_i \sum_j f_{ij} (y_{ij} - \bar{y})^2$$

$$= S_{ey}^2 + S_{my}^2;\ \text{where, } S_{my}^2 = \frac{1}{N} \sum_i n_i (\bar{y}_i - \bar{y})^2$$

$$\Rightarrow S_y^2 = S_{ey}^2 + S_{my}^2 \Rightarrow 1 = \frac{S_{ey}^2}{S_y^2} + \frac{S_{my}^2}{S_y^2} \Rightarrow \eta_{yx}^2$$

$$= \frac{S_{my}^2}{S_y^2} = \frac{\left[\frac{1}{N} \sum n_i (\bar{y}_i - \bar{y})^2\right]}{S_y^2} = \frac{\left[\sum \frac{T_i^2}{n_i} - \frac{T^2}{N}\right]}{\sum_i \sum_j f_{ij} y_{ij}^2 - \frac{T^2}{N}}.$$

One can substitute the values of the above quantities from the table to get η_{yx}^2.

Example 8.27. The following table gives the frequency distribution of milk yield (y) in kg/day and the age of the cows (x) in years. Find the correlation ratio of yield on the age of cows.

	Age of cows in year					
Yield (kg/day) y	3–4	5–6	7–8	9–10	11–12	13–14
4–7	**2**	**3**	**3**	-	-	-
8–11	-	**6**	**7**	**8**	-	-
12–15	-	**6**	**10**	**15**	**12**	**2**
16–19	-	**1**	**5**	**10**	**19**	**4**
20–23	-	-	-	**10**	**15**	**10**
24–27	-	-	-	-	**5**	**4**

Solution. From the given information, we frame the following table:

		Age of cows (years) (x)						
		3–4	5–6	7–8	9–10	11–12	13–14	
		Mid values (x)						
Milk yield (kg/day) (y)	Mid values (y)	3.5	5.5	7.5	9.5	11.5	13.5	Total
4–7	5.5	2	3	3	0	0	0	8
8–11	9.5	0	3	7	8	0	0	21
12–15	13.5	0	6	10	15	12	2	45
16–19	17.5	0	1	5	10	19	4	39
20–23	21.5	0	0	0	10	15	10	35
24–27	25.5	0	0	0	0	5	4	9
$n_i = \sum_{j=1}^{6} f_{ij}$		2	16	25	43	51	20	157
$T_i = \sum_{j=1}^{6} f_{ij} y_{ij}$		11	172	305.5	668.5	944.5	414	2,515.5
$\frac{T_i^2}{n_i}$		60.5	1,849	3,733.21	10,392.84	17,491.769	8,569.8	42,097.122
$\sum_{j=1}^{6} f_{ij} y_{ij}^2$		60.5	2,032	4,076.25	11,140.75	18,190.75	8,813	44,313.25

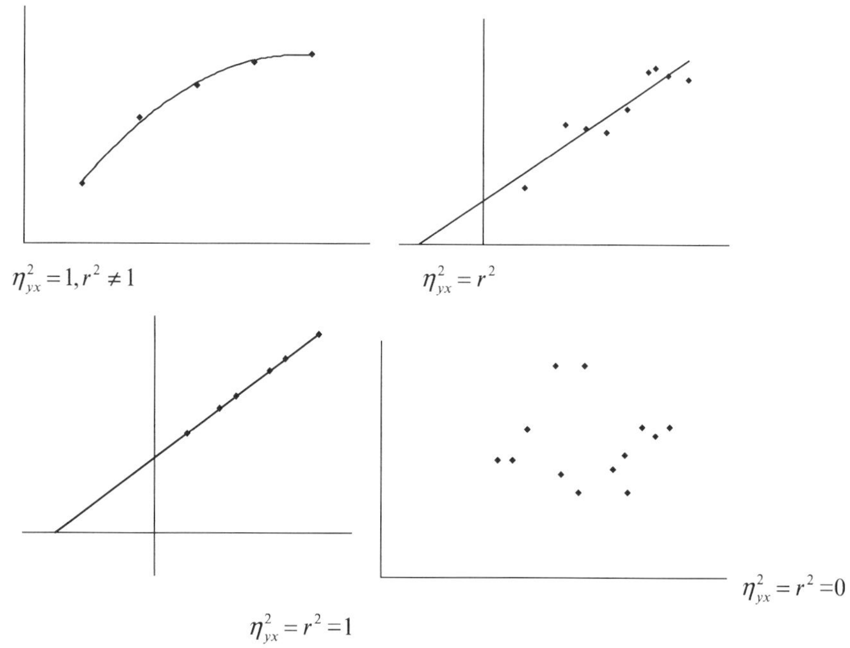

Fig 8.20 Graphical presentation of correlation coefficient and correlation ratio

One can get the $\eta^2_{yx} = 1 - \frac{S^2_{ey}}{S^2_y}$ or $\eta^2_{yx} = \frac{S^2_{my}}{S^2_y}$.

We have

$$NS^2_y = \sum_i \sum_j f_{ij}(y_{ij} - \bar{y})^2 = \sum_i \sum_j f_{ij} y^2_{ij} - N\bar{y}^2$$

$$= \sum_i \sum_j f_{ij} y^2_{ij} - \frac{T^2}{N} = 44313.25 - \frac{(2515.5)^2}{157}$$

$$= 44,313.25 - 40,304.077 = 4,009.173.$$

Again,

$$NS^2_{my} = \sum_i n_i(\bar{y}_i - \bar{y})^2 = \sum \frac{T^2_i}{n_i} - \frac{T^2}{N}$$

$$= 42,097.122 - 40,304.077 = 1,793.045.$$

So, $\eta^2_{yx} = \frac{S^2_{my}}{S^2_y} = \frac{1,793.045}{4,009.173} = 0.447.$

Properties of Correlation Ratio:

1. $r^2 \leq \eta^2_{yx} \leq 1$.
2. η^2_{yx} is independent of change of origin and scale.
3. $r_{xy} = r_{yx}$, but η^2_{yx} may or may not be equal to η^2_{xy}.
4. $\eta^2_{yx} = 1$, if $S^2_{ey} = 0$ i.e. $y_{ij} = \bar{y}_i$.
   That means all observations in any array coincide with their mean, and there is a functional relationship between y and x which of course need not be linear.
5. $\eta^2_{yx} = r^2$ if $\bar{y}_i = \hat{y}_{ij} = Y_i$; that is, array means lie on a straight line. Thus, $\eta^2_{yx} - r^2$ is the departure of regression from linearity.
6. If $r^2 = 1$, that is, the relationship between y and is linear, then η^2_{yx} must be unity and the relationship is to be linear.
7. $\eta_{yx} = 0$ if $\bar{y}_i = \bar{y}$; that is, all the array means are equal to a constant, that is, the overall mean, then r must be zero and there is no relationship (Fig. 8.20).

8.5 Association of Attributes

Qualitative characters like gender, religion, color, aroma, taste, education standard, and economic status cannot be measured as such in any numeric scale; rather, these could be grouped or categorized. Thus, the above-discussed measures of association may not be applicable. To have an association between the two attributes grouped into different categories, one can use Spearman's rank correlation coefficient and Yule's coefficient of attributes.

8.5.1 Rank Correlation

As discussed above, there are other groups of variables, which are not arithmetically measurable, rather they can be grouped or ranked into different groups or ranks. Even sometimes some measurable characters cannot be measured appropriately because of constraints in time of measurement, cost involvement (may be in the form of the cost of the measuring instrument), etc. In all these cases, we can arrange the individuals according to their assigned ranks. For example, in case of tea, the aroma and taste of different teas can be ranked rather than measured. Similarly, the susceptibility of the different varieties of scented rice towards pest and diseases can be ranked; it is very difficult to measure the susceptibility exactly. Now, if we want to have a degree of association between aroma and susceptibility, then one can use the *Spearman's rank correlation coefficient*.

Suppose we have n individuals whose ranks according to a character A are $x_1, x_2, x_3, \ldots, x_n$ and according to character B are $y_1, y_2, y_3, \ldots, y_n$. Assuming that no two individuals are tied in either the classification mentioned, wherein both x and y are all integral values from 1 to n, then Spearman's rank correlation coefficient is defined as $r_R = 1 - \frac{6\sum d_i^2}{n(n^2-1)}$, where $d_i (i = 1, 2, \ldots, n)$ are the difference between the ranks obtained in two different characters by the ith individual.

If more than one individual have the same rank, then the above formula is redefined as follows:

$$r_R = 1 - \frac{6\left\{\sum d^2{}_i + \frac{p(p^2-1)}{12} + \frac{q(q^2-1)}{12}\right\}}{n(n^2-1)},$$ where

p and q are the number of individuals involved in tied ranks for characters A and B, respectively.

Range of rank correlation coefficient: $-1 \leq r_R \leq +1$.

Example 8.28. Ten different types of teas were examined for aroma as well as for taste. The ranks are shown as follows. Find the association between aroma and pest susceptibility:

	Variety									
	T_1	T_2	T_3	T_4	T_5	T_6	T_7	T_8	T_9	T_{10}
Aroma:	3	1	6	7	8	4	2	5	9	10
Susceptibility:	6	2	9	10	4	5	1	7	8	3

The differences between the rankings in aroma and susceptibility are −3, −1, −3, −3, 4, −1, 1, −2, 1, and 7.

So, $\sum_{i=1}^{n} d_i^2 = 9 + 1 + 9 + 9 + 16 + 1 + 1 + 4 + 1 + 49 = 100$

$$\Rightarrow r_R = 1 - \frac{6.\sum_{i=1}^{10} d_i^2}{10(10-1)} = 1 - \frac{6.100}{10.99} = 1 - \frac{20}{33} = \frac{13}{33}.$$

Thus, there is a low association between the aroma and the pest susceptibility in scented rice.

Example 8.29. Farmers are ranked for their educational status and motivation index as follows. Find the association between the standard of education and the motivation index of the farmers:

	Farmers									
	F_1	F_2	F_3	F_4	F_5	F_6	F_7	F_8	F_9	F_{10}
Education	6	10	3.5	5	7	3.5	2	1	9	8
Motivation index	2	9	7	4.5	6	1	4.5	3	8	10

In education, there are two tied ranks 3.5 and in awareness, there are also two tied ranks 4.5. So the correction in education series is $\frac{2(4-1)}{12}=\frac{1}{2}$ and that in awareness series is also 1/2.

Now,

$$d_i = 4,\ 1, -3.5,\ 0.5,\ 1,\ 2.5, -2.5, -2,\ 1, -2$$

$$\Rightarrow \sum_{i=1}^{10} d_i^2 = 4^2 + 1^2 + (-3.5)^2 + (0.5)^2 + 1^2 + (2.5)^2 + (-2.5)^2 + (-2)^2 + 1^2 + (-2)^2$$

$$= 16 + 1 + 12.25 + 0.25 + 1 + 6.25 + 6.25 + 4 + 1 + 4 = 52$$

$$r_R = 1 - \frac{6\left[\sum_{i=1}^{10} d_i^2 + \sum^{2} \frac{m(m^2-1)}{12}\right]}{n(n^2-1)} = 1 - \frac{6.\left[52 + \frac{1}{2} + \frac{1}{2}\right]}{10.99} = 1 - 0.3212 = 0.6788.$$

So the education and motivation index of the farmers are substantially associated.

8.5.2 Yule's Coefficient

The main idea of Yule's coefficient of association is that the two attributes are associated if they appear in greater number/frequency than what is expected if these are independent. Let there be two attributes A and B. The occurrence of A or B is denoted by (A) and (B), respectively, whereas the absence of these is denoted by (a) and (b), respectively. Thus, we have (Table 8.13)

According to Yule's formula of association, between the attributes A and B

$$Q_{\mathrm{AB}} = \frac{(AB)(ab) - (Ab)(aB)}{(AB)(ab) + (Ab)(aB)}$$

where

(AB) is the frequency of the class AB, that is, the class in which both A and B are present

(Ab) is the frequency of the class Ab, that is, the class in which A is present but B is absent

(aB) is the frequency of the class aB, that is, the class in which A is absent but B is present

(ab) is the frequency of the class ab, that is, the class in which both a and b are absent

The value of Yule's coefficient $-1 \le Q_{AB} \le +1$.

Table 8.13 2×2 Distribution of two attributes

A\B	B	b
A	(AB)	(Ab)
a	(aB)	(ab)

8.5.3 Coefficient of Colligation

Using the same idea discussed above in the case of Yule coefficient, another coefficient of association between two attributes has been defined as coefficient of colligation using the following formula: $Q_{AB} = \frac{\left\{1 - \frac{(Ab)(aB)}{(AB)(ab)}\right\}}{\left\{1 + \frac{(Ab)(aB)}{(AB)(ab)}\right\}}$ when $Q = 0$ $\Rightarrow \gamma = 0$, $Q = -1 \Rightarrow \gamma = -1$, $Q = +1 \Rightarrow \gamma = +1$, and $Q = \frac{2\gamma}{1+\gamma^2}$.

Besides the above types of correlation, intraclass correlation, grade correlation, Kendall's rank correlation, serial correlation, tetrachoric correlation, etc., are also used.

8.6 Regression Analysis

Correlation coefficient measures the degree of linear association between any of the two given variables. Once after getting a good degree of association, our objective is to find out the actual relationship between or among the variables. The technique by which we can analyze the relationship among the correlated variable is known as regression analysis in the theory of statistics. In different agricultural and socioeconomic studies, different demographic, social, economical, educational, etc., parameters are studied to find out the dependence of the ultimate variables, say yield, adoption index, awareness, empowerment status, etc., on these parameters. Regression analysis is a technique by virtue of which one can study the relationship.

Thus, the main objective of regression analysis is to estimate and/or predict the average value of the dependent variable given the values for independent/explanatory variables which can be represented in the form of $Y = f(X_i, u_i)$.

It should emphatically be noted that statistical regression analysis does not imply cause and effect relationship between the explanatory variables and the dependent variable. To know more about causation, one should consider Granger test of causality.

The regression equation can also be used as prediction equation under the assumption that the trend of change in Y (the dependent variables) corresponding to the change in X (or the X_is) (the independent variables) remains the same. Once the constants are estimated from a given set of observations, the value of the dependent variable corresponds to any value of X (or a set of values of X_is) within the range of X (or the X_is can be worked out). To some extent, the prediction can be made for Y for the value(s) of X (X_is) beyond the range but not too far beyond the values taken for calculation.

Type of Regression:

Regression analysis may be of two main types: (1) *linear regression* and (2) *nonlinear regression*.

Again depending upon the number of variables present in the regression equations both the above-mentioned two types of regressions can be categorized into (a) *simple regression* and (b) *multiple regression*.

If only two variables are present in the regression equation, then it is termed as simple regression equation, for example, $Y = 12.5 + 1.25\,X$, where X is the independent variable and Y is the dependent variable. On the other hand, when a regression equation has more than two variables, then it is termed as multiple regression equation, for example, $Y = 10.0 + 0.89\,X_1 + 1.08\,X_2 - 0.76\,X_3 + \ldots$, where Xs are the independent variables and Y is the dependent variable.

Simple Linear Regression Analysis:

$Y = \alpha + \beta X$, where α and β are the parameters, known as the intercept constant and regression coefficient of Y on X. The intercept constant signifies the value of the dependent variable at initial stage, and the regression coefficient measures the change in a dependent variable to per unit change in the independent variable X.

$$\Rightarrow \beta = \frac{\mathrm{Cov}(X,Y)}{V(X)} = \frac{\rho\sigma_{xy}}{\sigma_x^2} = \rho\frac{\sigma_y}{\sigma_x} \text{ and}$$

$$\alpha = \mu_y - \rho\frac{\sigma_y}{\sigma_x}\mu_x$$

where ρ is the correlation coefficient between X and Y, μ_x and μ_y are the means of X and Y, respectively,

σ_{xy} is the convariance between X and Y, and σ_x^2 and σ_y^2 are the cariances of X and Y, respectively,

Using these expressions, the equation of the line may be written

$$\left(Y - \mu_y\right) = \rho\frac{\sigma_y}{\sigma_x}(X - \mu_x).$$

Properties of Regression Coefficient:

1. $b.b_1 = r\frac{s_Y}{s_X}.r\frac{s_X}{s_Y} = r^2$, where $b = \frac{s_{xy}}{s_x^2}$, $b_1 = \frac{s_{xy}}{s_y^2}$ are the regression coefficient of Y on X and X on Y, respectively.
2. Regression coefficients are independent of change of origin but not of scale.

   Let X and Y be the two variables having the mean and variances $\overline{X}$, $\overline{Y}$ and s_X^2, s_Y^2, respectively. Let us construct two more variables such that

   $$U = \frac{X - A}{C} \Rightarrow X = A + CU$$

   and

   $$V = \frac{Y - B}{D} \Rightarrow Y = B + DV$$

   where A, B, C, and D are constants.
   Now,

   $$\mathrm{Cov}(X,Y) = C.D.\mathrm{Cov}(U,V)$$

   and

   $$s_X^2 = C^2 s_U^2.$$

   Similarly,

   $$s_Y^2 = D^2 s_V^2$$

   $$b = \frac{\mathrm{Cov}(X,Y)}{s_X^2} = \frac{C.D.\mathrm{Cov}(U,V)}{C^2 s_U^2} = \frac{D}{C}b_{VU}.$$

   Similarly, it can be proved that $b_1 = \frac{C}{D}b_{UV}$ $\beta_{XY} = \frac{D}{C}\beta_{UV}$.
3. Let $(x_1, y_1), (x_2, y_2), \ldots, (x_n, y_n)$ be the n pair of observations.

Table 8.14 Comparison between the correlation coefficient and the regression coefficient

Sl no.	Correlation coefficient	Regression coefficient
1	Correlation coefficient (r_{XY}) measures the degree of linear association between any two given variables	Regression coefficient (b_{XY}, b_{YX}) measures the change in dependent variable due to per unit change in independent variable when the relation is linear
2	Correlation coefficient does not consider the dependency between the variables	One variable is dependent on the other variable
3	Correlation coefficient is a unit-free measure	Regression coefficient (b_{XY}, b_{YX}) has the unit depending upon the units of the variables under consideration
4	Correlation coefficient is independent of the change of origin and scale	Regression coefficient (b_{XY}, b_{YX}) does not depend on the change of origin but depends on the change of scale
5	$-1 \leq r_{XY} \leq +1$	b_{XY}, b_{YX} does not have any limit

Now, the total sum of squares (y)

$$\text{TSS} = \sum_{i=1}^{n} (y_i - \bar{y})^2 = \sum_{i=1}^{n} \left(y_i - \hat{y} + \hat{y} - \bar{y}\right)^2$$

$$= \sum_{i=1}^{n} (y_i - \hat{y})^2 + \sum_{i=1}^{n} (\hat{y} - \bar{y})^2,$$

where $\hat{y}$ is the estimate of Y

= Residual sum of squares + Regression sum of squares

= RSS + R_gSS

$$\therefore \text{TSS} = \text{RSS} + \text{R}_\text{g}\text{SS} \Rightarrow 1 = \frac{\text{RSS}}{\text{TSS}} + \frac{\text{R}_\text{g}\text{SS}}{\text{TSS}}$$

$$= \frac{\text{RSS}}{\text{TSS}} + r^2.$$

Now, we observe that $\frac{\text{R}_\text{g}\text{SS}}{\text{TSS}} = r^2$ and $\frac{\text{RSS}}{\text{TSS}} = 1 - r^2$.

If we get the calculated value of r^2 as 0.55, we mean that 55% variations are explained by the regression line, and the remaining 45% variations are unexplained (Table 8.14).

Example 8.30. The following table gives the panicle length (X in cm) and the corresponding yield (Y in q/ha) of paddy in ten different varieties to find out the degree of linear association between the yield and the panicle length of paddy and to find out the linear relationship of yield on the panicle length of paddy.

X	18	20	20	22	22	25	26	28	35	40
Y	14	14	15	15	16	18	17	20	22	21

Solution. Let us make the following table:

	X	Y	X^2	Y^2	XY
	18	14	324	196	252
	20	14	400	196	280
	20	15	400	225	300
	22	15	484	225	330
	22	16	484	256	352
	25	18	625	324	450
	26	17	676	289	442
	28	20	784	400	560
	35	22	1,225	484	770
	40	21	1,600	441	840
Mean	25.6	17.2			
Variance	44.84	7.76			
Covariance	17.28				
Correlation coefficient		0.926			

We have a no. of observations "n" = 10, and the mean and variances and covariance are calculated as per the formula given in Chap. 3. So the correlation coefficient between the yield and the panicle length is given by r = Cov(X,Y)/sd(X).sd(Y). Now, the regression equation of yield (Y) on the panicle length (X) is given by

$$(Y - \overline{Y}) = b_{yx}(X - \overline{X})$$

$$\Rightarrow (Y - 17.2) = \frac{\text{Cov}(X.Y)}{s_X^2}(X - 25.6)$$

$$= \frac{17.28}{44.84}(X - 25.6)$$

$$= 0.385(X - 25.6) = 0.385x - 9.865$$

$$\Rightarrow Y = 17.2 + 0.385X - 9.865$$

$$\Rightarrow Y = 7.335 + 0.385X.$$

Example 8.31. The following data give the number of insects per hill (X) and the corresponding yield (Y) of nine different varieties of rice. Work out the linear relationship between the yield and the number of insects per hill.

Variety	V1	V2	V3	V4	V5	V6	V7	V8	V9
X	16	16	22	23	24	24	25	28	29
Y	55	54	48	47	45	40	44	37	40

Solution. Let us make the following table:

Variety	x	y	$(x-\bar{x})$	$(y-\bar{y})$	$(x-\bar{x})^2$	$(y-\bar{y})^2$	$(x-\bar{x})(y-\bar{y})$
1	16	55	−7	9.44	49	89.2	−66.11
2	16	54	−7	8.44	49	71.31	−59.11
3	22	48	−1	2.44	1	5.98	−2.44
4	23	47	0	1.44	0	2.09	0
5	24	45	1	−0.56	1	0.31	−0.56
6	24	40	1	−5.56	1	30.86	−5.56
7	25	44	2	−1.56	4	2.42	−3.11
8	28	37	5	−8.56	25	73.2	−42.78
9	29	40	6	−5.56	36	30.86	−33.33
Mean	23	45.56					
Variance	18.44	34.02					
Covariance	−23.67						

The regression equation of yield on the number of insects is given by $(Y-\overline{Y}) = b_{yx}(X-\overline{X}) \Rightarrow (Y-45.56) = \frac{\text{Cov}(X.Y)}{s_X^2}(X-23) = \frac{-23.67}{18.44}(X-23) = -1.284(X-23) = -1.284X + 29.532 \;\Rightarrow Y = 75.092 - 1.284X$.

The above example can very well be worked out using MS Excel software as follows:

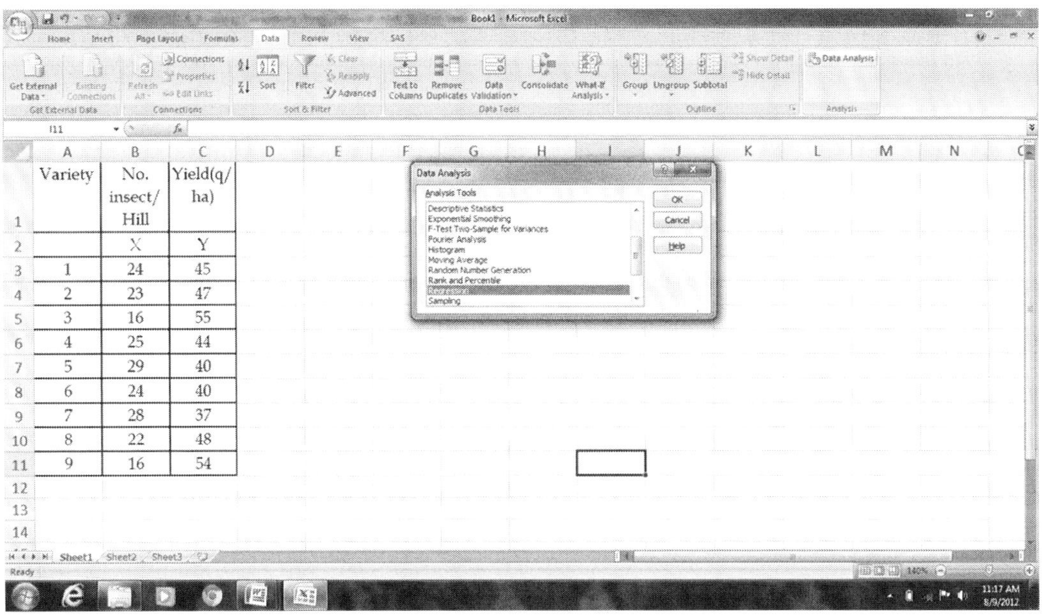

Slide 8.25: Step 1 showing data entry and the selection of the proper Analysis menu using MS Excel

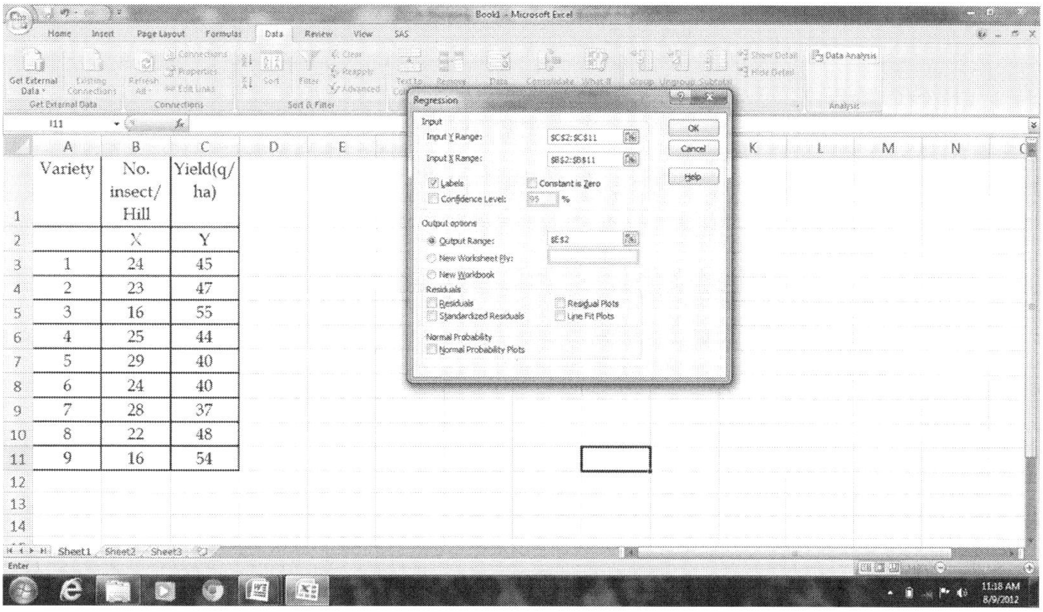

Slide 8.26: Step 2 showing data and the selection of proper data range and other submenus in regression analysis using MS Excel

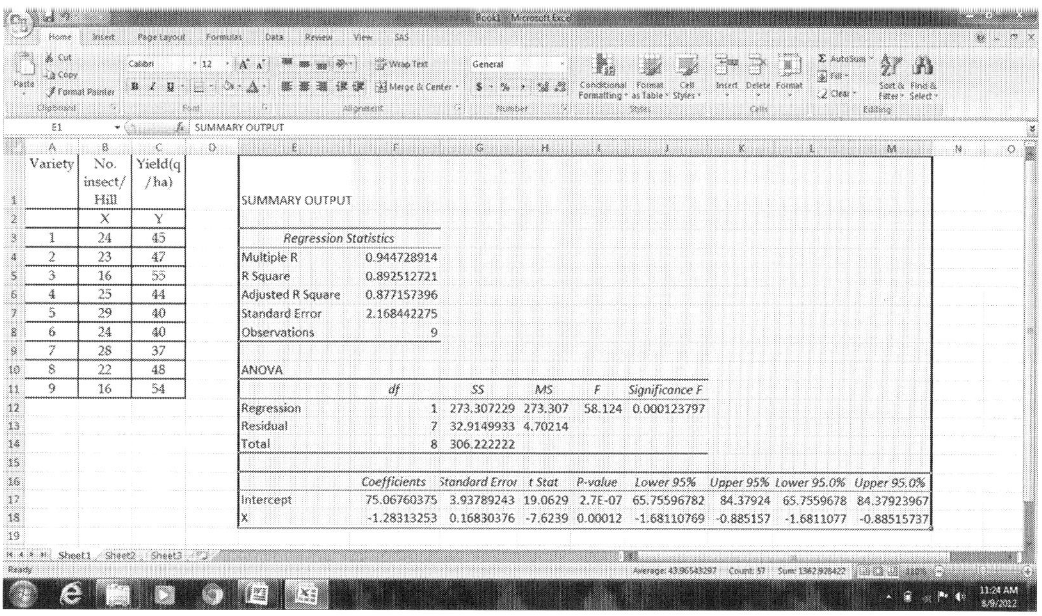

Slide 8.27: Step 3 showing the output generated in regression analysis using MS Excel

	No. of flowers per plant (*Y*)							
	Y	1–3	3–5	5–7	7–9	9–11	11–13	
Plant ht (*X*)	*X*	2	4	6	8	10	12	Total (f_i)
25–35	30	0	1	0	0	0	0	1
35–45	40	1	2	0	0	0	0	3
45–55	50	0	3	2	1	0	0	6
55–65	60	1	3	2	1	1	0	8
65–75	70	2	2	3	1	0	0	8
75–85	80	0	1	2	2	2	0	7
85–95	90	0	2	3	3	1	0	9
95–105	100	0	1	4	3	2	1	11
105–115	110	0	0	5	4	3	1	13
115–125	115	0	2	5	3	2	0	12
	Total ($f_{.j}$)	4	17	26	18	11	2	78

$\overline{X} = \frac{1}{N}\sum_{i=1}^{10} f_{i.}\, X_i = 86.795$ and $\overline{Y} = \frac{1}{N}\sum_{i=1}^{10} f_{.j} Y_j = 6.538$

$s_X^2 = \frac{1}{N}\sum_{i=1}^{10} f_{i.} X_i^2 - \overline{X}^2 = 574.343,\ s_Y^2 = \frac{1}{N}\sum_{j=1}^{6} f_{.j} Y_j^2 - \overline{Y}^2 = 5.505$

$s_{XY} = \frac{1}{N}\sum_{i=1}^{10}\sum_{j=1}^{6} f_{ij} X_i Y_j - \overline{XY} = 19.29\ r_{XY} = 0.343.$

Readers can verify whether the results in both cases are almost same. Additionally, some statistical properties of the coefficients as well as that of the whole regression exercise are also provided. While taking up multiple regression analysis during multivariate analysis, we would show how this analysis could also be performed using SPSS and SAS software.

Example 8.32. Given below are the data on plant height and the number of flowers per plant to find out the correlation coefficient and the relationship between the plant height and the number of flowers per plant

Now, the regression equation of the number of flowers per plant (*Y*) on plant height (*X*) is given by

$$Y - \overline{Y} = r\frac{s_Y}{s_X}(X - \overline{X})$$
$$\Rightarrow Y = \overline{Y} + r\frac{s_Y}{s_X}\left(X - \overline{X}\right)$$
$$\Rightarrow Y = 6.538 + 0.343\frac{2.346}{23.965}(X - 86.795)$$
$$= 6.538 + 0.034X - 2.915$$
$$\Rightarrow Y = 3.623 + 0.034X.$$

The above regression equation of the number of flowers per plant on plant height of okra indicates that for every centimeter change in plant height, the number of flowers per plant will increase by 0.034 numbers.

9 Formulation and Testing of Hypothesis

The main objective of the researchers is to study the population behavior to draw the inferences about the population, and in doing so, in most of the cases the researcher uses sample observations. As samples are part of the population, there are possibilities of difference in sample behavior from that of population behavior. Thus, the process/technique is of knowing accurately and efficiently the unknown population behavior from the statistical analysis of the sample behavior—known as *statistical inference*.

Problems in statistical inference:

(a) *Estimation*: Scientific assessment of the population characters from sample observations.
(b) *Testing of hypothesis*: Some information or some hypothetical values about the population parameter may be known or available but it is required to be tested how far these information or hypothetical values are acceptable or not acceptable in the light of the information obtained from the sample supposed to have been drawn from the same population.

9.1 Estimation

Let $(x_1, x_2, x_3, \ldots, x_n)$ be a random sample drawn from a population having density $f(x/\theta)$ where θ is an unknown parameter. Estimation is the problem of estimating the value θ with the help of the sample observations. There are mainly two types of estimation, namely, *point estimation* and *interval estimation*.

9.1.1 Point Estimation

Suppose $x_1, x_2, x_3, \ldots, x_n$ is a random sample from a density $f(x/\theta)$, where θ is an unknown parametric. Let t be a function of $x_1, x_2, x_3, \ldots, x_n$ and is used to estimate θ, then t is called a point estimator of θ. Among the many estimators based on sample observations, a good estimator is one which is (a) *unbiased*, (b) *consistent*, and (c) *efficient and sufficient*.

The details of these properties and their estimation procedures are beyond the scope of this book.

9.1.2 Interval Estimation

In contrast to the point estimation method, in interval estimation, we are always in search of a probability statement about the unknown parameter θ of the population from which the sample has been drawn. If $x_1, x_2, x_3, \ldots, x_n$ be a random sample drawn from a population, we shall be always in search of two functions u_1 and u_2 such that the probability of θ remains in the interval (u_1, u_2) is given by a value say $1 - \alpha$. Thus, $P(u_1 \leq \theta \leq u_2) = 1 - \alpha$. This type of interval, if exists, is called confidence interval

P.K. Sahu, *Research Methodology: A Guide for Researchers in Agricultural Science, Social Science and Other Related Fields*, DOI 10.1007/978-81-322-1020-7_9, © Springer India 2013

for the parameter θ. u_1 and u_2 are known as confidence limits and $1 - \alpha$ is called the confidence coefficient.

9.2 Testing of Hypothesis

A statistical hypothesis is an assertion about the population distribution of one or more random variables belonging to a particular population. In other words statistical hypothesis is a statement about the probability distribution of population characteristics which are to be verified on the basis of sample information. On the basis of the amount of information provided by a hypothesis, a statistical hypothesis is either (1) *simple* or (2) *composite*. Given a random variable from a population, a *simple hypothesis* is a statistical hypothesis which specifies all the parameters of the probability distribution of the random variable; otherwise it is a *composite hypothesis*. For example, if we say that the height of the Indian population is distributed normally with mean 5.6′ and variance 15.6, then it is a simple hypothesis. On the other hand, if we say that (1) the height of the Indian population is distributed normally with mean 5.6′ or (2) the height of the Indian population is distributed normally with variance 15.6 is composite in nature. Because to specify the distribution of height which is distributed normally, we require two information on mean and variance of the population. Depending upon the involvement or noninvolvement of the population parameter in the statistical hypothesis, it is again divided into *parametric* or *nonparametric* hypothesis, and corresponding tests are either *parametric test or nonparametric test*. Thus, a statistical hypothesis $\mu = \mu_0$ is a parametric hypothesis, but testing whether a time series data is random or not, that is, to test "the series is random," is a nonparametric hypothesis.

Any unbiased/unmotivated statistical assertion (hypothesis) whose validity is to be tested for possible rejection on the basis of sample observations is called *null hypothesis*. And the statistical hypothesis which contradicts the null hypothesis is called the *alternative hypothesis*. For example, against the null hypothesis H_0: $\mu = \mu_0$, (1) $\mu \neq \mu_0$, (2) $\mu > \mu_0$, and (3) $\mu < \mu_0$ are the alternative hypotheses.

In testing of hypothesis, a *test statistic* is a function of sample observations whose computed value when compared with the probability distribution, it follows, leads us to take final decision with regard to acceptance or rejection of null hypothesis.

9.2.1 Qualities of Good Hypothesis

1. Hypothesis should be clearly stated.
2. Hypothesis should be precise and stated in simple form as far as possible.
3. Hypothesis should state the relationships among the variables.
4. Hypothesis should be in the form of being tested.
5. Hypothesis should be such that it can be tested within the research area under purview.

Critical region: Let a random sample $x_1, x_2, x_3, \ldots, x_n$ be represented by a point x in n-dimensional sample space Ω and ω being a subset of the sample space, defined according to a prescribed test such that it leads to the rejection of the null hypothesis on the basis of the given sample if the corresponding sample point x falls in the subset ω; the subset ω is known as the *critical region* of the test. As it rejects the null hypothesis, it is also known as the *zone of rejection*. The complementary region to the critical region of the sample space, that is, ω' or $\overline{\omega}$, is known as the *region of acceptance*. Two boundary values of the critical region are also included in the region of acceptance.

Errors in decision: In statistical inference, drawing inference about the population parameter based on sample observation may lead to the following situations, and out of the four situations in two situations, one can commit errors. (Table 9.1)

9.2.2 Level of Significance

The probability of committing type I error (α) is called the level of significance. If the calculated

Table 9.1 Table of types of decision in statistical inference

Null	Decision taken	
hypothesis (H_0)	Reject H_0	Accept H_0
True	Incorrect decision (type I error)	Correct decision
False	Correct decision	Incorrect decision (type II error)

value of a test statistic lies in the critical region, the null hypothesis is said to be rejected at α level of significance. Generally, the level of significance is taken as 5 or 1%. This depends on the objective of the study. Sometimes we may have to opt for 0.01 or 0.001% level of significance, particularly in relation to medical studies. The researcher has the freedom to select the appropriate level of significance depending upon the objective of the study. Nowadays in many statistical software, the actual level of significance is provided instead of saying whether a particular null hypothesis is rejected or accepted at a particular prefixed level of significance or not.

9.2.3 Types of Test

Based on the nature of alternative hypothesis, a test is *one sided (one tailed) or both sided (two tailed)*. For example, if we are to test H_0: $\mu = 5.6'$, then the test against alternative hypothesis (1) $\mu \neq 50$ is a both-sided or two-tailed test while the test H_1: against either H_1: $\mu > 50$ or $\mu < 50$ is a one-sided or one-tailed test.

9.2.4 Degrees of Freedom

Degree of freedom is *actually the number of observations less the number of restrictions*. Generally, the degrees of freedom are $n - 1$ for a sample with n number of observations, but it should be kept in mind that the degrees of freedom are not necessarily be $n - 1$ for n observations. Depending upon the restrictions, the degrees of freedom are worked out, for example, in calculating χ^2 value from an $r \times s$ table, the degrees of freedom will be $(r - 1)(s - 1)$.

9.2.5 Steps in Testing of Statistical Hypothesis

Holistic and meticulous approach is required in each and every step of testing of statistical hypothesis so as to draw fruitful and correct decision about the population parameter based on sample observations. Some of the important steps to be followed in testing of statistical hypothesis are as follows:

(a) *Objective of the study*: There should not be any ambiguity in fixing the objective of the study, that is, to test whether the equality of two population means from two different samples drawn from the same parent population or from two different populations or to test whether population mean can be taken as μ_0 (a specified value) or not and so on.
(b) *Knowledge about population*: Extensive information or knowledge about the population and its distribution for which the parameters under study have been taken is essential and useful.
(c) *Hypothesis*: Commensurating with the objective of the study and the knowledge about the population under study selection or fixing of appropriate null hypothesis and alternative hypothesis will lead us to the type of test (i.e., one sided or two sided) to be performed.
(d) *Selection of test statistic*: While deciding the appropriate sampling distribution vis-a-vis the suitable test statistic, one should be very careful to select such a statistic which can best reflect the probability of the null hypothesis and the alternative hypothesis based on available information under study. A faulty selection of test statistic may lead to the wrong conclusion about the population parameter under study. Similarly, instead of a small sample test, if we go for a large sample test in a case where the number of observations is low, we may lead to an erroneous conclusion.

(e) *Selection of level of significance*: Depending upon the objective of the study, type of parameter, type of study object, precision required, etc., the appropriate level of significance is required to be decided. Selection of suitable level of significance (i.e., the region of rejection) is a must before the testing of hypothesis actually takes place.

(f) *Critical values*: Depending upon the type of test (one sided or both sided), test statistic, and its distribution, level of significance values separating the zone of rejection and the zone of acceptance are to be worked out from the table of specific corresponding distribution. Critical region should be selected in such a way that its power be maximum. A test is significant (i.e., rejection of null hypothesis) or not significant (acceptance of null hypothesis) depending upon the values of the calculated value of the test statistic and the table value of the statistic at prefixed level of significance. Fault in selection of the critical values may lead to a wrong conclusion.

(g) *Calculation of test statistic*: From the given information, one needs to accurately calculate the value of the test statistic under null hypothesis.

(h) *Comparison* of calculated value of the statistic at prefixed level of significance with that of the corresponding table value.

(i) *Decision* with respect to rejection or acceptance of null hypothesis.

(j) *Drawing conclusion* in line with the questions put forward in the objective of the study.

9.3 Statistical Test Based on Normal Population

1. Test for specified value of population mean with known population variance
2. Test for specified value of population mean with unknown population variance
3. Test of significance for specified population variance with known population mean (μ)
4. Test of significance for specified population variance with unknown population mean (μ)
5. Test of equality of two population variances from two normal populations with known population means
6. Test of equality of two population variances from two normal populations with unknown population means
7. Test for equality of two population means with known variances
8. Test for equality of two population means with equal but unknown variance
9. Test for equality of two population means with unequal and unknown variances—Fisher–Berhans problem
10. Test equality of two population means under the given condition of unknown population variances from a bivariate normal population-paired t-test
11. Test for significance of population correlation coefficient, that is, H_0: $\rho = 0$ against H_1: $\rho \neq 0$
12. Test for equality of two population variances from a bivariate normal distribution
13. Test for specified values of intercept and regression coefficient in a simple linear regression
14. Test for significance of the population multiple correlation coefficient
15. Test for significance of population partial correlation coefficient

The above tests are few to maintain but are not exhaustive.

A. Tests Based on a Normal Population

Depending upon the type of test (i.e., one sided or both sided) and the level of significance, the limiting values of the critical regions (i.e., region of acceptance and region of rejection) under standard normal probability curve are given as follows (Table 9.2):

Table 9.2 Critical values under standard normal probability curve

Type of test	Level of significance $\alpha = 0.05$	Level of significance $\alpha = 0.01$
Both-sided test	1.960	2.576
One sided (left tailed)	−1.645	−2.330
One sided (right tailed)	1.645	2.330

1. *Test for Specified Values of Population Mean with Known Population Variance from a Normal Population*

To test H_0: $\mu = \mu_0$ against the alternative hypotheses H_1: (1) $\mu \neq \mu_0$, (2) $\mu > \mu_0$, and (3) $\mu < \mu_0$ with known population variance σ^2. The test statistic under the given null hypothesis is

$$\tau = \frac{\bar{x} - \mu_0}{\sigma/\sqrt{n}},$$

where $\bar{x}$ is the sample mean and this τ follows a standard normal distribution.

Conclusion:

(a) For H_1: $\mu \neq \mu_0$, reject H_0 if the calculated value of $|\tau| < \tau_{\alpha/2}$, where $\tau_{\alpha/2}$ is the table value of τ at upper $\alpha/2$ level of significance (e.g., 1.96 or 2.576 at 5 or 1% level of significance, respectively); otherwise accept it.

(b) For H_1: $\mu > \mu_0$, reject H_0 if the calculated value of $\tau > \tau_\alpha$, where τ_α is the table value of τ at upper α level of significance (e.g., 1.645 or 2.33 at 5 or 1% level of significance, respectively); otherwise accept it.

(c) For H_1: $\mu < \mu_0$, reject H_0 if the calculated value of $\tau > \tau_{1-\alpha}$, where $\boldsymbol{\tau}_{1-\alpha}$ is the table value of τ at lower α level of significance (e.g., -1.645 or -2.33 at 5 or 1% level of significance, respectively); otherwise accept it.

Example 9.1. A random sample drawn from ten students of a college found that the average weight of the students is 45 kg. Test whether this average weight of the students be taken as 50 kg at 1% level of significance or not, given that the weight of the students follows a normal distribution with variance 9 kg^2.

Solution. Given that (1) the population is normal with variance 9, (2) sample mean ($\bar{x}$) is 45 and (3) population hypothetical mean is 50.

To test H_0: Population mean $\mu = 50$ against H_1: $\mu \neq 50$.

This is a both-sided test. As the sample has been drawn from a normal population with known variance, the appropriate test statistic will be

$$\tau = \frac{\bar{x} - \mu_0}{\sigma_X/\sqrt{n}}.$$

For this problem,

$$\tau = \frac{45 - 50}{\sqrt{9/10}} = \frac{-5}{0.949} = -5.269.$$

From the table of the standard normal variate, we have $\tau_{0.01}(= 2.576) \prec |\tau|_{cal}(= 5.269)$. So the test is significant at 1% level of significance. Hence, we reject the null hypothesis, that is, H_0: $\mu = \mu_0$. So the average student weight is not 50 kg.

2. *Test for Specified Value of Population Mean with Unknown Population Variance from a Normal Population*

To test H_0: $\mu = \mu_0$ against the alternative hypotheses H_1: (1) $\mu \neq \mu_0$, (2) $\mu > \mu_0$, and (3) $\mu < \mu_0$ with unknown population variance σ^2.

The test statistic under the given null hypothesis is $t = x - \mu_0/(s/\sqrt{n})$, where $(n - 1)$ is the degrees of freedom, and $\bar{x}$ and s^2 are the sample mean and sample mean square, $s^2 = \frac{1}{n-1}\sum_{i=1}^{n}(xi - \bar{x})^2$ (an unbiased estimator of population variance σ^2).

Conclusion

(a) For H_1: $\mu \neq \mu_0$, reject H_0 if the calculated value of $t > t_{\alpha/2,n-1}$ or cal $t < t_{(1-\alpha)/2,n-1} = -t_{\alpha/2,n-1}$ i.e. cal $|t| > t_{\frac{\alpha}{2},n-1}$ where $t_{\alpha/2,n-1}$ is the table value of t at upper $\alpha/2$ level of significance with $(n - 1)$ d.f.; otherwise accept it.

(b) For H_1: $\mu > \mu_0$, reject H_0 if the calculated value of $t > t_{\alpha,n-1}$, where $\tau_{\alpha,n-1}$ is the table value of t at upper α level of significance with $(n - 1)$d.f.; otherwise accept it.

(c) For H_1: $\mu < \mu_0$, reject H_0 if the calculated value of $t < t_{1-\alpha,n-1}$, where $t_{1-\alpha,n-1}$ is the table value of t at lower α level of significance with $(n - 1)$d.f.; otherwise accept it.

Example 9.2. Given below are the lengths (inches) of ten randomly selected panicles of particular wheat variety. If the length of panicle assumed to follow a normal distribution with unknown variance, can we say that the average length (inches) of panicle for the wheat variety be 12 in.?

Panicle length (inches): 10, 12, 6, 15, 14, 13, 13, 7, 14, 6

Solution. Given that the (1) length of panicle follows a normal distribution with unknown variance and (2) population hypothetical mean is 12 in.

To test H_0: Population mean $\mu = 12$ against H_1: $\mu \neq 12$. The test statistic under the null hypothesis is $t = \frac{\bar{x}-\mu}{s/\sqrt{n}}$ with $(n - 1)$ d.f. and the test is a both-sided test. We have sample mean,

$$\bar{x} = \frac{1}{n}\sum_{i=1}^{n} x_i = \frac{1}{10}[10 + 12 + \cdots + 6] = 11 \text{ in.}$$

and

$$s^2 = \frac{1}{n-1}\sum_{i=1}^{n}(x_i - \bar{x})^2 = \frac{1}{9}\left[\sum_{1}^{10} x_i^2 - 10.(11)^2\right]$$
$$= 12.22 \text{ in}^2.$$

So

$$t = \frac{11 - 12}{\sqrt{\frac{12.22}{10}}} = \frac{-1}{1.106} = -0.904.$$

The table value of t at 9 d.f. at 2.5% level of significance is 2.262 which is greater than the absolute value of the calculated value of t, that is, $|t| < t_{0.025,9}$. So the test is nonsignificant and the null hypothesis cannot be rejected. That means that the length of panicle for the given wheat variety may be taken as 12 in.

3. *Test for Significance for Specified Population Variance with Known Population Mean from a Normal Population*

If $x_1, x_2, x_3, \ldots, x_n$ be a random sample drawn from a normal population with mean μ and variance σ^2, that is, $N(\mu, \sigma^2)$ to test $H_0 : \sigma^2 = \sigma_0^2$ where σ_0^2, is any specified value for the population variance. Against the alternative hypotheses H_1: (1) $\sigma^2 \neq \sigma_0^2$, (2) $\sigma^2 > \sigma_0^2$, and (3) $\sigma^2 < \sigma_0^2$, the test statistic is

$$\chi_n^2 = \frac{\sum_{i=1}^{n}(x_i - \mu)^2}{\sigma_0^2}$$

with n d.f.

1. For $H_1 : \sigma^2 \neq \sigma_0^2$, the null hypothesis is rejected if the calculated value of χ_n^2 is greater than the table value of χ_n^2 at upper $\alpha/2$ level of significance and at n degrees of freedom, that is, cal $\chi_n^2 >$ tab $\chi^2_{\alpha/2,n}$ or calculated $\chi_n^2 <$ tabulated $\chi^2_{(1-\alpha)/2,n}$.
2. For $H_1 : \sigma^2 > \sigma_0^2$, the null hypothesis is rejected if the calculated value of χ_n^2 is greater than the table value of χ_n^2 at upper α level of significance and at n degrees of freedom, that is, cal $\chi_n^2 >$ tab $\chi^2_{\alpha,n}$.
3. For $H_1 : \sigma^2 < \sigma_0^2$, the null hypothesis is rejected if the calculated value of χ_n^2 is less than the table value of χ_n^2 at lower α level of significance and at n degrees of freedom, that is, cal $\chi_n^2 <$ tab $\chi^2_{(1-\alpha),n}$.

Example 9.3. From the following data on number of flowers per plant of rose, find out whether the variance of rose per plant can be taken as 6, given that the number of flowers per plant of rose follows a normal distribution with mean number of flowers per plant as 12.

No. of flowers per plant: 12,11,10,9,13,12, 8,15,16,13

Solution. Given that (1) the population mean is 12 (known), (2) the sample has been drawn from a normal population, (3) sample size is 10.

To test H_0: $\sigma^2 = 6$ against $H_1 : \sigma^2 \neq 6$.

Under the H_0, the test statistics is $\chi^2_{10} = \frac{\sum_{i=1}^{n}(x_i-\mu)^2}{\sigma^2} = \frac{[0+1+4+9+1+0+16+9+16+1]}{6}$
$= 57/6 = 9.5.$

From the table, we have a value of $\chi^2_{0.025,10} = 20.483$ and $\chi^2_{0.975,10} = 3.25$. The calculated value of χ^2, that is, 9.5, lies between these two values. So the null hypothesis cannot be rejected. That means we can conclude that the variance of no. of flowers per plant in rose can be taken as 6.

4. *Test of Significance for Hypothetical Population Variance with Unknown Population Mean from a Normal Population*
   If $x_1, x_2, x_3, \ldots, x_n$ be a random sample drawn from a normal population with unknown mean μ and variance σ^2 to test $H_0 : \sigma^2 = \sigma_0^2$ where σ_0^2 is any specified value for the population variance. Against the alternative hypotheses H_1: (1) $\sigma^2 \neq \sigma_0^2$, (2) $\sigma^2 > \sigma_0^2$, and (3) $\sigma^2 < \sigma_0^2$, the test statistic,

$$\chi^2_{(n-1)} = \frac{\sum_{i=1}^{n}(x_i - \bar{x})^2}{\sigma_0^2}$$

   with $(n - 1)$ degrees of freedom where $\bar{x}$ is the sample mean $= \frac{1}{n}\sum_{i=1}^{n} x_i$. The decision rule is as usual for the test mentioned in 3.

Example 9.4. A random sample of 25 trees of jackfruit gives average fruits per plant as 40 with variance 220 from a normal population. Test whether the variance of no. of fruits per plant of jackfruit can be taken as 200 given that the no. of fruits per plant follows a normal distribution.

Solution. Given that (1) sample size $n = 25$.

(2) Sample mean ($\bar{x} = 40$).

(3) Sample variance = 220.

(4) The sample has been drawn from a normal population with unknown mean.

To test H_0: $\sigma^2 = 200$ against H_1: $\sigma^2 \neq 200$, the test is a both-sided test, and under the null hypothesis, the test statistic is

$$\chi^2_{24} = \frac{\sum_{i=1}^{n}(x_i - \bar{x})^2}{\sigma^2} = \frac{nS^2}{\sigma^2} = \frac{25 \times 220}{200} = 27.5,$$

where S^2 is the sample variance.

From the table, we have $\chi^2_{0.995,24} = 9.89$ and $\chi^2_{0.005,24} = 45.558$. Since $9.89 <$ cal $\chi^2 < 45.56$, H_0 cannot be rejected. That means we can accept that the variance of number of fruits per tree may be 200.

5. *Test of Equality of Two Population Variances from Two Normal Populations with Known Population Means*
   Let two independent random samples be drawn from two normal populations $(x_1, x_2, x_3, \ldots, x_{n_1})$ and $(y_1, y_2, y_3, \ldots, y_{n_2})$. To test H_0: $\sigma_1^2 = \sigma_2^2$ under the condition, we have the of known means test statistic

$$F = \frac{\sum_{i=1}^{n_1}(x_i - \mu_1)^2/n_1}{\sum_{i=1}^{n_2}(y_i - \mu_2)^2/n_2}$$

   with n_1 and n_2 d.f.
   (a) When we are going to test $H_0 : \sigma_1^2 = \sigma_2^2$ against the alternative hypothesis $H_1 : \sigma_1^2 \neq \sigma_2^2$, we reject H_0 if cal $F >$ tab $F_{\alpha/2,(n_1,n_2)}$ or cal $F < F_{(1-\alpha/2),(n_1,n_2)}$.
   (b) When we are going to test $H_0 : \sigma_1^2 = \sigma_2^2$ against the alternative hypothesis $H_1 : \sigma_1^2 > \sigma_2^2$, we reject H_0 if cal $F >$ tab $F_{\alpha/2,(n_1,n_2)}$.
   (c) When we are going to test $H_0 : \sigma_1^2 = \sigma_2^2$ against the alternative hypothesis $H_1 : \sigma_1^2 < \sigma_2^2$, we reject the H_0 if cal $F < F_{(1-\alpha/2),(n_1,n_2)}$.

Example 9.5. The following figures are the weights of two batches of primary students assumed to follow a normal distribution with $(N(\mu_1, \sigma_1^2)$ and $N(\mu_2, \sigma_2^2)$ with respective mean weights of 24 and 26 kg, respectively. Can we assume from the data that both the populations have the same variance?

Yield in t/ha
Batch 1: 16,18,20,22,24,30
Batch 2: 19,22,23,27,29,31,33

Solution. Null hypothesis $H_0 : \sigma_1^2 = \sigma_2^2$ against the alternative hypothesis $H_1 : \sigma_1^2 \neq \sigma_2^2$, given that the populations are normal with mean 24 t/ha and 26 t/ha, respectively.

Under the given circumstances, the test statistic will be

$$F = \frac{\sum_{i=1}^{n_1} (x_i - \mu_1)^2 / n_1}{\sum_{i=1}^{n_2} (y_i - \mu_1)^2 / n_2}$$

with n_1 and n_2 d.f. the given alternative Under hypothesis, the test is a two-tailed test, given that $n_1 = 6$, $n_2 = 7$, $\mu_1 = 24$ kg, and $\mu_2 = 26$ kg.

From the given information, we have $\sum_{i=1}^{6} (x_i - \mu_1)^2 / n_1 = 26$

$$\sum_{i=1}^{7} (y_i - \mu_1)^2 / n_2 = 22.57, \ F = 1.152.$$

From the table, we have $F_{0.025;6,7} = 5.99$ and $F_{0.975;6,7} = 1/F_{0.025;7,6} = 1/6.98 = 0.1754$. Since $0.1754 <$ cal $F < 5.12$, so the null hypothesis cannot be rejected, that means we can conclude that both populations have the same variance.

6. *Test of Equality of Two Population Variances from Two Normal Populations with Unknown Population Means*

   Let two independent random samples be drawn from two normal populations (x_1, x_2, $x_3, \ldots, x_{n_1}$) and (y_1, y_2, $y_3, \ldots, y_{n_2}$). To test H_0: $\sigma_1^2 = \sigma_2^2$ under the unknown means condition, we have the test statistic $F = \frac{S_1^2}{S_2^2}$ with ($n_1 - 1$, $n_2 - 1$) d.f. where $S_1^2 \,\&\, S_2^2$ are the sample mean square of the respective populations.

   (a) When we are going to test $H_0 : \sigma_1^2 = \sigma_2^2$ against the alternative hypothesis $H_1 : \sigma_1^2 \neq \sigma_2^2$, we reject H_0 if cal $F >$ tab $F_{\alpha/2,}\ (n_1 - 1, n_2 - 1)$ or cal $F < F_{(1-\alpha/2),}\ (n_1 - 1, n_2 - 1)$.

   (b) When we are going to test $H_0 : \sigma_1^2 = \sigma_2^2$ against the alternative hypothesis $H_1 : \sigma_1^2 > \sigma_2^2$, we reject H_0 if cal $F >$ tab $F_{\alpha/2,(n_1-1,n_2-1)}$.

   (c) When we are going to test $H_0 : \sigma_1^2 = \sigma_2^2$ against the alternative hypothesis $H_1 : \sigma_1^2 < \sigma_2^2$, we reject the H_0 if cal $F < F_{(1-\alpha/2),(n_1-1,n_2-1)}$.

Example 9.6. The following figures are pertaining to lactation period (in months) for two random samples of cows, drawn from two normal populations fed with the same feed. Find whether the variability in lactation period of both the populations can be taken equal or not.

Lactation period (month)										
Sample 1	4.5	3	5	7	6.5	5.5	6	8	7.5	4
Sample 2	6	7	6.5	5.5	7.5	6.5	8	10	8.5	9

Solution. Given that (1) the populations are normal with unknown means.

(2) Sample sizes in each sample is 10, that is, $n_1 = 10$, $n_2 = 10$

To test H_0: $\sigma_1^2 = \sigma_2^2$ against H_1: $\sigma_1^2 \neq \sigma_2^2$. Under the given conditions, the test statistic is $F = \frac{s_1^2}{s_2^2}$ where s_1^2 and s_2^2 are the sample mean squares.

Let the level of significance be $\alpha = 0.05$.

We have

$$s_1^2 = \frac{1}{n_1 - 1} \sum_{i=1}^{10} \left(x_{1i} - x_1 \right)^2 \text{ and}$$

$$s_2^2 = \frac{1}{n_2 - 1} \sum_{i=1}^{10} \left(x_{2i} - x_2 \right)^2 \text{ and}$$

$$\bar{x}_1 = \frac{1}{n_1} \sum_{i=1}^{10} (x_{1i}), \ \bar{x}_2 = \frac{1}{n_2} \sum_{i=1}^{10} (x_{2i}).$$

From the given data, we have

$$\bar{x}_1 = 5.70, \ \bar{x}_2 = 7.45$$
$$\text{and } s_1^2 = 2.638 \text{ and } s_2^2 = 2.025.$$

So, $F = \frac{s_1^2}{s_2^2} = 1.302$. with (9,9) d.f.

From the table, we have $F_{0.025;9,9} = 4.03$ and $F_{0.975;9,9} = 1/F_{0.025;9,9} = 1/4.03 = 0.2481$. Since $0.2481 <$ cal $F < 4.03$, so the test is non-significant and the null hypothesis cannot be rejected, that means we can conclude that both populations have the same variance.

7. *Test for Equality of Two Population Means with Known Variances*

To test the H_0: $\mu_1 = \mu_2$ for $(x_1, x_2, x_3, \ldots, x_{n_1})$ and $(y_1, y_2, y_3, \ldots, y_{n_2})$ two independent random samples drawn from two normal populations $N(\mu_1, \sigma_1^2)$ and $N(\mu_2, \sigma_2^2)$, respectively, and under the condition of known variances σ_1^2 and σ_2^2, the test statistic is

$$\tau = \frac{\bar{x} - \bar{y}}{\sqrt{\frac{\sigma_1^2}{n_1} + \frac{\sigma_2^2}{n_2}}},$$

when τ is in a standard normal variate.

Example 9.7. The following data are pertaining to grain yield of two varieties of paddy under the same management. If the grain yield follows a normal distribution with known variances 36 and 64, respectively, for two varieties test whether these two varieties differ significantly with respect to grain yield (q/ha).

(Grain yield q/ha)
Variety A: 25, 30, 31, 35, 24, 14, 32, 25, 29, 31, 30
Variety B: 30, 32, 28, 36, 42, 18 ,16 , 20 , 22, 40

Solution. Assumption: grain yield follows a normal distribution. Population variances are known. Let the level of significance be $\alpha = 0.05$.

So the null hypothesis is H_0: $\mu_A = \mu_B$ (under known population variances) against alternative hypothesis $H_1 : \mu_A \neq \mu_B$.

Under the given condition, the test statistic is $\tau = \frac{\bar{A} - \bar{B}}{\sqrt{\frac{\sigma_A^2}{n_1} + \frac{\sigma_B^2}{n_2}}}$ which follows as a standard normal variate.

From the sample observations, we have $n_A = 11$ and $n_B = 10$ and $\bar{A} = 27.82$ and $\bar{B} = 28.4$

$$\tau = \frac{27.82 - 28.4}{\sqrt{\frac{36}{11} + \frac{64}{10}}} = \frac{-0.58}{\sqrt{3.277 + 6.4}} = \frac{-0.58}{\sqrt{9.677}}$$
$$= \frac{-0.58}{3.109} = -0.186.$$

We know that at $\alpha = 0.05$, the value of $\tau_{\alpha/2} = 1.96$, as the calculated $|\tau| < 1.96$. So we cannot reject the null hypothesis. We conclude that the varieties do not differ significantly with respect to grain yield.

8. *Test for Equality of Two Population Means with Equal but Unknown Variance*

   To test the H_0: $\mu_1 = \mu_2$ against $H_1 : \mu_1 \neq \mu_2$ for $(x_1, x_2, x_3, \ldots, x_{n_1})$ and $(y_1, y_2, y_3, \ldots, y_{n_2})$ two independent random samples drawn from two normal populations $N(\mu_1, \sigma_1^2)$ and $N(\mu_2, \sigma_2^2)$, respectively, under the condition that $\sigma_1^2 = \sigma_2^2 = \sigma^2$ (unknown). The test statistic under the given H_0 is

$$t = \frac{\bar{x} - \bar{y}}{s\sqrt{\frac{1}{n_1} + \frac{1}{n_2}}}$$

   with $(n_1 + n_2 - 2)$ degrees of freedom where $s^2 = \frac{(n_1-1)s_x^2 + (n_2-1)s_y^2}{n_1+n_2-2}$ and s_x^2 and s_y^2 are the sample mean squares for two samples, respectively.

   If the calculated absolute value of t is greater than the table value of t at upper $\alpha/2$ level of significance and at $(n_1 + n_2 - 2)$ d.f., then the test is significant and the null hypothesis is rejected, that means the two population means are unequal.

   Note: This test statistic is known as two-sample t statistic or Fisher's t statistic. Before performing the test we are to test first $H_0 : \sigma_1^2 = \sigma_2^2$ by F-test as already discussed. If it is accepted, then we perform t-test statistic; otherwise not. In that situation we perform the following test.

Example 9.8. Given below are the two samples about the vase life of two varieties of tuberose. Is it possible to draw inference that the vase life of variety A is more than that of the variety B, assuming that vase life behaves like a normal population?

	Sample size	Sample mean	Sample mean square
Variety A	7	72 h	36 h^2
Variety B	8	64 h	25 h^2

Solution. H_0: Both the varieties of tuberose have the same vase life, that is, H_0: $\mu_1 = \mu_2$ against the alternative hypothesis, H_1: Variety

A has more vase life than variety B, that is, $H_1: \mu_1 > \mu_2$. Let us select the level of significance, $\alpha = 0.05$. According to H_1, the test is a one-sided test. We assume that $\sigma_A^2 = \sigma_B^2 = \sigma^2$(unknown). The test statistic, under the given null hypothesis and unknown variance but equal, is

$$t = \frac{\overline{x} - \overline{y}}{s\sqrt{\frac{1}{n_1} + \frac{1}{n_2}}},$$

with $(n_1 + n_2 - 2)$ d.f. where $\overline{x}, \overline{y}$ are the sample mean vase life of two varieties and s^2 is the composite sample mean square and given by

$$s^2 = \frac{(n_1 - 1)s_x^2 + (n_2 - 1)s_y^2}{n_1 + n_2 - 2},$$

s_x^2 and s_y^2 by the sample mean squares.

First we test $H_0 : \sigma_A^2 = \sigma_B^2$ by $F = \frac{s_x^2}{s_y^2}$ with $(n_1 - 1, n_2 - 1)$ d.f. against the $H_1 : \sigma_A^2 \neq \sigma_B^2$.

Thus, $F = 36/25 = 1.44$ with (6,7) d.f.

From the table, we have $F_{0.025;6,7} = 5.12$ and $F_{0.975;6,7} = 1/F_{0.025;7,6} = 1/5.70 = 0.18$. Since $0.18 <$ cal $F < 5.12$, $H_0 : \sigma_A^2 = \sigma_B^2$ cannot be rejected. So we can perform the t-test.

$$s^2 = \frac{(n_1 - 1)s_1^2 + (n_2 - 1)s_2^2}{n_1 + n_2 - 2} = \frac{(7 - 1)6^2 + (8 - 1)5^2}{7 + 8 - 2} = \frac{216 + 175}{13} = 391/13 = 30.07.$$

$$t = \frac{72 - 64}{\sqrt{30.07\left(\frac{1}{7} + \frac{1}{8}\right)}} = \frac{8}{\sqrt{\left(30.07 \times \frac{15}{56}\right)}} = 2.82.$$

From the table, we have $t_{0.05,13} = 1.771$. Since cal $t > 1.771$, the test is significant and we reject the null hypothesis, that is, we accept $\mu_1 > \mu_2$. That means the vase life of variety A is more than that of the vase life of variety B.

9. *Test for Equality of Two Population Means with Unequal and Unknown Variances*

   The problem for test of significance of equality of two population means under unknown and unequal population variances (i.e., $\sigma_1^2 \neq \sigma_2^2$) is known as *Fisher–Berhans problem*. Under the null hypothesis H_0: $\mu_1 = \mu_2$ against $H_1 : \mu_1 > \mu_2$, the test statistic

$$t = \frac{\overline{x} - \overline{y}}{\sqrt{\frac{s_1^2}{n_1} + \frac{s_2^2}{n_2}}}$$

does not follow t distribution, and as such ordinary t-table will not be sufficient for comparing. According to Cochran's approximation, the calculated value of above t statistic is compared with

$$t^* = \frac{t_1 s_1^2/n_1 + t_2 s_2^2/n_2}{s_1^2/n + s_2^2/n_2},$$

where t_1 and t_2 are the table values of t distribution at $(n_1 - 1)$ and $(n_2 - 1)$ degrees of freedom, respectively, with upper α level of significance for acceptance or rejection of H_0.

Example 9.9. Given below are the two samples about the body weight (kg) of two breeds of cows. Is it possible to draw inference that the body weight of breed A is greater than that of breed B, assuming that body weight behaves like a normal population?

Variety	Sample size	Sample mean	Sample mean square
Breed A	7	185	100
Breed B	8	175	64

Solution. We are to test the null hypothesis H_0: $\mu_A = \mu_B$ (under unknown and unequal population variances) against alternative hypothesis $H_1 : \mu_A > \mu_B$.

Let the level of significance $\alpha = 0.05$.

Under the given conditions we apply Cochran's approximation to Fisher–Berhans problem. Thus, the test statistic is given by

$$t = \frac{\overline{x} - \overline{y}}{\sqrt{\frac{s_1^2}{n_1} + \frac{s_2^2}{n_2}}},$$

which is then compared with the value of

$$t^* = \frac{t_1 s_1^2/n_1 + t_2 s_2^2/n_2}{s_1^2/n + s_2^2/n_2}$$

and appropriate decision is taken.

We have

$$t = \frac{\bar{x} - \bar{y}}{\sqrt{\frac{s_1^2}{n_1} + \frac{s_2^2}{n_2}}} = \frac{185 - 175}{\sqrt{\frac{100}{7} + \frac{64}{8}}} = \frac{10}{4.721} = 2.118.$$

The table values of t at upper 5% level of significance with $(n_1 - 1 =)6$ d.f. and $(n_2 - 1 =)$ 7 d.f. are 1.943 and 1.895, respectively. So

$$t^* = \frac{1.943 \times 100/7 + 1.895 \times 64/8}{100/7 + 64/8} = \frac{42.917}{22.286}$$

$$= 1.926.$$

Now the $t_{cal} > t^*$, hence we can reject the null hypothesis, that is, H_1 is accepted. That means we conclude that the body weight of breed A is greater than that of breed B.

10. *To Test Equality of Two Population Means Under the Given Condition of Unknown Population Variances from a Bivariate Normal Population*

Let (x_1,y_1), (x_2,y_2), (x_3,y_3), …, (x_n,y_n) be n pairs of observations in a random sample drawn from a bivariate normal distribution with parameters $\mu_X, \mu_y, \sigma_X^2, \sigma_y^2$ and ρ where μ_X and μ_y are the means and σ_X^2, σ_y^2 are the variances and ρ is the population correlation coefficient between X and Y.

So to test the null hypothesis $H_0 : \mu_X = \mu_y$ i.e. $H_0 : \mu_X - \mu_y = \mu_d = 0$, we have the test statistic

$$t = \frac{\bar{d}}{\frac{s_d}{\sqrt{n}}} \text{ with } (n-1)\text{d.f., where } \bar{d} = \frac{1}{n}\sum_{i=1}^{n} d_i$$

$$= \frac{1}{n}\sum_{i=1}^{n}(x_i - y_i) \text{ and } s_d{}^2$$

$$= \frac{1}{n-1}\sum_{i=1}^{n}\left(d_i - \bar{d}\right)^2.$$

This t is known as *paired* t-test statistic and the test is known as *paired t-test*.

The table value of t at $(n - 1)$ d.f. for α level of significance will be compared with the calculated value of t for arriving at definite conclusion according to the nature of alternative hypothesis.

Example 9.10. To test whether a specific artificial hormonal spray has effect on fruit weight of papaya, initial and final weights from a sample of eight plants of a particular variety of papaya were taken at an interval of 15 days of spray.

Initial wt (g)	114	113	119	116	119	116	117	118
Final wt(g)	220	217	226	221	223	221	218	224

Solution. Let x represent the initial weight and y the final weight. So $x - y = d$. Assuming that X and Y follow a bivariate normal distribution with parameters μ_x, μ_y, σ_x, σ_y, and ρ_{xy}, we want to test

$H_0 : \mu_x = \mu_y$ against

$H_1 : \mu_x < \mu_y$. The test statistic under H_0 is t

$$= \frac{\bar{d}}{\mathrm{sd}/\sqrt{\mathrm{n}}} \text{ with } (n-1)\text{d.f.}$$

So $\bar{d} = -104.75$ and

Initial wt (g) X	114	113	119	116	119	116	117	118
Final wt (g) Y	220	217	226	221	223	221	218	224
$X - Y$ (d)	−106	−104	−107	−105	−104	−105	−101	−106

$$S_d = \sqrt{\frac{1}{(8-1)}\left[\sum d_i^2 - 8.\bar{d}^2\right]}$$
$$= \sqrt{\frac{1}{(7)}[87804.0 - 87780.5]}$$
$$= \sqrt{\frac{1}{7} \times 23.5} = \sqrt{33.571} = 1.832.$$

$$\text{So } t = \frac{\bar{d}}{\text{sd}/\sqrt{n}} = \frac{-104.75}{1.832/\sqrt{8}} = \frac{-104.75}{0.648}$$
$$= -161.651.$$

From the table value, we have $t_{0.95,7} = -1.895$ and $t_{.99,7} = -2.998$. The calculated value of t is less than the table value at both the level of significance. Hence, the test is significant at 1% level of significance. So we reject the null hypothesis $H_0 : \mu_x = \mu_y$ and accept $H_1 : \mu_x < \mu_y$. That is, there was significant effect of hormone spray on fruit weight of papaya.

11. *To Test for Significance of Population Correlation Coefficient, That Is, H_0: $\rho = 0$ Against $H_1 : \rho \neq 0$*
    The test statistic under H_0 will be $t = r\sqrt{n-2}/\sqrt{1-r^2}$ at $(n - 2)$ d.f. where r is the sample correlation coefficient between X and Y, the two variables considered.
    If the calculated value of $|t|$ is less than the tabulated value of t at $(n - 2)$ d.f. for upper $\alpha/2$ level of significance, we cannot reject the null hypothesis; that is, sample correlation coefficient is not significantly different from zero or the variables are uncorrelated. Otherwise we reject the null hypothesis, and sample correlation coefficient is significant to the population correlation coefficient b not zero and the variables have significant correlation between them.

Example 9.11. The correlation coefficient between the age and lactation duration of 20 breeds of Indian cows is found to be 0.850. Test for the existence of correlation between these two characters in Indian cows using 1% level of significance.

Solution. Under the given information, we want to test $H_0 : \rho = 0$ against $H_1 : \rho \neq 0$. The test statistic is given by

$$t = \frac{r\sqrt{n-2}}{\sqrt{1-r^2}} \text{ with } (n-2)\text{d.f.}$$

Given that $r = 0.850$, $n = 20$, so

$$t = \frac{0.85\sqrt{(20-2)}}{\sqrt{1-(0.85)^2}} = \frac{0.85 \times \sqrt{18}}{\sqrt{0.2775}} = 6.846.$$

The table value of $t_{0.005,18} = 2.878$. Since $|t| >$ table $t_{0.005,18}$, that is, $6.846 > 2.878$, so the null hypothesis $H_0 : \rho = 0$ is rejected, we accept $H_1 : \rho \neq 0$. So we can conclude that the age and lactation period in Indian cows are correlated.

12. *Test for Equality of Two Population Variances from a Bivariate Normal Distribution*
    The null hypothesis for testing the equality of two variances is $H_0 : \sigma_X^2 = \sigma_y^2$.
    Let us derive two new variables U and V such that $U = X + Y$ and $V = X - Y$. So the $\text{Cov}(U,V) = \text{Cov}(X + Y, \ X - Y) = \sigma_X^2 - \sigma_y^2$. Under the null hypothesis $H_0 : \sigma_X^2 = \sigma_y^2$, $\text{Cov}(U,V) = 0$, and thus U and V are two normal variates with correlation coefficient $\rho_{UV} = 0$ when H_0 is true.
    Hence, H_0: $\sigma_X^2 = \sigma_y^2$ is equivalent to test H_0: $\rho_{UV} = 0$.
    As usual the test statistic is given by $t = r_{uv}\sqrt{(n-2)}/\sqrt{1-r_{uv}{}^2}$ with $(n - 2)$ d.f, where r_{uv} is the sample correlation coefficient between u and v.

Example 9.12. To test whether a specific artificial hormonal spray has effect on variability of fruit weight of papaya, initial and final weights from a sample of eight plants of a particular variety of papaya were taken at an interval of 15 days of spray.

Initial wt (g)	114	113	119	116	119	116	117	118
Final wt (g)	220	217	226	221	223	221	218	224

Solution. To work out the significant differences in variability of fruit weights before and after spraying, the null hypothesis would be $H_0 : \sigma_x = \sigma_y$ against $H_0 : \sigma_x \neq \sigma_y$. This is equivalent to test $\rho_{uv} = 0$ where u and v are $x + y$ and $x - y$, respectively. The test statistic for the same will be

$$t = \frac{r_{uv}}{\sqrt{1 - r_{uv}^2}}\sqrt{n-2} \quad \text{with } (n-2) \text{ d.f.}$$

We have

X:	114	113	119	116	119	116	117	118
Y:	220	217	226	221	223	221	218	224
$U = (X + Y)$:	334	330	345	337	342	337	335	342
$V = (X - Y)$:	−106	−104	−107	−105	−104	−105	−101	−106

$$\therefore r_{uv} = \frac{\sum uv - n\overline{uv}}{\sqrt{\left(\sum u^2 - n\bar{u}^2\right)\left(\sum v^2 - n\bar{v}^2\right)}} = -0.4647.$$

$$\text{Thus, } t = \frac{-0.4647\sqrt{8-2}}{\sqrt{1-(-0.4647)^2}} = -1.2855 \text{ at 6 d.f.}$$

From the table we get $t_{0.025,6} = 2.447$. Thus, the observed value of $|t|$ is less than the table value. That is, $|t_{cal}| < t_{tab}$. So we cannot reject the null hypothesis of equality of variances.

13. *Test for Significance and Regression Coefficient in a Simple Linear Regression.*
    To test the H_0: $b = 0$, we have the test statistic $t = b/SE(b)$ with $n - 1$ d.f.: if the calculated value of $H \leq \text{tab}_{(n-1)}$ d.f. at $\frac{\alpha}{2}$ level of significance, then it cannot be rejected; otherwise t is rejected.

Example 9.13. The relation between yield and expenditure on plant protection measure was found to be $Y = 15.6 + 0.8$ PP with standard error estimates for the intercept and the regression coefficients being 6.2 and 0.625, respectively, from ten varieties of paddy. Test for significance of regression coefficient of plant protection.

Solution. To rest the H_0: $b = 0$ the test statistic is $t = b/\text{SE}(b) = 0.8/0.625 = 1.28$. The table value of t statistic at 5% level of significance is greater than the table value. Hence, the test is nonsignificant and the null hypothesis cannot be rejected. We conclude that population regression coefficient may be zero.

14. *Test for Significance of the Population Multiple Correlation Coefficients*
    Let there be p variables $(x_1, x_2, \ldots x_p)$ following a p-variate normal distribution and the multiple correlation of x_1 on $x_2, x_3, \ldots x_p$ be given by $\rho_{12.34,\ldots,p}$ and the corresponding sample multiple correlation coefficient being $R_{1.2,3,\ldots,p}$ based on a random sample of size n. The problem is to test whether the population multiple correlation is zero or not, that is, to test $H_0 : \rho_{1.2,3,\ldots,p} = 0$ against the alternative hypothesis $H_1 : \rho_{1.\,2,\,3,\ldots,p} > 0$. Under H_0 the test statistic is

$$F = \frac{R^2{}_{1.23\ldots p}/(p-1)}{\left(1 - R^2{}_{1.23\ldots p}\right)/(n-p)}$$
with $(p - 1, n - p)$d.f.

If cal $F > F_{\alpha;\,p-1,n-p}$, then H_0 is rejected, else it is accepted}.
It is noted that in this problem, alternative hypothesis is always one sided (right) because $\rho_{12.\,3,4,\ldots,p}$ cannot be negative.

Example 9.14. In a yield component analysis, the relationship of yield with five-yield component was worked out for 20 plants. The multiple correlation coefficients were found to be 0.98. Test for significance of multiple correlation coefficients.

Solution. Under the given situation, we are to test $H_0: \rho_{1.2,3,\ldots,p} = 0$ against $H_1: \rho_{1.2,3,\ldots,p} > 0$.

Given that $n = 20$ and $p = 6$, so the appropriate test statistic would be

$$F = \frac{R^2_{1.23\ldots p}/(p-1)}{(1-R^2_{1.23\ldots p})/(n-p)} = \frac{0.98^2/(6-1)}{(1-0.98^2)/(20-6)}$$
$$= 67.2 \text{ with } (6-1, 20-6)\text{d.f.}$$

Cal $F > F_{0.05;\,5,14}$, H_0 is rejected.

15. *Test for Significance of Population Partial Correlation Coefficient*
With the above variable consideration of multiple correlations, let $\rho_{21.34\ldots p}$ be the partial correlation coefficient of x_1 and x_2 after eliminating the effect of $x_3, x_4, \ldots x_p$, and the corresponding sample partial correlation coefficient from a random sample of size n is given by $r_{12.34\ldots p}$.
The test statistic for $H_0: \rho_{12.34\ldots p} = 0$ is

$$t = \frac{r_{12.34\ldots p}\sqrt{n-p}}{\sqrt{1-r^2_{12.34\ldots p}}} \text{ with } (n-p)\text{d.f.}$$

Decision on rejection/acceptance of null hypothesis is taken on the basis of comparison of the calculated value of t with that of table value of t at $(n-p)$ d.f. for α level of significance.

Example 9.15. During yield component analysis in ginger, the partial correlation coefficient of yield (X_1) with the weight of primary finger (X_3) was found to be 0.609, eliminating the effects of no. of tillers (X_2), no. of secondary finger (X_4), and finger (X_5). Again the multiple correlation coefficient of yield (X_1) on all other four components was found to be 0.85 from a sample of 40 plants drawn at random. Test for significant difference for both partial correlation coefficient and multiple correlation coefficients from zero.

Solution. Given that $n = 40$, number of variables $(p) = 5$, $r_{13.245} = 0.609$, and $R_{1.2345} = 0.85$. Let us take $\alpha = 0.01$.

(a) The test statistic for $H_0: \rho_{13.245} = 0$ against $H_0: \rho_{13.245} \neq 0$ is $t = \frac{r_{13.245}\sqrt{n-p}}{\sqrt{1-r^2_{13.245}}} = \frac{0.609\sqrt{35}}{\sqrt{1-0.609^2}} = 4.543$with $(40-5=)$ 35 d.f.
From the table, we have $t_{0.005,35} = 2.725$. Since the calculated $|t| > 2.725$, so the null hypothesis of zero partial correlation coefficient between yield and weight of primary figure is rejected.

(b) The test statistic under $H_0: \rho_{1.2345} = 0$ against $H_0: \rho_{1.2345} > 0$ is given by

$$F = \frac{R^2_{1.2345}/(p-1)}{(1-R^2_{1.2345})/(n-p)}$$
$$= \frac{(0.85)^2/4}{(1-0.85^2)/35} = \frac{0.181}{0.00793} = 22.8247.$$

The calculated value of F is greater than the table value of $F_{0.01;4,35}$, so the test is significant and the null hypothesis of zero multiple correlation coefficient is rejected. That means the population multiple correlation coefficient differs significantly from zero.

9.4 Large Sample Test

Large sample tests are based on the followings facets:

1. Any sample having sample size more than 30 is treated as large.
2. If a random sample of size n is drawn from an arbitrary population with mean μ and variance σ^2 and any statistic be t with mean $E(t)$ and variance $V(t)$, then t is asymptotically normally distributed with mean $E(t)$ and variance $V(t)$ as n gets larger and larger, that is, $t \sim N(E(t), V(t))$ as $n \to \infty$. The standard normal variate is

$$\tau = \frac{t - E(t)}{\sqrt{V(t)}} \sim N(0,1).$$

Let us now discuss some of the important and mostly used tests under this large sample test.

1. *Test for specified value of population mean*
2. *Test of significance between two means:*
   (a) *Under the given condition* $\sigma_1{}^2 = \sigma_2{}^2 = \sigma^2$ (unknown).
   (b) *Under the given condition of* $\sigma_1{}^2 \neq \sigma_2{}^2$ *and both being unknown.*
3. *Test for significance of specified population standard deviation*
4. *Test of significant difference between two standard deviations*
5. *Test for significance of specified population proportion*
6. *Test for equality of two population proportions*
7. χ^2 *test*

1. *Test for Specified Value of Population Mean*
   Let a sample $x_1, x_2, \ldots, x_n$ be drawn at random from a population with mean μ and variance σ^2 such that the size of the sample is $n \geq 30$. To test that whether the population mean $\mu = \mu_0$, a specified value under (1) the population variance σ^2 known and (2) the unknown population variance σ^2 is the test statistics are
   (a) $\tau = \frac{\bar{x}-\mu_0}{\sigma/\sqrt{n}}$, when σ^2 is known and
   (b) $\tau = \frac{\bar{x}-\mu_0}{s_n/\sqrt{n}}$, when σ^2 is unknown,
   where $s_n^2 = \frac{1}{n}\sum_{i=1}^{n}(x_i - \bar{x})^2$ is the sample variance.
   For acceptance or rejection of H_0, we have to compare the calculated value of τ with the appropriate table value, keeping in view the alternative hypotheses.

Example 9.16. A manufacturing company claims that the average monthly salary of the company is \$3,000. To test this claim, a sample of 64 employees were examined and found that the average monthly salary was \$2,890 with variance 256 h. Test whether the claim is justified or not using 5% level of significance.

Solution. Given that (1) the sample is large $n\,(= 64 > 30)$.

(2) Sample mean $(\bar{x}) = 2{,}890$ and sample variance = 256.

(3) Population variance is unknown.

We want to test

H_0: $\mu = 3{,}000$ h against

H_1: $\mu \neq 3{,}000$ h.

The approximate test statistic under the given condition is

$$\tau = \frac{\bar{x}-\mu_0}{\frac{s_n}{\sqrt{n}}} = \frac{2,890 - 3,000}{\frac{16}{\sqrt{64}}} = \frac{-110}{16/8} = -55.00.$$

Thus, the calculated value of $|\tau|$, that is, 55.00, is greater than the table value of $\tau_{0.025} = 1.96$; the test is significant and the null hypothesis is rejected. So we conclude that the claim of the manufacturer is not justified.

Example 9.17. The manufacturer of certain brand of water pump claims that the pump manufactured by this company will fill a 1,000-l water tank within 8 min with a variance of 9 min^2. To test the claim, 36 pumps were examined and found that the average time taken to fill tank is 10 min with s.d. 2.5 min. Conclude whether the claim is justified or not.

Solution. Given that (1) population mean (μ) = 8 min and variance (σ^2) – 9 min^2.

(2) Sample size $(n) = 36$, $\bar{x} = 10$ min, and $s_n = 2.5$ min.

Thus, under the given condition, the null hypothesis is

H_0: $\mu = 8$ min against

H_1: $\mu > 8$ min.

The test statistic is

$$\tau = \frac{\bar{x}-\mu_0}{\sigma/\sqrt{n}}$$

which follows a standard normal distribution. The calculated value of

$$\tau = \frac{10-8}{3/\sqrt{36}} = \frac{2}{3/6} = 4.$$

The calculated value of $|\tau| = 4.00$ is greater than $\tau_{0.025} = 1.96$. So the test is significant and the null hypothesis is rejected. Thus, we reject the claim that the pumps manufactured by the company fill a 1,000-l water tank in 8 min.

2. *Test of Significance Between Two Means*
   Two large samples $(x_{11}, x_{12}, \ldots, x_{1n_1})$ and $(x_{21}, x_{22}, \ldots, x_{2n_2})$ of sizes n_1 and n_2 with means $\bar{x}_1$ and $\bar{x}_2$ and variances $s^2_{n_1}$ and $s^2_{n_2}$ are drawn independently at random from two populations with means μ_1 and μ_2 and variances $\sigma_1{}^2$ and $\sigma_2{}^2$, respectively.
   As two large samples are drawn independently from two population so $\bar{x}_1 \sim N\left(\mu_1, \frac{\sigma_1{}^2}{n_1}\right)$ and $\bar{x}_2 \sim N\left(\mu_2, \frac{\sigma_2{}^2}{n_2}\right)$.
   So to test $H_0 : \mu_1 = \mu_2$, the test statistic is

$$\tau = \frac{(\bar{x}_1 - \bar{x}_2)}{\sqrt{\frac{\sigma_1{}^2}{n_1} + \frac{\sigma_2{}^2}{n_2}}},$$

which follows the distribution of a standard normal variate with mean zero and unit variance.

When population variance are unknown, then these are replaced by respective sample variances s^2_{n1} and s^2_{n2}. If both the population variances are equal and unknown, then estimate of common population variance is given by

$$\hat{\sigma}^2 = \frac{n_1 s_{n_1}{}^2 + n_2 s_{n_2}{}^2}{n_1 + n_2}.$$

Thus, the test statistic under $\sigma_1{}^2 = \sigma_2{}^2 = \sigma^2$ (unknown) comes out to be

$$\tau = \frac{\bar{x}_1 - \bar{x}_2}{\sqrt{\hat{\sigma}^2\left(\frac{1}{n_1} + \frac{1}{n_2}\right)}}$$

which follows $N(0, 1)$.

Example 9.18. A flower exporter is to select large quantity of cut flowers from two lots of flowers type A and B. To test the vase life of cut flowers he took 100 flower sticks from each type and found that the average vase life of two types of flowers sticks were 72 and 68 h with 8 and 9 h standard deviations, respectively. Assuming that (1) both the types of flowers have the same variability and (2) different variabilities, test whether the vase life of type A is better than type B or not.

Solution. Given that:

Type	Mean	S.D.	Sample size
A	72 h	8 h	100
B	68 h	9 h	100

(a) *Under the given condition* $\sigma_1{}^2 = \sigma_2{}^2 = \sigma^2$(unknown) and the null hypothesis H_0: vase life of two type of flowers are equal against H_1 : vase life of type A > vase life of type B. So the test statistic under H_0 is

$$\tau = \frac{\bar{x}_1 - \bar{x}_2}{\sqrt{\hat{\sigma}^2\left(\frac{1}{n_1} + \frac{1}{n_2}\right)}} \sim N(0, 1),$$

where

$$\hat{\sigma}^2 = \frac{n_1 s_{n_1}{}^2 + n_2 s_{n_2}{}^2}{n_1 + n_2} = \frac{100 \times 8^2 + 100 \times 9^2}{100 + 100} = \frac{100(64 + 81)}{200} = \frac{145}{2}.$$

So,

$$\tau = \frac{72 - 68}{\sqrt{\frac{145}{2}\left(\frac{1}{100} + \frac{1}{100}\right)}} = \frac{4}{\sqrt{1.45}} = 3.322.$$

Let the level of significance be $\alpha = 0.05$. This is a one-tailed test. Since the cal $\tau > \tau_{0.05} = 1.645$, the test is significant, we reject the null hypothesis and accept the alternative hypothesis H_1 : Vase life of type A is more than that of type B.

(b) *Under the given condition of* $\sigma_1{}^2 \neq \sigma_2{}^2$ *and both being unknown*. The null hypothesis and the alternative hypothesis remain the same, that is,

H_0 : Vase life of two types of flowers are same.
H_1 : Vase life of type A > vase life of type B, that is, $H_0: \mu_1 = \mu_2$
$H_1: \mu_1 > \mu_2$.

Let the level of significance be $\alpha = 0.05$; being a one-sided test, the critical value is 1.645 for standard normal variate τ. The test statistic is

$$\tau = \frac{\bar{x}_1 - \bar{x}_2}{\sqrt{\frac{{s_{n_1}}^2}{n_1} + \frac{{s_{n_2}}^2}{n_2}}} \sim N(0, 1).$$

Thus,

$$\tau = \frac{72 - 68}{\sqrt{\frac{64}{100} + \frac{81}{100}}} = \frac{4}{\sqrt{\frac{145}{100}}} = \frac{4}{\sqrt{1.45}} = \frac{4}{1.204} = 3.322.$$

Since the cal $\tau > 1.645$, so the test is significant, we reject the null hypothesis and accept the alternative hypothesis H_1 : vase life of type A is more than that of type B.

3. *Test for Significance of Specified Population Standard Deviation*
   Let $(x_1, x_2, \ldots, x_n)$ be a large sample of size n drawn randomly from a population with variance σ^2. The sampling distribution of the sample s.d. s_n follows an approximately normal distribution with mean $E(s_n) = \sigma$ and variance $V(s_n) = \frac{\sigma^2}{2n}$ as the sample size n tends to infinity. Thus, as $n \to \infty$, $s_n \sim N\left(\sigma, \frac{\sigma^2}{2n}\right)$.

Our objective is to test $H_0: \sigma = \sigma_0$.

To test the above null hypothesis, we have the following approximate test statistic

$$\tau = \frac{s_n - \sigma_0}{\sqrt{\frac{\sigma_0^2}{2n}}}$$

which is a standard normal variate with mean zero and variance unity.

For acceptance or rejection of H_0, we have to compare the calculated value of τ with the appropriate table value keeping in view the alternative hypotheses.

Example 9.19. To test the variability in vase life of a cut flower, a sample of 100 cut flowers gives s.d. 8.5 h. Can we conclude that the vase life of the cut flower has the variability of 65 h^2?

Solution. Given that the sample is large of size (n) 100. The population variance is assumed to be 65 h^2, and s.d. of the sample is 8.5 h. Under the given condition, the null hypothesis is the population s.d. $= \sqrt{65}$ h.

That is, $H_0: \sigma = 8.06$ against $H_1: \sigma' \neq 8.06$.

The test statistic for the above null hypothesis is

$$\tau = \frac{s_n - \sigma_0}{\sqrt{\frac{\sigma_0^2}{2n}}} = \frac{8.5 - 8.06}{\sqrt{\frac{65}{(2 \times 100)}}} = \frac{0.44}{0.57} = 0.77.$$

Let the level of significance be $\alpha = 0.05$, and the corresponding critical value of τ (standard normal variate) is 1.96. So the calculated value of $|\tau| <$ tabulated value of τ (1.96). The test is nonsignificant and the null hypothesis cannot be rejected. Hence, the population variance of vase life of the cut flower under consideration can be taken as 65 h^2.

4. *Test of Significant Difference Between Two Standard Deviations*
   Let us draw two independent samples $(x_{11}, x_{12}, \ldots, x_{1n_1})$ and $(x_{21}, x_{22}, \ldots, x_{2n_2})$ of sizes n_1 and n_2 with means $\bar{x}_1, \bar{x}_2$ and variance ${s_{n_1}}^2$, ${s_{n_2}}^2$ from two populations with variances ${\sigma_1}^2$ and ${\sigma_2}^2$, respectively.
   For a large sample, both s_{n_1} and s_{n_2} are distributed as $s_{n_1} \sim N\left(\sigma_1, \frac{{\sigma_1}^2}{2n_1}\right)$ and $s_{n_2} \sim N\left(\sigma_2, \frac{{\sigma_2}^2}{2n_2}\right)$.
   Now, $E(s_{n_1} - s_{n_2}) = E(s_{n_1}) - E(s_{n_2}) = \sigma_1 - \sigma_2$ and

$$\mathrm{SE}(s_{n_1} - s_{n_2}) = \sqrt{V(s_{n_1} - s_{n_2})} = \sqrt{\frac{\sigma_1^2}{2n_1} + \frac{\sigma_2^2}{2n_2}}.$$

As the samples are large, so $(s_{n_1} - s_{n_2}) \sim N\left(\sigma_1 - \sigma_2, \frac{\sigma_1^2}{2n_1} + \frac{\sigma_2^2}{2n_2}\right)$.

To test the equality of two standard deviations σ_1 and σ_2, that is, for null hypothesis $H_0 : \sigma_1 = \sigma_2$.

The test statistic is

$$\tau = \frac{(s_{n_1} - s_{n_2})}{\sqrt{\frac{\sigma_1{}^2}{2n_1} + \frac{\sigma_2{}^2}{2n_2}}}.$$

In practice, $\sigma_1{}^2$ and $\sigma_2{}^2$ are unknown, and for large samples, $\sigma_1{}^2$ and $\sigma_2{}^2$ are replaced by the corresponding sample variances. Hence, the test statistic becomes

$$\tau = \frac{(s_{n1} - s_{n2})}{\sqrt{\left(\frac{s_{n_1}^2}{2n_1} + \frac{s_{n_2}^2}{2n_2}\right)}},$$

which is a standard normal variate.

Example 9.20. The following table gives the features of two independent random samples drawn from two populations. Test whether the variability of the two are the same or not.

Sample	Size	Mean yield (q\ha)	S.D.
1	56	25.6	5.3
2	45	20.2	4.6

Solution. Let the variability be measured in terms of standard deviation. So under the given condition we are to test

H_0 : The standard deviations are equal against

H_1 : The standard deviations are not equal, that is,

$H_0 : \sigma_1 = \sigma_2$ against

$H_1 : \sigma_1 \neq \sigma_2$.

Let the level of significance be$\alpha = 0.05$.

Under the above null hypothesis, the test statistic is

$$\tau = \frac{s_{n_1} - s_{n_2}}{\sqrt{\frac{s_{n_1}{}^2}{2n_1} + \frac{s_{n_2}{}^2}{2n_2}}} \sim N(0,1), = \frac{5.3 - 4.6}{\sqrt{\frac{5.3^2}{112} + \frac{4.6^2}{90}}}$$

$$= \frac{0.7}{0.6971} = 1.004.$$

So the test is nonsignificant and the null hypothesis cannot be rejected. We conclude that the variabilities (measured in terms of s.d.) in two populations are the same.

5. *Test for Significance of Specified Population Proportion*

   To test the hypothesis $H_0 : P = P_0$ (where P_0 is the specified population proportion value with respect to a particular character), the test statistics would be

$$\tau = \frac{P - P_0}{\sqrt{\frac{P_0(1 - P_0)}{n}}} \sim N(0,1).$$

Example 9.21. Assumed that 10% of the students fail in mathematics. To test the assumption, randomly selected sample of 50 students were examined and found that 44 of them are good. Test whether the assumption is justified or not at 5% level of significance.

Solution. The null hypothesis will be that the proportion of passed students is 0.9, that is, $H_0 : P = 0.9$, against the alternative hypothesis $H_1 : P \neq 0.9$. The test statistic is

$$\tau = \frac{P - P_0}{\sqrt{\frac{P_0(1 - P_0)}{n}}} = \frac{0.88 - 0.90}{\sqrt{\frac{0.90 \times 0.10}{50}}} = -0.41716.$$

The calculated value of $|\tau|$ is less than the tabulated value of τ at upper 2.5% level of significance. So the test is nonsignificant and we cannot reject the null hypothesis. That means we conclude that the 10% of the students failed in mathematics.

6. *Test for Equality of Two Population Proportions*

   To test the equality of two population proportions, that is, $H_0 : P_1 = P_2 = P$ (known), the appropriate test statistic is

$$\tau = \frac{(p_1 - p_2)}{\sqrt{\frac{P(1 - P)}{n_1} + \frac{P(1 - P)}{n_2}}}.$$

If the value of P is unknown, we replace P by its unbiased estimator $\hat{P} = \frac{n_1 p_1 + n_2 p_2}{n_1 + n_2}$ based on both the samples. Thus,

$$\tau = \frac{(p_1 - p_2)}{\sqrt{\hat{P}\left(1 - \hat{P}\right)\left(\frac{1}{n_1} + \frac{1}{n_2}\right)}}.$$

Example 9.22. From two large samples of 500 and 600 of electric bulbs, 30 and 25%, respectively, are found to be defective. Can we conclude that the proportions of defective bulbs are equal in both the lots?

Solution. Under the given condition, the null hypothesis is $H_0 : P_1 = P_2$ against the alternative hypothesis $H_1 : P_1 \neq P_2$.

That means there exists no significant difference between the two proportions against the existence of significant difference.

Let the level of significance be $\alpha = 0.05$, and the test statistic for the above null hypothesis is

$$\tau = \frac{(p_1 - p_2)}{\sqrt{\hat{P}(1 - \hat{P})\left(\frac{1}{n_1} + \frac{1}{n_2}\right)}}, \text{ where } \hat{P} = \frac{n_1 p_1 + n_2 p_2}{n_1 + n_2} = \frac{500 \times 0.3 + 600 \times 0.25}{500 + 600} = \frac{3}{11}$$

$$\backslash\tau = \frac{(0.3 - 0.25)}{\sqrt{\frac{3}{11} \times \frac{8}{11}\left(\frac{1}{500} + \frac{1}{600}\right)}} = \frac{0.05}{0.02697} = 1.8539.$$

Since the calculated value of $|\tau|$ is less than the tabulated value of τ (1.96), we have to accept the null hypothesis and conclude that proportions of damaged bulbs are equal in both the lots.

7. χ^2 *Test*

The χ^2 test is one of the most important and widely used tests in testing of hypothesis. It is used both in parametric and in nonparametric tests. Some of the important uses of χ^2 test are testing equality of proportions, testing homogeneity or significance of population variance, test for goodness of fit, tests for association between attributes, etc.

(a) χ^2 *Test for Equality of* $k(\geq 2)$ *Population Proportions*

Let $X_1, X_2, X_3, \ldots, X_k$ be independent random variables with $X_i \sim B(n_i, P_i)$, $i = 1, 2, 3, \ldots, k$; $k \geq 2$. The random variable

$$\sum_{i=1}^{k} \left\{ \frac{X_i - n_i P_i}{\sqrt{n_i P_i (1 - P_i)}} \right\}^2$$

is distributed as χ^2 with k d.f. as $n_1, n_2, n_3, \ldots, n_k \rightarrow \infty$.

To test $H_0 : P_1 = P_2 = P_3 = \cdots = P_k = P$ (known) against all alternatives, the test statistic is

$$\chi^2 = \sum_{i=1}^{k} \left\{ \frac{x_i - n_i P}{\sqrt{n_i P (1 - P)}} \right\}^2 \text{ with } k \text{ d.f.},$$

and if cal $\chi^2 > \chi^2_{\alpha,k}$, we reject H_o.

In practice P will be unknown. The unbiased estimate of P is

$$\hat{P} = \frac{x_1 + x_2 + \cdots + x_k}{n_1 + n_2 + \cdots + nk}.$$ Then the statistic

$$\chi^2 = \sum_{i=1}^{k} \left(\frac{x_i - n_i \hat{P}}{\sqrt{n_i \hat{P}(1 - \hat{P})}} \right)^2$$

is asymptotically χ^2 with $(k - 1)$ d.f. If cal $\chi^2 > \chi^2_{\alpha,k-1}$, we reject $H_0 : P_1 = P_2 = P_3 = \cdots = P_k = P$ (unknown).

In particular when $k = 2$, $\chi^2 = \dfrac{(x_1 - n_1\hat{P})^2}{n_1\hat{P}(1-\hat{P})} + \dfrac{(x_2 - n_2\hat{P})^2}{n_2\hat{P}(1-\hat{P})}$.

Since $\hat{P} = \dfrac{x_1 + x_2}{n_1 + n_2}$, on simplification, we get

$\chi^2 = \dfrac{(p_1 - p_2)^2}{\hat{P}(1-\hat{P})\left(\dfrac{1}{n_1} + \dfrac{1}{n_2}\right)}$ with 1 d.f. which is equivalent to

$\tau = \dfrac{p_1 - p_2}{\sqrt{\hat{P}(1-\hat{P})\left(\dfrac{1}{n_1} + \dfrac{1}{n_2}\right)}}$ for testing $H_0 : P_1 = P_2 = P$ (unknown) provided in test 6.

(b) χ^2 *Bartlett's Test for Homogeneity of Variances*

Proposed by Bartlett also known as Bartlett's test for homogeneity of variances. When experiments are repeated in several seasons/places/years, pooling the data becomes necessary; test for homogeneity of variance is necessary before pooling the data of such experiments. When more than two sets of experimental data are required to be pooled, then homogeneity of variance test through F statistic as mentioned earlier does not serve the purpose. Bartlett's test for homogeneity of variances is essentially a χ^2 test. We suppose that we have k samples ($k > 2$) and that for the ith sample of size n_i, the observed values are x_{ij} ($j = 1, 2, 3, 4, \ldots, n_i$; $i = 1, 2, 3, \ldots, k$). We also assume that all the samples are independent and come from the normal populations with means $\mu_1, \mu_1, \mu_1, \ldots, \mu_k$ and variances $\sigma_1^2, \sigma_2^2, \sigma_3^2, \ldots, \sigma_k^2$. To test the null hypothesis $H_0 : \sigma_1^2 = \sigma_2^2 = \cdots = \sigma_k^2 (= \sigma^2$, say) against the alternative hypothesis H_1 is that these variances are not all equal, the approximate test statistic is

$$\chi^2{}_{k-1} = \frac{\left[\sum_{i=1}^{k} (n_i - 1)\log_e \frac{s^2}{s_i{}^2}\right]}{\left[1 + \frac{1}{3(k-1)}\left\{\sum_{i=1}^{k} \frac{1}{(n_i - 1)} - \frac{1}{\sum_{i=1}^{k}(n_i - 1)}\right\}\right]}$$ where

$s_i^2 = \dfrac{1}{n_i - 1}\sum_{j=1}^{n_i}(x_{ij} - \overline{x_i})^2$ = sample mean square for the ith sample ($i = 1, 2, 3, \ldots k$),

and $s^2 = \dfrac{\sum_{i=1}^{k}(n_i - 1)(s_i)^2}{\sum_{i=1}^{k}(n_i - 1)}$.

If cal $\chi^2 > \chi^2_{\alpha,k-1}$, H_0 is rejected.

Example 9.23. An experiment with the same protocol was practiced in three consecutive seasons for a particular variety of wheat. The following table gives the error mean squares along with degrees of freedom for three seasons. Test whether pooling of data from the three experiments can be done or not.

Season	d.f.	Error mean square
Season 1	15	380
Season 2	12	320
Season 3	12	412

Solution. The error mean squares are the estimates of the variances. So the null hypothesis for the present problem is H_0: Variances of the three experiments conducted in three seasons are homogeneous in nature, against the alternative hypothesis H_1: Variances of the three experiments conducted in three seasons are not homogeneous in nature.

The test statistic is

$$\chi^2_{k-1} = \frac{\left[\sum_{i=1}^{k}(n_i-1)\log\frac{s^2}{s_i^2}\right]}{\left[1+\frac{1}{3(k-1)}\left\{\sum_{i=1}^{k}\frac{1}{(n_i-1)}-\frac{1}{\sum_{i=1}^{k}(n_i-1)}\right\}\right]} = \frac{\left[\sum_{i=1}^{3}(n_i-1)\log\frac{s^2}{s_i^2}\right]}{\left[1+\frac{1}{3(3-1)}\left\{\sum_{i=1}^{3}\frac{1}{(n_i-1)}-\frac{1}{\sum_{i=1}^{3}(n_i-1)}\right\}\right]} \text{ at 2 d.f.}$$

Let the level of significance $\alpha = 0.05$.

$n_i - 1$	s_i^2	$(n_i-1)\,s_i^2$	$(n_i-1)\log_e\frac{s^2}{s_i^2}$	$\frac{1}{n_i-1}$
15	380	5,700	−0.3440	0.0667
12	320	3,840	1.7870	0.0833
12	412	4,944	−1.2454	0.0833
Total		14,484	0.1976	0.2333

$$\text{Now } s^2 = \frac{1}{\sum_{i=1}^{3}(n_i-1)}\sum_{i=1}^{3}(n_i-1)(s_i)^2 = \frac{14484}{39} = 371.385.$$

$$\chi_2^2 = \frac{0.1976}{1+\frac{1}{6}\left[0.2333-\frac{1}{39}\right]} = \frac{0.1976}{1.0346} = 0.1910 \text{ with 2 d.f.}$$

From the table, we have $\chi^2_{0.05,2} = 5.99$.

The calculated value of χ^2 (0.1910) is less than the table value of $\chi^2_{0.05,2}$, hence we cannot reject the null hypothesis. That means the variances are homogeneous in nature. So we can pool the information of three experiments.

(c) χ^2 *Test for Heterogeneity*

When a null hypothesis is to be tested for a large amount of test materials (treatments), it becomes sometimes difficult to accommodate all the treatments (testing materials) in one experiment. For example, when we test hundreds of germplasms to test a particular null hypothesis, then it becomes almost impossible to accommodate all these materials in one experiment even in one experimental station. So we are forced to lay down the experiment in different plots, sometimes in different experimental stations having different environmental conditions and sometimes being tested over different seasons. For example, we may be interested to test the Mendel's monohybrid cross ratio (1:2:1) of F_1 generation under different experiments using χ^2 test. Now the question is whether all these information from different experiments testing the same null hypothesis can be pooled/added to test for the theoretical ratio. χ^2 test for heterogeneity of variances based on additive property of χ^2 variates can be used here.

Let us suppose we have p experimental setup providing $\chi_1^2, \chi_2^2, \chi_3^2, \ldots \chi_p^2$ each with $(k - 1)$ d.f. The individual χ^2 values from p setup are added to get χ_s^2 with $p(k - 1)$ d.f. Then a χ^2 value known as χ_p^2 is calculated from the pooled sample of all the p experiments with $(k - 1)$ d.f. Now the heterogeneity χ^2 is calculated as $\chi_h^2 = \chi_s^2 - \chi_p^2$ with $\{p(k-1) - (k-1)\} = (p-1)(k-1)$ d.f.

Decisions are taken by comparing the calculated value of χ^2 with table value of χ^2 at upper level of significance with $(p - 1)(k - 1)$ degrees of freedom.

Example 9.24. Three sets of experiments were conducted at three different places, and the following information are collected to check whether the data follow Mendel's monohybrid cross ratio of 1:2:1 or not.

	Fruit shape		
Places	Long	Medium	Small
Place A	70	120	60
Place B	45	70	30
Place C	15	25	20

Solution. Under the given conditions, we are to test (1) the validity of monohybrid cross ratio with respect to fruit shape in all the three places separately and (2) whether we can pool the χ^2 values from different location to judge the validity of monohybrid cross ratio for fruit shape for the given population.

1. The null hypothesis for the first problem is H_0: The ratio of different fruit shape in F_1 generation follow 1:2:1 ratio in each location against the alternative hypothesis H_1: The ratio of different fruit shape in F_1 generation does not follow 1:2:1 ratio in each location.
   Under the given null hypothesis, the test statistic is

$$\chi_2^2 = \sum_{i=1}^{3} \frac{(\text{Obs} - \text{Exp})^2}{\text{Exp}} \quad \text{for each place.}$$

   Let the level of significance be $\alpha = 0.05$.
   The χ_2^2 value for each place are:

Place	χ^2 value	d.f.
Place A	1.2	2
Place B	3.276	2
Place C	1.666	2

   The χ_2^2 table value at 5% level of significance is 5.991. In all the places, the χ^2 value obtained is less than the table value of $\chi^2_{0.05,2}$, so we can conclude that in all the places, the null hypothesis cannot be rejected. That means the experimental F_1 generation data follow the theoretical ratio 1:2:1.
2. Our next objective is to work out the χ^2 value for pooled data, that is, 130:215:110 :: long:medium:small and to test whether homogeneity of variances is true for the whole experiment or not. Thus, the null hypothesis is that H_0: Homogeneity of variance exists against the alternative hypothesis that homogeneity of variance does not exist.
   Let the level of significance be $\alpha = 0.05$.
   The calculated value of χ_p^2 for whole data is 3.426, and the sum of all the χ^2 values obtained in different places is 6.142.
   Thus,

$$\begin{aligned}\chi_h^2 &= \chi_s^2 - \chi_p^2 \text{with } (p-1)(k-1) \text{ d.f.}\\ &= 6.142 - 3.426 \text{ with } 2 \times 2 = 4 \text{ d.f.}\\ &= 2.715.\end{aligned}$$

At 0.05 level of significance, the tabulated χ^2 value at 4 d.f. is 9.488. So the calculated value of χ^2 is less than the tabulated value of $\chi^2_{0.05,4}$, hence the test is nonsignificant and the null hypothesis cannot be rejected. We conclude that the results obtained in different places are homogeneous in nature.

(d) *χ^2 Test for Goodness of Fit*

In biological, agricultural, and other applied sciences, there are several laws; verification of these theoretical laws is taken up in practical field. χ^2 test for goodness of fit is a most widely used test. It is to test the agreement between the theory and practical. Now the agreement or disagreement between the theoretical values (expected) and observed values is being judged through χ^2 test for goodness of fit. For example, to test whether the different types of frequencies are in agreement with Mendel's respective theoretical frequencies or not, different observed frequencies are in agreement with the expected frequencies in accordance with different probability distributions or not. Suppose a population is classified into k mutually exclusive nominal like "male" or "female" or may be interval of the domain of some measured variable like height groups. To test whether the cell frequencies are in agreement with the expected frequencies or not. In all these cases, χ^2 known as frequency χ^2 or Pearsonian chi-square is most useful.

1. *χ^2 Test for Goodness of Fit when Population Is Completely Specified*

The problem is to test $H_0: P_1 = P_1^0, P_2 = P_2^0, P_3 = P_3^0, \ldots, P_k = P_k^0$, where $P_i^0 (i = 1, 2, 3, \ldots k)$ are specified values. The test statistic under H_0 is

$$\sum_{i=1}^{k} \frac{(O_i - nP_i^0)^2}{nP_i^0} = \sum_{i=1}^{k} \frac{(O_i - e_i)^2}{e_i} = \sum_{i=1}^{k} \frac{(O_i)^2}{nP_i^0} - n$$

where e_i = expected frequency of the ith class $= nP_i^0 (i = 1, 2, 3, \ldots, k)$ and which is (in the limiting for as $n \to \infty$) distributed as χ^2 with $(k - 1)$ d.f. If α be the level of significance, then we reject H_0 if cal $\chi^2 \geq \chi^2_{\alpha,k-1}$; otherwise we accept it.

Example 9.25. Two different types of tomato (red and yellow) varieties were crossed, and in F_1 generation, 35, red; 40, yellow; and 70, mixed (neither red nor yellow) types of plants are obtained. Do the data agree with the theoretical expectation of 1:2:1 ratio at 5% level of significance?

Solution. Total frequency 35 + 40 + 70 = 145. The expected frequency for both red and yellow $\frac{145}{4} \times 1 = 36.2536$, and mixed $\frac{145}{4} \times 2 = 72.573$.

H_0: The observed frequencies support the theoretical frequencies, that is, $P_1 = \frac{1}{4}, P_2 = \frac{2}{4}, P_3 = \frac{1}{4}$

against

H_1: The observed frequencies do not follow the theoretical frequencies, that is, $P_1' \neq \frac{1}{4},\ P_2' \neq \frac{2}{4},\ P_3' \neq \frac{1}{4}$.

Under H_0 the test statistic is

$$\chi^2 = \sum_{i=1}^{k} \frac{(O_i - e_i)^2}{e_i} \text{ with } k - 1\text{d.f.}$$

$$= \frac{(35 - 36)^2}{36} + \frac{(70 - 73)^2}{73} + \frac{(40 - 36)^2}{36}$$

$$= \frac{1}{36} + \frac{9}{73} + \frac{16}{36} = 0.596 \text{ with 2 d.f.}$$

From the table, we have $\chi^2_{0.05,2} = 5.991$. So the calculated value of χ^2 is less than the table value of χ^2. Hence, we cannot reject the null hypothesis. That means the data agree with the theoretical ratio.

2. *χ^2 Test for Goodness of Fit When Some Parameters of the Hypothetical Population Are Unspecified*

An observed frequency distribution in a sample may often be supposed to arise from a true binomial, Poisson, normal, or some other known type of distribution in the population. In most cases in practice, the hypothesis that we want to test is H_0: The sample has been drawn from a parent population of certain type.

This hypothesis may be tested by comparing the observed frequencies in various classes with those which would be given by the assumed theoretical distribution. Usually the parameters of this distribution may not be known but will have to be estimated from the sample. The test statistic under H_0 is

$$\chi^2 = \sum_{i=1}^{k} \frac{(O_i - e_i)^2}{e_i} = \sum_{i=1}^{k} \frac{\left(O_i - n\hat{P}_i\right)^2}{n\hat{P}_0}$$

which is χ^2 distribution with $(k - s - 1)$ d.f. where $s(<k - 1)$ is the number of parameters of this distribution to be estimated from the sample and $\hat{P}_i$ is the estimated probability that a single item in the sample falls in the ith class which is a function of the estimated parameters. We reject H_0 if cal $\chi^2 > \chi^2_{\alpha,k-s-1}$, and otherwise we accept it.

Example 9.26. While analyzing a set of data following frequency (observed and expected) are obtained by a group of students. Test whether the data fitted well with the expected distribution or not.

Class	Observed frequency	Expected frequency
2–4	2	1
4–6	4	7
6–8	26	28
8–10	60	56
10–12	62	58
12–14	29	31
14–16	4	9
16–18	4	1

Solution.

H_0: The data fit well with the theoretical expectation against

H_1: The data do not fit well with the theoretical expectation.

Let the level of significance be $\alpha = 0.05$. The test statistic is given by

$$\chi^2 = \sum_{i=1}^{n} \frac{(O_i - E_i)^2}{E_i} \text{ at } (k - 2 - 1) \text{ d.f.}$$
$$= 14.896 \text{ at } 5 \text{ d.f.}$$

Class	Obs.	Exp.	(Obs. − exp.)^2/exp.
1	2	1	1
2	4	7	1.285714
3	26	28	0.142857
4	60	56	0.285714
5	62	58	0.275862
6	29	31	0.129032
7	4	9	2.777778
8	4	1	9

From the table, we have $\chi^2_{0.05,5} = 11.07$, that is, the calculated value of χ^2 is greater than the table value of χ^2. So the test is significant and the data do not fit well with the expected distribution.

(e) *χ^2 Test for Independence of Attributes*

A large number of statistical theories are available which deal with quantitative characters. χ^2 test is one of the few available statistical tools dealing with qualitative characters (attributes). χ^2 test for independence of attributes work out the association or independence between the qualitative characters like color, shape, and texture which cannot be measured but grouped. χ^2 test for independence of attributes helps in finding the dependence or independence between such qualitative characters from their joint frequency distribution.

Let us suppose two attributes A and B grouped into r $(A_1, A_2, \ldots, A_r)$ and

Table 9.3 "$r \times s$" contingency table

B A	B_1	B_2	...	B_j	...	B_s	Total
A_1	(A_1B_1)	(A_1B_2)		(A_1B_j)		(A_1B_s)	(A_1)
A_2	(A_2B_1)	(A_2B_2)		(A_2B_j)		(A_2B_s)	(A_2)
...							
A_i	(A_iB_1)	(A_iB_2)		(A_iB_j)		(A_iB_s)	(A_i)
...							
A_r	(A_rB_1)	(A_rB_2)		(A_rB_j)		(A_rB_s)	(A_r)
Total	(B_1)	(B_2)		(B_j)		(B_s)	N

s classes ($B_1, B_2, \ldots, B_s$). The various cell frequencies are expressed according to the following table format known as "$r \times s$" manifold contingency table where (A_i) is the frequency of the individual units possessing the attribute A_i, ($i = 1, 2, 3, \ldots, r$), (B_j) is the frequency of the individual units possessing the attribute B_j ($j = 1, 2, 3, \ldots, s$), and (A_iB_j) is the frequency of the individual units possessing both the attributes A_i and B_j with $\sum_{i=1}^{r} (A_i) = \sum_{j=1}^{s} (B_j) = N$, where N is the total frequency.

1. Table 9.3

   The problem is to test if the two attributes A and B under consideration are independent or not. Under the null hypothesis H_0: The attributes are independent, the theoretical cell frequencies are calculated as follows:

$$P[A_i] = \text{probability that an individual unit possesses the attribute } A_i = \frac{(A_i)}{N};\, i = 1, 2, \ldots, r;$$

$$P\left[B_j\right] = \text{probability that an individual unit possesses the attribute } B_j = \frac{(B_j)}{N};\, j = 1, 2, \ldots, s.$$

Since the attributes A_i and B_j are independent, under null hypothesis, using the theorem of compound probability (Chap. 6), we get

$$P\left[A_iB_j\right] = \text{probability that an individual unit possesses both the attributes } A_i \text{ and } B_j$$

$$= P[A_i]\, P\left[B_j\right] = \frac{(A_i)}{N} \cdot \frac{(B_j)}{N};\, i = 1, 2, 3, \ldots, r;\; j = 1, 2, 3, \ldots, s \text{ and}$$

$$\left(A_iB_j\right)_e = \text{expected number of individual units possessing both the attributes } A_i \text{ and } B_j$$

$$= N.\, P\left[A_iB_j\right] = \frac{(A_i)(B_j)}{N}$$

$$\left(A_iB_j\right)_e = \frac{(A_i)(B_j)}{N},\, (i = 1, 2, 3, \ldots, r; j = 1, 2, 3, \ldots, s)$$

By using this formula, we can find out the expected frequencies for each of the cell frequencies (A_iB_j) ($i = 1, 2, 3, \ldots, r$; $j = 1, 2, 3, \ldots, s$), under the null hypothesis of independence of attributes. The approximate test statistic for the test of independence of attributes is derived from the

χ^2 test for goodness of fit as given below:

$$\chi^2 = \sum_{i=1}^{r}\sum_{j=1}^{s}\left[\frac{\left\{(A_iB_j)_o - (A_iB_j)_e\right\}^2}{(A_iB_j)_e}\right]$$

$$= \sum_{i=1}^{r}\sum_{j=1}^{s}\frac{(f_{ij} - e_{ij})^2}{e_{ij}} : \chi^2_{(r-1)(s-1)}$$

where

f_{ij} = observed cell frequency of ith row and jth column combination and

e_{ij} = expected cell frequency of ith row and jth column combination.

2. "2×2" χ^2 Test

Under the special case, the $r \times s$ contingency table becomes a 2×2 contingency table, in which both the attributes are classified into two groups each. The test of independence of attributes can be performed in the similar way as in the case of $r \times s$ contingency table. Moreover, the value of χ^2 can also be calculated directly from the observed frequency table. Let us suppose we have the following 2×2 contingency Table 9.4 as given below: The formula for calculating χ^2 value is

$$\chi^2 = \frac{N(ad - bc)^2}{(a+c)(b+d)(a+b)(c+d)} \sim \chi^2_{(2-1)(2-1)} \sim \chi^2_1.$$

Now comparing the table value of χ^2 at 1 degree of freedom at a prefixed α level of significance with the calculated value of χ^2, one can arrive at a definite conclusion about the independence of the two attributes.

Table 9.4 "2×2" contingency table

	Attribute B		
Attribute A	B_1	B_2	Total
A_1	a	b	$a + b$
A_2	c	d	$c + d$
Total	$a + c$	$b + d$	$a + b + c + d = N$

Yates's Correction

Validity of the χ^2 depends on the condition that the expected frequency in each cell should be sufficiently large; generally the expected frequency is required to be at least five. One can overcome the problem of lower cell frequency by coalescing or merging consecutive rows or columns. In doing so, one has to compromise with the basic concept of the classification of the particular subject on hand. In a 2×2 table, the possibility of merging cell frequencies is ruled out, rather we adopt Yates's corrected formula for χ^2 as given below:

$$\chi^2_1 = \frac{\left(|ad - bc| - \frac{N}{2}\right)^2 N}{(a+b)(c+d)(a+c)(b+d)},$$

where N is the total frequency and a, b, c, and d are as usual.

Example 9.27. The following table gives the frequency distribution of education standard and type occupation for 280 persons. Test whether the type of occupation is independent of education standard or not.

	Cultivation	Teaching	Govt. service	Others
HS	65	5	12	8
Graduation	35	25	15	10
Postgraduation	25	40	25	15

Solution. Under the given condition

H_0: Educational standard and occupation are independent against

H_1: Educational standard and occupation are not independent.

Under the given H_0, the test statistic is

$$\chi^2_{(3-1)(3-1)} = \sum_{i=1}^{r}\sum_{j=1}^{s}\left[\frac{\left\{(A_iB_j)_o - (A_iB_j)_e\right\}^2}{(A_iB_j)_e}\right]$$

$$= \sum_{i=1}^{r}\sum_{j=1}^{s}\frac{(f_{ij} - e_{ij})^2}{e_{ij}}.$$

Let the level of significance be $\alpha = 0.05$

We calculate the expected frequencies as follows:

$$P[A_iB_j] = \text{probability that an individual unit possesses both the attributes } A_i \text{ and } B_j$$
$$= P[A_i]\, P[B_j]$$
$$= \frac{(A_i)}{N} \cdot \frac{(B_j)}{N};\, i = 1, 2, 3, \ldots, r;\; j = 1, 2, 3, \ldots, s, \text{ and}$$
$$(A_iB_j)_e = \text{expected number of individual units possessing both the attributes } A_i \text{ and } B_j$$
$$= N.\, P[A_iB_j] = \frac{(A_i)(B_j)}{N}.$$

As follows:

Groups	Cultivation	Teaching	Govt. service	Others	Total
HS	= 90 × 125/280	= 90 × 70/280	= 90 × 52/280	= 90 × 33/280	**90**
Graduation	= 85 × 125/280	= 85 × 70/281	= 85 × 52/281	= 85 × 33/281	**85**
Postgraduation	= 105 × 125/280	= 105 × 70/282	= 105 × 52/282	= 105 × 33/282	**105**
Total	**125**	**70**	**52**	**33**	**280**

The expected frequencies are as follows:

	Cultivation	Teaching	Govt. service	Others
HS	40.1786	22.50	16.7143	10.6071
Graduation	37.9464	21.25	15.7857	10.0178
Postgraduation	46.8750	26.25	19.500	12.3750

So,

$$\chi_6^2 = \sum_{i=1}^{r}\sum_{j=1}^{s}\frac{(f_{ij} - e_{ij})^2}{e_{ij}} = \frac{(65 - 40.1786)^2}{401786} + \frac{(5 - 22.5)^2}{22.5} + \cdots + \frac{(15 - 12.375)^2}{12.375} = 51.3642.$$

The table value of $\chi^2_{0.05,6} = 12.592$ is less than the calculated value of χ^2. Hence, the test is significant and the null hypothesis is rejected. We can conclude that the two attributes, educational standard and the type of occupation, are not independent of each other.

Example 9.28. The following table is pertaining to classification of rice varieties classified into grain shape and aroma. Test whether the taste and aroma are independent of each other or not.

	Grain shape	
Aroma	Long	Dwarf
Aromatic	40	50
Nonaromatic	45	65

Solution. To test H_0: Taste and aroma are independent of each other against the alternative hypothesis H_1: Grain shape and aroma are not independent of each other. The test statistic follows a χ^2 distribution with 1 d.f. Let the level of significance be 0.05.

So

$$\chi^2 = \frac{(ad - bc)^2N}{(a + b)(c + d)(a + c)(b + d)}.$$

We don'*t* have to go for Yates's correction. From the above given information, we have

$$\chi^2 = \frac{(40 \times 65 - 50 \times 45)^2 200}{90 \times 110 \times 85 \times 115} = 0.2532.$$

The calculated value of χ^2 is less than the tabulated value of $\chi^2_{0.05,1}$ (3.84). So the test is nonsignificant and the null hypothesis cannot be rejected. That means the two attributes, namely, the grain shape and aroma are independent of each other.

Example 9.29. The following 2 × 2 table gives the frequency distribution of rice varieties based on two attributes, namely, shape of grain and the aroma of rice. Test whether the two attributes are independent or not.

	Rice aroma	
Grain type	Scented	Non-scented
Elongated	15	30
Short	3	42

Solution. Under the given condition, we are to test whether the two attributes, namely, "aroma" and "grain shape" in rice are independent of each other or not.

So H_0: Aroma and grain shape in rice are independent of each other, against

H_1: Aroma and grain shape in rice are not independent of each other.

The test statistic for the problem will be χ^2 with 1 d.f. Seeing the data that a cell frequency will be less than five, so we are to adopt the formula for χ^2 with Yates's correction. Thus,

$$\chi_{\text{corr.}}^2 = \frac{\left(|ad - bc| - \frac{N}{2}\right)^2 N}{(a+b)(c+d)(a+c)(b+d)}$$

$$= \frac{\left(|15 \times 42 - 30 \times 3| - \frac{90}{2}\right)^2 90}{(30+15)(15+3)(3+42)(30+42)}$$

$$= 8.402.$$

Let the level of significance be 0.05 and the corresponding table value at 1 d.f. is 3.84. So the calculated value of $\chi_{\text{corr.}}^2$ is more than the table value of χ^2 at 1 d.f. at 5% level of significance. Hence, the test is significant and the null hypothesis is rejected. We can conclude that the two attributes, aroma and grain shape in rice, are not independent of each other.

9.5 Nonparametric Tests

In parametric tests discussed so far, there is one or more assumption about the population behavior which is supposed to be valid under the specific situation. In nonparametric tests instead of the normality assumption, one assumes the continuity of the distribution function and the probability density function of the variable and independence of the sample observations.

Nonparametric methods are synonymously used as "distribution-free" methods, etc., as mentioned above but is nonparametric if the parent distribution is dependent on some general assumption like continuity. A distribution-free method depends neither on the form nor on the parameters of the parent distribution.

Merits nonparametric methods are as follows:

1. Nonparametric statistical tests are exact irrespective of the nature of the population distribution.
2. For unknown population distribution and for very small (say, 6) sample size, then tests are useful.
3. Nonparametric tests are also available for a sample made up of observations from different populations.
4. Nonparametric tests are useful both for inherent by ranked/qualified data and for the data which are potential to be ranked from numerical figure.
5. Data measured in nominal scales, mostly in socioeconomic studies can also be put under nonparametric tests.

Demerits of nonparametric methods:

1. Probability of type II error is more in nonparametric method. If all the assumptions of the parametric model are valid, then a parametric test is superior.
2. Suitable non parametric method corresponding to interaction effect estimation through ANOVA is lacking.
3. Estimation of population parameters cannot be done by non-parametric method.
4. Disregard the actual scale of measurement.

We discuss below some of the nonparametric tests widely used in agriculture and allied field.

9.5.1 One-Sample Tests

1. *Sign Test*

For a random sample $x_1, x_2, x_3, \ldots x_n$ of size n from a distribution with unknown median θ, we are to test H_0: $\theta = \theta_0$ against the alternative hypotheses (1) $\theta \neq \theta_0$, (2) $\theta > \theta_0$, and (3)$\theta < \theta_0$.

If θ_0 be the median, then there will be equal number of observations below and above the value θ_0. We denote the observations greater than the θ_0 with plus (+) signs and the observations smaller than θ_0 with minus (−) signs and ignore the sample values equal to the median. Let item be r plus (+) sign and minus signs. Such that $r + s = m \leq n$. Distribution of r given $r + s = m$ is binomial with probability ½. Thus, the above null hypothesis becomes equivalent to testing H_0: $P = 1/2$, where $P = P(x > \theta_0)$. One-tailed cumulative binomial probabilities are given in Table A.9. For H_1: $\theta \neq \theta_0$ the critical region for α level of significance is given by $r \leq r'_{\alpha/2}$ and $r \geq r_{\alpha/2}$ where $r'_{\alpha/2}$ and $r_{\alpha/2}$ are the largest and smallest integer such that $\sum_{r=0}^{r'_{\alpha/2}} \binom{m}{r}^m \leq \alpha/2$ and $\sum_{r=\alpha/2}^{m} \binom{m}{r}\left(\frac{1}{2}\right)^m \leq \alpha/2$. For H_1: $\theta > \theta_0$, the critical region for α level of significance is given by $r \geq r_\alpha$, where r_α is the smallest integer such that $\sum_{r=r_\alpha}^{m} \binom{m}{r}\left(\frac{1}{2}\right)^m \leq \alpha$. For H_1: $\theta < \theta_0$, the critical region for ^α level of significance is given by $r \leq r'_\alpha$ where r'_α is the larger integer such that $\sum_{r=0}^{m} \binom{m}{r}\left(\frac{1}{2}\right) \leq \alpha$.

Example 9.30. Test whether the median long jump (θ) of a group of students is 10 ft or not at 5% level of significance from the following data on long jumps.

Long jumps (feet): 11.2, 12.2, 10.0, 9, 13, 12.3, 9.3, 9.8, 10.8, 11, 11.6, 10.6, 8.5, 9.5, 9.8, 11.6, 9.7, 10.9, 9.2, 9.8

Solution. Let us first assign the signs to each of the given observations as follows:

Long jump	10	9	11.2	12.2	10	9	13	12.3	10.8	11	9.3	9.8
Signs	−	−	+	+	−	−	+	+	+	+	−	−
Long jump	8.5	9.5	11.6	10.6	8.5	9.5	9.8	11.6	9.2	9.8	9.7	10.9
Signs	−	−	+	+	−	−	−	+	−	+	−	−

There are 10 (= r) plus (+) signs and 9 (= s) minus (−) signs, and one observation is equal to the median value and discarded. This r is a binomial variate with parameter ($m = r + s = 20 - 1 = 19$) and $P(= 1/2)$. Thus, testing of $H0 : \theta = 10$ ft against $H_1 : \theta \neq 10$ ft is equivalent to testing of H_0: P = ½ against H_1: $P \neq 1/2$.

The critical region for $\alpha = 0.05$ (two-sided test) is $r \geq r_{\alpha/2}$ and $r \leq r'_{\alpha/2}$ where r is the number of plus signs and $r_{\alpha/2}$ and $r'_{\alpha/2}$ are the smallest and largest integer, respectively, such that $\sum_{r_{\alpha/2}}^{19} \binom{19}{x}\left(\frac{1}{2}\right)^{19} \leq \alpha/2$ and $\sum_{0}^{r'_{\alpha/2}} \binom{19}{x}\left(\frac{1}{2}\right)^{19} \leq \alpha/2$.

From the table we get $r_{0.025} = 14$ and $r'_{0.025} = 4$ for 19 distinct observations at $p = 1/2$. For this example we have $4/r = 10 < 14$, which lies between 4 and 14, so we cannot reject the null hypothesis at 5% level of significance, that is, we conclude that the median of the long jump can be taken as 10 ft.

[If total number of signs, that is, plus (+) signs plus the minus (−) signs (i.e., $r + s = m$), is greater than 25, that is, $m = r + s > 25$, then normal approximation to binomial may be used, and accordingly the probability of r or fewer success will be tested with the statistic $\tau = \frac{r - \frac{r+s}{2}}{\sqrt{\frac{r+s}{4}}} = \frac{\frac{r-s}{2}}{\frac{\sqrt{r+s}}{2}} = \frac{r - s}{\sqrt{r + s}}$.]

2. *Test of Randomness*
   (a) *One Sample Run Test*

   In socioeconomic or time series analysis, the researchers often wants to know whether the data point or observations have changed following a definite pattern or in a haphazard manner. One-sample run test is used to test the hypothesis that a sample is random. In other form, run test is used to test the hypothesis that the given sequence/arrangement is random.

   A run is a sequence of letters (signs) of the same kind bounded by letters (signs) of other kind. Let n_1 = the number of elements of one kind and n_2 be the number of element of other kind. That is, n_1 might be the number of heads and n_2 be the number of tails, or n_1 might be the number of pluses and n_2 might be the number of minuses. n = total number of observed events, that is, $n = n_1 + n_2$. To use the one-sample run test, first we observe the n_1 and n_2 events in the sequence in which they occurred and determine the value of r, the number of runs.

Let $x_1, x_2, x_3, \ldots x_n$ be a sample drawn from a single population. At first we find the median of the sample and we denote observations below the median by a minus sign and observations above the median by plus signs. We discard the value equal to the median. Then we count the number of runs (r) of plus and minus signs. If both n_1 and n_2 are equal to or less than 20, *then Table A.12* gives the critical value of r under H_0 for $\alpha = 0.05$. These are critical values from the sampling distribution of r under H_0. If the observed value of r falls between the critical values, we accept H_0. For large n_1 and n_2 or both the number of runs below and above, the sample median value is a random variable with mean

$$E(r) = \frac{n}{2} + 1 \text{ and variance } \mathrm{Var}(r) = \frac{n(n-2)}{4(n-1)}.$$

This formula is exact when the n is even and the r is normally distributed as the number of observations n increases. That means $\tau = \frac{r - E(r)}{\sqrt{\mathrm{Var}(r)}} \sim N(0, 1)$and we can conclude accordingly.

Example 9.31. In an admission counter, the boys (B) and girls (G) were queued as follows:

B, G, B, G, B, B, B, G, G, B, G, B, B, G, B, B, B.

Test the hypothesis that the order of males and females in the queue was random.

Solution. H_0: The order of boys and girls in the queue was random against

H_1: The order of boys and girls in the queue was not random.

In this problem there are $n_1 = 11$ boys and $n_2 = 6$ girls. The data is B G B G B B B G G B G B B G B B B.

There are 11 runs in this series, that is, $r = 11$. Table A.12 given in the Appendix shows that for $n_1 = 11$and $n_2 = 6$, a random sample would be expected to contain more than 4 runs but less than 13. The observed $< r = 11 < r = 13$ fall in the region of acceptance for $\alpha = 0.05$. So we accept H_0, that is, the order of boys and girls in the queue was random.

Example 9.32a. The following figures give the production (million tons) of rice in certain state of India. To test whether the production of rice has changed randomly or followed a definite pattern.

Year	1971	1972	1973	1974	1975	1976	1977	1978	1979	1980
Production	13.12	13	13.3	14.36	10.01	12.22	13.99	8.98	13.99	11.86
Year	1981	1982	1983	1984	1985	1986	1987	1988	1989	1990
Production	13.55	13.28	15.8	12.69	14.45	14.18	13.03	13.81	13.48	13.1
Year	1991	1992	1993	1994	1995	1996	1997	1998	1999	2000
Production	15.91	15.52	15.44	14.6	18.2	17.22	15.45	19.64	21.12	19.78
Year	2001	2002	2003	2004	2005	2006	2007	2008	2009	2010
Production	20	19.51	21.1	21.03	19.79	21.79	21.18	21.18	22.27	21.81

Solution. We are to test the null hypothesis H_0: The series is random against the alternative hypothesis that H_1: The series is not random.

Let us first calculate the median and put plus signs to those values which are greater than the median value and minus signs to those values which are less than the median value; the information are provided in the table given below. Median = 15.02(m.t).

Year	1971	1972	1973	1974	1975	1976	1977	1978	1979	1980
Production (m.t)	13.12	13.00	13.30	14.36	10.01	12.22	13.99	8.98	13.99	11.86
Signs	–	–	–	–	–	–	–	–	–	–
Year	1981	1982	1983	1984	1985	1986	1987	1988	1989	1990
Production (m.t)	13.55	13.28	15.80	12.69	14.45	14.18	13.03	13.81	13.48	13.10
Signs	–	–	+	–	–	–	–	–	–	–
Year	1991	1992	1993	1994	1995	1996	1997	1998	1999	2000
Production (m.t)	15.91	15.52	15.44	14.60	18.20	17.22	15.45	19.64	21.12	19.78
Signs	+	+	+	–	+	+	+	+	+	+
Year	2001	2002	2003	2004	2005	2006	2007	2008	2009	2010
Production (m.t)	20.00	19.51	21.10	21.03	19.79	21.79	21.18	21.18	22.27	21.81
Signs	+	+	+	+	+	+	+	+	+	+

As the sample size is large, one should apply randomness test through normality test of the number of runs, r. Here $r = 6$, so the number of runs is found to be 6.

$$E(r) = \frac{n}{2} + 1 = 40/2 + 1 = 21 \text{ and } \mathrm{Var}(r) = \frac{n(n-2)}{4(n-1)} = \frac{40(40-2)}{4(40-1)} = \frac{380}{39} = 9.744$$

$$\tau = \frac{6 - E(r)}{\sqrt{\mathrm{Var}(r)}} = \frac{6-21}{\sqrt{9.744}} = \frac{-15}{3.122} = -4.806.$$

Let the level of significance be $\alpha = 0.05$. Thus, $|\tau| = 4.806 > 1.96$ (the critical value at $\alpha = 0.05$); hence, the null hypothesis of randomness is rejected. We conclude that the production of the state has not changed in random manner.

(b) *Test of Turning Points*

In the test of turning points to test the randomness of a set of observations, count peaks and troughs in the series. A "peak" is a value greater than the two neighboring values, and a "trough" is a value which is lower than of its two neighbors. Both the peaks and troughs are treated as turning points of the series. At least three consecutive observations are required to find a turning point, let U_1, U_2, and U_3. If the series is random then these three values could have occurred in any order, namely, in six ways. But in only four of these ways would there be a turning point. Hence, the probability of turning points in a set of three values is 4/6 = 2/3.

Let $U_1, U_2, U_3, \ldots, U_n$ be a set of observations and let us define a marker variable X_i by

$$X_i = 1 \text{ when } U_i < U_{i+1} > U_{i+2} \text{ and } U_i > U_{i+1} < U_{i+2}$$
$$= 0 \text{ otherwise } \forall,\ I = 1, 2, 3, \ldots, (n-2).$$

Hence, the number of turning points p is then $p = \sum_{i=1}^{n-2} x_i$,

then we have $E(p) = \sum_{i=1}^{n-2} E(x_i) = \frac{2}{3}(n-2)$

and $E(p^2) = E\left(\sum_{i=1}^{n-2} (x_i)\right)^2 = \frac{40n^2 - 144n + 131}{90}$

$\mathrm{Var}\ (p) = E(p^2) - (E(p))^2 = \frac{(16n-29)}{90}$. As n, the number of observations, increases the distribution of p tends to normality. Thus, for testing the null hypothesis H_0: Series is random the test statistic, $\tau = \frac{p - E(p)}{\sqrt{\mathrm{Var}(p)}} \sim N$ (0,1) and we can conclude accordingly.

Example 9.32b. The area ('000ha) under cotton crop in a particular state of India since 1971 is given below. Test whether the area under jute has changed randomly or not.

Year	1971	1972	1973	1974	1975	1976	1977	1978	1979	1980
Area	326	435	446	457	404	423	496	269	437	407
Year	1981	1982	1983	1984	1985	1986	1987	1988	1989	1990
Area	461	367	419	370	335	441	479	538	504	610
Year	1991	1992	1993	1994	1995	1996	1997	1998	1999	2000
Area	506	439	464	534	731	518	424	415	427	500
Year	2001	2002	2003	2004	2005	2006	2007	2008	200	2010
Area	573	493	475	508	516	620	642	642	614	613

From the given information, one has (1) number of observations = 40 and (2) number of turning points $p = 20$. The null hypothesis is given as H_0: The series is random.

We have the expectation of turning point (p), $E(p) = \frac{2}{3}(n-2) = \frac{2}{3}(40-2) = \frac{76}{3} = 25.33$ and the variance

$$\mathrm{var}(p) = \frac{16n-29}{90} = \frac{16 \times 40 - 29}{90} = \frac{611}{90} = 6.789.$$

Thus, the test statistic

$$\tau = \frac{p - E(p)}{\sqrt{\mathrm{Var}(p)}} = \frac{20 - 25.33}{\sqrt{6.789}} = \frac{-5.33}{2.606} = -2.045.$$

We know that the τ is standard normal variate, and the value of standard normal variate at $P = 0.05$ is 1.96. As the calculated value of $|\tau| > 1.96$, so the test is significant, we reject the null hypothesis. We conclude that at 5% level of significance, there is no reason to take that the area under jute in West Bengal has changed randomly since 1961.

Note: Data for the above two examples have been taken from the various issues of the economic review published by the Government of West Bengal.

3. *Kolmogorov–Smirnov One Sample Test*

   χ^2 test for goodness of fit is valid under certain assumptions like large sample size. The parallel test to the χ^2 tests which can also be used under small sample conditions is Kolmogorov–Smirnov one-sample test. We test the null hypothesis that the sample of observations x_1, $x_2, x_3, \ldots x_n$ has come from a specified population distribution against the alternative hypothesis that the sample has come from other distribution. Let $x_1, x_2, x_3, \ldots, x_n$ be a random sample from a population of distribution function $F(x)$, and the sample cumulative distribution function is given as $F_n(x)$ where $F_n(x)$ is defined as $F_n(x) = k/n$ where k is the number of observations equal to or less than x. Now for fixed value of x, $F_n(x)$ is a statistic since it depends on the sample, and it follows a binomial distribution with parameter (n, $F(x)$). To test both-sided goodness of fit for H_0: $F(x) = F_0(x)$ for all x against the alternative hypothesis H_1: $F(x) \neq F_0(x)$, the test statistic is $D_n = \mathrm{Sup}_x[|F_n(x) - F_0(x)|]$. The distribution of D_n does not depend on F_0 so long F_0 is continuous. Now if F_0 represents the actual distribution function of x, then one would expect very small value of D_n; on the other hand, a large value of D_n is an indication of the deviation of distribution function from F_0. The decision is taken with the help of *Table A.10*.

Example 9.33. The following data presents a random sample of the proportion of insects killed by an insecticide in ten different jars. Assuming the proportion of insect killing varying between (0,1), test whether the proportion of insect killed follow rectangular distribution or not.

Proportion of insect killed: 0.404, 0.524, 0.217, 0.942, 0.089, 0.486, 0.394, 0.358, 0.278, 0.572

Solution. If $F_0(x)$ be the distribution function of a rectangular distribution over the range [0,1], then H_0: $F(x) = F_0(x)$. We know that

$$\begin{aligned} F_0(x) &= 0 \text{ if } x < 0 \\ &= x \text{ if } 0 \leq x \leq 1 \\ &= 1 \text{ if } x > 1 \end{aligned}$$

for a rectangular distribution. We make the following table:

x	$F_0(x)$	$F_n(x)$	$\lvert F_n(x) - F_0(x)\rvert$
0.089	0.089	0.1	0.011
0.217	0.217	0.2	0.017
0.278	0.278	0.3	0.022
0.358	0.358	0.4	0.042
0.394	0.394	0.5	0.106
0.404	0.404	0.6	0.196
0.486	0.486	0.7	0.214
0.524	0.524	0.8	0.276
0.572	0.572	0.9	0.328
0.942	0.942	1.0	0.058

Let the level of significance $\alpha = 0.05$.

From the table we get for $n = 10$ the critical value of K–S statistic D_n at 5% level of significance is 0.409. Thus, the calculated value of $D_n = \operatorname{Sup}_x [\lvert F_n(x) - F_0(x)\rvert] = 0.328 <$ the table value 0.409, so we cannot reject the null hypothesis. That means we conclude that the given sample is from the rectangular parent distribution. This example is due to Sahu (2007).

4. *The Median Test*

Synonymous to that of the parametric test to test whether two samples have been taken from the same population or not, the median test is also used to test whether two groups have been taken from the same population or not. Actually in the median test, the null hypothesis tested is that whether two sets of scores differ significantly between them or not. Generally it is used to test whether there exists any significant difference between the experimental group and the control group or not. If the two groups have been drawn at random from the same population, it is quite obvious that half of the frequencies will lie above and below the median. The whole process is given stepwise:

(a) Arrange the ranks/scores of both the group units taken together.
(b) A common median is worked out.
(c) Scores of each group are then divided into two subgroups: (a) those above the common median and (b) those below the common median; generally the ranks equal to the medians are ignored.
(d) These frequencies are then put into 2×2 contingency table.
(e) χ^2 value is obtained using the usual norms.
(f) The decision rule is the same as that was for χ^2 test.

Example 9.34. The following table gives the scores of two groups of students. Test whether the two groups could be taken as one or not.

	Students											
Group	1	2	3	4	5	6	7	8	9	10	11	12
Group 1	9	8	12	6	8	9	5	10	9	7	11	10
Group2	8	11	4	9	7	3	10	9	5	6	12	8

Solution. Under the given condition, we can go for median test to test the equality of two groups with respect to their scores.

Taking both the groups as one can have the following arrangement:

3	4	5	5	6	6	7	7	8	8	8	**8**	**9**	9	9	9	9	10	10	10	11	11	12	12

From the above arrangements, we have the median score as the average of the 12th and 13th ordered observation, that is, $(8 + 9)/2 = 8.5$.

Now a 2×2 contingency table is prepared as follows:

	Below	Above
Group1	(a) 5	(b) 7
Group2	(c) 7	(d) 5

$$X_1^2 = \frac{(ad - bc)^2 N}{(a+b)(a+c)(b+d)(c+d)}$$
$$= \frac{24 \times 24}{12 \times 12 \times 12} = 0.66.$$

From this 2 × 2 contingency table using the formula, the calculated value of $\chi^2 = 0.66 <$ the table value of $\chi^2 = 3.841$ at 5% level of significance. So we have no reason to reject the null hypothesis that the two groups have come from the same population, that is, the two groups do not differ.

5. *The Mann–Whitney U-Test*

 Analogous to the parametric t-test, the Mann–Whitney U-test is aimed at determining whether two independent samples have been drawn from the same population or not. This test does not require any assumption except that the continuity. Steps involved in the process are given below:

 (a) Rank the data of both the samples taking them as one sample from low to high (generally low ranking is given to low value and so on).
 (b) For tied ranks take the average of the ranks for each of the tied values (e.g., if 4th, 5th, 6th, and 7th places are having the same value, then each of these identical values would be assigned ((4 + 5 + 6 + 7)/4=) 5.5 rank).
 (c) Find the sum of the ranks of both the samples separately (say, R_1 and R_2, respectively, for 1st and 2nd sample).
 (d) Calculate $U = n_1 \cdot n_2 + \frac{n_1(n_1+1)}{2} - R_1$, where n_1 and n_2 are the sizes of the 1st and 2nd sample, respectively.
 (e) If both n_1 and n_2 are sufficiently large (>8), the sampling distribution of U can be approximated to a normal distribution and usual critical values could be used to arrive at the conclusion. On the other hand, if the sample sizes are small, then the significance test can be worked out as per the table values of the Wilcoxon's unpaired distribution.

Example 9.35. The following table gives the scores in mathematics of two groups of students. Using Mann–Whitney U-test, can you say that these two groups of students are the same at 5% level of significance, that is, they have come from the same population?

	Students											
Group	1	2	3	4	5	6	7	8	9	10	11	12
Group A	59	48	52	45	32	56	98	78	70	85	43	80
Group B	39	45	55	87	36	67	96	75	65	38		

Solution. Using the above information we get the following table:

Given that $n_1 = 12$ and $n_2 = 10$, so

$$U = n_1 \cdot n_2 + \frac{n_1(n_1+1)}{2} - R_1$$
$$= 12.10 + \frac{12(12+1)}{2} - 143.5$$
$$= 120 + 78 - 143.5 = 54.5.$$

Since both n_1 and n_2 are greater than 8, so normal approximation can be used with mean ($\mu_U = n_1 \cdot n_2/2 = 60$) and variance

$$\sigma_U^2 = n_1 \cdot n_2 + \frac{n_1(n_1+n_2+1)}{12}$$
$$= 12.10 + \frac{12(12+10+1)}{12} = 120 + 23$$
$$= 143 \text{ and } \sigma_U = 11.958.$$

Original score		Ordered score			Rank	
Group	Score	Group	Score	Unified rank	Group A	Group B
A	59	A	32	1	1	
A	48	B	36	2		2
A	52	B	38	3		3
A	45	B	39	4		4
A	32	A	43	5	5	
A	56	A	45	6.5	6.5	
A	98	B	45	6.5		6.5
A	78	A	48	8	8	
A	70	A	52	9	9	
A	85	B	55	10		10
A	43	A	56	11	11	
A	80	A	59	12	12	
B	39	B	65	13		13
B	45	B	67	14		14
B	55	A	70	15	15	
B	87	B	75	16		16
B	36	A	78	17	17	
B	67	A	80	18	18	
B	96	A	85	19	19	
B	75	B	87	20		20
B	65	B	96	21		21
B	38	A	98	22	22	
Total				**253**	**143.5**	**109.5**

At 5% level of significance, the critical value of Z the standard normal variate is 1.96.

So the upper and lower confidential limits are

$$\begin{aligned}\mu_U + 1.96\sigma_U &= 60 + 1.96 \times 11.958 \\ &= 83.438 \text{ and } \mu_U - 1.96\sigma_U \\ &= 60 - 1.96 \times 11.958 = 36.562.\end{aligned}$$

As the observed value is within the acceptance zone, one can accept the null hypothesis to conclude that the two groups of students have come from the same population.

By calculating U using R_2, one get the value of U as

$$\begin{aligned}U &= n_1 \cdot n_2 + \frac{n_2(n_2+1)}{2} - R_2 \\ &= 12.10 + \frac{10(10+1)}{2} - 109.5 \\ &= 120 + 55 - 109.5 = 65.5\end{aligned}$$

which is also within the zone of acceptance. So it does not matter how we are calculating the U value, the inference remains the same.

6. *The Kruskal–Wallis Test*

The Kruskal–Wallis test is used to compare the average performance of many groups, analogous to the one-way analysis of variance in parametric method. But unlike the parametric method, it does not require the assumption that the samples have come from normal populations. This test is mostly an extension of the Mann–Whitney U-test; difference is that here more than two groups are compared, and the test statistic is approximated to χ^2 with the stipulation that none of the groups should have less than five observations. The steps for this test are given below:

(a) Rank the data of all the samples taking them as one sample from low to high (generally low ranking is given to low value and so on).

(b) For tied ranks take the average of the ranks for each of the tied values (e.g., if 4th, 5th, 6th, and 7th places are having the same value then each of these identical values would be assigned (4 + 5 + 6 + 7)/4= 5.5 rank).

(c) Find the sum of the ranks of all the samples separately (say, $R_1, R_2 \ldots R_j$, respectively, for 1st and 2nd … and jth sample).

(d) The test statistic H is calculated as follows: $H = \frac{12}{n(n+1)} \sum_{j=1}^{k} \frac{R_j^2}{n_j} - 3(n+1)$ where $n = n_1 + n_2 + n_3 + \cdots + n_j + \cdots n_k$ and n_j being the no. of observations in jth sample.

(e) Under the null hypothesis H_0: There is no differences among the sample means, the sampling distribution of H can be approximated with χ^2 distribution with $(k - 1)$ degrees of freedom provided none of the sample has got observation less than five. Based on the calculated value of H and corresponding table value of χ^2 at required level of significance, appropriate conclusion could be made.

Example 9.36. The following table gives the yield of paddy corresponding to five groups of farmers. Using Kruskal–Wallis test, test whether five groups could be taken as one or not.

Farmers group	Yield (q/ha)						
Group A	15	17	37	25	24	32	42
Group B	25	28	26	33	30	38	
Group C	35	37	28	36	44		
Group D	42	45	52	18	24	29	41
Group E	23	28	19	51	46	37	

Solution.

Original Scores		Ordered score			Rank				
Group	Score	Group	Score	Unified rank	Group A	Group B	Group C	Group D	Group E
A	15	A	15	1	1				
A	17	A	17	2	2				
A	37	D	18	3				3	
A	25	E	19	4					4
A	24	E	23	5					5
A	32	A	24	6.5	6.5				
A	42	D	24	6.5				6.5	
B	25	A	25	8.5	8.5				
B	28	B	25	8.5		8.5			
B	26	B	26	10		10			
B	33	B	28	12		12			
B	30	C	28	12			12		
B	38	E	28	12					12
C	35	D	29	14				14	
C	37	B	30	15		15			
C	28	A	32	16	16				
C	36	B	33	17		17			
C	44	C	35	18			18		
D	42	C	36	19			19		
D	45	A	37	21	21				
D	52	C	37	21			21		
D	18	E	37	21					21
D	24	B	38	23		23			
D	29	D	41	24				24	
D	41	A	42	25.5	25.5				
E	23	D	42	25.5				25.5	
E	28	C	44	27			27		
E	19	D	45	28				28	

(continued)

(continued)

Original Scores		Ordered score			Rank				
Group	Score	Group	Score	Unified rank	Group A	Group B	Group C	Group D	Group E
E	51	E	46	29					29
E	46	E	51	30					30
E	37	D	52	31				31	
R_j					80.5	85.5	97	132	101
n_j					7	6	5	7	6
$\frac{R_j^2}{n_j}$					925.75	1218.375	1881.8	2489.143	1700.167

$$
\begin{aligned}
H &= \frac{12}{n(n+1)} \sum_{j=1}^{k} \frac{R_j^2}{n_j} - 3(n+1) \\
&= \frac{12}{31(31+1)} [925.75 + 1218.375 + 1881.8 + 2489.143 + 1700.167] - 3(31+1) \\
&= \frac{12}{31 \times 32} [8215.235] - 3(32) = 99.378 - 96 = 3.378.
\end{aligned}
$$

As all the groups have five or more observations, so H is distributed as χ^2 with $(5 - 1 =)4$ d.f. The table value of χ^2 at 5% level of significance and 4 d.f. is 9.488 which is greater than the calculated value, so we cannot reject the null hypothesis of equality of means of five farmers groups.

7. *The McNemar Test*

Just like to that of paired t-test, McNemar test is applicable for related samples, particularly concerned with nominal data. Initially when the subjects are categorized into two groups, treatment is applied and again the responses are recorded and subjects are grouped. This test is used to record whether there has been any significant change in grouping due to treatment or not.

	After treatment	
Before treatment	Disliked	Liked
Liked	a	b
Disliked	c	d

Where a, b, c, and d are the respective cell frequencies. Clearly $a + d$ is the change in response due to treatment whereas $b + c$ does not indicate any change. Thus, under normal situation one can expect equal probability of changes in either direction; that means one should go for testing H_0: $P(a) = P(d)$ against the alternative H_1: $P(a) \neq P(b)$, and the appropriate test statistic is $\chi_1^2 = (|a - d| - 1)^2/(a + d)$.

Example 9.37. Eight hundred farmers were asked about the effectiveness of SRI (system rice intensification); then these farmers were demonstrated the practice. The following table gives the responses before and after the demonstration. Examine whether the demonstration has got any impact on farmers' attitude towards SRI or not.

	After demonstration	
Before demonstration	Disliked	Liked
Liked	125	225
Disliked	75	375

Solution. Under the given condition, we are to test H_0: P(like, dislike) $=$ P(dislike, like) against H_1: P(like, dislike) $\neq$ P(dislike, like). The appropriate test statistic is $\chi_1^2 = (|a - d| - 1)^2/(a + d)$; here $a = 125$, $d = 375$, so the

$$
\begin{aligned}
\chi_1^2 &= \frac{(|a-d|-1)^2}{(a+d)} = \frac{(|125-375|-1)^2}{(125+375)} \\
&= \frac{(249)^2}{(500)} = 124.002.
\end{aligned}
$$

Table 9.5 $N \times K$ table for Kendall's coefficient of concordance test

	Individuals/objects				
Judges	1	2	......................*j*......................	$N-1$	N
Judge1	2	4	....	1	3
Judge2	$N-2$	N	....	4	2
:	:	:	....	:	:
:	:	:	....	:	:
Judge $(K-1)$	:	:	....	:	:
Judge K	:	:	....	:	:
Total rank (R_j)	R_1	R_2	......................R_j....................	R_{N-1}	R_N
$(R_j-\overline{R}_j)^2$	$(R_1-\overline{R}_j)^2$	$(R_2-\overline{R}_j)^2$	................$(R_j-\overline{R}_j)^2$..........	$(R_{N-1}-\overline{R}_j)^2$	$(R_N-\overline{R}_j)^2$

The calculated value of $\chi^2 > \chi_1^2 = 3.84$, so the test is significant and the null hypothesis of equality of probabilities is rejected. So one can conclude safely that there has been significant change in the attitude of the farmers towards SRI after demonstration.

8. *The Kendall's Coefficient of Concordance*

In social/sports or other branches of sciences, frequently a set of objects are required to be assessed by a group of experts based on both quantitative and qualitative characters. In these cases, for unbiased judgment, it is assumed that there is no association among several sets of ranking and the individuals or objects under consideration. In this aspect, Kendall's coefficient of concordance plays vital role, the special case of this test is being the Spearman's rank correlation coefficient already discussed in Chap. 8.

Let us suppose N no. of individuals has been ranked by K no. of judges. Thus, these rankings can be put into $N \times K$ matrix form as follows (Table 9.5):

$$\overline{R}_j = \frac{R_1 + R_2 + \ldots. + R_j + \ldots. + R_N}{N} = \frac{1}{N}\sum_{j=1}^{N} R_j, \text{ and}$$

$$S = (R_1-\overline{R}_j)^2 + (R_2-\overline{R}_j)^2 + \cdots + (R_j-\overline{R}_j)^2 + \cdots + (R_N-\overline{R}_j)^2 = \sum_{j=1}^{N}(R_j-\overline{R}_j)^2.$$

Kendall' coefficient of concordance is given by

$$W = \frac{S}{\frac{1}{12}K^2N(N^2-1)}$$

where R_j is the total rank of jth object/individual.

For Tied Ranks

Just like to that of Spearman's rank correlation coefficient, in Kendall's coefficient of concordance, if there exists tied ranks, then a correction factor is computed for each of the K. Judges as follows:

$$T = \frac{\sum_{j=1}^{N}\left(t_j^3 - t_j\right)}{12}$$ where t is no of ties in jth object/individual, $0 \le t < N$.

Next totals of all Ts from each of the judges are obtained, and the Kendall's coefficient of concordance W is calculated as follows:

$$W = \frac{S}{\frac{1}{12}K^2N(N^2-1) - K\sum T}.$$

This correction is generally used when there are large number of ties.

To test the null hypothesis that Kendall's coefficient of concordance W is equal to zero or not, the no. of objects/individuals (N) plays a vital role. If $N > 7$, then W is distributed as

$$\chi^2 = K(N-1)\cdot W \text{ with } (N-1)\text{d.f.},$$

and the conclusion is drawn accordingly. But if the $N \leq 7$, then a table has been prepared for different critical values of S associated with W. If the observed $S \geq$ the value shown in table, then H_0: K sets of ranking are independent may be rejected at the particular level of significance.

Example 9.38. In a heptathlon event, eight athletes were judged by six judges, and the following table shows the rank provided by each judge to all the athletes. Test whether the judges independently worked or not.

	Athletes							
	1	2	3	4	5	6	7	8
Judges	Ranks							
Judge 1	2	4	3	5	7	8	1	6
Judge 2	3	2	4	5	6	7	1	8
Judge 3	4	3	5	6	7	8	2	1
Judge 4	5	4	3	7	6	8	2	1
Judge 5	4	5	6	8	7	3	2	1
Judge 6	6	4	8	7	5	1	3	2

Solution. There are six sets of ranking for eight objects. Kendall's coefficient of concordance can very well be worked out followed by the test of H_0: $W = 0$. For this purpose we make the following table:

	Athletes							
Judges	1	2	3	4	5	6	7	8
Judge 1	2	4	3	5	7	8	1	6
Judge 2	3	2	4	5	6	7	1	8
Judge 3	4	3	5	6	7	8	2	1
Judge 4	5	4	3	7	6	8	2	1
Judge 5	4	5	6	8	7	3	2	1
Judge 6	6	4	8	7	5	1	3	2
Total rank (R_j)	24	22	29	38	38	35	11	19
$(R_j - \bar{R}_j)^2$	9	25	4	121	121	64	256	64

$$\bar{R}_j = \frac{R_1 + R_2 + \cdots + R_8}{8} = \frac{1}{N}\sum_{j=1}^{N} R_j = 27.$$

$$\begin{aligned} S &= (R_1 - \bar{R}_j)^2 + (R_2 - \bar{R}_j)^2 + \cdots + (R_8 - \bar{R}_j)^2 \\ &= (24 - 24)^2 + (22 - 27)^2 + \cdots + (19 - 27)^2 \\ &= 664. \end{aligned}$$

$$W = \frac{S}{\frac{1}{12}K^2N(N^2 - 1)} = \frac{664}{\frac{1}{12} \times 6^2 \times 8(8^2 - 1)} = 0.439.$$

As $N > 7$, we can use the approximate $\chi^2 = K(N - 1) \cdot W$ with $(N - 1)$d.f. Here $N = 8$, $K = 6$, and $W = 0.439$, $\chi^2 = K(N - 1) \cdot W = 6 \times (8 - 1) \times 0.439 = 18.438$; table value of $\chi^2_{0.05,9} = 16.919 < \chi^2_{\text{cal}}$. Thus, the null hypothesis of independence of judgment is rejected, and we conclude that W is significant at 5% level of significance.

9.5.2 Two-Sample Test

1. *Paired-Sample Sign Test*
   The sign test for one sample mentioned above can be easily modified to apply to sampling from a bivariate population. Let a random sample of n pairs (x_1,y_1), (x_2,y_2), (x_3,y_3), ..., (x_n,y_n) be drawn from a bivariate population. Let $d_i = x_i - y_i$ $(i = 1, 2, 3, \ldots, n)$. It is assumed that the distribution function of difference, d_i, is also continuous. We want to test H_0: Med$(D) = 0$, that is, $P(D > 0) = P(D < 0) = 1/2$. It is to be noted that Med(D) is not necessarily equal to Med (X) – Med(Y), so that H_0 is not that Med(X) = Med(Y), but the Med$(D) = 0$. Like the one-sample sign test, we assign plus (+) and minus (−) signs to the difference values which are greater and lesser than zero, respectively. We perform the one-sample sign test as given in the previous section and conclude accordingly.

Example 9.39. Ten students were subjected to physical training. Their body weights before and after the training were recorded as given below. Test whether there is any significant effect of training on body weight or not.

		1	2	3	4	5	6	7	8	9	10
Body weight (kg)	Before	55	48	45	62	48	59	37	40	48	52
	After	56	51	48	60	52	57	45	46	49	55

Solution. Let d_i ($i = 1, 2, 3, \ldots, 10$) be the gain in weight of ten students due to training. The null hypothesis under the given condition is $H_0 : \theta = 0$ against the alternative hypothesis $H_1 : \theta > 0$ where θ is the median of the distribution of the differences.

		1	2	3	4	5	6	7	8	9	10
Body weight (kg)	Before (y)	55	48	45	62	48	59	37	40	48	52
	After (x)	56	51	48	60	52	57	45	46	49	55
Difference in weight	($x - y$)	1	3	3	−2	4	−2	8	6	1	3
		+	+	+	−	+	−	+	+	+	+

We have 8 plus signs and 2 minus signs, and we know that under the null hypothesis, the number of plus signs follow a binomial distribution with parameter n and p. In this case the parameters are $n = 10$ and $p = ½$. The critical region ω is given by

$r \geq r_\alpha$, where r_α is the smallest integer value such that

$$P(r \geq r_\alpha/H_0) = \sum_{x=0}^{10} \binom{10}{x} \left(\frac{1}{2}\right)^{10} \leq \alpha = 0.05, \text{ that is, } 1 - \sum_{x=r0}^{r_\alpha - 1} \binom{10}{x} \left(\frac{1}{2}\right)^{10} \leq 0.05$$

$$\Rightarrow \sum_{x=0}^{r_\alpha - 1} \binom{10}{x} \left(\frac{1}{2}\right)^{10} \geq 1 - 0.05 = 0.95.$$

From the table we have $r\,\alpha = 9$, corresponding to $n = 10$ at 5% level of significance, but for this example, we got $r = 8$ which is less than the table value. So we cannot reject the null hypothesis.

Alternatively, we can calculate the probability of getting $r \geq 8$, from the fact that r follows binomial distribution under the given null hypothesis. Thus, $P(r \geq 8/H_0)$ is given by

$P(r = 8) + P(r = 9) + P(r = 10)$, that is,

$$\binom{10}{8}\left(\frac{1}{2}\right)^{10} + \binom{10}{9}\left(\frac{1}{2}\right)^{10} + \binom{10}{10}\left(\frac{1}{2}\right)^{10}$$

$$= \frac{\frac{10.9}{2} + 10 + 1}{2^{10}} = \frac{56}{2^{10}} = \frac{7}{128}$$

$$= 0.056 > 0.05.$$

Thus, the null hypothesis cannot be rejected.

2. *Two-Sample Run Test*

Sometimes we want to test the null hypothesis that whether two samples drawn at random and independently have come from the same population distribution. In testing the above hypothesis, we assume that the population distributions are continuous. The procedure is as follows.

We have two random and independent samples $x_1, x_2, x_3, \ldots, x_{n_1}$ and $y_1, y_2, y_3, \ldots, y\ y_{n_1}$ of sizes n_1 and n_2, respectively. This $n_1 + n_2 = N$ number of values are then arranged either in ascending or descending order which (may) give rise to the following sequence: $x\,x\,y\,x\,x\,x\,y\,y\,x\,x\,y\,y\,y\,y \ldots$. Now we count the “runs.” A run is a sequence of values coming from one sample surrounded by the values from the other sample. Let the number of runs in total $n_1 + n_2 = N$ arranged observations be r. Number of runs r is expected to be very high if the two samples are thoroughly mixed; that means the two samples are coming from the identical distributions, otherwise the number of runs will be very small. Table for critical values of r for given values of n_1 and n_2 are provided in Table A.10 (Appendix) for both n_1 and n_2 less than 20. If the calculated r value is greater than the critical value of run for a given set of n_1 and n_2, then we cannot reject the null hypothesis; otherwise any value of calculated r is less than or equal to the critical value of r for a given set n_1 and n_2, the test is significant and the null hypothesis is rejected.

For large n_1 and n_2 (say, >10) or any one of them is greater than 20, the distribution of r is asymptotically normal with

$$E(r) = \frac{2n_1n_2}{N} + 1 \text{ and } \mathrm{Var}(r) = \frac{2n_1n_2(2n_1n_2 - N)}{N^2(N-1)},$$

and we can perform an approximate test statistic as

$$\tau = \frac{r - E(r)}{\sqrt{\mathrm{Var}(r)}} \sim N(0, 1).$$

Example 9.40. Two independent samples of 9 and 8 plants of coconuts were selected randomly, and the number of nuts per plant was recorded. On the basis of nut characters, we are to decide whether the two samples came from the same coconut population or not.

Sample	Plant 1	Plant 2	Plant 3	Plant 4	Plant 5	Plant 6	Plant 7	Plant 8	Plant 9
Sample 1	140	135	85	90	75	110	112	95	100
Sample 2	80	125	95	100	112	90	105	108	

Solution. Null hypothesis is H_0: Samples have come from identical distribution against the alternative hypothesis that they have come from different populations.

We arrange the observations as follows:

Nut/plant	75	80	85	90	90	95	95	100	100
Sample	1	2	1	1	2	1	2	1	2
Nut/plant	105	108	110	112	112	125	135	140	
Sample	2	2	1	1	2	2	1	1	

The value of r counted from the above table is 11 and the table value of r corresponding to 9 and 8 is 5. Thus, we cannot reject the null hypothesis of equality of distributions. Hence, we conclude that the two samples have been drawn from the same population.

3. *Two-Sample Median Test*

The test parallel to equality of two means test in parametric procedure is the two-sample median tests in nonparametric method. The objective of this test is to test that the two independent samples drawn at random having the same or different sample sizes are from identical distributions against the alternative hypotheses that they have different location parameters (medians).

Let us draw two random independent samples of sizes n_1 and n_2 from two populations. Make an ordered combined sample of size $n_1 + n_2 = N$ and get the median $(\hat{\theta})$ of the combined sample. Next, we count the number of observations below and above the estimated median value $\hat{\theta}$ for all the two samples, which can be presented as follows (Table 9.6):

Table 9.6 Frequency distribution table for two-sample median test

	Number of observations		
	$<\hat{\theta}$	$\geq \hat{\theta}$	Total
Sample 1	n_{11}	n_{12}	n_1
Sample 2	n_{21}	n_{22}	n_2
Total	n_1	n_2	N

If n_1, n_2, n_1, and n_2 are small, we can get the exact probability of the above *table* with fixed marginal frequencies as follows:

$$P = \frac{n_1!n_2!n_1!n_2!}{n_{11}!n_{12}!n_{21}!n_{22}!N!}.$$

On the other hand, if the fixed marginal frequencies are moderately large, we can use the χ^2 statistic for a 2×2 contingency table using the formula

$$\chi^2 = \frac{(n_{11}n_{22} - n_{12}n_{21})^2 N}{n_1 \times n_2 \times n_1 \times n_2}.$$

Example 9.41. The following table gives the distribution of nuts per palm for two different samples of coconut. Test whether the samples have been drawn from the same coconut population or not.

Sample1	80	85	90	110	78	86	92	115	120	118
Sample2	110	82	89	114	122	128	130	126	128	

Solution. 78, 80, 82, 85, 86, 89, 90, 110, 110, 114, 115, 118, 120, 122, 126, 128, 128, 130

From the arranged data, we get the median of combined samples as 110 and we construct the following table:

Sample	<110	≥110	Total
1	6	4	10
2	2	7	9
Total	8	11	19

The exact probability of getting likely distribution of the above table is given by

$$P = \frac{n_1!n_2!n_1!n_2!}{n_{11}!n_{12}!n_{21}!n_{22}!N!} = \frac{10!9!8!11!}{6!4!2!7!19!} = 0.3.$$

The probability of getting less likely distribution than the observed one is obtained from the following table:

Sample	<112	≥112	Total
1	8	2	10
2	1	8	9
Total	9	10	19

Probability = 0.00438

Thus, the probability of getting the observed table or more extreme table in one direction is = 0.3 + 0.00438 = 0.30438.

By symmetry, we have the exact probability of getting the observed distribution or less likely ones in either direction is $2 \times 0.05108 = 0.10216$. As $0.10216 > 0.05$, we cannot reject the null hypothesis that the two distributions are identical against both-sided alternatives that they are not identical.

4. *Kolmogorov–Smirnov Two-Sample Test*

In χ^2 test, we have come across with the test procedure to test the homogeneity of two distributions in the previous sections of this chapter. But we know that χ^2 test is valid under certain assumptions like a large sample size. The parallel test to the above-mentioned χ^2 tests which can also be used under small sample conditions is Kolmogorov–Smirnov two-sample test. Kolmogorov–Smirnov two-sample test is the test for homogeneity of two populations. We draw two random independent samples $(x_1, x_2, x_3, \ldots, x_m)$ and $(y_1, y_2, y_3, \ldots, y_n)$ from two continuous cumulative distribution functions F and G, respectively. The empirical distribution functions of the variable are given by

$$\begin{aligned} F_m(x) &= 0 \text{ if } x < x_{(1)} \\ &= i/m \text{ if } x_{(i)} \leq x < x_{(i+1)} \\ &= 1 \text{ if } x \geq x_{(m)} \text{ and} \\ G_n(x) &= 0 \text{ if } x < y_{(1)} \\ &= i/m \text{ if } y_{(i)} \leq x < y_{(i+1)} \\ &= 1 \text{ if } x \geq y_{(n)} \end{aligned}$$

where $x_{(i)}$ and $y_{(i)}$ are the ordered values of the two samples, respectively. In a combined ordered arrangements of m x's and n y's, F_m and G_n represent the respective proportions of x and y values that do not exceed x. Thus, we are interested to test whether the two distribution functions are identical or not, that is, to test H_0: $F(x) = G(x)$ against the alternative hypothesis H_1: $F(x) \neq G(x)$; the test statistic is $D_{m,n} = \underset{x}{\text{Sup}}\,[|F_m(x) - G_n(x)|]$. Now if the null hypothesis is true, then one would expect very small value of $D_{m,n}$; on the other hand, a large value of $D_{m,n}$ is an indication that the parent distributions are not identical. Table A.11 given in the Appendix gives the critical values of D for different sample sizes (n_1,n_2) at different level of significance. If the calculated value of $D <$ critical value $D_{m,n};\alpha$ we accept H_0, that is, the parent distributions are identical.

Example 9.42. Two samples of ten each are taken independently at random for the number of grains per panicle in wheat. Test whether the two samples belong to the identical parent population distribution or not.

Sample	1	2	3	4	5	6	7	8	9	10
S1	70	80	75	85	100	110	90	115	120	95
S2	73	82	78	83	98	102	97	114	118	116

Solution. Here the problem is to test whether the two samples have come from the same parent population or not, that is, H_0: $Fm = Gn$ against the alternative hypothesis H_1: $Fm \neq Gn$ where Fm and Gn are the two distributions from which the above two samples have been drawn independently at random. For this example, $m = n$. Under the given null hypothesis, we apply K–S two-sample test having the statistic $D_{m,n} = \text{Sup}_x[|F_m(x) - G_n(x)|]$. Let the level of significance be $\alpha = 0.05$. We make the following table:

We make a cumulative frequency distribution for each sample of observations using the same intervals for both distributions.

	Frequency	
Class interval	Sample 1	Sample 2
70–75	2	1
76–81	1	1
82–87	1	2
88–93	1	0
94–99	1	2
100–105	1	1
106–111	1	0
112–117	1	2
118–123	1	1

For the calculation of $D_{m,n}$, we make the following table:

	70–75	76–81	82–87	88–93	94–99	100–105	106–111	112–117	118–123
$F_{10}(x)$	2/10	3/10	4/10	5/10	6/10	7/10	8/10	9/10	10/10
$F_{10}(x)$	1/10	2/10	4/10	4/10	6/10	7/10	7/10	9/10	10/10
$F_{10}(x) - F_{10}(x)$	1/10	1/10	0	1/10	0	0	1/10	0	0

Appendix

Table A.1 Table of random numbers

	00–04	05–09	10–14	15–19	20–24	25–29	30–34	35–39	40–44	45–49
00	54463	22662	65905	70639	79365	67382	29085	69831	47058	08186
01	15389	85205	18850	39226	42249	90669	96325	23248	60933	26927
02	85941	40756	82414	02015	13858	78030	16269	65978	01385	15345
03	61149	69440	11286	88218	58925	03638	52862	62733	33451	77455
04	05219	81619	10651	67079	92511	59888	84502	72095	83463	75577
05	41417	98326	87719	92294	46614	50948	64886	20002	97365	30976
06	28357	94070	20652	35774	16249	75019	21145	05217	47286	76305
07	17783	00015	10806	83091	91530	36466	39981	62481	49177	75779
08	40950	84820	29881	85966	62800	70326	84740	62660	77379	90279
09	82995	64157	66164	41180	10089	41757	78258	96488	88629	37231
10	96754	17676	55659	44105	47361	34833	86679	23930	53249	27083
11	34357	88040	53364	71726	45690	66334	60332	22554	90600	71113
12	06318	37403	49927	57715	50423	67372	63116	48888	21505	80182
13	62111	52820	07243	79931	89292	84767	85693	73947	22278	11551
14	47534	09243	67879	00544	23410	12740	02540	54440	32949	13491
15	98614	75993	84460	62846	59844	14922	48730	73443	48167	34770
16	24856	03648	44898	09351	98795	18644	39765	71058	90368	44104
17	96887	12479	80621	66223	86085	78285	02432	53342	42846	94771
18	90801	21472	42815	77408	37390	76766	52615	32141	30268	18106
19	55165	77312	83666	36028	28420	70219	81369	41843	41366	41067
20	75884	12952	84318	95108	72305	64620	91318	89872	45375	85436
21	16777	37116	58550	42958	21460	43910	01175	87894	81378	10620
22	46230	43877	80207	88877	89380	32992	91380	03164	98656	59337
23	42902	66892	46134	01432	94710	23474	20423	60137	60609	13119
24	81007	00333	39693	28039	10154	95425	39220	19774	31782	49037
25	68089	01122	51111	72373	06902	74373	96199	97017	41273	21546
26	20411	67081	89950	16944	93054	87687	96693	87236	77054	33848
27	58212	13160	06468	15718	82627	76999	05999	58680	96739	63700
28	70577	42866	24969	61210	76046	67699	42054	12696	93758	03283
29	94522	74358	71659	62038	79643	79769	44741	05437	39038	13163
30	42626	86819	85651	88678	17401	03252	99547	32404	17918	62880
31	16051	33763	57194	16752	54450	19031	58580	47629	54132	60631
32	08244	27647	33851	44705	94211	46716	11738	55784	95374	72655
33	59497	04392	09419	89964	51211	04894	72882	17805	21896	83864
34	97155	13428	40293	09985	58434	01412	69124	82171	59058	82859
35	98409	66162	95763	47420	20792	61527	20441	39435	11859	41567
36	45476	84882	65109	96597	25930	66790	65706	61203	53634	225571
37	89300	69700	50741	30329	11658	23166	05400	66669	48708	03887
38	50051	95137	91631	66315	91428	12275	24816	6809	71710	33258
39	31753	85178	31310	89642	98364	02306	24617	09607	83942	22716
40	79152	53829	77250	20190	56535	18760	69942	77448	33278	48805
41	44560	38750	83750	56540	64900	42912	13953	79149	18710	68618
42	68328	83378	63369	71381	39564	05615	42451	64559	97501	65747
43	46939	38689	58625	08342	30459	85863	20781	09284	26333	91777
44	83544	86141	15707	96256	23068	13782	08467	89469	93469	55349
45	91621	00881	04900	54224	46177	55309	17852	27491	89415	23466
46	91896	67126	04151	03795	59077	11848	12630	98375	52068	60142
47	55751	62515	21108	80830	02263	29303	37204	96926	30506	09808
48	85156	87689	95493	88842	00664	55017	55539	17771	69448	87530
49	07521	56898	12236	60277	39102	62315	12239	07105	11844	01117

Table A.2 Table of area under standard normal curve

.00	.3989423	.5039894						
.01	.3989223	.5039894	.51	.3502919	.6949743	1.01	.2395511	.8437524
.02	.3988625	.5079783	.52	.3484925	.6984682	1.02	.2371320	.8461358
.03	.3987628	.5119665	.53	.3466677	.7019440	1.03	.2307138	.8484950
.04	.3986233	.515%34,	.54	.3448080	.7054015	1.04	.2322970	.8508300
.05	.3984439	.5199388	.55	.3429439	.7088403	1.05	.2298821	.8531409
.06	.3982248	.5239222	.56	.3410458	.7122603	1.06	.22746%	.8554277
.07	.3979661	.5279032	.57	.3391243	.7156612	1.07	.2250599	.8576903
.08	.3976677	.5318814	.58	.3371799	.7190427	1.08	.2226535	.8599289
.09	.3973298	.5358564	.59	.3352132	.7224047	1.09	.2202508	.8621434
.10	.3%9525	.5398278	.60	.5332246	.7257469	1.10	.2178522	.8643339
.11	.3%5360	.5437853	.61	.3312147	.7290691	1.11	.2154582	.8665005
.12	.3960802	.5477584	.62	.3291840	.7323711	1.12	.2130691	.8686431
.13	.3955854	.5517168	.63	.3271330	.7356527	1.13	.2106856	.8707619
.14	.3950517	.5556700	.64	.3250623	.7389137	1.14	.2083078	.8727568
.15	.3944793	.5596177	.65	.3229724	.7421539	1.15	.2059363	.8749281
.16	.3938684	.5635595	.66	.3208638	.7453731	1.16	.2035714	.8769756
.17	.3932190	.5"674949	.67	.3187371	.7485711	1.17	.2012135	.8789995
.18	.3925315	.5714237	.68	.3165929	.7517478	1.18	.1988631	.8809999
.19	.3918060	.5753454	.69	.3144317	.7549029	1.19	.1965205	.8829768
.20	.3910427	.5792597	.70	.3122539	.7580363	1.20	.1941861	.8849303
.21	.3902419	.5831662	.71	.3100603	.7611479	1.21	.1918602	.8868605
.22	.3893038	.5870644	.72	.3078513	.7642375	1.22	.1895432	.8887676
.23	.3885286	.5909541	.73	.3056274	.7673049	1.23	.1872354	.8906514
.24	.3876166	.5948349	.74	.3033893	*.7703500*	1.24	.1849373	.8925123
.25	.3866681	.5987063	.75	.3011374	.7733726	1.25	.1826491	.8943502
.26	.3856834	.6025681	.76	.2988724	.7763727	1.26	.1803712	.8%1653
.27	.3846627	.6064199	.77	.2965948	.7793501	1.27	.1781038	.8979577
.28	.3836063	.6102612	.78	.2943050	.7823046	1.28	*.175IW74*	.8997274
.29	.3625146	.6140919	.79	.2920038	.7852361	1.29	.1736022	.9014747
.30	.3813878	.6178114	.80	.2896916	.7881446	1.30	.1713686	.9031995
.31	.3802264	.6217195	.81	.2873689	.7910299	1.31	.1691468	.9049021
.32	.3790305	.6255158	.82	.2850364	.7938919	1.32	.1669370	.9065825
.33	.3778007	.6293000	.83	.2826945	.7967306	1.33	.1647397	.9082409
.34	.3765372	.6330717	.84	.2803438	.7995458	1.34	.1625551	.9098773
.35	.3752403	.6368307	.85	.2779849	.8023375	1.35	.1603833	.9114920
.36	.3739106	.6405764	.86	.2755182	.805155	1.36	.1582248	.9130850
.37	.3725483	.7443088	.87	.2732444	.8078498	1.37	.1560797	.9146565
.38	.3711539	.6480273	.88	.2708640	.8105703	1.38	.1539483	.9162067
.39	.3692277	.6517317	.89	.2684774	.8132671	1.39	.1518308	.9177356
.40	.3682701	.6554217	.90	.2660852	.8159399	1.40	.1497227	.9192433
.41	.3667817	.6590970	.91	.2636880	.8185887	1.41	.1476385	.9207302
.42	.3652627	.6627573	.92	.2612863	.8212136	1.42	.1455741	.9221962
.43	.3637136	.6664022	.93	.2588805	.8238145	1.43	.1435046	.9236415
.44	.3621346	.6700314	.94	.2564713	.8263912	1.44	.1414600	.9250663
.45	.3605270	.6736448	.95	.2540591	.8289439	1.45	.1394306	.9264707
.46	.3588903	.6772419	.96	.2516443	.8314724	1.46	.1374165	.9278550
.47	.3572253	.6808225	.97	.2492277	.8339768	1.47	.1354181	.9292191
.48	.3555325	.6843863	.98	.2468095	.8364568	1.48	.1334353	.9305634
.49	.3538124	.6879331	.99	.2443095	.8389129	1.49	.1314684	.9318879
.50	.3520653	.6914625	1.00	.2419707	.8413447	1.50	.1295176	.9331928
1.51	.1275830	.9344783	2.01	.0529192	.9777844	2.51	.0170947	.9939634
1.52	.1256646	.9357445	2.02	.0518636	.9783083	2.52	.0166701	.9941323
1.53	.1237628	.9369916	2.03	.0508239	.9788217	2.53	.0162545	.9942969
1.54	.1218775	.9382198	2.04	.0498001	.9793248	2.54	.0158476	.9944574

(continued)

Table A.2 (continued)

1.55	.1200090	.9394292	2.05	.0487920	.9798178	2.55	.0154493	.9946139
1.56	.1181573	.9406201	2.06	0477996	.9803007	2.56	.0150596	.9947664
1.57	.1163225	.9417924	2.07	.0468226	.9807738	2.57	~0146782	.9949151
1.58	.1145048	.9429466	2.08	.0458611	.9812372	2.58	.0143051	.9950600
1.59	.1127042	.9440826	2.09	.0449148	.9816911	2.59	.0139401	.9952012
1.60	.1109208	.9452007	2.10	.0439836	.9821356	2.60	.0135830	.9953388
1.61	.1091548	.9463011	2.11	.0430674	.9825708	2.61	.0132337	.9954729
1.62	.1074061	.9473839	2.12	.0421661	.9829970	2.62	.0128921	.9956035
1.63	.1056748	.9484493	2.13	.0412795	.9834142	2.63	.0125581	.9957308
1.64	.1039611	.9494974	2.14	.0404076	.9838226	2.64	.0122315	.9958547
1.65	.1022649	.9505285	2.15	.0395500	.9842224	2.65	.0119122	.9959754
1.66	.1005864	.9515428	2.16	.0387069	.9846137	2.66	.0116001	.9960930
1.67	.0989255	9525403	2.17	.0378779	.9849966	2.67	.0112951	.9962074
1.68	.0972823	.9535213	2.18	.0370629	.9853713	2.68	.0109969	.9963189
1.69	.0956568	.9544860	2.19	.0362619	.9857379	2.69	.0107056	.9964274
1.70	.0940491	.9554345	2.20	.0354746	.9860966	2.70	.0104209	.9965330
1.71	.0924591	.9563671	2.21	.0347009	.9864474	2.71	.0101428	.9966358
1.72	.0908870	.9572838	2.22	.0339408	.9867906	2.72	.0098712	.9967359
1.73	.0893326	.9581849	2.23	.0331939	.9871263	2.73	.0096058	.9968333
1.74	.0877961	.9590705	2.24	.0324603	.9874545	2.74	.0093466	.9969280
1.75	.0862773	.9599408	2.25	.0317397	.9877755	2.75	.0090936	.9970202
1.76	.0847764	.9607961	2.26	.0310319	.9880894	2.76	.0068465	.9971099
1.77	.0832932	.9616364	2.27	.0303370	.9883962	2.77	.0086052	.9971972
1.78	.0818278	.9624620	2.28	.02%546	.9886962	2.78	.0083697	.9972821
1.79	.0803801	.9632730	2.29	.0289647	.9889893	2.79	.0081398	.9973646
1.80	.0789502	.9640697	2.30	.0283270	.9892759	2.80	.0079155	.9974449
1.41	.0775379	.9648521	2.31	.0276816	.9895559	2.81	.0076965	.9975229
1.82	.0761433	.9656205	2.32	.0270481	.98982%	2.82	.0074829	.9975983
1.83	.0741663	.9663750	2.33	.0264265	.9900969	2.83	.0072744	.9976726
1.84	.()1.34068	.9671159	2.34	.0258166	.9903581	2.84	.0070711	.9977443
1.85	.0720649	4678432	2.35	.0252182	.9906153	2.85	.0068728	.9978140
1.86	.0707404	.9685572	2.36	.0246313	.9908625	2.86	.0066793	.9978818
1.87	.0694333	.9692581	2.37	.0240556	.9911060	2.87	.0064907	.9979476
1.83	.0681436	.9699460	2.38	.4234910	.9913437	2.88	.0063067	.9980116
1.89	.0668711	.9706210	2.39	.0229374	.9915758	2.89	.0061274	.998073J~
1.90	.0656158	9712834	2.40	.0223945	.9918025	2.90	.0059525	.9981342
1.91	.0643777	.9719334	2.41	.0218624	.9920237	2.91	.0057821	.9981929
1.92	.0631566	.9725711	2.42	.0213407	.9922397	2.92	.0056160	.9982498
1.93	.0619524	.9731966	2.43	.0208294	.9924506	2.93	.0054541	.9983052
1.94	.0607652	.9738102	2.44	.0203284	.9926564	2.94	.0052963	.9983589
1.95	4595947	.9744119	2.45	4198374	9928572	2.95	.0051426	.9984111
1.%	.0584409	.9750021	2.46	.0193563	.9930531	2. %	.0049929	.9984618
1.97	.0573038	.9755808	2.47	.0188850	.9932443	2.97	.0048470	.9985110
1.98	.0561831	.9761482	2.48	.0184233	.9934309	2.98	.0047050	.9985583
1.99	.0550789	9767045	2.49	.0179711	.9936128	2.99	.0045666	.9986051
2.00	.0539910	.9772499	2.50	.0175283	.9937903	3.00	.0044318	.9986501

Table A.3 Table of *t* statistics

	Probability of a larger value, sign ignored								
Degree of freedom	0.500	0.400	0.200	0.100	0.050	0.025	0.010	0.005	0.001
1	1.000	1.376	3.078	6.314	12.706	25.452	63.657		
2	0.816	1.061	1.886	2.920	4.303	6.205	9.925	14.089	31.598
3	.765	0.978	1.638	2.535	3.182	4.176	5.841	7.453	12.941
4	.741	.941	1.533	2.132	2.776	3.495	4.604	5.598	8.610
5	.727	.920	1.476	2.015	2.571	3.163	4.032	4.773	6.859
6	.718	.906	1.440	1.943	2.447	2.969	3.707	4.317	5.956
7	.711	.896	1.415	1.895	2.365	2.841	3.499	4.029	5.405
8	.706	.889	1.397	1.860	2.306	2.752	3.355	3.832	5.041
9	.703	.883	1.383	1.833	2.262	2.685	3.250	3.690	4.781
10	.700	.879	1.372	1.812	2.228	2.634	3.199	3.581	4.587
11	.697	.876	1.363	1.796	2.201	2.593	3.106	3.497	4.437
12	.695	.873	1.356	1.782	2.179	2.560	3.055	3.428	4.318
13	.694	.870	1.350	1.771	2.160	2.533	3.012	3.372	4.221
14	.692	.868	1.345	1.761	2.145	2.510	2.977	3.326	4.410
15	.691	.866	1.341	1.753	2.131	2.490	2.947	3.286	4.073
16	.690	.865	1.337	1.746	2.120	2.473	2.921	3.252	4.015
17	.689	.863	1.333	1.740	2.110	2.458	2.898	3.222	3.965
18	.688	.862	1.330	1.734	2.101	2.445	2.878	3.197	3.922
19	.688	.861	1.328	1.729	2.093	2.433	2.861	3.174	3.883
20	.687	.860	1.325	1.725	2.086	2.423	2.845	3.153	3.850
21	.686	.859	1.323	1.721	2.080	2.414	2.831	3.135	3.819
22	.686	.858	1.321	1.717	2.074	2.406	2.819	3.119	3.792
23	.685	.858	1.319	1.714	2.069	2.398	2.807	3.104	3.767
24	.685	.857	1.318	1.711	2.064	2.391	2.797	3.090	3.745
25	.684	.856	1.316	1.708	2.060	2.385	2.787	3.078	3.725
26	.684	.856	1.315	1.706	2.056	2.379	2.779	3.067	3.707
27	.684	.855	1.314	1.703	2.052	2.373	2.771	3.056	3.690
28	.683	.855	1.313	1.701	2.048	2.368	2.763	3.047	3.674
29	.683	.854	1.311	1.699	2.045	2.364	2.756	3..038	3.659
30	.683	.854	1.310	1.697	2.042	2.360	2.750	3.030	3.646
35	.682	.852	1.306	1.690	2.030	2.342	2.724	2.996	3.591
40	.681	.851	1.303	1.684	2.021	2.329	2.704	2.971	3.551
45	.680	.850	1.301	1.680	2.014	2.319	2.690	2.952	3.520
50	.680	.849	1.299	1.676	2.008	2.310	2.678	2.937	3.486
55	.679	.849	1.297	1.673	2.004	2.304	2.669	2.925	3.476
60	.679	.848	1.296	1.661	2.000	2.299	2.660	2.915	3.460
70	.678	.847	1.294	1.667	1.994	2.290	2.648	2.899	3.435
80	.678	.847	1.293	1.665	1.989	2.284	2.638	2.887	3.416
90	.678	.846	1.291	1.662	1.986	2.279	2.631	2.878	3.402
100	.677	.846	1.290	1.661	1.982	2.276	2.625	2.871	3.390
120	.677	.845	1.289	1.658	1.980	2.270	2.617	2.860	3.373
∞	.6745	.8416	1.2816	1.6448	1.9600	2.2414	2.5758	2.8070	3.2905

Table A.4 Table of F statistic (5% level of significance)

n_1 / n_2	1	2	3	4	5	6	8	12	24	∞
1	161.40	199.50	215.70	224.60	230.20	234.00	238.90	243.90	249.00	254.30
2	18.51	19.00	19.16	19.25	19.30	19.33	19.37	19.41	19.45	19.50
3	10.13	9.55	9.28	9.12	9.01	8.94	8.84	8.74	8.64	8.53
4	7.71	6.94	6.59	6.39	6.26	6.16	6.04	5.91	5.77	5.63
5	6.61	5.79	5.41	5.19	5.05	4.95	4.82	4.68	4.53	4.36
6	5.99	5.14	4.76	4.53	4.39	4.28	4.15	4.00	3.84	3.67
7	5.59	4.74	4.35	4.12	3.97	3.87	3.73	3.57	3.41	3.23
8	5.32	4.46	4.07	3.84	3.69	3.58	3.44	3.28	3.12	2.93
9	5.12	4.26	3.86	3.63	3.48	3.37	3.23	3.07	2.90	2.71
10	4.96	4.10	3.71	3.48	3.33	3.22	3.07	2.91	2.74	2.54
11	4.84	3.98	3.59	3.86	3.20	3.09	2.95	2.79	2.61	2.40
12	4.75	3.88	3.49	3.26	3.11	3.00	2.85	2.69	2.50	2.30
13	4.67	3.80	3.41	3.18	3.02	2.92	2.77	2.60	2.42	2.21
14	4.60	3.74	3.34	3.11	2.96	2.85	2.70	2.53	2.35	2.13
15	4.54	3.68	3.29	3.06	2.90	2.79	2.64	2.48	2.29	2.07
16	4.49	3.63	3.24	3.01	2.85	2.74	2.59	2.42	2.24	2.01
17	4.45	3.59	3.20	2.96	2.81	2.70	2.55	2.38	2.19	1.96
18	4.41	3.55	3.16	2.93	2.77	2.66	2.51	2.34	2.15	1.92
19	4.38	3.52	3.13	2.90	2.74	2.63	2.48	2.31	2.11	1.88
20	4.35	3.49	3.10	2.87	2.71	2.60	2.45	2.28	2.08	1.84
21	4.32	3.47	3.07	2.84	2.68	2.57	2.42	2.25	2.05	1.81
22	4.30	3.44	3.05	2.82	2.66	2.55	2.40	2.23	2.03	1.78
23	4.28	3.42	3.03	2.80	2.64	2.53	2.38	2.20	2.00	1.76
24	4.26	3.40	3.01	2.78	2.62	2.51	2.36	2.18	1.98	1.73
25	4.24	3.38	2.99	2.76	2.60	2.49	2.34	2.16	1.96	1.71
26	4.22	3.37	2.98	2.74	2.59	2.47	2.32	2.15	1.95	1.69
27	4.21	3.35	2.96	2.73	2.57	2.46	2.30	2.13	1.93	1.67
28	4.20	3.34	2.95	2.71	2.56	2.44	2.29	2.12	1.91	1.65
29	4.18	3.33	2.93	2.70	2.54	2.43	2.28	2.10	1.90	1.64
30	4.17	3.32	2.92	2.69	2.53	2.42	2.27	2.09	1.89	1.62
40	4.03	3.23	2.84	2.61	2.45	2.34	2.18	2.00	1.79	1.51
60	4.00	3.16	2.76	2.52	2.37	2.25	2.10	1.92	1.70	1.39
120	3.92	3.07	2.68	2.45	2.29	2.17	2.02	1.83	1.61	1.25
∞	3.84	2.99	2.60	2.37	2.21	2.09	1.94	1.75	1.52	1.00

Table A.5 Table of F statistic (1% level of significance)

n_1 / n_2	1	2	3	4	5	6	8	12	24	∞
1	4052	4999	5403	5625	5764	5859	5981	6106	6234	6366
2	98.49	99.01	99.17	99.25	99.30	99.33	99.36	99.42	99.46	99.50
3	34.12	30.81	29.46	28.71	28.24	27.91	27.49	27.05	26.60	26.12
4	21.20	18.00	16.69	15.98	15.52	15.21	14.80	14.37	13.93	13.46
5	16.26	13.27	12.06	11.39	10.97	10.67	10.27	9.89	9.47	9.02
6	13.74	10.92	9.78	9.15	8.75	8.47	8.10	7.72	7.31	6.88
7	12.25	9.55	8.45	7.85	7.46	7.19	6.84	6.47	6.07	5.65
8	11.26	8.65	7.59	7.01	6.63	6.37	6.03	5.67	5.28	4.86
9	10.56	8.02	6.99	6.42	6.06	5.80	5.47	5.11	4.73	4.31
10	10.04	7.56	6.55	5.99	5.64	5.39	5.06	4.71	4.33	3.91
11	9.65	7.20	6.22	5.67	5.32	5.07	4.74	4.40	4.02	3.60
12	9.33	6.93	5.95	5.41	5.06	4.82	4.50	4.16	3.78	3.36
13	9.07	6.70	5.74	5.20	4.86	4.62	4.30	3.96	3.59	3.16
14	8.86	6.51	5.56	5.03	4.69	4.46	4.14	3.80	3.43	3.00
15	8.68	6.36	5.42	4.89	4.56	4.32	4.00	3.67	3.29	2.87
16	8.53	6.23	5.29	4.77	4.44	4.20	3.89	3.55	3.18	2.75
17	8.40	6.11	5.18	4.67	4.34	4.10	3.79	3.45	3.08	2.65
18	8.28	6.01	5.09	4.58	4.25	4.01	3.71	3.37	3.00	2.57
19	8.18	5.93	5.01	4.50	4.17	3.94	3.63	3.30	2.92	2.49
20	8.10	5.85	4.94	4.43	4.10	3.87	3.56	3.23	2.86	2.42
21	8.02	5.78	4.87	4.37	4.04	3.81	3.51	3.17	2.80	2.36
22	7.94	5.72	4.82	4.31	3.99	3.76	3.45	3.12	2.75	2.31
23	7.88	5.66	4.76	4.26	3.94	3.71	3.41	3.07	2.70	2.26
24	7.82	5.61	4.72	4.22	3.90	3.67	3.36	3.03	2.66	2.21
25	7.77	5.57	4.68	4.18	3.86	3.63	3.32	2.99	2.62	2.17
26	7.72	5.53	4.64	4.14	3.82	3.59	3.29	2.96	2.58	2.13
27	7.68	5.49	4.60	4.11	3.78	3.56	3.26	2.93	2.55	2.10
28	7.64	5.45	4.57	4.07	3.75	3.53	3.23	2.90	2.52	2.06
29	7.60	5.42	4.54	4.04	3.73	3.50	3.20	2.87	2.49	2.03
30	7.56	5.39	4.51	4.02	3.70	3.47	3.17	2.84	2.47	2.01
40	7.31	5.18	4.31	3.83	3.51	3.29	2.99	2.66	2.29	1.80
60	7.08	4.98	4.13	3.65	3.34	3.12	2.82	2.50	2.12	1.60
120	6.85	4.79	3.95	3.48	3.17	2.96	2.66	2.34	1.95	1.38
∞	6.64	4.60	3.78	3.32	3.02	2.80	2.51	2.18	1.79	1.00

Table A.6 Table of cumulative distribution of χ^2

Degree of freedom	Probability of greater value 0.995	0.990	0.975	0.950	0.900	0.750	0.500	0.250	0.100	0.050	0.025	0.010	0.005
1	....	....	....	....	0.02	0.10	0.45	1.32	2.71	3.84	5.02	6.63	7.88
2	0.01	0.02	0.05	0.10	0.21	0.58	.39	2.77	4.61	5.99	7.38	9.21	10.60
3	0.07	0.11	0.22	0.35	0.58	1.21	2.37	4.11	6.25	7.81	9.35	11.34	12.84
4	0.21	0.30	0.48	0.71	1.06	1.92	3.36	5.39	7.78	9.49	11.14	13.28	14.86
5	0.41	0.55	0.83	1.15	1.61	2.67	4.35	6.63	9.24	11.07	12.83	15.09	16.75
6	0.68	0.87	1.24	1.64	2.20	3.45	5.35	7.84	10.64	12.59	14.45	16.81	18.55
7	0.99	1.24	1.69	2.17	2.83	4.25	6.35	9.04	12.02	14.07	16.01	18.48	20.28
8	1.34	1.65	2.18	2.73	3.49	5.07	7.14	10.22	13.36	15.51	17.53	20.09	21.96
9	1.73	2.09	2.70	3.33	4.17	5.90	8.34	11.39	14.68	16.92	19.02	21.67	23.59
10	2.16	2.56	3.25	3.94	4.87	6.74	9.34	12.55	15.99	18.31	20.48	23.21	25.19
11	2.60	3.05	3.82	4.57	5.58	7.58	10.34	13.70	17.28	19.68	21.92	24.72	26.76
12	3.07	3.57	4.40	5.23	6.30	8.44	11.34	14.85	18.55	21.03	23.34	26.22	28.30
13	3.57	4.11	5.01	5.89	7.04	9.30	12.34	15.98	19.81	22.36	24.74	27.69	29.82
14	4.07	4.66	5.63	6.57	7.79	10.17	13.34	17.12	21.06	23.68	26.12	29.14	31.32
15	4.60	5.23	6.27	7.26	8.55	11.04	14.34	18.25	22.31	25.00	27.49	30.58	32.80
16	5.14	5.81	6.91	7.96	9.31	11.91	15.34	19.37	23.54	26.30	28.85	32.00	34.27
17	5.70	6.41	7.56	8.67	10.09	12.79	16.34	20.49	24.77	27.59	30.19	33.41	35.72
18	6.26	7.01	8.23	9.39	10.86	13.68	17.34	21.60	25.99	28.87	31.53	34.81	37.16
19	6.84	7.63	8.91	10.12	11.65	14.56	18.34	22.72	27.20	30.14	32.85	36.19	38.58
20	7.43	8.26	9.59	10.85	12.44	15.45	19.34	23.83	28.41	31.41	34.17	37.57	40.00
21	8.03	8.90	10.28	11.59	13.24	16.34	20.34	24.93	29.62	32.67	35.48	38.93	41.40
22	8.64	9.54	10.98	12.34	14.04	17.24	21.34	26.04	30.11	33.92	36.78	40.29	42.80
23	9.26	10.20	11.69	13.09	14.85	18.14	22.34	27.14	32.01	35.17	38.08	41.64	44.18
24	9.89	10.86	12.40	13.85	15.66	19.04	23.34	28.24	33.20	36.42	39.36	42.98	45.56
25	10.52	11.52	13.12	14.61	16.47	19.94	24.34	29.34	34.38	37.65	40.65	44.31	46.93
26	11.16	12.20	13.84	15.38	17.29	20.84	25.34	30.43	35.56	38.89	41.92	45.64	48.29
27	11.81	12.88	14.57	16.15	18.11	21.75	26.34	31.53	36.74	40.11	43.19	46.96	49.64
28	12.46	13.56	15.31	16.93	18.94	22.66	27.34	32.62	37.92	41.34	44.46	48.28	50.99
29	13.12	14.26	16.05	17.71	19.77	2.1.57	28.34	33.71	39.09	42.56	45.72	49.59	52.34
30	13.79	14.95	16.79	18.49	20.60	24.48	29.34	34.80	40.26	43.77	46.98	50.89	53.67
40	20.71	22.16	24.43	26.51	29.05	33.66	39.34	45.62	51.80	55.76	59.34	63.69	66.77
50	27.99	29.71	32.36	34.76	37.69	42.94	49.33	56.33	63.17	67.50	71.42	76.15	79.49
60	35.53	37.48	40.48	43.19	46.46	52.29	59.33	66.98	74.40	79.08	83.30	88.38	91.95
70	43.28	43.44	48.76	51.74	55.33	61.70	69.33	77.58	85.53	90.53	95.02	100.42	104.22
80	51.17	53.54	57.15	60.39	64.28	71.14	79.33	88.13	96.58	101.88	106.63	112.33	116.32
90	59.20	61.75	65.65	69.13	73.29	80.62	89.33	98.64	107.56	113.14	118.14	124.12	128.30
100	67.33	70.06	74.22	77.93	82.36	90.13	99.33	09.14	118.50	124.34	129.56	135.81	140.17

Table A.7 Table of critical values of simple and partial correlation coefficients

	Two-sided		
d.f	5%	1%	0.1%
1	.969	.988	.988
2	.960	.900	.900
3	.878	.959	.911
4	.811	.917	.974
5	.754	.875	.951
6	.707	.834	.925
7	.606	.798	.898
8	.632	.765	.872
9	.602	.735	.847
10	.576	.708	.823
11	.553	.684	.801
12	532	.661	.780
13	.514	.041	.760
14	.497	.623	.742
15	.482	.606	.725
16	468	.590	.708
17	.456	.575	.093
18	.444	.561	.679
19	.433	.549	.065
20	.423	.537	.652
21	.413	.526	.640
22	.404	.615	.629
23	.396	.505	.618
24	.388	.496	.607
25	.381	.487	.697
26	.374	.478	.588
27	.367	.470	.578
28	.361	.463	.570
29	.355	.456	.562
30	.349	.449	.654
40	.304	.393	.490
50	.273	.354	.443
60	.250	.325	.408
70	.232	.302	.380
80	.117	.283	.357
100	.196	.254	.321
160	.159	.208	.263
200	.138	.181	.230
260	.124	.162	.206
300	.113	.146	.188

Table A.8 Table for one-sided and both-sided K–S one sample statistic

One sided test:											
$\alpha =$	.10	.05	.025	.01	.005	$\alpha =$	.10	.05	.025	.01	.005
Two sided test:											
$\alpha =$	.20	.10	.05	.02	.01	$\alpha =$	.20	.10	.05	.02	.01
$n = 1$	.900	.950	.975	.990	.995	$n = 21$	.226	.59	.297	.321	.344
2	.684	.776	.842	.900	.929	22	.221	.253	.281	.314	.337
3	.565	.636	.708	.785	.829	23	.216	.247	.275	.307	.330
4	.493	.565	.624	.689	.734	24	.212	.242	.269	.301	.323
5	.447	.509	.563	.627	.669	25	.208	.238	.264	.295	.317
6	.410	.468	.519	.577	.617	26	.204	.233	.259	.291	.311
7	.381	.436	.483	.538	.576	27	.200	.229	.254	.284	.305
8	.358	.410	.454	.507	.542	28	.197	.225	.250	.27?	.300
9	.339	.387	.430	.480	.513	29	.193	.221	.246	.275	.295
10	.323	.369	.409	.457	.489	30	.190	.218	.242	.270	.290
11	.308	.352	.391	.437	.468	31	.187	.214	.238	.266	.285
12	.296	.338	.375	.419	.44	32	.184	.211	.234	.262	.281
13	.285	.325	.361	.404	.432	33	.182	.208	.231	.258	.277
14	.275	.314	.349	.390	.418	34	.179	.205	.227	.254	.273
15	.266	.304	.338	.377	.404	35	.177	.202	.224	.251	.269
16	.258	.295	.327	.366	.392	36	.174	.199	.221	.247	.265
17	.250	.286	.318	.355	.381	37	.172	.196	.218	.244	.262
18	.244	.279	.309	.346	.371	38	.170	.194	.215	.241	.258
19	.237	.271	.301	.337	.361	39	.168	.191	.213	.238	.255
20	.232	.265	.294	.329	.352	40	.165	.189	.210	.235	.252
			Approximation for $n > 40$				$\frac{1.07}{\sqrt{n}}$	$\frac{1.22}{\sqrt{n}}$	$\frac{1.36}{\sqrt{n}}$	$\frac{1.52}{\sqrt{n}}$	$\frac{1.63}{\sqrt{n}}$

Table A.9 Critical values for K–S two-sample test statistic

This table gives the values of $D^+_{m,n,a}$ and $D_{m,n,a}$ for which $\alpha \geq P\left\{D^+_{m,n}\ D^+_{m,n,a}\right\}$ for some selected values of N_1 = smaller sample size, N_2 = larger sample size, and α

One sided test:	$\alpha =$	.10	.05	.025	.01	.005
Two sided test:	$\alpha =$	.20	.10	.05	.02	.01
$N_1 = 1$	$N_2 = 9$	17/18				
	10	9/10				
$N_1 = 2$	$N_2 = 3$	5/6				
	4	3/4				
	5	4/5	4/5			
	6	5/6	5/6			
	7	5/7	6/7			
	8	3/4	7/4	7/8		
	9	7/9	8/9	8/9		
	10	7/10	4/5	9/10		
$N_1 = 3$	$N_2 = 4$	3/4	3/4			
	5	2/3	4/5	4/5		
	6	2/3	2/3	5/6		
	7	2/3	5/7	6/7	6/7	
	8	5/8	3/4	3/4	7/8	
	9	2/3	2/3	7/9	8/9	8/9
	10	3/5	7/10	4/5	9/10	9/10
	12	7/12	2/3	3/4	5/6	11/12
$N_1 = 4$	$N_2 = 5$	3/5	3/4	4/5	4/5	
	6	7/12	2/3	3/4	5/6	5/6
	7	17/28	5/7	3/4	6/7	6/7
	8	5/8	5/8	3/4	7/8	7/8
	9	5/9	2/3	3/4	7/9	8/9
	10	11/20	13/20	7/10	4/5	4/5
	12	7/12	2/3	2/3	3/4	5/6
	16	9/1	5/8	11/16	3/4	13/16
$N_1 = 5$	$N_2 = 6$	3/5	2/3	2/3	5/6	5/6
	7	4/7	23/35	5/7	29/35	6/7
	8	11/20	5/8	27/40	4/5	4/5
	9	5/9	3/5	31/45	7/9	4/5
	10	1/2	3/5	7/10	7/10	4/5
	15	8/15	3/5	2/3	77/15	11/15
	20	1/2	11/20	3/5	7/10	3/4
$N_1 = 6$	$N_1 = 7$	23/42	4/7	29/42	5/7	5/6
	8	1/2	7/12	2/3	3/4	3/4
	9	1/2	5/9	2/3	13/18	7/9
	10	1/2	17/30	19/30	7/10	1/15
	12	1/2	7/12	7/12	2/3	3/4
	18	4/9	5/9	11/18	2/3	2/3

(continued)

Table A.9 (continued)

One sided test:	$\alpha =$	.10	.05	.025	.01	.005
Two sided test:	$\alpha =$	.20	.10	.05	.02	.01
	24	11/24	1/2	7/12	5/8	2/3
$N_1 = 7$	$N_2 = 8$	27/56	33/56	5/8	41/56	3/4
	9	31/63	5/9	40/63	5/7	47/63
	10	33/70	39/70	43/70	7/10	5/7
	14	3/7	1/2	4/7	9/14	5/7
	28	3/7	13/28	15/28	17/28	9/14
$N_1 = 8$	$N_2 = 9$	4/9	13/24	5/8	2/3	3/4
	10	19/40	21/40	13/40	17/40	7/10
	12	11/24	1/2	7/12	5/8	2/3
	16	7/16	1/2	9/16	5/8	5/8
	32	13/32	7/16	1/2	9/16	19/32
$N_1 = 9$	$N_2 = 10$	7/15	1/2	26/45	2/3	31/45
	12	4/9	1/2	5/9	11/18	2/3
	15	19/45	22/45	8/15	3/5	29/45
	18	7/18	4/9	1/2	5/9	11/18
	36	13/36	5/12	17/36	19/36	5/9
$N_1 = 10$	$N_2 = 15$	2/5	7/15	1/2	17/30	19/30
	20	2/5	9/20	1/2	11/20	3/5
	40	7/20	2/5	9/20	1/2	
$N_1 = 12$	$N_2 = 15$	23/60	9/20	1/2	11/20	7/12
	16	3/8	7/16	23/48	13/24	7/12
	18	13/36	5/12	17/36	19/36	5/9
	20	11/30	5/12	7/15	31/60	17/30
$N_1 = 15$	$N_2 = 20$	7/20	2/5	13/30	29/60	31/60
$N_1 = 16$	$N_2 = 20$	27/80	31/80	17/40	19/40	41/40
Large-sample approximation		$1.07\sqrt{\frac{m+n}{mn}}$	$1.22\sqrt{\frac{m+n}{mn}}$	$1.36\sqrt{\frac{m+n}{mn}}$	$1.52\sqrt{\frac{m+n}{mn}}$	$1.63\sqrt{\frac{m+n}{mn}}$

Table A.10 Table for critical values of r in run test

n_2 n_1	2	3	4	5	6	7	8	9	10	11	12	13	14	15	16	17	18	19	20
2											2	2	2	2	2	2	2	2	2
3					2	2	2	2	2	2	2	2	2	3	3	3	3	3	3
4				2	2	2	3	3	3	3	3	3	3	3	4	4	4	4	4
5				2	2	3	3	3	3	4	4	4	4	4	4	4	5	5	5
6		2	2	3	3	3	3	4	4	4	4	5	5	5	5	5	5	6	6
7		2	2	3	3	3	4	4	5	5	5	5	5	6	6	6	6	6	6
8		2	3	3	3	4	4	5	5	5	6	6	6	6	6	7	7	7	7
9		2	3	3	4	4	5	5	5	6	6	6	7	7	7	7	8	8	8
10		2	3	3	4	5	5	5	6	6	7	7	7	7	8	8	8	8	9
11		2	3	4	4	6	5	6	6	7	7	7	8	8	8	9	9	9	9
12	2	2	3	4	4	5	6	6	7	7	7	8	8	8	9	9	9	10	10
13	2	2	3	4	5	5	6	6	7	7	8	8	9	9	9	10	10	10	10
14	2	2	3	4	5	5	6	7	7	8	8	9	9	9	10	10	10	11	11
15	2	3	3	4	5	6	6	7	7	8	8	9	9	10	10	11	11	11	12
16	2	3	4	4	5	6	6	7	8	8	9	9	10	10	11	11	11	12	12
17	2	3	4	4	5	6	7	7	8	9	9	10	10	11	11	11	12	12	13
18	2	3	4	6	5	6	7	8	8	9	9	10	10	11	11	12	12	13	13
19	2	3	4	5	6	6	7	8	8	9	10	10	11	11	12	12	13	13	13
20	2	3	4	5	6	6	7	8	9	9	10	10	11	12	12	13	13	13	14

n_2 n_1	2	3	4	5	6	7	8	9	10	11	12	13	14	15	16	17	18	19	20
2																			
3																			
4				9	9														
5			9	10	10	11	11												
6		9	10	11	12	12	13	13	13	13									
7			11	12	13	13	14	14	14	14	15	15	15						
8			11	12	13	14	14	15	15	16	16	16	16	17	17	17	17	17	
9				13	14	14	15	16	16	16	17	17	18	18	18	18	18	18	
10				13	14	15	16	16	17	17	18	18	18	19	19	19	20	20	
11				13	14	15	16	17	17	18	19	19	19	20	20	20	21	21	
12				13	14	16	16	17	18	19	19	20	20	21	21	21	22	22	
13					15	16	17	18	19	19	20	20	21	21	22	22	23	23	
14					15	16	17	18	19	20	20	21	22	22	23	23	23	24	
15					15	16	18	18	19	20	21	22	22	23	23	24	24	25	
16						17	18	19	20	21	21	22	23	23	24	25	25	25	
17						17	18	19	20	21	22	23	23	24	25	25	26	26	
18						17	18	19	20	21	22	23	24	25	25	26	26	27	
19						17	18	20	21	22	23	23	24	25	26	26	27	27	
20						17	18	20	21	22	23	24	25	25	26	27	27	28	

For one sample and two sample run test any value of r equal to or smaller than the value shown in Table A.12 is significant at 0.05 level. For one sample run test, any value of r greater than or equals to the values shown in Table A.12A is also significant at 0.05 level

Table A.11 Values of Wilcoxon's (unpaired) distribution

$[W_s - \text{Min } W_s]$ or $[\text{Max } W_l - W_l]$																								
s	l	Min W_s	Max W_l	0	1	2	3	4	5	6	7	8	9	10	11	12	13	14	15	16	17	18	19	20
2	2	3	7																					
	3	3	12	.100																				
	4	3	18	.067	.134																			
	5	3	25	.048	.095	.190																		
	6	3	33	.086	.071	.143																		
	7	3	42	.028	.056	.111																		
	8	3	52	.022	.044	.089	.133																	
3	2	6	9	.100																				
	3	6	15	.050	.100																			
	4	6	22	.029	.057	.114																		
	5	6	30	.018	.036	.071	.125																	
	6	6	39	.012	.024	.048	.083	.131																
	7	6	49	.008	.017	.033	.058	.092	.133															
	8	6	60	.006	.012	.024	.042	.067	.097	.139														
4	2	10	11	.007	[a]																			
	3	10	18	.029	.057	.114																		
	4	10	26	.014	.029	.057	.100																	
	5	10	35	.008	.016	.032	.056	.095	.143															
	6	10	45	.005	.010	.019	.033	.057	.086	.129														
	7	10	56	.003	.006	.012	.021	.036	.055	.082	.115													
	8	10	68	.002	.004	.008	.014	.024	.036	.055	.077	.107												
5	3	15	21	.018	.036	.071	.125	[a]																
	4	15	30	.008	.016	.032	.056	.095	.143															
	5	15	40	.004	.008	.016	.028	.048	.075	.111														
	6	15	51	.002	.004	.009	.015	.026	.041	.063	.089	.123												
	7	15	63	.001	.003	.005	.009	.015	.024	.037	.053	.074	.101											
	8	15	76	.001	.002	.003	.005	.009	.015	.023	.033	.047	.064	.085	.111									
6	3	21	24	.012	.024	[a]																		
	4	21	34	.005	.010	.019	.033	.057	.086	.129	[a]													
	5	21	45	.002	.004	.009	.015	.026	.041	.063	.089	.123												
	6	21	57	.001	.002	.004	.008	.013	.021	.032	.047	.066	.090	.120										
	7	21	70	.001	.001	.002	.004	.007	.011	.017	.026	.037	.051	.069	.090	.117								
	8	21	84	.000	.001	.001	.002	.004	.006	.010	.015	.021	.030	.041	.054	.071	.091	.114						

7	4	28	38	.003	.006	.012	.021	.036	.055	[a]														
	5	28	50	.001	.003	.005	.009	.015	.024	.037	053	.074	.101											
	6	28	63	.001	.001	.002	.004	.007	.011	.017	026	.037	.051	.069	.090	.117								
	7	28	77	.000	.001	.001	.002	.003	.006	.009	013	.019	.027	.036	.049	.064	.082	.104						
	8	28	92	.000	.000	.001	.001	.002	.003	.005	007	.010	.014	.020	.027	.036	.047	.060	.076	.095	.116			
8	4	36	42	.002	.004	.008	.014	[a]																
	5	36	55	.001	.002	.003	.005	.009	.015	.023	.033	.047	.064	[a]										
	6	36	69	.000	.001	.001	.002	.004	.006	.010	.015	.021	.030	.041	.054	.071	.091	.114						
	7	36	84	.000	.000	.001	.001	.002	.003	.005	.007	.010	.014	.020	.027	.036	.047	.060	.076	0.95	.116			
	8	36	100	.000	.000	.000	.001	.001	.001	.002	.003	.005	.007	.010	.014	.019	.025	.032	.041	0.52	.065	.080	.097	.117

[a]Indicates that the value at head of this column (and those values that are larger are not possible for the given values of s and l in this row)

Table A.12 Values of the Kendall's coefficient of concordance

	N					Some additional values for $N = 3$	
k	3	4	5	6	7	*k*	*s*
Values at 5% level of significance							
3			64.4	103.9	157.3	9	54.0
4		49.5	88.4	143.3	217.0	12	71.9
5		62.6	112.3	182.4	276.2	14	83.8
6		75.7	136.1	221.4	335.2	16	95.8
8	48.1	101.7	183.7	299.0	453.1	18	107.7
10	60.0	127.8	231.2	376.7	571.0		
15	89.8	192.9	349.8	570.5	864.9		
20	119.7	258.0	468.5	764.4	1158.7		
Values at 1% level of significance							
3			75.6	122.8	185.6	9	75.9
4		61.4	109.3	176.2	265.0	12	103.5
5		80.5	142.8	229.4	343.8	14	121.9
6		99.5	176.1	282.4	422.6	16	140.2
8	66.8	137.4	242.7	383.3	579.9	18	158.6
10	85.1	175.3	309.1	494.0	737.0		
15	131.0	269.8	475.2	758.2	1129.5		
20	177.0	364.2	641.2	1022.2	1521.9		

Table A.13 One-tailed cumulative binomial probabilities under H_0: $P = Q = 0.5$

x																
N	0	1	2	3	4	5	6	7	8	9	10	11	12	13	14	15
5	031	188	500	812	969	†										
6	016	109	344	656	891	984	†									
7	008	062	227	500	773	938	992	†								
8	004	035	145	363	637	855	965	996	†							
9	002	020	090	254	500	746	910	980	998	†						
10	001	011	055	172	377	623	828	945	989	999	†					
11		006	033	113	274	500	726	887	967	994	†	†				
12		003	019	073	194	387	613	806	927	981	997	†	†			
13		002	011	046	133	291	500	709	867	954	989	998	†	†		
14		001	006	029	090	212	395	605	788	910	971	994	999	†	†	
15			004	018	059	151	304	500	696	849	941	982	996	†	†	†
16			002	011	038	105	227	402	598	773	895	962	989	998	†	†
17			001	006	025	072	166	315	500	685	834	928	975	994	999	†
18			001	004	015	048	119	240	407	593	760	881	952	985	996	999
19				002	010	032	084	180	324	500	676	820	916	968	990	998
20				001	006	021	058	132	252	412	588	748	868	942	979	994
21				001	004	013	039	095	192	332	500	668	808	905	961	987
22					002	008	026	067	143	262	416	584	738	857	933	974
23					001	005	017	047	105	202	339	500	661	798	895	953
24					001	003	011	032	076	154	271	419	581	729	846	924
25						002	007	022	054	115	212	345	500	655	788	885

Note: Probability values are in decimal points

† Values are almost unity

10 Analysis of Variance and Experimental Designs

In Chap. 9, a discussion has been made as to how two sample means can be compared using τ or t-test. Problem arises when one wants to compare more than two populations at a time. One of the possible solutions to this problem is to take $^{m}C_{2}$ no. of pairs of samples and test these using the τ or t-test as applicable. Another important procedure is to use the analysis of variance technique. The analysis of variance technique, in short ANOVA, is a powerful technique used in the field of agriculture, social science, business, education, medicine, and several other fields. Using this tool the researcher can draw inference about the samples whether these have been drawn from the same population or they belong to different populations. Using this technique the researcher can establish whether a no. of varieties differ significantly among themselves with respect to their different characteristics like yield, susceptibility to diseases and pests, nutrient acceptability, and stress tolerance and efficiency of different salesmen; for example, one can compare different plant protection chemicals, different groups of people with respect to their innovativeness, and different drugs against a particular disease, different programs of poverty reduction, performances of different business houses, and so on.

In statistical studies variability and measures of variability are the major focal points of attention. The essence of ANOVA lies in partitioning the variance of a set of data into a number of components associated with the type, classification, and nature of data. *Analysis of variance is a systematic approach of partitioning the variance of a variable into assignable and non-assignable parts*. Through ANOVA one can investigate one or more factors (variety, type of irrigation, fertilizer doses, social groups, business houses, etc.) with varying levels (e.g., five or six varieties at a time; three or more irrigation types; different levels of NPK; social groups like innovators, early adopters, early majority, late majority, and laggards; more than one business house) that are hypothesized to influence the dependent variable. Depending upon the nature, type, and classification of data, the analysis of variance is developed for *one-way classified data, two-way classified data with one observation per cell, two-way classified data with more than one observation per cell, etc*. Before taking up the analysis of variance in detail, let us discuss about linear model, which is mostly being used in the analysis of variance.

10.1 Linear Models

Taken observation on any dependent variable Y, y_i can be assumed as $y_i = \alpha_i + e_i$, where α_i is its true value and e_i is the error part, which may be because of chance factor. This α_i again may be the linear combination of m unknown quantities $\gamma_1, \gamma_2, \ldots, \gamma_m$ as $\alpha_i = a_{i1}\gamma_1 + a_{i2}\gamma_2 + \cdots + a_{im}\gamma_m$, where a_{ij} $(j = 1, 2, \ldots, m)$ is the constant and take the values 0 or 1.

Thus, $y_i = a_{i1}\gamma_1 + a_{i2}\gamma_2 + \cdots + a_{im}\gamma_m + e_i$.

P.K. Sahu, *Research Methodology: A Guide for Researchers in Agricultural Science, Social Science and Other Related Fields*, DOI 10.1007/978-81-322-1020-7_10, © Springer India 2013

If $a_{ij} = 1$ for a particular value of j and for all i, then γ_j is termed general mean or general effect. A linear model in which all the γ_j's are unknown constants (known as parameters) is termed *fixed effect model*. On the other hand, a linear model in which γ_j's are random variables excepting the general mean or general effect is known as *variance component model* or *random effect model*. A linear model in which at least one γ_j is a random variable and at least one γ_j is a constant (other than general effect or general mean) is called a *mixed effect model*.

Assumptions in Analysis Variance

The analysis of variance is based on following assumptions:

1. The effects are additive in nature.
2. The observations are independent.
3. The variable concerned must be normally distributed.
4. Variances of all populations from which samples have been drawn must be the same. In other words, all samples should be drawn from normal populations having common variance with the same or different means.

The interpretation of analysis of variance is valid only when the assumptions are met. A larger deviation from these assumptions affects the level of significance and the sensitivity of F- and t-test. *Two independent factors are said to be additive in nature if the effect of one factor remains constant over the levels of the other factor. On the contrary when the effects of one factor remain constant by a certain percentage over the levels of other factors, then the factors are multiplicative or nonadditive in nature.* The models for additive and multiplicative effects may be presented as follows:

$$y_{ij} = \mu + \alpha_i + \beta_j + e_{ij} \text{ (additive model)}$$

$$y_{ij} = \mu + \alpha_i + \beta_j + (\alpha\beta)_{ij} + e_{ij} \text{ (multiplicative model)}$$

The second assumption is related to the error associated with the model. Precisely, this assumption means that the errors should be independently and identically distributed with mean 0 and constant (σ^2) variance.

Several normality test procedures are available in the literature. However, the nature of the functional relationship between the mean and the variance may help us in identifying whether a distribution is normal or non-normal. Homogeneity of error variance is one of the important assumptions in the analysis of variance. In practice we may not get exactly equal (homoscedastic) variances. The variances can be tested for homogeneity through *Hartley's test* or *Bartlett's* test.

10.2 One-Way ANOVA

In one-way analysis of variance, different samples for only one factor are considered. That means the whole data set comprised several groups with many observations based on one factor as shown below:

Let $y_{i1}, y_{i2}, y_{i3}, \ldots y_{in_i}$ be a random sample from an $N(\mu_i, \sigma^2)$ population and $i = 1, 2, 3, 4, \ldots, k$. The random samples are independent. Suppose there are $n = \sum_{i=1}^{k} n_i$ observations y_{ij} ($i = 1, 2, 3, \ldots, k$; $j = 1, 2, 3, 4, \ldots, n_i$), which are grouped into k classes in the following way (Table 10.1):

If we consider the fixed effect model, then it can be written as $y_{ij} = \mu_i + e_{ij}$, where μ_i is a fixed effect due to ith class and e_{ij}s are independently $N(0, \sigma^2)$. This μ_i can be regarded as the sum of two components, namely, μ, the overall mean, and a component due to the specific class. Thus, we can write

$$\mu_i = \mu + \alpha_i.$$

Thus, the mathematical model will be

$$y_{ij} = \mu + \alpha_i + e_{ij},$$

Table 10.1 One-way classification of data

1	2	..........*i*.........	*k*
y_{11}	y_{21}	y_{i1}	y_{k1}
y_{12}	y_{22}	y_{i2}	y_{k2}
:	:	:	:
j	:	y_{ij}	y_{jk}
:	:	:	:
y_{1n_1}	y_{2n_2}	y_{in_i}	y_{kn_k}

where

y_{ij} = response due to j^{th} observation of ith group,

μ = general mean effect,

α_i = additional effect due to i^{th} class group,

e_{ij} = errors component associated with j^{th} observation of i^{th}class group and are i.i.d.N $(0,\sigma^2)$,

$$\sum_{i=1}^{k} n_i\alpha_i = 0.$$

We want to test the equality of the population means, that is,

$H_0: \alpha_1 = \alpha_2 = \alpha_3 = \cdots = \alpha_k = 0$ against the alternative hypothesis

$$H_1 : \text{All } \alpha\text{'s are not equal.}$$

Let the level of significance chosen be equal to 0.05.

$$y_{ij} = \bar{y}_{00} + (\bar{y}_{i0} - \bar{y}_{00}) + (y_{ij} - \bar{y}_{i0}),$$

where $\bar{y}_{i0}$ = mean of the ith group/class,

$\bar{y}_{00}$ = grand mean,

$$\therefore y_{ij} - \bar{y}_{00} = (\bar{y}_{i0} - \bar{y}_{00}) + (y_{ij} - \bar{y}_{i0}).$$

Squaring and taking sum for both the sides over i and j, we have

$$\begin{aligned}\sum_i\sum_j (y_{ij} - \bar{y}_{00})^2 &= \sum_i\sum_j \left[(\bar{y}_{i0} - \bar{y}_{00}) + (y_{ij} - \bar{y}_{i0})\right]^2 \\ &= \sum_i\sum_j (\bar{y}_{i0} - \bar{y}_{00})^2 + \sum_i\sum_j (y_{ij} - \bar{y}_{i0})^2 + 2\sum_i\sum_j (\bar{y}_{i0} - \bar{y}_{00})(y_{ij} - \bar{y}_{i0}) \\ &= \sum_i\sum_j (\bar{y}_{i0} - \bar{y}_{00})^2 + \sum_i\sum_j (y_{ij} - \bar{y}_{i0})^2 + 2\sum_i (\bar{y}_{i0} - \bar{y}_{00})\sum_j (y_{ij} - \bar{y}_{i0}) \\ &= \sum_i n_i(\bar{y}_{i0} - \bar{y}_{00})^2 + \sum_i\sum_j (y_{ij} - \bar{y}_{i0})^2 \left[\because \sum_j (y_{ij} - \bar{y}_{i0}) = 0\right]\end{aligned}$$

SS(Total) = SS(class/group) + SS(error),

TSS = CSS + ErSS.

Thus, the total sum of squares is partitioned into sum of squares due to classes/groups and sum of squares due to error.

$n - 1$ d.f. is attributed to SS(total) or TSS because it is computed from the n quantities of the form $(y_{ij} - \bar{y})$ which is subjected to one linear constraint $\sum_j (y_{ij} - \bar{y}_i) = 0, i = 1, 2, 3, ..., k$. Similarly, $(k - 1)$ d.f. is attributed to CSS since $\sum_i n_i(\bar{y}_{i0} - \bar{y}_{00}) = 0$. Finally ErSS will have $(n - k)$ d.f. since it is based on quantities which are subjected to k linear constraints $\sum_j (\bar{y}_{ij} - \bar{y}_i) = 0, i = 1, 2, 3, \ldots, k$. It is to be noted that CSS and ErSS add up to TSS and the corresponding d.f. is also additive. Dividing SS by their respective d.f., we will get the respective mean sum of squares, that is, MSS due to class and error. Thus,

$$\text{CMS} = \frac{\text{CSS}}{k-1} = \text{mean sum of squares due to class/group},$$

$$\text{ErMS} = \frac{\text{ErSS}}{n-k} = \text{mean sum of squares due to error}.$$

Table 10.2 ANOVA table for one-way classification of data

Sources of variation (SOV)	d.f.	SS	MS	F-ratio
Class/group	$k-1$	CSS	MSC	MSC/ErMS
Error	$n-k$	ErSS	ErMS	
Total	$n-1$	TSS		

Since e_{ij} are independently and normally distributed with zero mean and common variance σ^2, the test statistic under H_0 is given by

$$F = \left(\frac{\text{CSS}}{\sigma^2}\cdot\frac{1}{k-1}\right) \div \left(\frac{\text{ErSS}}{\sigma^2}\cdot\frac{1}{n-k}\right) = \frac{\text{CMS}}{\text{ErMS}},$$

which has an F distribution with $(k-1)$ and $(n-k)$ d.f. If the calculated value of $F > F_{\alpha;k-1,n-k}$, then H_0 is rejected; otherwise, it is accepted.

So we have Table 10.2 as the following ANOVA table.

For all practical purposes, the various sums of squares are calculated using the following short-cut formulae:

$$G = \text{Grand total} = \sum_{i=1}^{k}\sum_{j=1}^{n_i} y_{ij};$$

$$\text{Correction factor (CF)} = n\bar{y}^2 = \frac{G^2}{n}, \text{ where } n = \sum_{i-1}^{k} n_i;$$

Total sum of squares (TSS)

$$= \sum_i\sum_j (y_{ij} - \bar{y})^2$$

$$= \sum_i\sum_j y_{ij}^2 - n\bar{y}^2 = \sum_i\sum_j y_{ij}^2 - \text{CF}$$

Class sum of squares (CSS)

$$= \sum_i n_i\hat{\alpha}_i^2 = \sum n_i(\bar{y}_i - \bar{y})^2 = \sum_i n_i\bar{y}_i^2 - n\bar{y}^2$$

$$= \sum_i n_i\left(\frac{\sum_j y_{ij}}{n_i}\right)^2 - \text{CF} = \sum_{i=1}^{k}\frac{y_{i0}^2}{n_i} - \text{CF}$$

where $y_{i0} = \sum_j y_{ij}$ is the sum of observations for ith class;

$\therefore$ Error sum of squares (ErSS) = TSS − CSS.

Example 10.1. The following figures give panicle lengths (cm) of 30 hills of 4 local cultivars of rice. Test (1) whether the panicle lengths of all the four varieties are equal or not and (2) the variety having the best panicle length.

Variety 1	Variety 2	Variety 3	Variety 4
12.5	13.2	11.5	14.5
13.4	13.2	11.6	14.2
12.6	13.4	11.7	14.3
12.8	12.9	11.8	14.2
12.9	13.7	12.2	14.6
12	13.8	11	15.2
12.2	11.9	15.6	
13		15.5	
12.4			

Solution. We are to test whether all the varieties are equal or not w.r.t. the panicle length, that is, to test the null hypothesis

$H_0: \alpha_1 = \alpha_2 = \alpha_3 = \alpha_4 = 0$ against the
$H_1: \alpha$'s are not all equal,
where α_i is the effect of ith $(i = 1,2,3,4)$ variety.

This is a fixed effect model and can be written as

$$y_{ij} = \mu + \alpha_i + e_{ij}.$$

From the above information, we calculate the following quantities:

Variety	Variety 1	Variety 2	Variety 3	Variety 4
Total(y_{i0})	113.8	80.2	97.2	102.6

$$\text{GT} = 12.5 + 13.4 + 12.6 + \cdots + 15.2 + 15.6 + 15.5 = 393.8$$

$$\text{CF} = \text{GT}^2/n = 393.8^2/30 = 5169.281$$

$$\text{TSS} = 12.5^2 + 13.4^2 + 12.6^2 + \cdots + 15.2^2 + 15.6^2 + 15.5^2 - \text{CF} = 43.9387$$

$$\text{SS(Variety)} = \frac{113.80^2}{9} + \frac{80.20^2}{6} + \frac{97.2^2}{8} + \frac{118.1^2}{7} - \text{CF} = 26.46597$$

$$\text{ErSS} = \text{TSS} - \text{SS(Variety)} = 5.2686$$

ANOVA for one-way analysis of variance

Source of variation	d.f.	SS	MS	F
Variety (between)	3	26.4659	8.8219	13.1274
Error (within)	26	17.4728	0.6720	
Total	29	43.9387		

The table value of $F_{0.05;3,26} = 2.975$, that is, $F_{\text{cal}} > F_{\text{tab.}}$ So the test is significant at 5% level of significance and we reject the null hypothesis. Thus, we can conclude that the panicle lengths of all the four varieties are not equal.

The next task is to find out which pair of variety differs significantly and which is the best or worst variety. To compare the schools, we calculate the means ($\bar{y}_i$) of the observations of four schools and the means are arranged in decreasing order. Thus, we get

Variety	V4	V2	V1	V3
Mean ($\bar{y}_i$)	14.66	13.37	12.64	12.15

Now we find the critical difference (CD) (also known as the least significant difference (LSD)) value at $\alpha/2$ level of significance, which is calculated as

$$\sqrt{\text{MSE}\left(\frac{1}{r_i}+\frac{1}{r_j}\right)} \times t_{0.025,\text{err.df.}}$$

$$= \sqrt{0.672\left(\frac{1}{r_i}+\frac{1}{r_j}\right)} \times t_{0.025,26.}$$

$$= \sqrt{0.672\left(\frac{1}{r_i}+\frac{1}{r_j}\right)} \times 2.056$$

where r_i and r_j are the number of observations of the two schools in comparison. Thus, for comparing the schools, we have the following critical difference and mean difference values among the schools:

CD/LSD values		Mean difference	
CD(0.05) (variety 1–variety 2)	0.888	Difference between variety 1 and variety 2	0.72
CD(0.05) (variety 1–variety 3)	0.818	Difference between variety 1 and variety 3	0.494
CD(0.05) (variety 1–variety 4)	0.849	Difference between variety 1 and variety 4	2.012
CD(0.05) (variety 2–variety 3)	0.910	Difference between variety 2 and variety 3	1.467
CD(0.05) (variety 2–variety 4)	0.938	Difference between variety 2 and variety 4	1.29
CD(0.05) (variety 3–variety 4)	0.872	Difference between variety 3 and variety 4	2.507

It is found from the above table that not all the values of the mean differences are greater than the respective critical difference values, except the first two pairs. Among the four varieties, variety 4 has significantly the best panicle length (14.657 cm).

The above analysis could be done using the data analysis module of MS Excel, as given below in a stepwise manner.

Step 1: Arrange the data in Ms Excel workout as shown below:

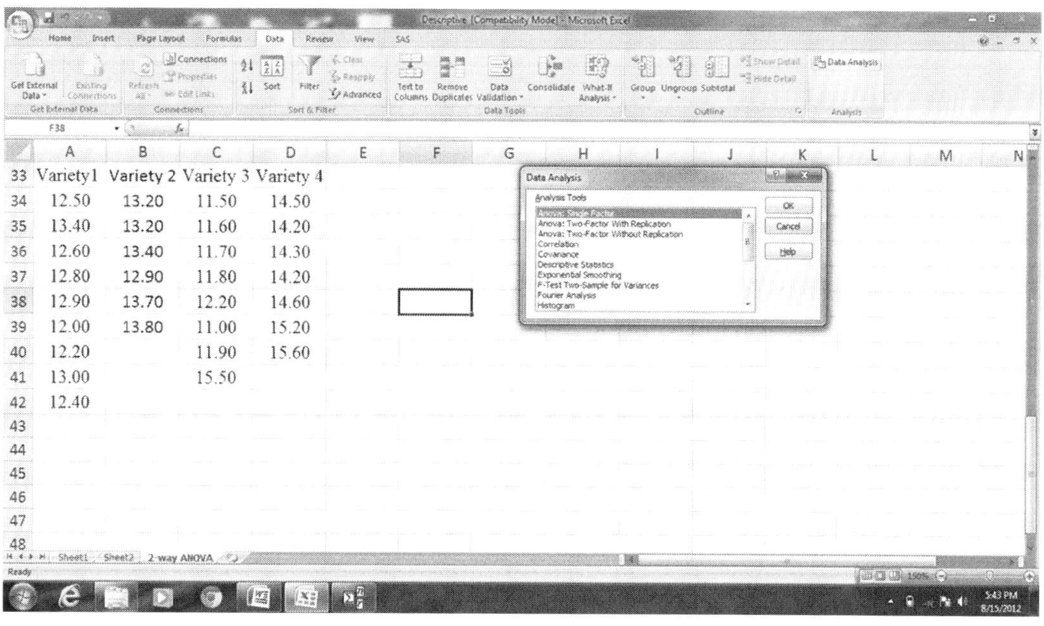

Slide 10.1: Arrangement of one-way classified data in MS Excel work sheet

Step 2: Go to Data menu and then Data Analysis add-in and select ANOVA: Single Factor as shown above.

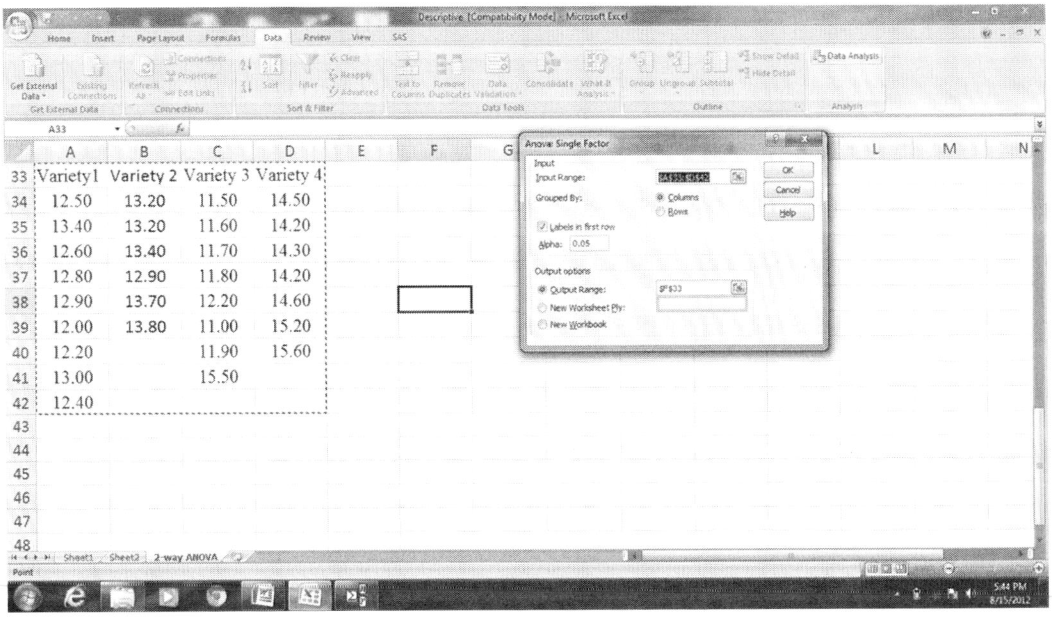

	A	B	C	D
33	Variety1	Variety 2	Variety 3	Variety 4
34	12.50	13.20	11.50	14.50
35	13.40	13.20	11.60	14.20
36	12.60	13.40	11.70	14.30
37	12.80	12.90	11.80	14.20
38	12.90	13.70	12.20	14.60
39	12.00	13.80	11.00	15.20
40	12.20		11.90	15.60
41	13.00		15.50	
42	12.40			

Slide 10.2: Selection of data range and other appropriate fields in analysis of one-way classified data using MS Excel

Step 3: Select the input range and click the appropriate boxes as shown above. Also select a cell when the output is to be written. Click OK to get the following output:

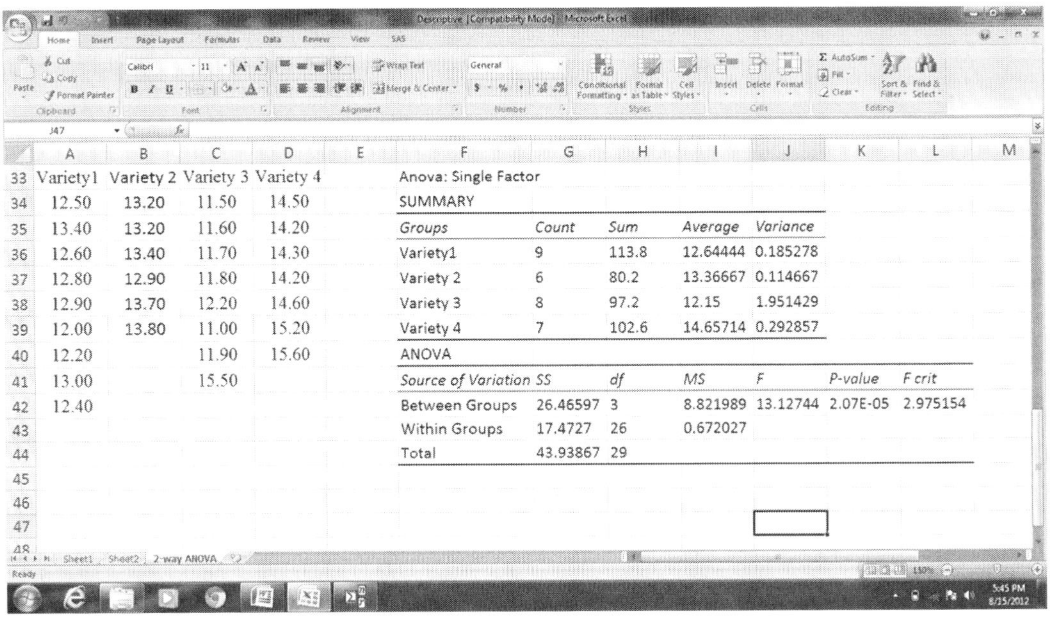

	A	B	C	D
33	Variety1	Variety 2	Variety 3	Variety 4
34	12.50	13.20	11.50	14.50
35	13.40	13.20	11.60	14.20
36	12.60	13.40	11.70	14.30
37	12.80	12.90	11.80	14.20
38	12.90	13.70	12.20	14.60
39	12.00	13.80	11.00	15.20
40	12.20		11.90	15.60
41	13.00		15.50	
42	12.40			

Anova: Single Factor

SUMMARY

Groups	Count	Sum	Average	Variance
Variety1	9	113.8	12.64444	0.185278
Variety 2	6	80.2	13.36667	0.114667
Variety 3	8	97.2	12.15	1.951429
Variety 4	7	102.6	14.65714	0.292857

ANOVA

Source of Variation	SS	df	MS	F	P-value	F crit
Between Groups	26.46597	3	8.821989	13.12744	2.07E-05	2.975154
Within Groups	17.4727	26	0.672027			
Total	43.93867	29				

Slide 10.3: Data along with output of analysis of one-way classified data using MS Excel

Now comparing the critical values with F-value, we are to decide whether the varieties differ significantly or not, that is, whether to accept or reject the null hypothesis. On rejection of the null hypothesis, CD/LSD value is required to be calculated to compare the treatment mean differences among the pairs of means, as discussed above.

10.3 Two-Way ANOVA

Two-way analysis of variance is the most powerful tool used in agriculture, social, educational, business, economics, medical, and other fields of research. In this analysis, instead of considering one factor at a time, one can consider two factors in the same experiment. Two-way analysis of variance is of two types: (1) *two-way analysis of variance with one observation per cell* and (2) *two-way analysis of variance with more than one observation per cell*. In the former case the interaction effect is not possible to be worked out.

Two-Way Analysis of Variance with One Observation per Cell

In two-way analysis of variance with one observation per cell, for each factor there will be the same or different number of classes/groups or levels. Considering the fixed effect model, if there are two factors A and B having m and n groups, respectively, then this can be represented in accordance with Table 10.3.

If we consider the fixed effect model and follow the same procedure of one-way classified data to test the null hypotheses $H_{01}: \alpha_1 = \alpha_2 = \alpha_3 = \cdots \alpha_m = 0$ and $H_{02}: \beta_1 = \beta_2 = \beta_3 = \cdots \beta_n = 0$ for the equality of the effects of different levels of A and of different levels of B, respectively, the appropriate model is

$$\begin{aligned} y_{ij} &= \mu_{ij} + e_{ij} \\ &= \mu + \alpha_i + \beta_j + e_{ij} \\ &= \mu + \alpha_i + \beta_j + e_{ij}, \end{aligned}$$
$$\text{where } i = 1, 2, \ldots, m; \quad j = 1, 2, \ldots, n$$

y_{ij} = value of the observation corresponding to the ith class/group or level of factor A and jth class/group or level of factor B

μ = general mean effect

α_i = additional effect due to ith class/group or level of factor A

β_j = additional effect due to jth class/group of factor B

e_{ij} = errors associated with ith class/group or level of factor A and jth class/group or level of factor B and are i.i.d. $N(0, \sigma^2)$ and $\sum_{i=1}^{m} \alpha_i = \sum_{j=1}^{n} \beta_j = 0$

Table 10.3 Two-way classification of data

	B_1	B_2	......	B_j	......	B_n	Total	Mean
A_1	y_{11}	y_{12}	......	y_{1j}	......	y_{1n}	y_{1o}	$\bar{y}_{10}$
A_2	y_{21}	y_{22}	......	y_{2j}	......	y_{2n}	y_{20}	$\bar{y}_{20}$
:	:	:		:		:	:	:
:	:	:		:		:	:	:
A_i	y_{i1}	y_{i2}	......	y_{ij}	......	y_{in}	y_{i0}	$\bar{y}_{i0}$
:	:	:		:		:	:	:
:	:	:		:		:	:	:
A_m	y_{m1}	y_{m2}	......	y_{mj}	......	y_{mn}	y_{m0}	$\bar{y}_{m0}$
Total	y_{01}	y_{02}	......	y_{0j}	......	y_{0n}	y_{00}	
Mean	$\bar{y}_{01}$	$\bar{y}_{02}$	......	$\bar{y}_{0j}$	......	$\bar{y}_{0n}$		$\bar{y}_{00}$

y_{ij} is the observation due to ith group of factor A and jth group of factor B.

and

$\hat{\mu} = \bar{y}_{00}$,
$\hat{\alpha}_i = \bar{y}_{i0} - \bar{y}_{00}$,
$\hat{\beta}_j = \bar{y}_{0j} - \bar{y}_{00}$,
with $\bar{y}_{00} = \frac{1}{mn}\sum_{i,j} y_{ij}, \bar{y}_{i0} = \frac{1}{n}\sum_j y_{ij}, \bar{y}_{0j} = \frac{1}{m}\sum_i y_{ij}$.

Thus, $y_{ij} = \bar{y}_{00} + (\bar{y}_{i0} - \bar{y}_{00}) + \left(\bar{y}_{0j} - \bar{y}_{00}\right) + \left(y_{ij} - \bar{y}_{i0} - \bar{y}_{0j} + \bar{y}_{00}\right)$. (The product terms vanish because of assumption of independence.)

$$\sum_i\sum_j (y_{ij} - \bar{y}_{00})^2 = n\sum_i (\bar{y}_{i0} - \bar{y}_{00})^2 + m\sum_j \left(\bar{y}_{0j} - \bar{y}_{00}\right) + \sum_i\sum_j \left(y_{ij} - \bar{y}_{i0} - \bar{y}_{0j} + \bar{y}_{00}\right)^2$$

or SS(total) = SS(factor A) + SS(factor B) + SS(error)
or TSS = SS(A) + SS(B) + ErSS.

The corresponding partitioning of the total d.f. is as follows:

$$mn - 1 = (m - 1) + (n - 1) + (m - 1)(n - 1).$$

Hence, the test statistic under H_{01} is given by

$$F = \left[\frac{\text{SS(A)}}{\sigma^2} \cdot \frac{1}{m-1}\right] \div \left[\frac{\text{ErSS}}{\sigma^2} \cdot \frac{1}{(m-1)(n-1)}\right] = \frac{\text{MS(A)}}{\text{ErMS}}$$

which follows F distribution with $[(m - 1), (m - 1)(n - 1)]$ d.f.

Thus, the null hypothesis H_{01} is rejected at α level of significance if

$$F = \frac{\text{MS(A)}}{\text{ErMS}} > F_{\alpha;(m-1),(m-1)(n-1)}\text{d.f.}$$

Similarly the test statistics under H_{02} is given by

$$F = \frac{\text{MS(B)}}{\text{ErMS}} \text{ with } \alpha; (m-1), (m-1)(n-1) \text{ d.f.}$$

For practical purposes the various sums of squares are calculated using the following formulae:

$$G = \text{grand total} = \sum_{i=1}^{m}\sum_{j=1}^{n} y_{ij};$$

$$\text{Correction factor (CF)} = \frac{G^2}{mn};$$

$$\text{Total sum of squares (TSS)} = \sum_i\sum_j (y_{ij} - \bar{y}_{00})^2 = \sum_i\sum_j y_{ij}^2 - \text{CF};$$

Sum of squares (A) = SS (A)

$$= n\sum_i (\bar{y}_{i0} - \bar{y}_{00})^2 = n\left[\sum_i \bar{y}_{i0}^2 - m\bar{y}_{00}^2\right] = n\sum_i \left(\frac{\sum_{j=1}^{n} \bar{y}_{ij}}{n}\right)^2 - nm\bar{y}_{00}^2 = \frac{1}{n}\sum_i y_{i0}^2 - \text{CF},$$

where, $y_{i0} = \sum_{j=1}^{n} y_{ij}$ is the sum of observations for ith level of factor A.

Sum of squares (B) = SS (B)

$$= m\sum_j \left(\bar{y}_{0j} - \bar{y}_{00}\right)^2 = \frac{1}{m}\sum_j y_{0j}^2 - \text{CF},$$

where $y_{0j} = \sum_{i=1}^{m} y_{ij}$ is the sum of observations for j^{th} level of factor B;

ErSS = TSS − SS (A) − SS (B)

Dividing this SSs by their respective degrees of freedom, we will get the corresponding mean sum of squares, that is, mean sum of squares due to classes and error mean sum of squares.

We have Table 10.4 as the ANOVA table for two-way classification with one observation per cell.

Table 10.4 ANOVA table for two-way classification of data

SOV	d.f.	SS	MS	F
Factor A	$m-1$	SS(A)	AMS = SS(A)/($m-1$)	AMS/ErMS
Factor B	$n-1$	SS(B)	BMS = SS(B)/($n-1$)	BMS/ErMS
Error	$(m-1)(n-1)$	ErSS	ErMS = ErSS/$(m-1)(n-1)$	
Total	$mn-1$	TSS		

Example 10.2. The following table gives the milk produced (liter/day) by four breeds of cows under five different feeds. Analyze the data to find out the best feed and best breed of cow with respect to production of milk.

	Breed			
Feed	Breed A	Breed B	Breed C	Breed D
Feed 1	17	10.6	47	30
Feed 2	19	10.8	46	28
Feed 3	21	10.9	45.5	29
Feed 4	23	11.2	48	25
Feed 5	25	11.3	45	26

Solution. Under the given condition we are to analyze the data per the two-way ANOVA with one observation per cell to test the null hypotheses:

$H_{01}: \alpha_1 = \alpha_2 = \alpha_3 = \alpha_4 = \alpha_5 = 0$ against

$H_{11}: \alpha$'s are not all equal and

$H_{02}: \beta_1 = \beta_2 = \beta_3 = \beta_4 = 0$ against

$H_{12}: \beta$'s are not all equal.

The appropriate model to test the above is $y_{ij} = \mu + \alpha_i + \beta_j + e_{ij}$, where $i = 5, j = 4$

y_{ij} = value of the observation corresponding to the ith feed and jth breed

μ = general mean effect

α_i = additional effect due to the ith feed

β_j = additional effect due to the jth breed

e_{ij} = errors associated with the ith feed and jth breed

For the purpose let us frame the following table:

Total number of observations = $N = mn = 20$,

	Breed					
Feed	Breed A	Breed B	Breed C	Breed D	Total	Mean
Feed 1	17	10.6	47	30	104.6	26.15
Feed 2	19	10.8	46	28	103.8	25.95
Feed 3	21	10.9	45.5	29	106.4	26.6
Feed 4	23	11.2	48	25	107.2	26.8
Feed 5	25	11.3	45	26	107.3	26.825
Total (y_{0j})	105	54.8	231.5	138	529.3	26.465
Mean $\left(\bar{y}_{0j}\right)$	21	10.96	46.3	27.6		

$$G = 17.0 + 10.6 + 47.0 + \cdots + 11.3 + 45 + 26 = 529.3$$

$$\mathrm{CF} = G^2/\mathrm{N} = 529.3^2/20 = 14007.92$$

$$\mathrm{TSS} = 17.0^2 + 10.6^2 + 47.0^2 + \cdots + 11.3^2 + 45^2 + 26^2 - \mathrm{CF} = 3388.27$$

$$\mathrm{FSS} = \mathrm{SS(Feed)} = \frac{104.6^2}{4} + \frac{103.8^2}{4} + \frac{106.4^2}{4} + \frac{107.2^2}{4} + \frac{107.3^2}{4} - 14007.92 = 2.498$$

$$\mathrm{BrSS} = \mathrm{SS(Breed)} = \frac{105^2}{5} + \frac{54.8^2}{5} + \frac{231.5^2}{5} + \frac{138^2}{5} - 14007.92 = 3324.934$$

$$\mathrm{ErSS} = \mathrm{TSS} - \mathrm{FSS} - \mathrm{BrSS} = 60.834$$

ANOVA table

SOV	d.f.	SS	MS	F
Feed	4	2.498	0.625	0.123
Breed	3	3324.934	1108.311	218.623
Error	12	60.834	5.069	
Total	19	3388.27		

Let the level of significance be $\alpha = 0.05$.

The table value of $F_{0.05;4,12} = 3.26$, that is, $F_{\text{cal}} < F_{\text{tab}}$. Therefore, the test is nonsignificant and we cannot reject the null hypothesis H_{01}. We conclude that there exists no significant difference among the effects of feeds with respect to milk production.

Next, let us check whether there exists any significant difference between milk production and different breeds of cows. The table value of $F_{0.05;3,12} = 3.49$, that is, $F_{\text{cal}} > F_{\text{tab}}$. Therefore, the test is significant and we reject the null hypothesis H_{02}, meaning there exists a significant difference among the breeds of cows w.r.t and milk production. Now our task is to find out which breed is significantly different from other and which breed is the best or worst with respect to milk production. To compare the breeds we calculate the critical difference (least significant difference) value, which is given as

$$\text{CD/LSD} = \sqrt{\frac{2\text{ErMS}}{m}} \times t_{0.025,\ \text{err.df.}}$$

$$= \sqrt{\frac{2 \times 5.07}{5}} \times t_{0.05,12.}$$

$$= \sqrt{\frac{2 \times 5.07}{5}} \times 2.179 = 3.103.$$

Arrangement of the breeds according to their milk production is given below:

$\bar{A}$	$\bar{B}$	$\bar{C}$	$\bar{D}$	$\Rightarrow \bar{C} > \bar{D} > \bar{A} > \bar{B}$
21	10.96	46.3	27.6	

The difference between pairs of means is greater than the CD/LSD value. So we conclude that all the breeds are significantly different from each other and that breed C has the highest milk production. Hence, breed C is the best breed with respect to milk production per day.

Using MS Excel program, in the following slides, the above calculation has been presented:

Step 1: Select "Anova: Two-Factor Without Replication," as shown below.

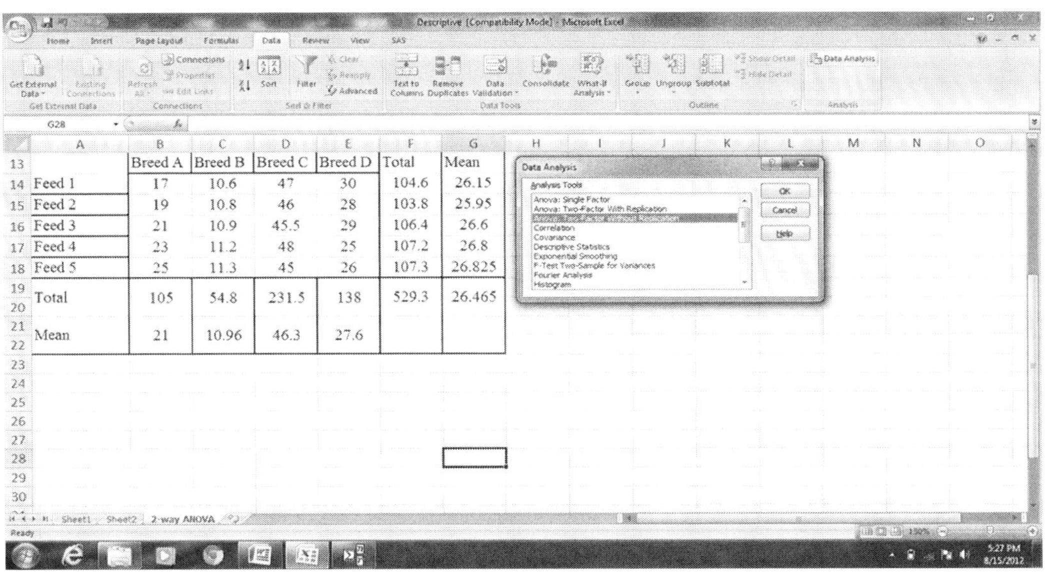

Slide 10.4: Arrangement of two-way classified data in MS Excel work sheet

Step 2: Select the input range and output range shown above. Click on other appropriate boxes followed by OK to get the following output, as shown below:

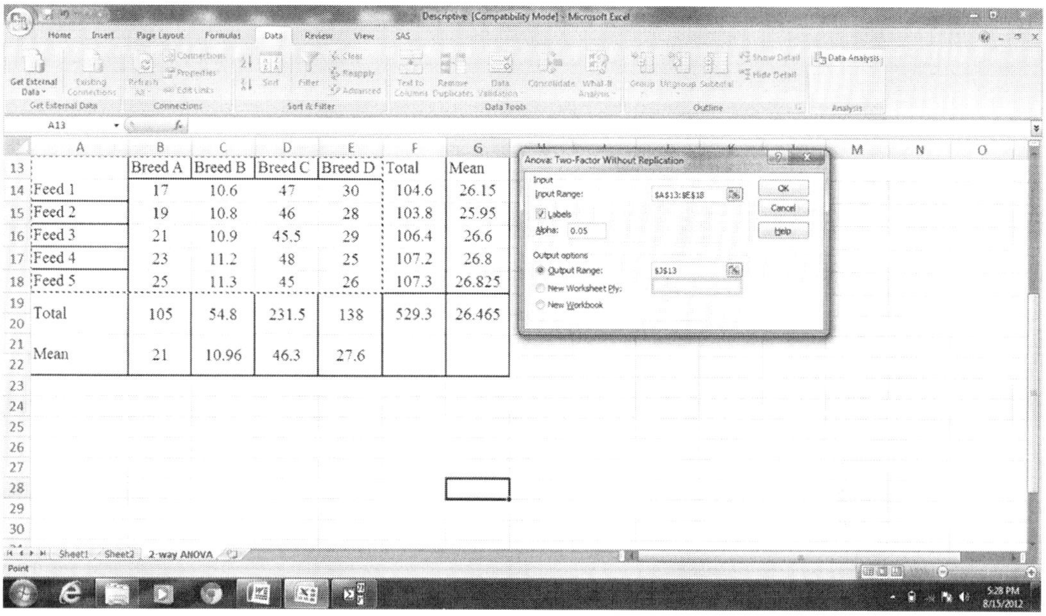

Slide 10.5: Step 2 showing the selection of data range and other appropriate fields in the analysis of two-way classified data using MS Excel

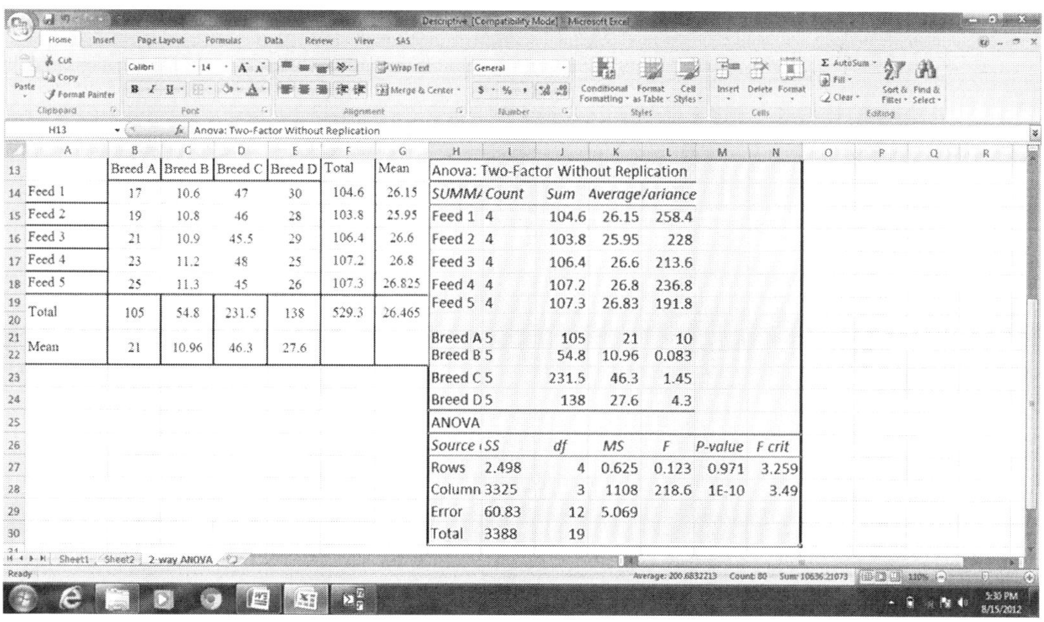

Slide 10.6: Step 3 showing the data along with output of analysis of two-way classified data using MS Excel

From the above calculation we are to take decision and calculate CD/LSD values required using the formula discussed during manual calculation. There are certain packages like SPSS where the pairwise mean comparison also comes out as output.

Two-Way Analysis of Variance with More Than One Observations per Cell

Working out the interaction effects of more than one factor from a single experiment is one of the major objectives of the two-way analysis of variance with more than one observation per cell. To test

$$H_{01} : \alpha_1 = \alpha_2 = \alpha_3 = \cdots \alpha_m = 0,$$
$$H_{02} : \beta_1 = \beta_2 = \beta_3 = \cdots \beta_n = 0,$$
$$H_{03} : \gamma_{ij} = 0 \text{ for all } i \text{ and } j,$$

the appropriate model for the purpose would be

$$y_{ijk} = \mu + \alpha_i + \beta_j + \gamma_{ij} + e_{ijk},$$

where $i = 1, 2, \ldots, m$; $j = 1, 2, \ldots, n$; $k = 1, 2, \ldots, l$

μ = general effect

α_i = additional effect due to the ith group of factor A

β_j = additional effect due to the jth group of factor B

γ_{ij} = interaction effect due to the ith group of factor A and the jth group of factor B

e_{ijk} = errors that are associated with the kth observation of the ith class of factor A and the jth class of factor B and are i.i.d. $N(0, \sigma^2)$; $\sum_i \alpha_i = \sum_j \beta_j = \sum_i \gamma_{ij} = 0,$

$$\text{where } \hat{\mu} = \bar{y}_{000}, \hat{\alpha}_i = \bar{y}_{i00} - \bar{y}_{000}, \hat{\beta}_j = \bar{y}_{0j0} - \bar{y}_{000}, \text{ and } \hat{\gamma}_{ij} = \bar{y}_{ij0} - \bar{y}_{i00} - \bar{y}_{0j0} + \bar{y}_{000}.$$

Thus, the linear model becomes

$$y_{ijk} = \bar{y}_{000} + (\bar{y}_{i00} - \bar{y}_{000}) + \left(\bar{y}_{0j0} - \bar{y}_{000}\right) + \left(\bar{y}_{ij0} - \bar{y}_{i00} - \bar{y}_{0j0} + \bar{y}_{000}\right) + \left(y_{ijk} - y_{ij0}\right).$$

Transferring y_{000} to the left, squaring both the sides, and summing over i, j, k, we get

$$\sum_i^m \sum_j^n \sum_k^l (y_{ijk} - \bar{y}_{000})^2 = nl \sum_i (\bar{y}_{i00} - \bar{y}_{000})^2 + ml \sum_j \left(\bar{y}_{0j0} - \bar{y}_{000}\right)^2 + l \sum_i \sum_j \left(\bar{y}_{ij0} - \bar{y}_{i00} - \bar{y}_{0j0} + \bar{y}_{000}\right)^2 + \sum_i \sum_j \sum_k \left(y_{ijk} - \bar{y}_{ij0}\right)^2$$

(Other product terms vanish because of the assumptions as usual.)

or SS(total) = SS(factor A) + SS(factor B) + SS(factor AB) + SS(error)

or TSS = SS(A) + SS(B) + SS(AB) + ErSS.

The corresponding partitioning of the total d.f. is as follows:

$$Lmn - 1 = (m - 1) + (n - 1) + (m - 1) \times (n - 1) + mn(l - 1),$$

where $\bar{y}_{000} = \frac{1}{mnl} \sum_i \sum_j \sum_k y_{ijk}$ = mean of all observations,

$\bar{y}_{i00} = \frac{1}{nl} \sum_j \sum_k y_{ijk}$ = mean of ith level of A,

$\bar{y}_{0j0} = \frac{1}{ml} \sum_i \sum_k y_{ijk}$ = mean of jth level of B,

$\bar{y}_{ij0} = \frac{1}{l} \sum_k y_{ijk}$ = mean of the observations for ith level of A and jth level of B.

Dividing the sum of squares by their corresponding d.f., we have the following:

Mean sum of squares due to factor A $= \frac{\text{SS(A)}}{m - 1} = \text{MS(A)},$

Mean sum of squares due to factor B $= \frac{\text{SS(B)}}{n - 1} = \text{MS(B)},$

Mean sum of squares due to factor AB $= \frac{\text{SS(AB)}}{(m - 1)(n - 1)} = \text{MS(AB)},$

Mean sum of squares due to error $= \frac{\text{ErSS}}{mn(1 - 1)} = \text{ErMS}.$

Table 10.5 ANOVA table for two-way classified data with m $l(>1)$ observations per cell

SOV	d.f.	SS	MS	F
Factor A	$m-1$	SS(A)	AMS = SS(A)/$(m-1)$	AMS/ErMS
Factor B	$n-1$	SS(B)	BMS = SS(B)/$(n-1)$	BMS/ErMS
Interaction (A × B)	$(m-1)(n-1)$	SS(AB)	ABMS = SS(AB)/ $(m-1)(n-1)$	ABMS/ErMS
Error	By subtraction = $mn(l-1)$	ErSS	ErMS = ErSS/$mn(l-1)$	
Total	$mnl-1$	TSS		

The mean sum of squares due to error (ErMS) always provides an unbiased estimate of σ^2, but MS(A), MS(B), and MS(AB) provide unbiased estimates of σ^2 under H_{01}, H_{02}, and H_{03}, respectively. They are also independent. The test statistics under H_{01}, H_{02}, and H_{03} are given by

$$F_A = \frac{\text{MS(A)}}{\text{ErMS}} \sim F_{m-1,mn(l-1)},$$
$$F_B = \frac{\text{MS(B)}}{\text{ErMS}} \sim F_{n-1,mn(l-1)},$$
$$F_{AB} = \frac{\text{MS(AB)}}{\text{ErMS}} \sim F_{(m-1)(n-1),mn(l-1)}, \text{ respectively.}$$

Thus, H_{03} is rejected at α level of significance if $F_{AB} > F_{\alpha;(m-1)(n-1),mn(l-1)}$; otherwise, it is accepted. If H_{03} is accepted, that is, interaction is absent, the tests for H_{01} and H_{02} can be performed. H_{01} is rejected

if $F_A > F_{\alpha;m-1,mn(l-1)}$. Similarly H_{02} is rejected if$F_B > F_{\alpha;n-1,mn(l-1)}$.

If H_{03} is rejected, that is, interaction is present, then we shall have to compare for each level of B at the different levels of A and for each level of A at the different levels of B (Table 10.5).

For practical purposes, different sums of squares are calculated using the following formulae:

$$\text{Grand total} = G = \sum_i^m \sum_j^n \sum_k^l y_{ijk};$$

$$\text{Correction factor} = \text{CF} = \frac{G^2}{mnl};$$

Treatment sum of squares = TrSS

$$= \sum_i^m \sum_j^n \sum_k^l (y_{ijk} - \bar{y}_{000})^2$$
$$= \sum_i^m \sum_j^n \sum_k^l (y_{ijk})^2 - \text{CF};$$

Sum of squares due to A = SS(A)

$$= nl\sum_i (\bar{y}_{i00} - \bar{y}_{000})^2 = nl\left[\sum_i \bar{y}_{i00}^2 - m\bar{y}_{000}^2\right]$$
$$= nl\sum_i \left(\frac{\sum_j\sum_k y_{ijk}}{nl}\right)^2 - \text{CF} = \frac{1}{nl}\sum_i y_{i00}^2 - \text{CF};$$

Sum of squares due to B = SS(B)

$$= ml\sum_j (\bar{y}_{0j0} - \bar{y}_{000})^2 = \frac{1}{ml}\sum_i y_{0j0}^2 - \text{CF};$$

Sum of squares due to (AB)

$$= \sum_i^m \sum_j^n \frac{y_{ij0}^2}{1} - \text{CF} - \text{SS (A)} - \text{SS (B)};$$

$$\therefore \text{ErSS} = \text{TSS} - \text{SS(A)} - \text{SS(B)} - \text{SS(AB)},$$

where y_{i00} = sum of the observations for ith level of A,
y_{0j0} = sum of the observations for jth level of B,
y_{ij0} = sum of the observations for ith level of A and jth level of B.

Example 10.3. The following table gives the yield (q/ha) of four different varieties of paddy in response to three different doses of nitrogen. Analyze the data to show whether:

(a) There exist significant differences in yield of four rice varieties.
(b) Three doses of nitrogen are significantly different with respect to production.
(c) The varieties have performed equally under different doses of nitrogen or not.

Doses of N	Variety 1	Variety 2	Variety 3	Variety 4
N1	13.0	17.0	15.5	20.5
	13.4	17.5	15.6	20.8
	13.5	17.4	15.5	20.4
N2	14.0	18.0	16.2	24.5
	14.2	18.5	16.1	25.6
	14.6	18.9	16.3	25.0
N3	25.0	22.0	18.5	32.0
	24.8	22.5	18.6	32.5
	24.3	22.3	18.7	32.0

Solution. The problem can be visualized as the problem of two-way analysis of variance with three observations per cell. The linear model is

$y_{ijk} = \mu + \alpha_i + \beta_j + \gamma_{ij} + e_{ijk}$,

where

α_i is the effect of ith level of nitrogen, $i = 3$;

β_j is the effect of jth level of variety, $j = 4$;

γ_{ij} is the interaction effect of ith level of nitrogen and jth level of variety.

We want to test the following null hypotheses:

$H_{01} : \alpha_1 = \alpha_2 = \alpha_3 = 0$ against H_{11} : α's are not all equal,

$H_{02} : \beta_1 = \beta_2 = \beta_3 = \beta_4 = 0$ against H_{12} : β's are not all equal,

$H_{03} : \gamma_{ij}$'s $= 0$ for all i, j against H_{13} : γ_{ij}'s are not all equal.

Let the level of significance be $\alpha = 0.05$.

Let us construct the following table of totals (y_{ij}):

Nitrogen	Variety 1	Variety 2	Variety 3	Variety 4	Total (y_{i00})	Mean
N_1	39.90	51.90	46.60	61.70	200.10	16.675
N_2	42.80	55.40	48.60	75.10	221.90	18.492
N_3	74.10	66.80	55.80	96.50	293.20	24.433
Total(y_{0j0})	156.800	174.100	151.000	233.300	715.200	
Mean	17.422	19.344	16.778	25.922		19.867

Total number of observations $= mnl = 3 \times 4 \times 3 = 36 = N$

$G = 13.0 + 17 + 15.5 + \cdots + 22.3 + 18.7 + 32. = 715.2,$

$\mathrm{CF} = \mathrm{GT}^2/N = 715.2^2/36 = 14208.64,$

$\mathrm{TSS} = 13.0^2 + 17^2 + 15.5^2 + \cdots + 22.3^2 + 18.7^2 + 32^2 - \mathrm{CF} = 975.52,$

$$\mathrm{NSS} = \frac{200.1^2}{12} + \frac{221.9^2}{12} + \frac{293.20^2}{12} - \mathrm{CF} = 395.182,$$

$$\mathrm{VSS} = \frac{156.8^2}{9} + \frac{174.10^2}{9} + \frac{151.00^2}{9} - \mathrm{CF} = 472.131,$$

$\mathrm{SS\ (N \times V)} = 1/3\left[39.9^2 + 51.9^2 + 46.6^2 + \cdots + 66.80^2 + 55.80^2 + 96.5^2\right] - \mathrm{CF} - \mathrm{NSS} - \mathrm{VSS} = 106.04,$

$\mathrm{ErSS} = \mathrm{TSS} - \mathrm{NSS} - \mathrm{VSS} - \mathrm{SS\ (N \times V)} = 975.52 - 395.182 - 472.131 - 106.04 = 2.17.$

ANOVA table for two-way analysis of variance with three observations per cell

SOV	d.f.	SS	MS	F
Variety	3	472.131	157.377	1743.253
Nitrogen	2	395.182	197.591	2188.698
Interaction (V × N)	4	106.041	17.673	195.767
Error	24	2.167	0.090	
Total	26	975.520		

Let the level of significance be $\alpha = 0.05$.

The table value of $F_{0.05;2,24} = 3.40$, $F_{0.05;3,24} = 3.01$, and $F_{0.05;6,24} = 2.51$, that is, $F_{\text{cal}} > F_{\text{tab,}}$ in all the cases, that is, nitrogen, variety, and nitrogen × variety interaction. So the tests are significant and we reject the null hypotheses and conclude that there exists significant difference among the effects of doses of nitrogen, varieties, and their interactions with respect to the yield of paddy.

Now, we are interested in identifying the best nitrogen, the best variety, and the nitrogen × variety combination providing the best yield.

To accomplish this task, we calculate the critical difference (CD)/LSD values for nitrogen, variety, and interaction separately as follows:

$$\begin{aligned}\text{CD}_{0.05}(\text{Nitrogen}) &= \sqrt{\frac{2\text{MSE}}{l \times v}} \times t_{0.025,\,\text{err.d.f.}}\\ &= \sqrt{\frac{2 \times 0.073}{3 \times 4}} \times t_{0.025,18.}\\ &= \sqrt{\frac{2 \times 0.090}{3 \times 4}} \times 2.064 = 0.252,\end{aligned}$$

$$\begin{aligned}\text{CD}_{0.05}(\text{Variety}) &= \sqrt{\frac{2\text{MSE}}{r \times \text{t}}} \times t_{0.025,\text{err.d.f.}}\\ &= \sqrt{\frac{2 \times 0.09}{3 \times 3}} \times t_{0.025,18.}\\ &= \sqrt{\frac{2 \times 0.09}{3 \times 3}} \times 2.064 = 0.292,\end{aligned}$$

$$\begin{aligned}\text{CD}_{0.05}(\text{Tr.} \times \text{V}) &= \sqrt{\frac{2\text{MSE}}{r}} \times t_{0.025,\,\text{err.d.f.}}\\ &= \sqrt{\frac{2 \times 0.090}{3}} \times t_{0.025,18.}\\ &= \sqrt{\frac{2 \times 0.090}{3}} \times 2.064 = 0.505.\end{aligned}$$

Comparing the three nitrogen means provided in the table of treatment totals, one can find that the difference between any two nitrogen means is greater than the corresponding critical difference value 0.252. So we can conclude that nitrogen doses significantly differ among themselves with respect to yield of paddy and the nitrogen dose giving significantly the highest yield (24.433q/ha) is treatment N_3.

As far as the response of four varieties of paddy is concerned, it is found that the difference between any two variety means is greater than the critical difference value 0.292. Thus, the varieties differ significantly among themselves with respect to yield. Among the varieties, variety 4 has produced significantly the highest yield, so this is the best variety.

After identifying the best variety and the best treatment separately, our task is now to identify the best nitrogen × variety combination for yield of paddy from the above information. Let us construct the following table of average interaction values of nitrogen and variety.

Mean table of N × V

Nitrogen	Variety 1	Variety 2	Variety 3	Variety 4
N1	13.300	17.300	15.533	20.567
N2	14.267	18.467	16.200	25.033
N3	24.700	22.267	18.600	32.167

From the above table, it can be seen that the difference between any two combinations is more than the corresponding critical difference value for nitrogen × variety interaction (0.505), so all the combinations produce significantly different yields of paddy. Among the nitrogen × variety combinations, it is noticed that variety 4–nitrogen 3 combination produced the highest yield; hence, this is the best combination, followed by V_4N_2, which is at par with V1N3. Thus, from the analysis we draw the following conclusions:

1. The treatments differ significantly among themselves and the best treatment is treatment T_3.
2. The varieties differ significantly among themselves and the best variety is V_2.
3. Treatment T_3 along with variety V_1, produces significantly higher yield than any other combination.

Let us try to know how the above analysis of variance could be done with MS Excel program available in majority of computers having MS Office.

Step 1: Go to Data Analysis of Tool menu.

Step 2: Select Anova: Two-Factor with Replication.

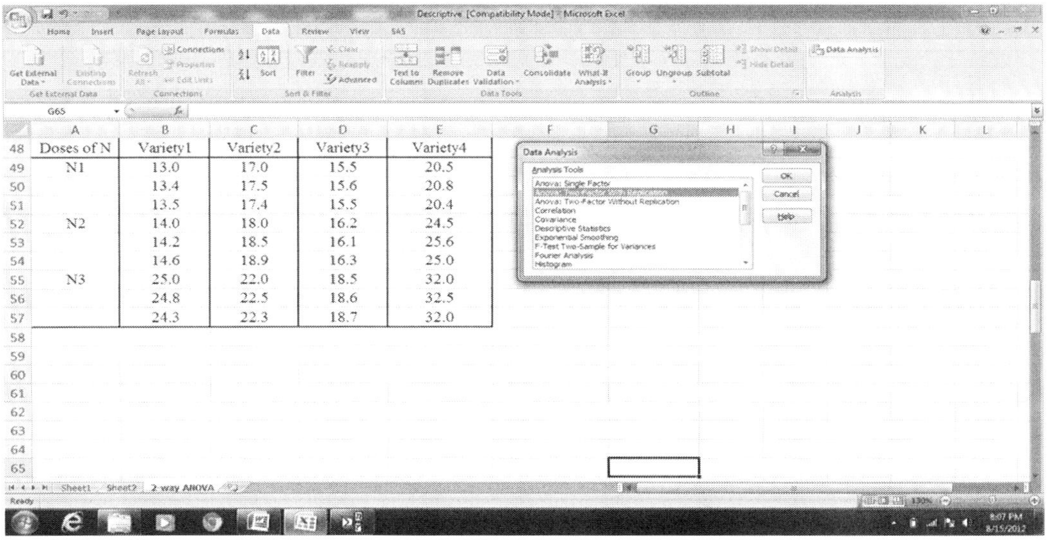

Slide 10.7: Arrangement of two-way classified data, with more than one observation per cell in MS Excel work sheet

Step 3: Select the options as shown below.

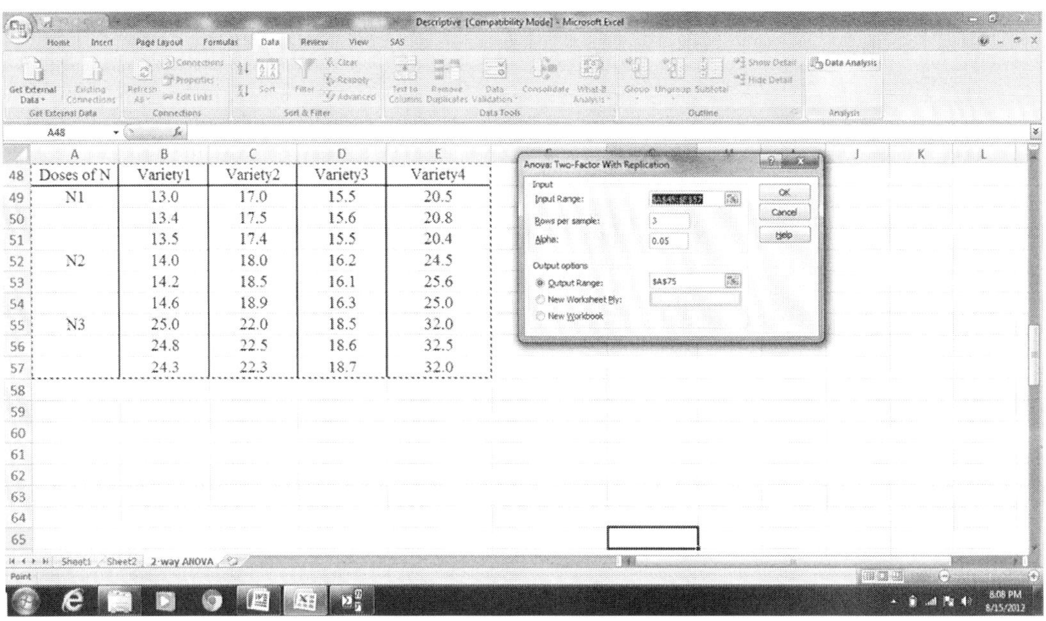

Slide 10.8: Step 2, showing the selection of data range and other appropriate fields in the analysis of two-way classified data, with more than one observation per cell using MS Excel

Table 10.6 Output of two-way classified data with m $l(>1)$ observations per cell using MS Excel

Anova: two-factor with replication						
Summary	Variety 1	Variety 2	Variety 3	Variety 4	Total	
N1						
Count	3	3	3	3	12	
Sum	39.900	51.900	46.600	61.700	200.100	
Average	13.300	17.300	15.533	20.567	16.675	
Variance	0.070	0.070	0.003	0.043	7.733	
N2						
Count	3	3	3	3	12	
Sum	42.800	55.400	48.600	75.100	221.900	
Average	14.267	18.467	16.200	25.033	18.492	
Variance	0.093	0.203	0.010	0.303	18.083	
N3						
Count	3	3	3	3	12	
Sum	74.100	66.800	55.800	96.500	293.200	
Average	24.700	22.267	18.600	32.167	24.433	
Variance	0.130	0.063	0.010	0.083	26.942	
Total						
Count	9	9	9	9		
Sum	156.800	174.100	151.000	233.300		
Average	17.422	19.344	16.778	25.922		
Variance	30.042	5.143	1.957	25.782		
ANOVA						
Source of variation	SS	d.f.	MS	*F*	*P* value	*F* crit
Sample(nitrogen)	395.182	2	197.591	2188.698	0.000	3.403
Columns (variety)	472.131	3	157.377	1743.253	0.000	3.009
Interaction (V × N)	106.041	6	17.673	195.767	0.000	2.508
Within	2.167	24	0.090			
Total	975.520	35				

The ultimate output will be as follows (Table 10.6):

Step 4: Get the result as given above. Here samples are the nitrogen, and columns are the varieties.

Step 5: From the above table we see that both the factors and their interaction effects are significant at $p = 0.05$. So we need to calculate the CD/LSD values to identify the pair of nitrogen means and pair of variety means which are significantly different from each other and also to identify the best type and the best combination of nitrogen and variety to produce the highest yield. Corresponding CD values are calculated per formulae given above.

10.4 Violation of Assumptions in ANOVA

Analysis of variance works under the assumption that the effects (treatments and the environmental) are additive in nature and experimental errors are i.i.d. $N(0, \sigma^2)$; failures to meet these assumptions adversely affect both the sensitivity of F- and t-tests and the level of significance. Thus, the data which are found to be drastically deviated from one or more of the assumptions are required to be corrected before taking up the analysis of variance. Data transformation is by far the most widely used procedure for data violating the assumptions of analysis of variance.

10.4.1 Data Transformation

Among the different types of transformation generally used to make the data corrected for analysis of variance, *logarithmic transformation, square root transformation, and angular transformation* are widely used.

10.4.1.1 Logarithmic Transformation

The number of plant insects, number of egg mass per unit area, number of larvae per unit area, etc., are typical examples wherein variance is proportional to the mean and logarithmic transformation can be used effectively. The procedure is to take simply the logarithm of each and every observation and carry out the analysis of variance following the usual procedure with the transformed data. However, if in the data set small values (less than 10) are recorded, then instead of taking $\log(x)$, it will be better to take $\log(x + 1)$. The final results or inference should be drawn on the basis of transformed mean values and on the basis of calculations made through transformed data. However, while presenting the mean table, it will be appropriate to recalculate the means by taking the antilog of the transformed data. In practice, the treatment means are calculated from the original data because of simplicity of calculations, but statistically the procedure of converting transformed mean to original form is more appropriate. If there is a mismatch in the two procedures, then the procedure of converting the transformed mean with the help of the antilog is preferred over the other procedure.

Example 10.4. The following data give the number of fruit set after the application of eight different hormonal treatments in jackfruit. Use suitable transformation and analyze the data using the suitable model.

Hormone	Replication 1	Replication 2	Replication 3	Replication 4
H1	2	9	8	2
H2	4	5	6	3
H3	5	7	12	6
H4	3	11	8	13
H5	19	26	15	22
H6	24	34	24	27
H7	3	1	1	5
H8	15	11	13	15

Solution. Because this experiment is related with count data, violation of the assumption of ANOVA is suspected and the analysis of the data is required to be made on the basis of the transformed data. The appropriate transformation would be logarithmic transformation.

The following table presents the transformed data. It may be noted that as some of the observations are below 10, we apply $\log(X + 1)$ transformation instead of $\log(X)$.

Hormone	H1	H2	H3	H4	H5	H6	H7	H8
Replication 1	0.4771	0.6990	0.7782	0.6021	1.3010	1.3979	0.6021	1.2041
Replication 2	1.0000	0.7782	0.9031	1.0792	1.4314	1.5441	0.3010	1.0792
Replication 3	0.9542	0.8451	1.1139	0.9542	1.2041	1.3979	0.3010	1.1461
Replication 4	0.4771	0.6021	0.8451	1.1461	1.3617	1.4472	0.7782	1.2041
Average	0.7271	0.7311	0.9101	0.9454	1.3246	1.4468	0.4956	1.1584

Using MS Excel program with the transformed data, we analyze the above data to get the following ANOVA table:

ANOVA with transformed data						
Summary	Count		Sum	Average		Variance
Replication 1	8		7.061452	0.88268156		0.130069
Replication 2	8		8.116066	1.014508192		0.148587
Replication 3	8		7.916744	0.989593054		0.107176
Replication 4	8		7.861564	0.982695553		0.128056
H1	4		2.908485	0.727121255		0.083682
H2	4		2.924279	0.731069822		0.010964
H3	4		3.640283	0.910070657		0.021079
H4	4		3.781612	0.945402946		0.058717
H5	4		5.298242	1.324560395		0.009283
H6	4		5.787106	1.446776523		0.004745
H7	4		1.982271	0.495567808		0.055628
H8	4		4.633549	1.158387312		0.003536
ANOVA						
Source of variation	SS	d.f.	MS	*F*	*P* value	*F* crit
Replication	0.080983	3	0.026994341	0.856422	0.478938	3.072467
Hormone	2.935294	7	0.419327722	13.30358	0.000	2.487578
Error	0.661918	21	0.031519908			
Total	3.678195	31				

The LSD value at 5% level of significance is given by

$$\mathrm{LSD}_{(0.05)} = \sqrt{\frac{2\mathrm{ErMS}}{r}} \times t_{0.025,21}$$
$$= \sqrt{\frac{2 \times 0.0315}{4}} \times 2.08 = 0.261.$$

By arranging the treatment means in descending order, we find that hormone 6 is the best treatment, recording the highest number of fruit set per plant, and hormone 5 is statistically at par with it.

Hormone	Transformed mean
H6	1.4468
H5	1.3246
H8	1.1584
H4	0.9454
H3	0.9101
H2	0.7311
H1	0.7271
H7	0.4956

Let us examine the analysis of the above problem without transforming the data using MS Excel and following the steps discussed above.

ANOVA with original data

Summary	Count	Sum	Average	Variance
H1	4	21	5.2500	14.2500
H2	4	18	4.5000	1.6667
H3	4	30	7.5000	9.6667
H4	4	35	8.7500	18.9167
H5	4	82	20.5000	21.6667
H6	4	109	27.2500	22.2500
H7	4	10	2.5000	3.6667
H8	4	54	13.5000	3.6667
Replication 1	8	75	9.3750	74.5536
Replication 2	8	104	13.0000	125.4286
Replication 3	8	87	10.8750	47.5536
Replication 4	8	93	11.6250	85.6964

ANOVA

Source of variation	SS	d.f.	MS	*F*	*P* value	*F* crit
Hormone	2100.21875	7	300.0313	27.1105	0.0000	2.4876
Replication	54.84375	3	18.2813	1.6519	0.2078	3.0725
Error	232.40625	21	11.0670			
Total	2387.46875	31				

The LSD value at 5% level of significance with original data is given by

$$\mathrm{LSD}_{(0.05)} = \sqrt{\frac{2\mathrm{ErMS}}{r}} \times t_{0.025,21}$$

$$= \sqrt{\frac{2 \times 11.067}{4}} \times 2.08 = 4.893.$$

By arranging the treatment means in descending order, we find that hormone 6 is the best treatment, recording the highest number of fruit set per plant, and all other hormonal effects are statistically different with it. *Thus, the conclusion on the basis of original data clearly differs from that of transformed data.*

Hormone	Original average
H6	27.25
H5	20.5
H8	13.5
H4	8.75
H3	7.5
H1	5.25
H2	4.5
H7	2.5

10.4.1.2 Square Root Transformation

For count data consisting of small whole numbers and percentage data where the data ranges either between 0 and 30% or between 70 and to 100%, that is, the data in which the variance tends to be proportional to the mean, the square root transformation is used. Data obtained from counting rare events like the number of death per unit time, the number of infested leaf per plant, the number of call received in a telephone exchange, or the percentage of infestation (disease or pest) in a plot (either 0–30% or 70–100%) are examples where square root transformation can be useful before taking up analysis of variance to draw a meaningful conclusion or inference. If most of the values in a data set are small (<10) coupled with the presence of 0 values, instead of using $\sqrt{x}$ transformation, it is better to use $\sqrt{(x+0.5)}$. Then analysis of variance to be conducted with the transformed data and the mean table should be made from the transformed data instead of taking the mean from the original data because of the facts stated earlier.

Example 10.5. The following information pertains to the number of insects per 20 insects that remained alive after the application of six different insecticides. Analyze the data to work out the most efficient insecticide.

Insecticides	Replication 3	Replication 2	Replication 1	Replication 4
I1	3	5	1	3
I2	6	3	2	3
I3	6	7	5	5
I4	3	4	4	5
I5	9	8	7	8
I6	10	9	12	11

Solution. Information given in this problem is in the form of small whole numbers, so violation of assumption of ANOVA is suspected.

Insecticides	Replication 1	Replication 2	Replication 3	Replication 4	Average
I1	1	2	3	2	2
I2	2	5	4	4	3.75
I3	2	7	6	6	5.25
I4	1	5	4	6	4
I5	4	12	9	8	8.25
I6	12	6	10	11	9.75

Hence, a square root transformation will be appropriate before taking up the analysis of variance.

So from the given data table, we first make the following table by taking the square roots of the given observations. It may be noted that the data set contains small whole number (<10); instead of $\sqrt{X}$, we have taken $\sqrt{(X + 0.5)}$.

Insecticides	Replication 1	Replication 2	Replication 3	Replication 4
I1	1.22474487	1.58113883	1.87083	1.58114
I2	1.58113883	2.34520788	2.12132	2.12132
I3	1.58113883	2.738612788	2.54951	2.54951
I4	1.22474487	2.34520788	2.12132	2.54951
I5	2.12132034	3.535533906	3.08221	2.91548
I6	3.53553391	2.549509757	3.24037	3.39116

Using MS Excel (as described in the previous example) with the transformed data, we analyze the above data to get the following ANOVA table:

ANOVA						
Source of variation	d.f.	SS	MS	F	P value	F crit
Replication	3	1.8011	0.6004	4.1221	0.0256	3.2874
Insecticides	5	7.2037	1.4407	9.8917	0.0002	2.9013
Error	15	2.1848	0.1457			
Total	23	11.1896				

The LSD value at 5% level of significance is given by

$$LSD_{(0.05)} = \sqrt{\frac{2ErMS}{r}} \times t_{0.025,21}$$
$$= \sqrt{\frac{2 \times 0.1457}{4}} \times 2.131 = 0.5751.$$

Insecticides	Transformed means
I1	1.5645
I2	2.0422
I3	2.0602
I4	2.3547
I5	2.9136
I6	3.1791

Arranging the treatment means in ascending order, we find that insecticide 1 is the best treatment, recording the lowest number of insects per 20 insects, and insecticides 2 and 3 are at par with insecticide 1. On the other hand, insecticide 6 has produced maximum input per 20 insects.

$$LSD_{(0.05)} = \sqrt{\frac{2ErMS}{r}} \times t_{0.025,21}$$
$$= \sqrt{\frac{2 \times 1.686}{4}} \times 2.131 = 1.957.$$

ANOVA with original data				
Summary	Count	Sum	Average	Variance
I1	4	12	3.0000	2.6667
I2	4	14	3.5000	3.0000
I3	4	23	5.7500	0.9167
I4	4	16	4.0000	0.6667
I5	4	32	8.0000	0.6667
I6	4	42	10.5000	1.6667
Replication 1	6	31	5.1667	15.7667
Replication 2	6	36	6.0000	5.6000
Replication 3	6	37	6.1667	8.5667
Replication 4	6	35	5.8333	9.7667

ANOVA						
Source of variation	SS	d.f.	MS	F	P value	F crit
Rows	173.208	5.000	34.642	20.545	0.000	2.901
Columns	3.458	3.000	1.153	0.684	0.576	3.287
Error	25.292	15.000	1.686			
Total	201.958	23.000				

Arranging the treatment means in ascending order, we find that insecticide 1 is the best insecticide treatment, recording the lowest number of insects per 20 insects, and insecticides 2 and 4 are at par with insecticide 1. On the other hand, insecticide 6 has retained insect weeds per 20 insects. But with transformed data I1, I2, and I3 are at par.

Insecticide	Average
I1	3.0000
I2	3.5000
I4	4.0000
I3	5.7500
I5	8.0000
I6	10.5000

10.4.1.3 Angular Transformation

Data on proportions/percentages arising out of count data require angular transformation or the arcsine transformation. It may be noted emphatically that the percentage data (e.g., percentage of carbohydrate, protein, and sugar; percentage of marks; and percentage of infections) which are not arising out of count data $\left(\frac{m}{n} \times 100\right)$ should not be put under arcsine transformation. Moreover, all the percentage data arising out of count data need not be subjected to arcsine transformation before analysis of variance:

(a) For percentage data ranging either between 0 and 30% or 70 and 100% but not both, the square root transformation should be used.
(b) For percentage data ranging between 30 and 70%, no transformation is required.
(c) Percentage data which overlaps the above two situations should only be put under arcsine transformation.

Thus, a data set having 0 to more than 30%, less than 70–100%, and 0–100% should be put under arcsine transformation.

The values of 0% should be substituted by $\frac{1}{4n}$ and the values of 100% by $100 - \frac{1}{4n}$, where n is the number of counts on which percentages are worked out before transformation of data following arcsine rule of transformation. The essence of arcsine transformation is to convert the percentage data into angles measured in degrees, that is, to transform 0–100% data into 0–90° angles. The actual procedure is to convert the percentage data into proportions and transform it into $\sin^{-1}\sqrt{p}$, where p is the percentage data measured in proportions. Ready-made tables are available for different percentage values with their corresponding transformed values. However, in MS Excel with the following functional form, a percentage data can be directly converted to arcsine transformed data.

= degrees (asin(sqrt(p/100))), where p is the percentage value.

Analysis of variance is taken up on the transformed data and inferences are made accordingly with the transformed means. However, to get original means retransformation of transformed means is preferred over the means from original data because of the fact stated earlier.

Example 10.6. The following table gives the germination percentage of eight varieties of wheat in a field trial with three replications. Analyze the data to find out the best variety w.r.t. germination.

Variety	Replication 3	Replication 1	Replication 2
V1	62	55	48
V2	28	23	32
V3	70	75	72
V4	85	86	88
V5	92	90	95
V6	18	15	17
V7	20	16	18
V8	68	65	66

Solution. From the given information it is clear that (a) the data can be analyzed per the analysis of two-way classified data and (b) the data relates to percentage data arising out of count data and the percentage ranges from 15 to 95%. Thus, there is a chance of heterogeneous variance.

So we opt for arcsine transformation before taking up the analysis of variance.

The transformed data are presented in the following table:

Variety	Replication 2	Replication 3	Replication 1
V1	43.85	51.94	47.87
V2	34.45	31.95	28.66
V3	58.05	56.79	60.00
V4	69.73	67.21	68.03
V5	77.08	73.57	71.57
V6	24.35	25.10	22.79
V7	25.10	26.57	23.58
V8	54.33	55.55	53.73

The analysis of variance with transformed data per the analysis of two classified data given in the analysis of variance table at 5% level of significance will be as follows:

ANOVA						
SOV	d.f.	SS	MS	F	P value	F critical
Replication	2	11.4088	5.7044	1.1204	0.3537	3.7389
Variety	7	7814.2733	1116.3248	219.2512	0.0000	2.7642
Error	14	71.2815	5.0915			
Total	23	7896.9635				

Thus, as the calculated value of F (219.2512) exceeds the critical value (2.7642) of F-statistic at 5% level of significance, the test is significant and the null hypothesis of equality of varietal effects w.r.t. germination percentage is rejected. That means the varieties differ significantly among themselves w.r.t. germination percent.

To identify the best variety w.r.t. germination percentage, the LSD (CD) value at 5% level of significance is worked out as given below:

$$\text{LSD}_{(0.05)} = \sqrt{\frac{2\text{ErMS}}{r}} \times t_{0.025,21}$$

$$= \sqrt{\frac{2 \times 5.0915}{3}} \times 2.069 = 3.812.$$

We arrange the varietal means in descending order and compare them in relation to the above CD value:

Variety	Mean germination %
V5	74.071
V4	68.324
V3	58.280
V8	54.537
V1	47.889
V2	31.685
V7	25.082
V6	24.080

Thus, from the above table, it is clear that variety 5 is the best variety, having maximum germination percentage, followed by variety 4, variety 3, and so on. On the other hand, the lowest germination percentage is recorded in variety 6, which is at par with variety 7. Hence, variety 5 is the best and varieties 6 and 7 are the worst varieties w.r.t. germination percent.

Let us see how this analysis differs from the analysis with original data. The following is the result of analysis of data without transforming the same.

ANOVA, without transforming the data

ANOVA						
Source of variation	SS	d.f.	MS	F	P value	F crit
Rows	19159.333	7	2737.0476	231.6494	0.0000	2.7642
Columns	20.583	2	10.2917	0.8710	0.4400	3.7389
Error	165.417	14	11.8155			
Total	19345.333	23				

Variety	Average germination %
V5	92.333
V4	86.333
V3	72.333
V8	66.333
V1	55.000
V2	27.667
V7	18.000
V6	16.667

It is found from the above two analyses that there is no difference in results and conclusion.

Thus, from the above three analyses based on transformation principle, it can be inferred that results and conclusion may or may not remain the same. But it is always wise to take care of necessary transformation, wherever required.

Effect of Change in Origin and Scale in Analysis of Variance:

Sometimes in real-life situation, we come across with data which are very large or very small in nature, thereby posing difficulty in taking up statistical analysis further. Analysis of variance, as has been mentioned earlier, is nothing but the partitioning of variance due to assignable causes and non-assignable causes. Thus, it is essentially an exercise with variance. We know that variance does not depend on the change of origin but depends on the change of scale. So if we add or subtract a constant quantity to each and every observation of the data set to be subjected under analysis of variance, there should not be any effect of such mathematical manipulation (change of origin). But definitely there will be effect of change of scale on analysis of variance. The following examples demonstrate the above two factors.

1. Analysis of variance with change of origin
2. Analysis of variance with change of scale
3. Analysis of variance with change of both origin and scale

Let us take the following example to demonstrate the above.

Example 10.7. The following table gives the yield (kg/plot) of four different varieties of onion in three different irrigation treatments. Analyze the data (a) with change of origin to 100 kg/plot and (b) with change of origin to 100 kg/plot and scale to 8 kg/plot, and comment on the effect of change of origin and scale on the results of ANOVA.

(A) Analysis with original data

	Variety 1	*Variety 2*	*Variety 3*	*Variety 4*
I 1	181	195	185	110
	184	197	184	125
	182	198	187	115
I 2	224	245	228	145
	225	240	227	135
	223	247	229	142
I 3	200	210	205	130
	201	208	202	132
	202	207	204	131

Summary

	Variety 1	*Variety 2*	*Variety 3*	*Variety 4*	*Total*
I1	182.333	196.667	185.333	116.667	170.250
I2	224.000	244.000	228.000	140.667	209.167
I3	201.000	208.333	203.667	131.000	186.000
Average	202.444	216.333	205.667	129.444	

(continued)

(continued)

(A) Analysis with original data

	Variety 1	Variety 2	Variety 3	Variety 4	
ANOVA					
SOV	d.f.	SS	MS	*F*	*P* value
Irrigation	2	9197.056	4598.528	486.903	0.000
Variety	3	42762.528	14254.176	1509.266	0.000
Interaction (IxV)	6	688.722	114.787	12.154	0.000
Error	24	226.667	9.444		
Total	35	52874.972			

(B) Analysis of data with origin shifted to 100, that is, observation −100

	Variety 1	Variety 2	Variety 3	Variety 4
I 1	81	95	85	10
	84	97	84	25
	82	98	87	15
I 2	124	145	128	45
	125	140	127	35
	123	147	129	42
I 3	100	110	105	30
	101	108	102	32
	102	107	104	31

Summary

	Variety 1	Variety 2	Variety 3	Variety 4	Total
I1	82.333	96.667	85.333	16.667	70.250
I2	124.000	144.000	128.000	40.667	109.167
I3	101.000	108.333	103.667	31.000	86.000
Average	102.444	116.333	105.667	29.444	

ANOVA					
SOV	d.f.	SS	MS	*F*	*P* value
Irrigation	2	9197.056	4598.528	486.903	0.000
Variety	3	42762.528	14254.176	1509.266	0.000
Interaction (I × V)	6	688.722	114.787	12.154	0.000
Error	24	226.667	9.444		
Total	35	52874.972			

(C) ANOVA with change of origin

	Variety 1	Variety 2	Variety 3	Variety 4
I 1	10.1	11.9	10.6	1.3
	10.5	12.1	10.5	3.1
	10.3	12.3	10.9	1.9
I 2	15.5	18.1	16.0	5.6
	15.6	17.5	15.9	4.4
	15.4	18.4	16.1	5.3
I 3	12.5	13.8	13.1	3.8
	12.6	13.5	12.8	4.0
	12.8	13.4	13.0	3.9

Summary

	Variety 1	Variety 2	Variety 3	Variety 4	Total
I 1	10.292	12.083	10.667	2.083	8.781
I 2	15.500	18.000	16.000	5.083	13.646
I 3	12.625	13.542	12.958	3.875	10.750
Average	12.806	14.542	13.208	3.681	

(continued)

(continued)

(C) ANOVA with change of origin

	Variety 1	*Variety 2*	*Variety 3*	*Variety 4*	
ANOVA					
SOV	d.f.	SS	MS	*F*	*P* value
Irrigation	2	143.704	71.852	486.903	0.000
Variety	3	668.164	222.721	1509.266	0.000
Interaction (I × V)	6	10.761	1.794	12.154	0.000
Error	24	3.542	0.148		
Total	35	826.171			

From the above analyses the following points may be noted:

1. *F*-ratios and the significance level do not change under the above three cases.
2. Mean values change in the second and third cases.
3. Sum of squares and mean sum of squares values do not change in the first and second cases.
4. As error mean square remains the same in the first and second cases, critical difference values also remain the same.

Thus, with the change of origin and/or scale, the basic conclusion from ANOVA does not change, but while comparing the means, care should be taken to adjust the mean values and the critical difference values accordingly.

10.5 Experimental Reliability

Reliability of experimental procedure, in its simplest form, can be verified with the help of coefficient of variations. The coefficient of variation is defined as

$$CV = \frac{\text{S.D}}{\text{Grand Mean}} \times 100.$$

Here the positive square root of error mean square in analysis of variance is taken as an estimate of the standard deviation. Thus, from the table of analysis of variance, one can work out the CV% as follows:

$$CV = \frac{\sqrt{\text{ErMS}}}{\text{Grand Mean}} \times 100.$$

Now the question is, what should be the range or the value of CV% for the information from an experiment to be taken reliable? There is no hard and fast rule to determine the cutoff value of CV % for an experiment is reliable; it depends on the condition of the experiment (laboratory condition, field condition, etc.), type of materials/treatments tested in the experiment, desired precision from the experiment, etc. Generally the CV% should be less in experiments conducted under laboratory conditions compared to field experiments. Similarly CV% also depends on the type of field crop, size and shape of experimental units, etc. However, by and large, a CV% less than 20% is regarded as an indication for the reliability of the experiments. If the CV value is more than 20%, there is a need to verify the experimental procedure and need for emphasis on the reduction of experimental error.

10.6 Comparison of Means

The *F*-test in analysis of variance indicates the acceptance or the rejection of the null hypothesis. The *F-test in analysis of variance can be significant even if only one pair of means among several pairs of means is significant. If the null hypothesis is rejected, we need to find out the means which are significantly different from each other and resulting the significance of F-test.* Thus, comparison of treatment means becomes essential. The comparison of treatment means can be done either through *pair comparison* or through *group comparison*. In group comparison, two or more than two treatment means are involved in the process. Under the group comparison there are *between-group comparison, within-group comparison, trend comparison, and factorial comparison*. Pair comparison is the simplest and most commonly used method.

In the following section we will discuss pair comparison.

10.6.1 Pair Comparison

Quite a good number of test procedures are available for the purpose of comprising pair of treatment means. Among the methods *the least significant difference test (LSD), also known as critical difference (CD) test; student Newman–Keuls test, Duncan multiple range test (DMRT), and Tukey's honestly significant difference test* are commonly used. Again, among the four test procedures, the CD and the DMRT are mostly used.

10.6.1.1 CD/LSD Test

The essence of LSD test is to provide a single value at specified level of significance which serves as the limiting value of significant or non-significant difference between two treatment means. Thus, two treatment means are declared significantly different at the specified level of significance if the difference between the treatment means exceeds the LSD/CD value; otherwise they are not.

Calculation of LSD Values

LSD = standard error of difference between two treatment means under comparison multiplied by the table value of t distribution at error degrees of freedom with specific level of significance.

Thus, for both-sided tests $\text{LSD}_\alpha/\text{CD}_\alpha = \text{SE(difference)} \times t_{\frac{\alpha}{2};\text{error d.f}}$.

The standard error of difference is calculated as

$$\text{SE}_d = \sqrt{\text{ErMS}\left(\frac{1}{r_i}+\frac{1}{r_{i'}}\right)},$$

where r_i and $r_{i'}$ are the replications/number of observations for the ith and the i'th treatment under comparison. For $r_i = r_{i'}$ (in case of designs like RBD LSD), the CD values are calculated as

$$\text{CD}_d = \sqrt{\frac{2\text{ErMS}}{r}} \times t_{\frac{\alpha}{2};\text{error d.f}}.$$

The LSD/CD test is most appropriate for comparing planned pair; in fact it is not advisable to use LSD for comparing all possible pairs of means. When the number of treatment is large, the number of possible pair of treatment means increases rapidly, and thereby, increasing the probability of significance of at least one pair of means will have the difference exceeding the LSD value. As a result LSD test is used only when the F-test for treatment effects is significant and the number of treatments is not too many, preferably less than six. LSD test is best used in comparing the control treatment with that of the other treatments.

10.6.1.2 Duncan Multiple Range Test

In the experiment, very large numbers of treatment means are compared in number, as has been mentioned already, and thus, the LSD test is not suitable. The fact that LSD can be used only when the F-test is significant in the analysis of variance has prompted the use of Duncan multiple range test. *Duncan multiple range test can be used irrespective of the significance or nonsignificance of the F-test in the analysis of variance.* The essence of Duncan multiple range test calculation is that instead of using standard error of difference between the two means, we use standard error of mean in this case. The standard error of mean is multiplied by r_p values for different values of p (the number of treatment means involved in the comparison) from the statistical tables for significant studentized ranges at $p = 2, 3, \ldots, t$ treatments with different error degrees of freedom.

Thus, if $\text{SE}_m = \sqrt{\frac{\text{ErMS}}{r}}$, then $r_p \times \text{SE}_m$ gives us R_p value. This R_p is subtracted from the largest mean arranged in either ascending or descending order. All the means less than the above subtracted values are declared as significantly different from the largest mean. Other treatments whose values are larger than the above difference, these are compared with appropriate R_p values depending upon the number of treatment means to be compared. That means for three means to be compared with R_3, two means remain to be compared with R_2 and so on. Let us demonstrate the procedure with the help of the following example.

Example 10.8. Five treatments were used to test the superiority measured in terms of harvest index. The following information is noted. Find out the best treatment using DMRT.

Treatment means are as follows: 2.6, 2.1, 2.5, 1.0, and 1.2

ErMS = 0.925, error d.f. = 21, no. of observations per mean is 5.

Solution. Five treatment means are 2.6, 2.1, 2.5, 1.0, and 1.2 with error mean square = 0.925; replication, 5; and error degrees of freedom = 21.

At the first instance, arrange the mean in descending order, that is, 2.6, 2.5, 2.1, 1.2, and 1.0.

$$\text{So the SE}_m = \sqrt{\frac{\text{ErMS}}{5}} = \sqrt{\frac{0.925}{5}} = 0.43.$$

The R_p values for different p values at error degrees of freedom = 21 are read from the studentized range table and are as follows:

p	R_p (0.05)	R_p
2	2.94	$2.94 \times 0.43 = 1.26$
3	3.09	$3.09 \times 0.43 = 1.329$
4	3.18	$3.18 \times 0.43 = 1.367$
5	3.24	$3.24 \times 0.43 = 1.39$

From the largest mean (t_1) 2.6, we subtract the R_p value for the largest $p = 5$, that is, 1.39, to get $2.6 - 1.39 = 1.21$.

Declare all the mean values which are less than this difference value as significantly different from the largest mean. In this case t_4 and t_5 are significantly different from t_1.

The difference between treatments t_1 and t_2 is $2.6 - 2.1 = 0.5$. Since there are three treatment means not declared significantly different from t_1, we are to compare the above difference with R_3 value, that is, $R_3 = 1.329$. The difference is less than the R_3 value. So these treatments are not different significantly. Thus, the conclusion is that t_1, t_2, and t_3 are statistically at par while t_4 and t_5 are also statistically at par. The largest treatment mean is for treatment t_1. In the next few sections, some basic and complex experimental designs have been discussed.

10.7 Completely Randomized Design (CRD)

Among all experimental designs, completely randomized design (CRD) is the simplest one where only two principles of design of experiments, that is, replication and randomization, have been used. The principle of local control is not used in this design. CRD is being analyzed as per the model of one-way ANOVA. The basic characteristic of this design is that the whole experimental area (1) should be homogeneous in nature and (2) should be divided into as many numbers of experimental units as the sum of the number of replications of all the treatments. Let us suppose there are five treatments A, B, C, D, and E replicated 5, 4, 3, 3, and 5 times, respectively, then according to this design, we require the whole experimental area to be divided into 20 experimental units of equal size. Under laboratory condition, completely randomized design is the most accepted and widely used design.

Analysis

Let there be t number of treatments with $r_1, r_2, r_3, \ldots, r_t$ number of replications, respectively, in a completely randomized design. So the model for the experiment will be $y_{ij} = \mu + \alpha_i + e_{ij}$, $i = 1, 2, 3, \ldots, t; j = 1, 2, \ldots, r_i$,

where

y_{ij} = response corresponding to the jth observation of the ith treatment

μ = general effect

α_i = additional effect due to the ith treatment and $\sum r_i\alpha_i = 0$

e_{ij} = errors that are associated with the jth observation of the ith treatment and are i.i.d. $N(0, \sigma^2)$

Assumption of the Model

The above model is based on the assumptions that the effects are additive in nature and the error components are identically, independently distributed as normal variate with mean zero and constant variance.

Hypothesis to Be Tested

$H_0 : \alpha_1 = \alpha_2 = \alpha_3 = \cdots = \alpha_i = \cdots = \alpha_t = 0$

against the alternative hypothesis

H_1 : All $\alpha's$ are not equal

Table 10.7 Treatment structure of CRD

	Treatment					
Observations	1	2	………	i	………	t
1	y_{11}	y_{21}	………	y_{i1}	………	y_{t1}
2	y_{12}	y_{22}	………	y_{i2}	………	y_{t2}
:	:	:	………	:	………	:
:	:	:		:		:
:	:	y_{2r2}		:		:
			………	:	………	:
r_i			………	y_{iri}	………	:
:	y_{1r1}					:
:						y_{tr_t}
Total	y_{10}	y_{20}	………	y_{i0}	………	y_{t0}
Mean	$\bar{y}_{10}$	$\bar{y}_{20}$		$\bar{y}_{i0}$		$\bar{y}_{t0}$

Table 10.8 ANOVA table for completely randomized design

Sources of variation	d.f.	SS	MS	F-ratio
Treatment	$t-1$	TrSS	$\text{TrMS} = \frac{\text{TrSS}}{t-1}$	$\frac{\text{TrMS}}{\text{ErMS}}$
Error	$n-t$	ErSS	$\text{ErMS} = \frac{\text{ErSS}}{n-t}$	
Total	$n-1$	TSS		

Let the level of significance be α. Let the observations of the total $n = \sum_{i=1}^{t} r_i$ experimental units be as follows (Table 10.7):

The analysis for this type of data is the same as that of one-way classified data discussed in Chap. 1, Sect. 1.2. From the above table we calculate the following quantities (Table 10.8):

$$\text{Grand total} = \sum_{i=1}^{t}\sum_{j}^{r_i}(\text{observation}) = y_{11} + y_{21} + y_{31} + \cdots + y_{trt} = \sum_{i=1}^{t}\sum_{j=1}^{r_t} y_{ij} = G.$$

$$\text{Correction factor} = \frac{G^2}{n} = \text{CF}.$$

$$\text{Total sum of squares(TSS)} = \sum_{i=1}^{t}\sum_{j}^{r_i}(\text{observation})^2 - \text{CF} = \sum_{i=1}^{t}\sum_{j=1}^{r_t} y_{ij}^2 - \text{CF}.$$

$$= y_{11}^2 + y_{21}^2 + y_{31}^2 + \cdots + y_{tr_t}^2 - \text{CF}.$$

$$\text{Treatment sum of squares(TrSS)} = \sum_{i=1}^{t}\frac{{y_{i0}}^2}{r_i} - \text{CF, where } y_{i0} = \sum_{j=1}^{r_i} y_{ij}$$

$$= \text{sum of the observations for the } i\text{th treatment.}$$

$$= \frac{y_{10}^2}{r_1} + \frac{y_{20}^2}{r_2} + \frac{y_{30}^2}{r_3} + \cdots \frac{y_{i0}^2}{r_i} + \cdots \frac{y_{t0}^2}{r_t} - \text{CF}.$$

$$\text{Error sum of squares(by subtraction)} = \text{TSS} - \text{TrSS} = \text{Erss}.$$

If the calculated value of F-ratio corresponding to treatment is greater than the table value at α level of significance with $(t-1),(n-t)$ degrees of freedom, that is, $F_{\text{cal}} > F_{\text{tab}\ \alpha;(t-1),(n-t)}$, the null hypothesis is rejected; otherwise, one cannot reject the null hypothesis. When the test is significant, one should find out which pair of treatments is significantly different and which treatment is either the best or the worst with respect to the particular characters under consideration. This is accomplished with the help of the least significant difference (critical difference) value per the given formula below:

$\text{LSD}_\alpha = \sqrt{\text{ErMS}\left(\frac{1}{r_i}+\frac{1}{r_{i'}}\right)} \times t_{\frac{\alpha}{2},(n-t)}$, where i and i' refer to the treatments involved in comparison and t is the table value of t distribution at α level of significance with $(n-t)$ d.f. and $\sqrt{\text{ErMS}\left(\frac{1}{r_i}+\frac{1}{r_{i'}}\right)}$ is the standard error of difference (SE_d) between the means for treatments i and i'. Thus, if the absolute value of the difference between the treatment means exceeds the corresponding CD value, then the two treatments are significantly different and the better treatment is adjudged based on the mean values commensurate with the nature of the character under study.

Example 10.9. An experiment was conducted with five varieties of wheat. The following data gives the weight of grains per hill (g) recorded from different varieties:

Varieties	Grain weight(gm)/hill							
A	150		148		145		145	140
B	162		158		150	155	148	146
C	140	142	144	146	149	140	152	150
D	130		134		135		132	
E	145		135	134	140	142	137	138

Analyze the data and draw the necessary conclusion.

This is a problem of completely randomized design with unequal replications. The fixed effect model for the purpose would be $y_{ij} = \mu + \alpha_i + e_{ij}$, where α_i is the effect of the ith variety, $i = 1, 2, 3, 4, 5$. That means the problem is to test

$H_0: \alpha_1 = \alpha_2 = \alpha_3 = \alpha_4 = \alpha_5$ against
$H_1: \alpha_i$'s are not all equal.

Let the level of significance be $\alpha = 0.05$.

Varieties					
	A	B	C	D	E
	150	162	140	130	145
	148	158	142	134	135
	145	150	144	135	134
	145	155	146	132	140
	140	148	149		142
		146	140		137
			152		138
			150		
Sum (y_{i0})	728	919	1,163	531	971
Average ($\bar{y}_{i0}$)	145.6	153.167	145.375	132.75	138.714

$$\text{Grand total(GT)} = 150 + 148 + 145 + \cdots + 142 + 137 + 138 = 4312$$

$$\text{Correction factor(CF)} = \frac{\text{GT}^2}{n} = \frac{4312^2}{30} = 619778.1$$

$$\text{Total sum of squares(TSS)} = 150^2 + 148^2 + 145^2 + \cdots + 142^2 + 137^2 + 138^2 - \text{CF} = 1737.867$$

$$\text{Variety sum of squares(VSS)} = \frac{728^2}{5} + \frac{919^2}{6} + \frac{1163^2}{8} + \frac{531^2}{4} + \frac{971^2}{7} - \text{CF} = 1231.78$$

$$\text{Error sum of squares(ErSS)} = \text{TSS} - \text{VSS} = 1737.867 - 1231.78 = 506.087$$

ANOVA				
SOV	d.f.	SS	MS	F
Variety	4	1231.780	307.945	15.212
Error	25	506.087	20.243	
Total	29	1737.867		

The table value of $F_{0.05;4,25} = 4.18$.

Thus, $F(\text{cal}) > F(\text{tab})_{0.05;4,25}$, so the test is significant and reject the null hypothesis of equality of grain weight per hill. We conclude that there exist significant differences among the five varieties of wheat with respect to grain weight per hill.

So, the next objective is to find out the varieties which differ significantly among themselves and the variety/varieties having significantly the highest weight per hill.

To compare the varieties we calculate the critical difference value, which is given as

$$\sqrt{\text{MSE}\left(\frac{1}{r_i}+\frac{1}{r_{i'}}\right)} \times t_{0.025,\ \text{err.d.f.}}$$

$$= \sqrt{20.243\left(\frac{1}{r_i}+\frac{1}{r_{i'}}\right)} \times t_{0.025,25}.$$

$$= \sqrt{20.243\left(\frac{1}{r_i}+\frac{1}{r_{i'}}\right)} \times 2.06,$$

where r_i and $r_{i'}$ are the number of observations of the two varieties under comparison. Thus, for comparing the varieties, we have the following critical difference and mean difference values among the varieties:

Comparison	CD (0.05) values	Mean difference $\lvert y_{i0} - y_{i'0}\rvert$	Conclusion
(Variety A–variety B)	5.612	7.567	Varieties are different
(Variety A–variety C)	5.284	0.225	Varieties are the same
(Variety A–variety D)	6.217	12.85	Varieties are different
(Variety A–variety E)	5.427	6.886	Varieties are different
(Variety B–variety C)	5.006	7.792	Varieties are different
(Variety B–variety D)	5.983	20.417	Varieties are different
(Variety B–variety E)	5.156	14.452	Varieties are different
(Variety C–variety D)	5.676	12.625	Varieties are different
(Variety C–variety E)	4.797	6.661	Varieties are different
(Variety D–variety E)	5.809	5.964	Varieties are different

Thus, variety B is the variety giving significantly the highest grain weight per hill followed by variety A and variety C, which are statistically at par.

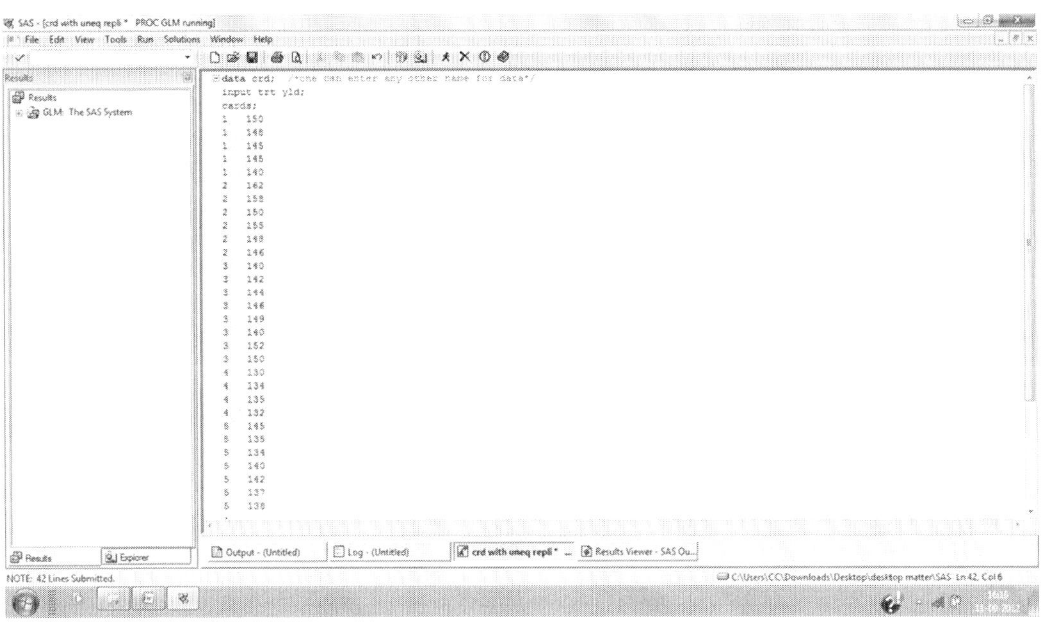

Slide 10.9: Step 1 showing the data input for analysis of CRD using SAS

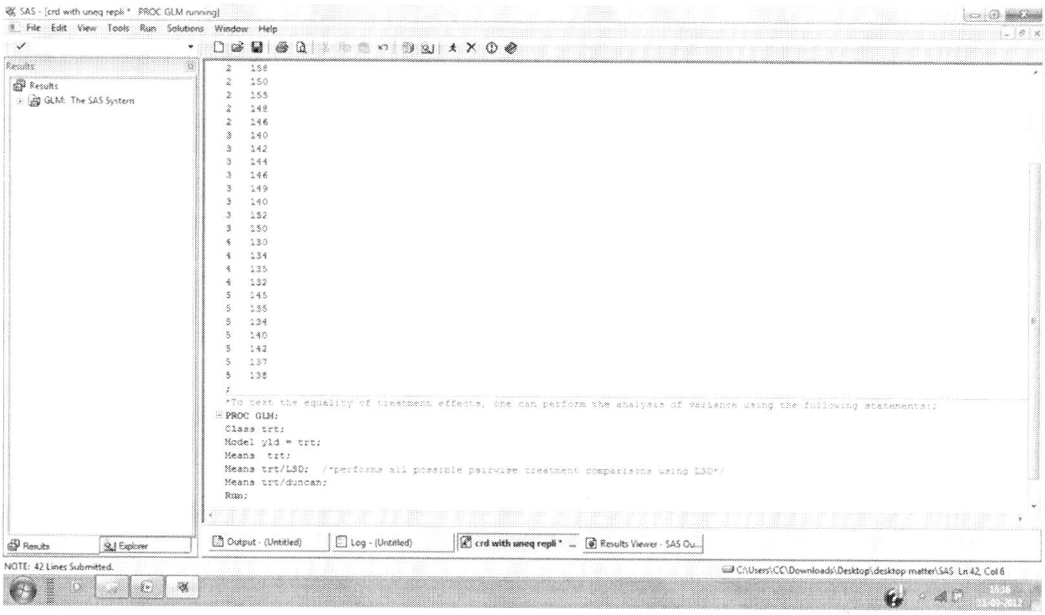

Slide 10.10: Step 2 showing the data input along with commands for analysis of CRD using SAS

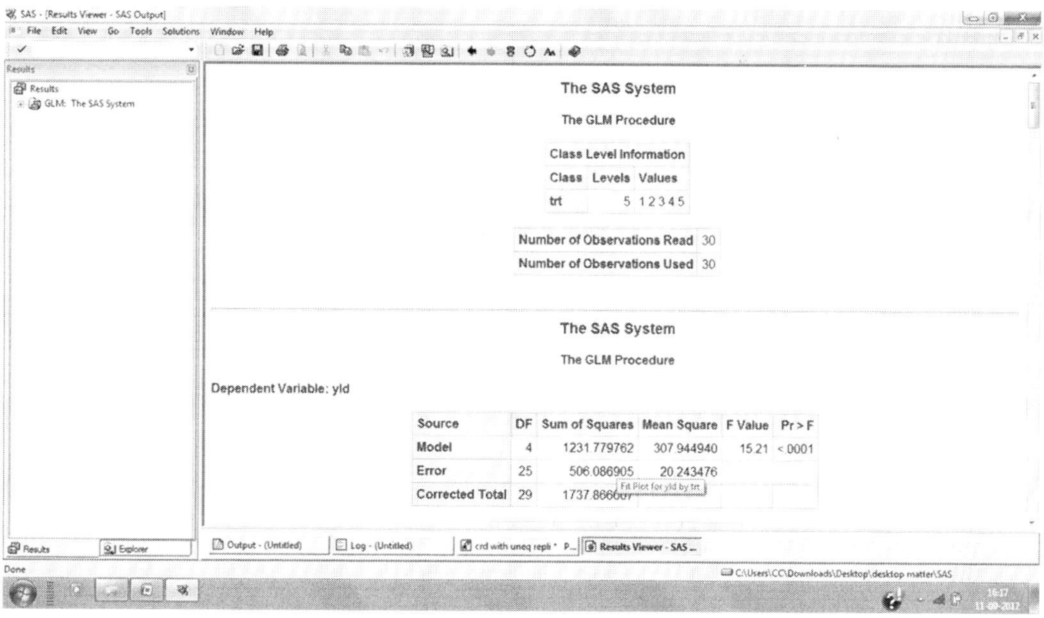

Slide 10.11: Step 3 showing the portion of output for analysis of CRD using SAS

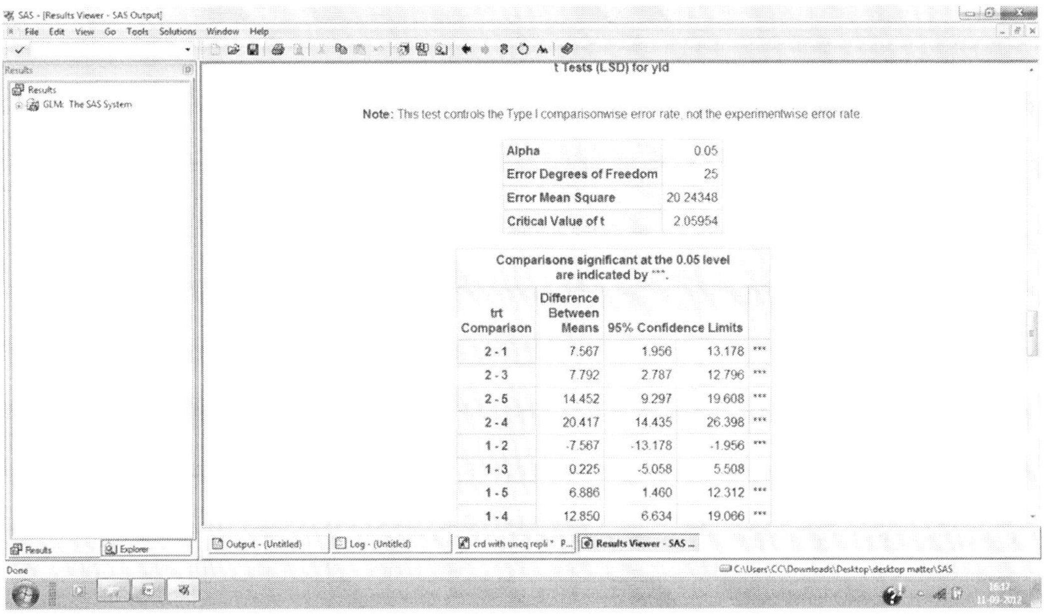

t Tests (LSD) for yld

Note: This test controls the Type I comparisonwise error rate, not the experimentwise error rate.

Alpha	0.05
Error Degrees of Freedom	25
Error Mean Square	20.24348
Critical Value of t	2.05954

Comparisons significant at the 0.05 level are indicated by ***.

trt Comparison	Difference Between Means	95% Confidence Limits		
2 - 1	7.567	1.956	13.178	***
2 - 3	7.792	2.787	12.796	***
2 - 5	14.452	9.297	19.608	***
2 - 4	20.417	14.435	26.398	***
1 - 2	-7.567	-13.178	-1.956	***
1 - 3	0.225	-5.058	5.508	
1 - 5	6.886	1.460	12.312	***
1 - 4	12.850	6.634	19.066	***

Slide 10.12: Step 3 showing the portion of output for analysis of CRD using SAS

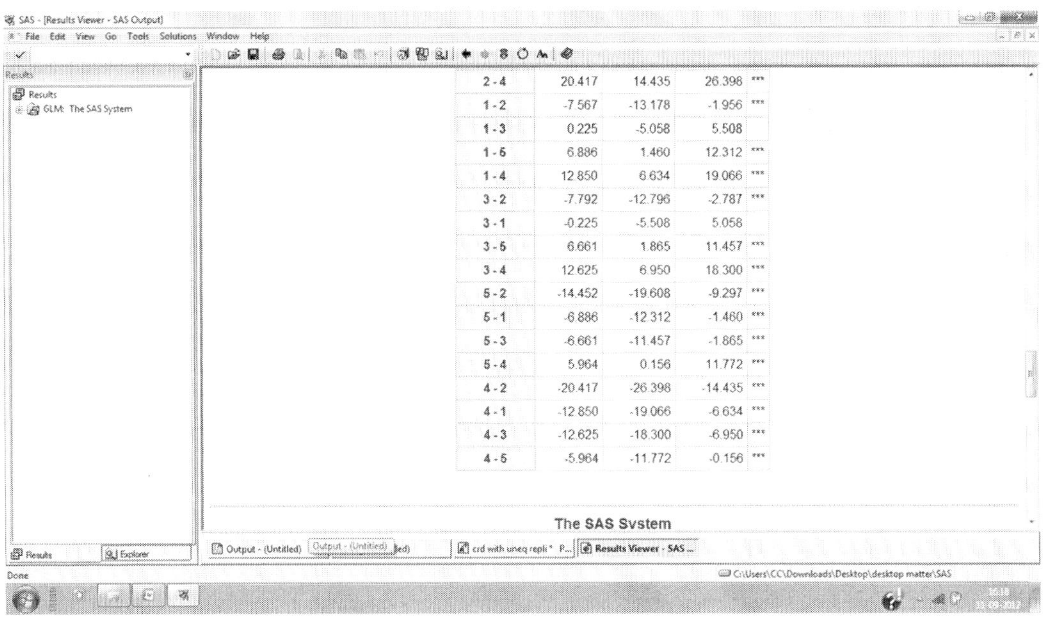

trt Comparison	Difference Between Means	95% Confidence Limits		
2 - 4	20.417	14.435	26.398	***
1 - 2	-7.567	-13.178	-1.956	***
1 - 3	0.225	-5.058	5.508	
1 - 5	6.886	1.460	12.312	***
1 - 4	12.850	6.634	19.066	***
3 - 2	-7.792	-12.796	-2.787	***
3 - 1	-0.225	-5.508	5.058	
3 - 5	6.661	1.865	11.457	***
3 - 4	12.625	6.950	18.300	***
5 - 2	-14.452	-19.608	-9.297	***
5 - 1	-6.886	-12.312	-1.460	***
5 - 3	-6.661	-11.457	-1.865	***
5 - 4	5.964	0.156	11.772	***
4 - 2	-20.417	-26.398	-14.435	***
4 - 1	-12.850	-19.066	-6.634	***
4 - 3	-12.625	-18.300	-6.950	***
4 - 5	-5.964	-11.772	-0.156	***

The SAS System

Slide 10.13: Step 3 showing the portion of output for analysis of CRD using SAS

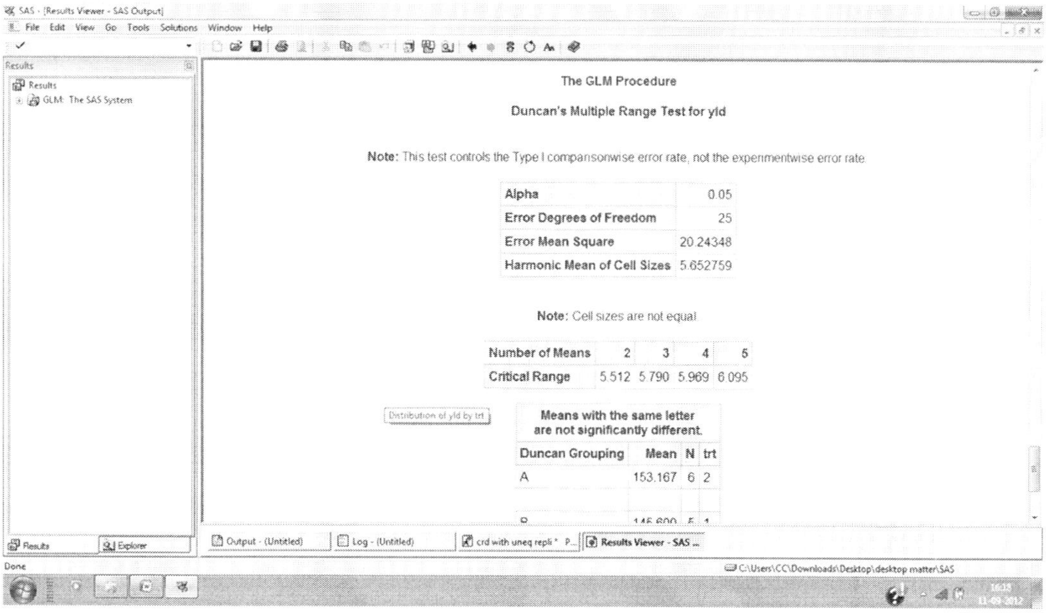

Slide 10.14a: Step 3 showing the portion of output for analysis of CRD using SAS

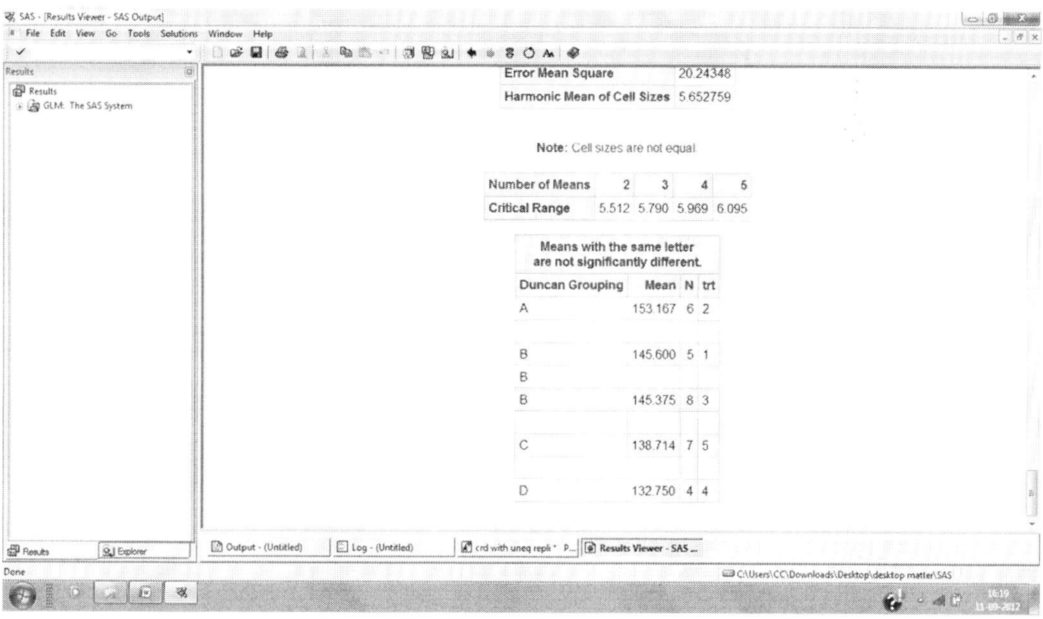

Slide 10.14b: Step 3 showing the portion of output for analysis of CRD using SAS

Example 10.10. Seven varieties of wheat were tested for yield using CRD with 5 plots per variety. The following table gives the yields of grain in quintal per acre. Test whether the 7 varieties differ significantly with respect to yield or not.

Variety A	Variety B	Variety C	Variety D	Variety E	Variety F	Variety G
18	19	17	18	21	20	19
14	19.5	15	14	21.5	16	13
16	20	14	19	22	17	17
17	22	20	21	23	19	18
15	18	16	14	22	15	14

Solution. The statement shows that the experiment was laid out in completely randomized design with seven varieties each replicated five times.

The model for the purpose is $y_{ij} = \mu + \alpha_i + e_{ij}$, $i = 7, j = 5$

where

y_{ij} = jth observation for the ith variety

μ = general effect

α_i = additional effect due to the ith variety

e_{ij} = errors that are associated with jth observation in the ith variety and are i.i.d. $N(0, \sigma^2)$

So the problem is to test

$H_0 : \alpha_1 = \alpha_2 = \alpha_3 = \alpha_4 = \alpha_5 = \alpha_6 = \alpha_7$ against

$H_1 : \alpha_i$'s are not all equal.

Let the level of significance be $\alpha = 0.05$. Let us construct the following table:

	A	B	C	D	E	F	G
	18	19	17	18	21	20	19
	14	19.5	15	14	21.5	16	13
	16	20	14	19	22	17	17
	17	22	20	21	23	19	18
	15	18	16	14	22	15	14
Total (y_{i0})	80	98.5	82	86	109.5	87	81
Average (y_{i0})	16	19.7	16.4	17.2	21.9	17.4	16.2

$$\text{Grand total(GT)} = 18 + 14 + 16 + \cdots + 17 + 18 + 14 = 624.00$$

$$\text{Correction factor(CF)} = \frac{\text{GT}^2}{n} = \frac{624^2}{35} = 11125.03$$

$$\text{Total sum of squares(TSS)} = 18^2 + 14^2 + 16^2 + \cdots + 17^2 + 18^2 + 14^2 - \text{CF} = 268.471$$

$$\text{Variety sum of squares(VSS)} = \frac{80^2}{5} + \frac{98.5^2}{5} + \frac{82^2}{5} + \frac{86^2}{5} + \frac{109.5^2}{5} + \frac{87.00^2}{5} + \frac{81.00^2}{5} - \text{CF} = 143.471$$

$$\text{Error sum of squares(ErSS)} = \text{TSS} - \text{VSS} = 268.471 - 143.471 = 125.00$$

ANOVA

SOV	d.f.	SS	MS	F
Variety	6	143.471	23.912	5.356
Error	28	125.000	4.464	
Total	34	268.471		

The table value of $F_{0.05;6,28} = 2.45$.

Thus, $F(\text{cal}) > F(\text{tab})_{0.05;6,28}$, so the test is significant and we reject the null hypothesis of equality of yields.

So, the next objective is to find out the varieties which differ significantly among themselves and the variety/varieties having significantly highest yield.

To compare the varieties we calculate the critical difference value, which is given as

$$\sqrt{\text{MSE}\left(\frac{1}{r_i}+\frac{1}{r_{i'}}\right)} \times t_{0.025,\,\text{err.d.f.}}$$

$$=\sqrt{4.464\left(\frac{1}{r}+\frac{1}{r}\right)} \times t_{0.025,\,28}$$

$$=\sqrt{4.464\left(\frac{2}{5}\right)} \times 2.048$$

$$=2.737,$$

where r_i and $r_{i'}$ are the number of observations of the two varieties under comparison and for this problem all are equal to 5.

We arrange the mean values corresponding to different varieties in descending order and compare the differences with CD value as follows:

Variety	Avg. yield (q/acre)
Variety E	21.9
Variety B	19.7
Variety F	17.4
Variety D	17.2
Variety C	16.4
Variety G	16.2
Variety A	16

Varieties joined by the same line are statistically at par; that is, they are not significantly different among themselves. Thus, varieties E and B; varieties B, F, and D; and varieties F, D, C, G, and A are statistically at par. From the above, it is clear that variety E is by far the best variety giving highest yield followed by variety B, which are statistically at par, and variety A is the lowest yielder.

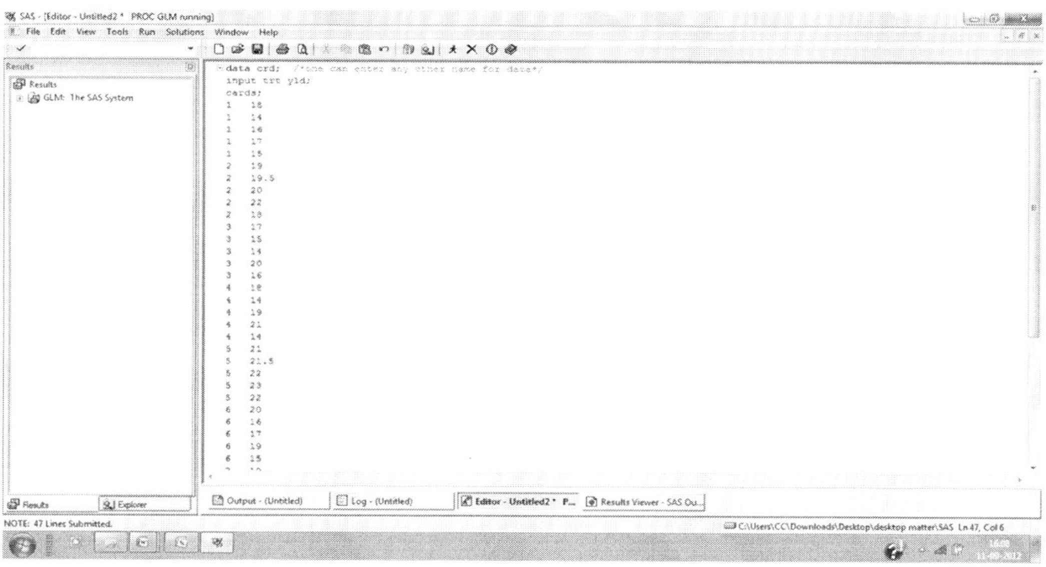

Slide 10.15: Step 1 showing the data for analysis of CRD using SAS

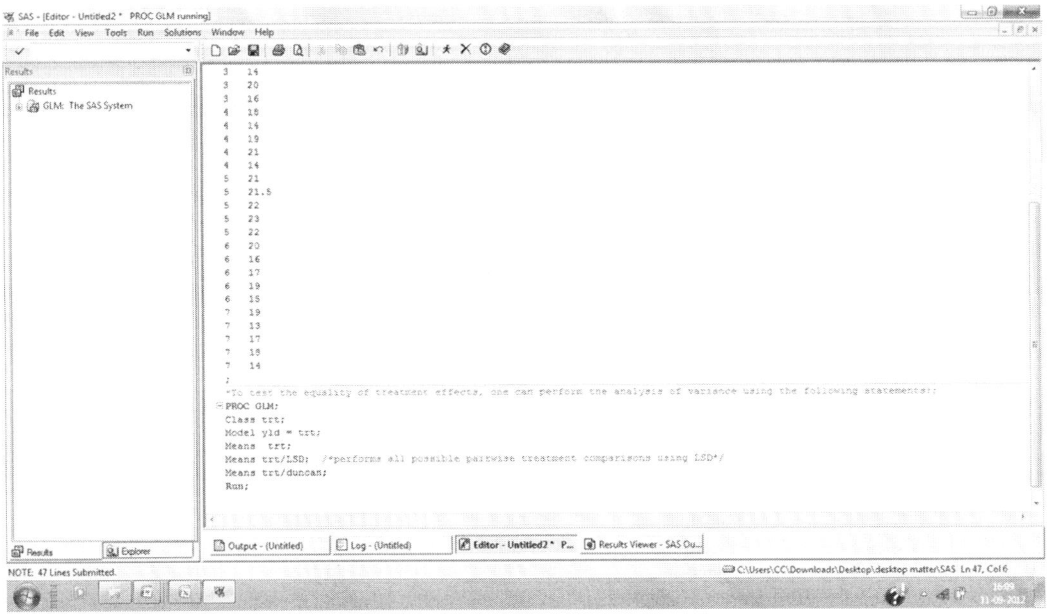

Slide 10.15A: Step 2 showing the portion of data and commands for analysis of CRD using SAS

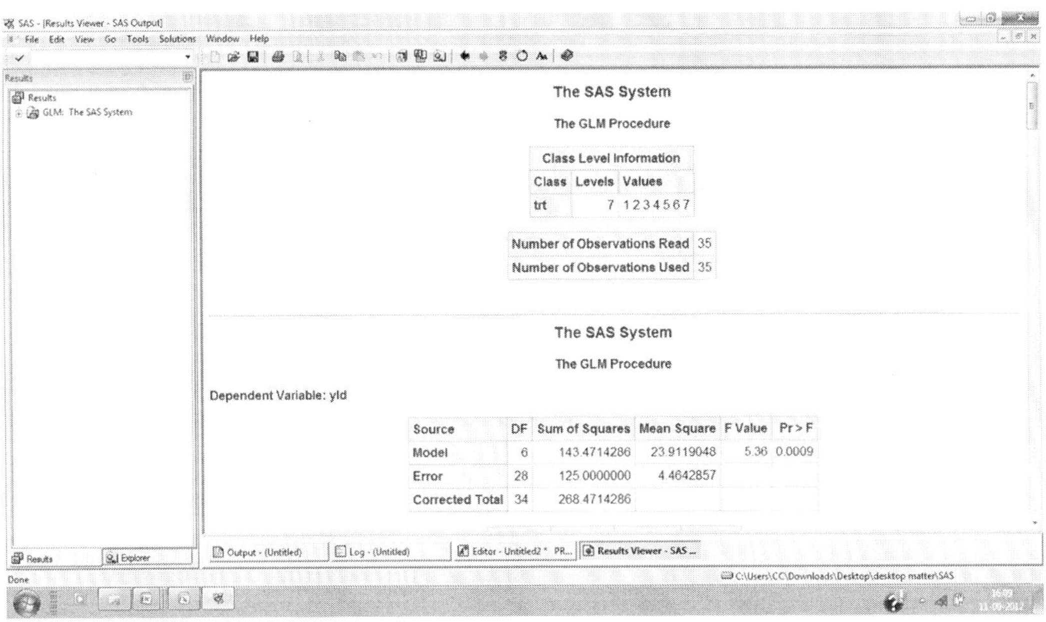

Slide 10.16: Step 3 showing the portion of output for analysis of CRD using SAS

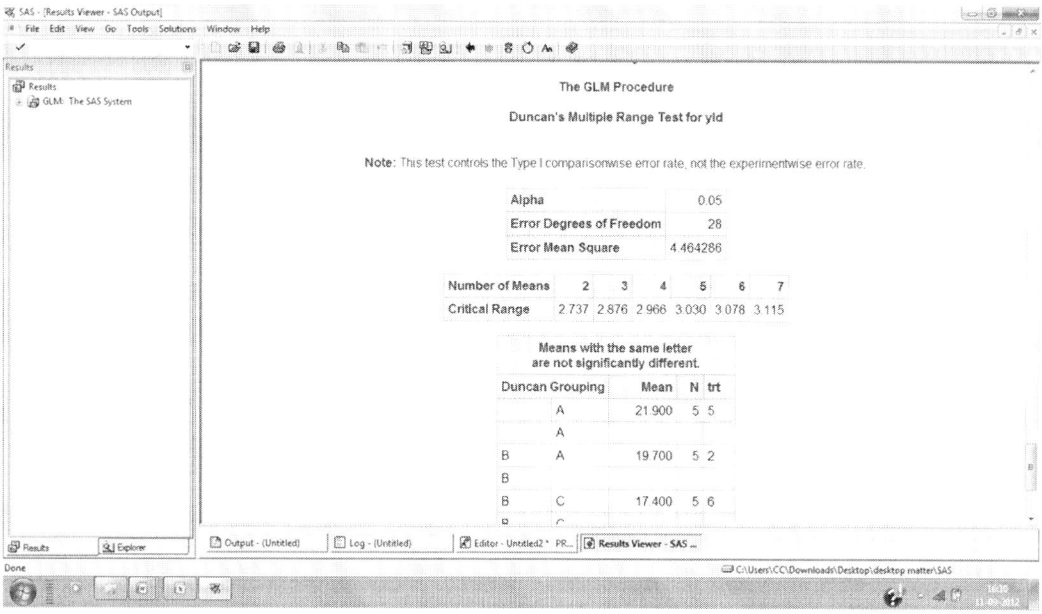

Slide 10.17: Step 3 showing the portion of output for analysis of CRD using SAS

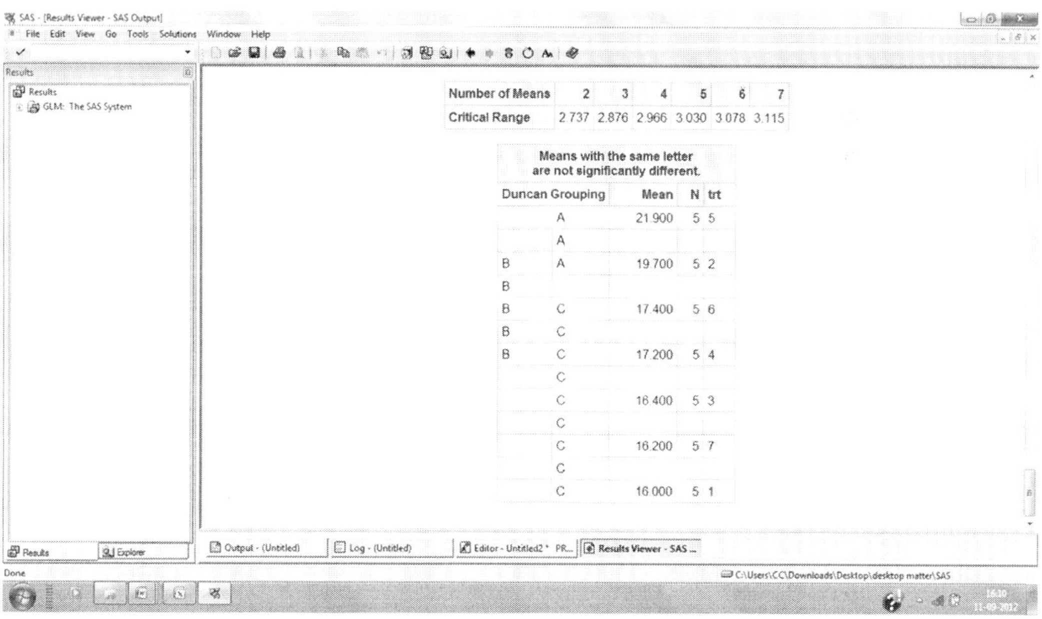

Slide 10.18: Step 3 showing the portion of output for analysis of CRD using SAS

10.8 Randomized Block Design (RBD)

Randomized block design uses all the three basic principles of experimental design, namely, (1) replication, (2) randomization, and (3) local control. Randomized complete block design (RCBD) or simply randomized block design (RBD) takes into account the variability among the experimental units in one direction. The whole experimental area is divided into a number of homogeneous blocks, each having a number of experimental units as the number of treatments. Blocking is done in such a way that the variations among the experimental units within the block are minimum (homogeneous). In doing so, blocking is done perpendicular to the direction of the variability of the experimental area. Thus, if the variability (may be fertility gradient) is in east–west direction, then blocking should be done in north–west direction.

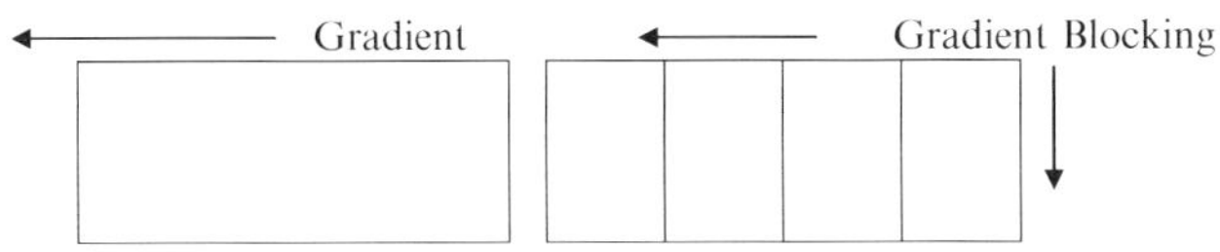

Let there be t number of treatments, each to be replicated r number of times. So according to randomized block design, we should have r number of blocks each having t number of experimental units as follows (Table 10.9):

Thus, the total number of experimental units (plots) in randomized block design is "number of treatments (t)" multiplied by "number of replications (r)," that is, $rt = n$.

Analysis

Let us suppose that we have t number of treatments, each being replicated r number of times. The appropriate statistical model for RBD will be

$$y_{ij} = \mu + \alpha_i + \beta_j + e_{ij}, i = 1, 2, 3, \cdots, t; \quad j = 1, 2, \cdots, r$$

where

y_{ij} = response corresponding to jth replication/block of the ith treatment

μ = general effect

α_i = additional effect due to the ith treatment and $\sum \alpha_i = 0$

β_j = additional effect due to the jth replication/block and $\sum \beta_j = 0$

e_{ij} = errors associated with jth replication/block of ith treatment and are i.i.d. $N(0, \sigma^2)$

The above model is based on the assumptions that the effects are additive in nature and the error components are identically, independently distributed as normal variate with mean zero and constant variance.

Let the level of significance be α.

Hypothesis to Be Tested

The null hypotheses to be tested are

Table 10.9 Treatments and blocks in RBD

Replication/blocks					
Experimental unit	1	2	....	$r - 1$	r
	1	$t + 1$	....	$(r - 2)t + 1$	$(r - 1)t + 1$
	2	$t + 2$	....	$(r - 2)t + 2$	$(r - 1)t + 2$
	:	:	....	:	:
	:	:	....	:	:
	t	$2t$	....	$(r - 1)t$	rt

Table 10.10 Summary of observations from RBD

Replications/blocks								
Treatments	1	2	....	j	....	r	Total	Mean
1	y_{11}	y_{12}	....	y_{1j}	....	y_{1r}	y_{10}	$\bar{y}_{10}$
2	y_{21}	y_{22}	....	y_{2j}	....	y_{2r}	y_{20}	$\bar{y}_{20}$
:	:	:	:	:	:	:	:	:
i	y_{i1}	y_{i2}	....	y_{ij}	....	y_{ir}	y_{i0}	$\bar{y}_{i0}$
:	:	:	:	:	:	:	:	:
t	y_{t1}	y_{t2}	....	y_{tj}	....	y_{tr}	y_{t0}	$\bar{y}_{t0}$
Total	y_{01}	y_{02}	....	y_{0j}	....	y_{0r}	y_{00}	
Mean	$\bar{y}_{01}$	$\bar{y}_{02}$	....	$\bar{y}_{0j}$	....	$\bar{y}_{0r}$		

Table 10.11 ANOVA table for RBD

SOV	d.f.	SS	MS	F-ratio	Tabulated F (0.05)	Tabulated F (0.01)
Treatment	$t-1$	TrSS	$\text{TrMS} = \dfrac{\text{TrSS}}{t-1}$	$\dfrac{\text{TrMS}}{\text{ErMS}}$		
Replication (block)	$r-1$	RSS	$\text{RMS} = \dfrac{\text{RSS}}{r-1}$	$\dfrac{\text{RMS}}{\text{ErMS}}$		
Error	$(t-1)(r-1)$	ErSS	$\text{ErMS} = \dfrac{\text{ErSS}}{(t-1)(r-1)}$			
Total	$rt-1$	TSS				

$$H_0 :(1)\ \alpha_1 = \alpha_2 = \cdots = \alpha_i = \cdots = \alpha_t = 0,$$
$$(2)\ \beta_1 = \beta_2 = \cdots = \beta_j = \cdots = \beta_r = 0$$

against the alternative hypotheses

$$H_1 :(1)\ \alpha\text{'s are not equal},$$
$$(2)\ \beta\text{'s are not equal}.$$

Let the observations of these $n = rt$ units be as follows (Table 10.10):

The analysis of this design is the same as that of two-way classified data with one observation per cell discussed in Chap. 1, Sect. 4.2.3.2.

From the above table, we calculate the following quantities (Table 10.11):

$$\text{Grand total} = \sum_{i,j} y_{ij} = y_{11} + y_{21} + y_{31} + \cdots + y_{tr} = G$$

$$\text{Correction factor} = \frac{G^2}{rt} = \text{CF}$$

$$\text{Total sum of squares(TSS)} = \sum_{i,j} y_{ij}^2 - \text{CF} = y_{11}^2 + y_{21}^2 + y_{31}^2 + \cdots + y_{tr}^2 - \text{CF}$$

$$\text{Treatment sum of squares(TrSS)} = \frac{\sum_{i=1}^{t} y_{i0}^2}{r} - \text{CF} = \frac{y_{10}^2}{r} + \frac{y_{20}^2}{r} + \frac{y_{30}^2}{r} + \cdots \frac{y_{i0}^2}{r} + \cdots \frac{y_{t0}^2}{r} - \text{CF}$$

$$\text{Replication sum of squares(RSS)} = \frac{\sum_{j=1}^{r} y_{0j}^2}{t} - \text{CF} = \frac{y_{01}^2}{t} + \frac{y_{02}^2}{t} + \frac{y_{03}^2}{t} + \cdots \frac{y_{0j}^2}{t} + \cdots \frac{y_{0r}^2}{t} - \text{CF}$$

$$\text{Error sum of squares(by subtraction)} = \text{TSS} - \text{TrSS} - \text{RSS}$$

If the calculated values of F-ratio corresponding to treatment and replication is greater than the corresponding table value at the α level of significance with $(t-1)$, $(t-1)(r-1)$ and $(r-1)$, $(t-1)(r-1)$ degrees of freedom, respectively, then the corresponding null hypothesis is rejected. Otherwise, we conclude that there exist no significant differences among the treatments/replications with respect to the particular character under consideration; all treatments/replications are statistically at par.

Upon rejection of the null hypothesis, corresponding LSD/CD is calculated using the following formula:

Replication:

$$\mathrm{LSD}_\alpha/(\mathrm{CD}_\alpha) = \sqrt{\frac{2\mathrm{ErMS}}{t}} \times t_{\frac{\alpha}{2};(t-1)(r-1)} \quad \text{and}$$

Treatment:

$$\mathrm{LSD}_\alpha(\mathrm{CD}_\alpha) = \sqrt{\frac{2\mathrm{ErMS}}{r}} \times t_{\frac{\alpha}{2};(t-1)(r-1)},$$

where t is the number of treatments and $t_{\frac{\alpha}{2};(t-1)(r-1)}$ is the table value of t at α level of significance and $(t-1)(r-1)$ degrees of freedom.

By comparing the mean difference among the replication/treatment mean with corresponding LSD/CD value, appropriate decisions are taken.

Example 10.11. An experiment was conducted with six varieties of paddy. The following table gives the layout and corresponding yield (q/ha). Analyze the data and find out the best variety of paddy.

Rep-1	Rep-2	Rep-3	Rep-4
V2 28.8	V4 14.5	V3 34.9	V1 27.8
V3 22.7	V6 23.5	V5 17.7	V5 16.2
V5 17.0	V3 36.8	V2 31.5	V2 30.6
V6 22.5	V2 40.0	V4 15.0	V4 16.2
V4 16	V5 15.4	V1 28.5	V6 22.3
V1 27.3	V1 38.5	V6 22.8	V3 27.7

Solution. From the layout of the experiment, it is clear that the experiment has been laid out in randomized block design with six varieties in four replications.

So the model for RBD is given by $y_{ij} = \mu + \alpha_i + \beta_j + e_{ij}$, $i = 6, j = 4$

where,

y_{ij} = effect due to the ith variety in jth replicates
μ = general effect
α_i = additional effect due to the ith variety
β_j = additional effect due to the jth replicate
e_{ij} = errors associated with ith variety in jth replicate and are i.i.d. $N(0, \sigma^2)$

The hypotheses to be tested are

$H_0 : \alpha_1 = \alpha_2 = \alpha_3 = \alpha_4 = \alpha_5 = \alpha_6$ against
$H_1 : \alpha_i$'s are not all equal and
$H_0 : \beta_1 = \beta_2 = \beta_3 = \beta_4$ against
H_1 : All β_j's are not all equal.

Let the level of significance be $\alpha = 0.05$.

We shall analyze the data in the following steps:

Make the following table from the given information.

Replication	V1	V2	V3	V4	V5	V6	Total (y_{0j})	Mean $\left(\bar{y}_{0j}\right)$
R1	27.30	28.80	22.70	16.00	17.00	22.50	134.30	22.38
R2	38.50	40.00	36.80	14.50	15.40	23.50	168.70	28.12
R3	28.50	31.50	34.90	15.00	17.70	22.80	150.40	25.07
R4	27.80	30.60	27.70	16.20	16.20	22.30	140.80	23.47
Total y_{i0}	122.10	130.90	122.10	61.70	66.30	91.10	**594.20**	
Mean $\bar{y}_{i0}$	30.53	32.73	30.53	15.43	16.58	22.78		

Calculate the following quantities:

$$\begin{aligned}
\text{CF} &= \frac{G^2}{n} = \frac{594.20^2}{6\times 4} = 14711.40167;\\
\text{TSS} &= \sum \text{Obs.}^2 - \text{CF} = 27.3^2 + 28.8^2 + \cdots + 16.2^2 + 22.3^2 - 14711.40167\\
&= 16156.72 - 14711.40167 = 1445.318;\\
\text{RSS} &= \frac{1}{6}\sum_{j=1}^{4} y_{0j}^2 - \text{CF} = \frac{1}{6}\left[134.3^2 + 168.7^2 + 150.4^2 + 140.8^2\right] - 14711.40167\\
&= \frac{88940.98}{6} - 14711.40167 = 112.095;\\
\text{TrSS} &= \frac{1}{4}\sum_{i=1}^{6} y_{i0}^2 - \text{CF} = \frac{1}{4}\left[122.1^2 + 130.9^2 + 122.1^2 + 61.7^2 + 66.3^2 + 91.1^2\right] - 14711.40167\\
&= \frac{63453.42}{4} - 14711.40167 = 1151.953;\\
\text{ErSS} &= \text{TSS} - \text{RSS} - \text{TrSS} = 1445.318 - 112.095 - 1151.953 = 181.27,
\end{aligned}$$

where TSS, RSS, TrSS, and ErSS are the total, replication, treatment, and error sum of squares, respectively.

Construct the ANOVA table as given below.

ANOVA				
SOV	d.f.	SS	MS	F
Replication	3	112.095	37.365	3.091935
Variety	5	1151.953	230.3907	19.06471
Error	15	181.270	12.08467	
Total	23	1445.318		

The table value of $F_{0.05;3,15} = 3.29$ and $F_{0.05;5,15} = 2.90$. Thus, we find that the test corresponding to replication is not significant, but the test corresponding to the effect of varieties is significant. So the null hypothesis of equality of replication effects cannot be rejected; that means there is no significant difference among the effects of the replications; they are statistically at par. On the other hand, the null hypothesis of equality of varietal effect is rejected; that means there exist significant differences among the varieties. So we are to identify the varieties, which varieties are significantly different from each other, and the best variety.

Calculate critical difference value at $\alpha = 0.05$ using the following formula:

$$\begin{aligned}
\text{CD}_{0.05}(\text{Variety}) &= \sqrt{\frac{2\text{MSE}}{r}} \times t_{0.025,\text{err.d.f.}}\\
&= \sqrt{\frac{2\times 12.085}{4}} \times t_{0.025,15}\\
&= \sqrt{\frac{2\times 12.085}{4}} \times 2.131 = 5.238.
\end{aligned}$$

Arrange the varietal mean values in descending order and compare the difference between any two treatment mean differences with that of the critical difference value. If the critical difference value be greater than the difference of two varietal means, then the treatments are statistically at par; there exists no significant difference among the means under comparison.

Variety	Mean yield (bushels/acre)	
V2	32.725	
V1	30.525	difference < CD(0.05)
V3	30.525	
V6	22.775	
V5	16.575	
V4	15.425	difference < CD(0.05)

Variety 2 is the best variety having highest yield, but variety 1 and variety 3 are also at par with that of variety 2. Variety 6 is significantly different from all other varieties. Variety 5 and variety 4 are statistically at par and are the lowest yielders among the varieties.

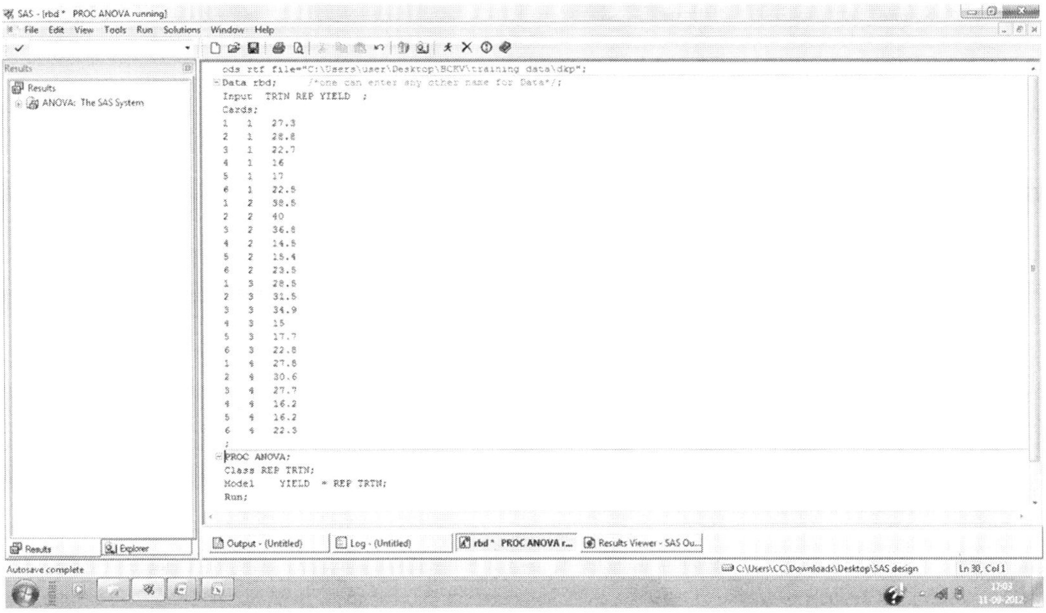

Slide 10.19: Step 1 showing data input and commands for RBD analysis using SAS

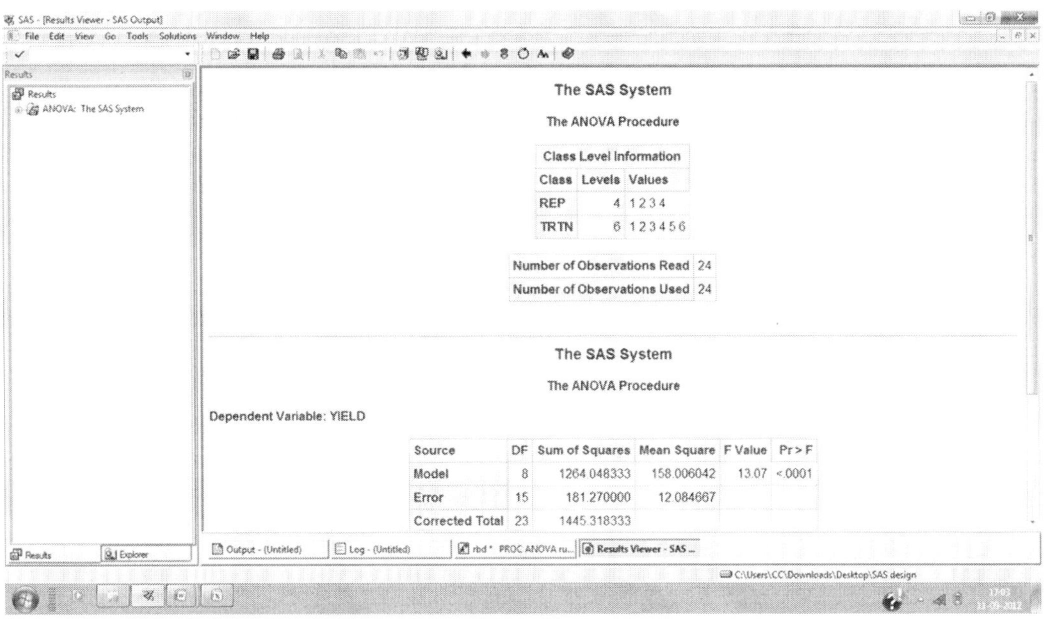

Slide 10.20: Step 2 showing a portion of output for RBD analysis using SAS

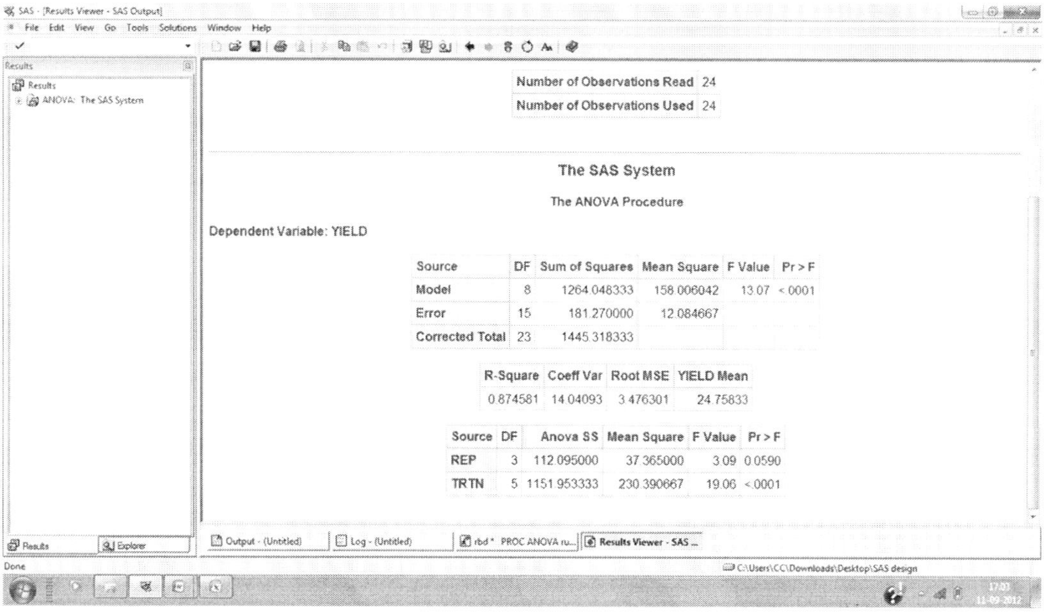

Slide 10.21: Step 2 showing a portion of output for RBD analysis using SAS

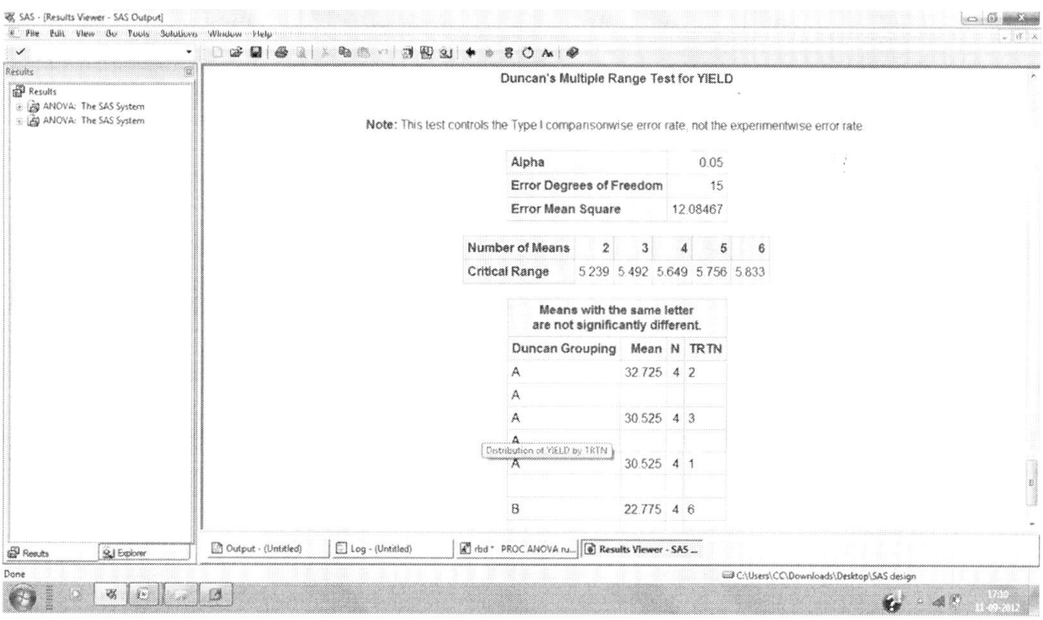

Slide 10.22: Step 2 showing a portion of output for RBD analysis using SAS

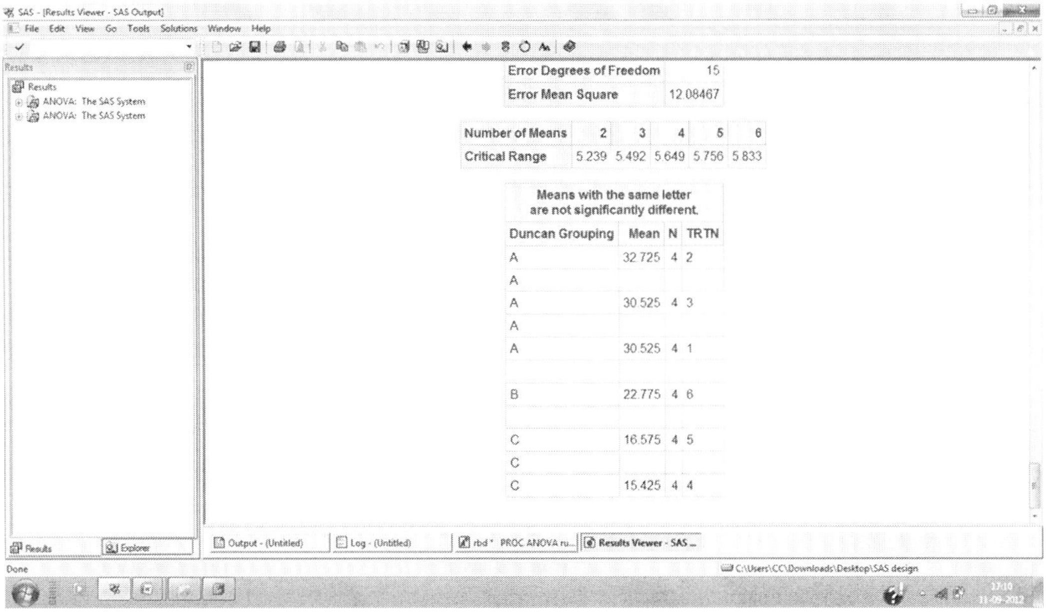

Slide 10.23: Step 2 showing a portion of output for RBD analysis using SAS

10.9 Latin Square Design (LSD)

LSD is a design in which known sources of variation in two perpendicular directions, that is, north to south and east to west or south to north and west to east, could be taken into consideration. In this type of field, we require framing of blocks into perpendicular directions which take care of the heterogeneity in both directions. A Latin square is an arrangement of treatments in such a way that each treatment occurs once and only once in each row and each column. If t is the number of treatments, then the total number of experimental units needed for this design is $t \times t$. These t^2 units are arranged in t rows and t columns. This type of experiments is rare in laboratory condition but can be conducted or useful in field conditions or greenhouse conditions. The two perpendicular sources of variations in greenhouse may be the difference among the rows of the plot and their distances from the greenhouses.

So while selecting an LSD design, an experimenter faces two problems: how to maintain the minimum replication and how to accommodate the maximum number of treatments in the experiment. This makes the LSD design's applicability limited in field experimentations. The number of treatments in LSD design should generally lie in between four and eight. All these limitations have resulted in the limited use of Latin square design in spite of its high potentiality for controlling experimental errors.

Analysis

Let there be t treatments, so there should be t rows and t columns. So we need a field of $t \times t$ experimental units. The appropriate statistical model for the analysis of the information from a $t \times t$ LSD can be given as follows:

$$y_{ijk} = \mu + \alpha_i + \beta_j + \upsilon_k + e_{ijk}, i = j = k = t$$

Where

μ = general effect

α_i = additional effect due to the ith treatment and $\sum \alpha_i = 0$

β_j = additional effect due to the jth row and $\sum r_j = 0$

υ_k = additional effect due to the kth treatment and $\sum c_k = 0$

e_{ijk} = errors that are associated with the ith treatment in the jth row and the kth column and are i.i.d. $N(0, \sigma^2)$

Table 10.12 ANOVA table for LSD

Sources of variation	d.f.	SS	MS	F-ratio
Treatment	$t-1$	TrSS	$\text{TrMS} = \frac{\text{TrSS}}{t-1}$	$\frac{\text{TrMS}}{\text{ErMS}}$
Row	$t-1$	RSS	$\text{RMS} = \frac{\text{RSS}}{t-1}$	$\frac{\text{RMS}}{\text{ErMS}}$
Column	$t-1$	CSS	$\text{RMS} = \frac{\text{CSS}}{t-1}$	$\frac{\text{CMS}}{\text{ErMS}}$
Error	$(t-1)(t-2)$	ErSS	$\text{ErMS} = \frac{\text{ErSS}}{(t-1)(t-2)}$	
Total	t^2-1	TSS		

Hypothesis to Be Tested

$$H_0: (1)\ \alpha_1 = \alpha_2 = \cdots = \alpha_i = \cdots = \alpha_t = 0,$$
$$(2)\ \beta_1 = \beta_2 = \cdots = \beta_j = \cdots = \beta_t = 0,$$
$$(3)\ \gamma_1 = \gamma_2 = \cdots = \gamma_k = \cdots = \gamma_t = 0$$

against

$$H_1: (1)\ \alpha\text{'s are not equal},$$
$$(2)\ \beta\text{'s are not equal},$$
$$(3)\ \gamma\text{'s are not equal}.$$

Analysis

From least square estimates, we have

$$\sum\sum\sum_{i,j,k}(y_{ijk}-\bar{y}_{000})^2 = t\sum_i(\bar{y}_{i00}-\bar{y}_{000})^2 + t\sum_j(\bar{y}_{0j0}-\bar{y}_{000})^2 + t\sum_k(\bar{y}_{00k}-\bar{y}_{000})^2$$
$$+\sum_{i,j,k}(y_{ijk}-\bar{y}_{i00}-\bar{y}_{0j0}-\bar{y}_{00k}+2\bar{y}_{000})^2,$$

TSS = Tr SS + RSS + CSS + ErSS,

where

$\hat{\mu} = \bar{y}_{000},\ \hat{\alpha}_i = \bar{y}_{i00} - \bar{y}_{000}, \hat{\beta}_j = \bar{y}_{0j0} - \bar{y}_{000},$

$\hat{\gamma}_k = \bar{y}_{00k} - \bar{y}_{000},$

$\bar{y}_{000}$ = mean of all t^2 observations,

$\bar{y}_{i00}$ = mean of t observations from ith treatment,

$\bar{y}_{0j0}$ = mean of t observations from jth row,

$\bar{y}_{00k}$ = mean of t observations from kth column.

Various sums of squares are calculated using the following formulae (Table 10.12):

$$\text{Grand total} = \sum_{(i,j,k)\in D} y_{ijk} = G$$

$$\text{Correction factor} = \frac{(G)^2}{t^2} = \text{CF}$$

$$\text{Total sum of squares(TSS)} = \sum_{(i,j,k)\in D} {y_{ijk}}^2 - \text{CF}$$

$$\text{Treatment sum of squares(TrSS)} = \frac{\sum_{i=1}^{t} {y_{i00}}^2}{t} - \text{CF}$$
$$= \frac{y_{100}^2}{t} + \frac{y_{200}^2}{t} + \frac{y_{300}^2}{t} + \cdots \frac{y_{i00}^2}{t} + \cdots + \frac{y_{t00}^2}{t} - \text{CF}$$

$$\text{Row sum of squares(RSS)} = \frac{\sum_{j=1}^{t} {y_{0j0}}^2}{t} - \text{CF}$$
$$= \frac{y_{010}^2}{t} + \frac{y_{020}^2}{t} + \frac{y_{030}^2}{t} + \cdots \frac{y_{0j0}^2}{t} + \cdots + \frac{y_{0t0}^2}{t} - \text{CF}$$

$$\text{Column sum of squares(CSS)} = \frac{\sum_{k=1}^{t} {y_{00k}}^2}{t} - \text{CF}$$
$$= \frac{y_{001}^2}{t} + \frac{y_{002}^2}{t} + \frac{y_{003}^2}{t} + \cdots \frac{y_{00k}^2}{t} + \cdots + \frac{y_{00t}^2}{t} - \text{CF}$$

Error sum of squares(by subtraction)

$$= \text{TSS} - \text{TrSS} - \text{RSS} - \text{CSS}$$

The null hypotheses are rejected at α level of significance if the calculated values of F-ratio corresponding to treatment or row or column are greater than the corresponding table values at the same level of significance with the same $(t-1)$, $(t-1)(t-2)$ degrees of freedom, respectively. When any one of the three null hypotheses is rejected, then LSD for treatments/rows/columns is calculated $\sqrt{\frac{2\text{ErMS}}{t}} \times t_{\frac{\alpha}{2};(t-1)(t-2)}$, where $t_{\frac{\alpha}{2};(t-1)(t-2)}$ is the table value of t at α level of significance with $(t-1)(t-2)$ degrees of freedom.

If the absolute value of the difference between any pair of means of row/column/treatment is more than the critical difference value, as calculated above, then the row/column/treatment means are significantly different from each other; otherwise, they are not.

Example 10.12. Five varieties of wheat were tested using the Latin square design. The following information pertains to the yield data (q/ha) along with the layout of the above experiment. Analyze the data and find out the best variety.

Layout along with yield data (q/ha)				
D 34.0	A 19.1	E 21.1	B 32.0	C 37.2
E 16.2	B 33.1	A 19.0	C 34.3	D 28.1
C 30.6	E 28.5	B 33.1	D 35.8	A 19.2
A 25.8	C 26.1	D 41.7	E 23.7	B 39.9
B 39.3	D 24.6	C 36.1	A 21.3	E 19.4

Solution. The model for LSD is $y_{ijk} = \mu + \alpha_i + \beta_j + \gamma_k + e_{ijk},\ i = j = k = 5$

where

y_{ijk} = effect due to the ith variety in the jth row and the kth column

μ = general effect

α_i = additional effect due to the ith variety, $\sum_i \alpha_i = 0$

β_j = additional effect due to the jth row, $\sum_j \beta_j = 0$

γ_k = additional effect due to the kth column, $\sum_k \gamma_k = 0$

e_{ijk} = errors that are associated with the ith variety in the jth row and kth column and are i.i.d. $N(0, \sigma^2)$

The hypotheses to be tested are

$$H_0 : \alpha_1 = \alpha_2 = \alpha_3 = \alpha_4 = \alpha_5,$$
$$\beta_1 = \beta_2 = \beta_3 = \beta_4 = \beta_5,$$
$$\gamma_1 - \gamma_2 - \gamma_3 - \gamma_4 - \gamma_5$$

against

H_1 : α_i's are not all equal,

β_j's are not all equal,

γ_k's are not all equal.

Let the level of significance be $\alpha = 0.05$.

Make the following two tables from the given information:

	C1	C2	C3	C4	C5	Total (y_{0j0})	Average ($\bar{y}_{0j0}$)
R1	34	19.1	21.1	32	37.2	143.4	28.68
R2	16.2	33.1	19	34.3	28.1	130.7	26.14
R3	30.6	28.5	33.1	35.8	19.2	147.2	29.44
R4	25.8	26.1	41.7	23.7	39.9	157.2	31.44
R5	39.3	24.6	36.1	21.3	19.4	140.7	28.14
Total (y_{00k})	145.9	131.4	151	147.1	143.8	719.2	
Average ($\bar{y}_{00k}$)	29.18	26.28	30.2	29.42	28.76		

	Varieties				
	A	B	C	D	E
	25.8	39.3	30.6	34	16.2
	19.1	33.1	26.1	24.6	28.5
	19	33.1	36.1	41.7	21.1
	21.3	32	34.3	35.8	23.7
	19.2	39.9	37.2	28.1	19.4
Total (y_{ioo})	104.4	177.4	164.3	164.2	108.9
Average ($\bar{y}_{ioo}$)	20.88	35.48	32.86	32.84	21.78

Calculate the following quantities:

$$\mathrm{CF} = \frac{G^2}{t \times t} = \frac{719.20^2}{5 \times 5} = 20689.9;$$

$$\mathrm{TSS} = \sum \mathrm{Obs.}^2 - \mathrm{CF} = 34.0^2 + 19.1^2 + \cdots + 21.3^2 + 19.4^2 - 20689.9 = 1385.91;$$

$$\mathrm{RSS} = \frac{1}{5}\sum_{j=1}^{5} y_{0j0}^2 - \mathrm{CF} = \frac{1}{5}\left[143.4^2 + 130.7^2 + \cdots + 140.7^2\right] - 20689.9456 = 74.498;$$

$$\mathrm{CSS} = \frac{1}{5}\sum_{k=1}^{5} y_{00k}^2 - \mathrm{CF} = \frac{1}{5}\left[145.9^2 + 131.4^2 + \cdots + 143.80^2\right] - 20689.9456 = 44.178;$$

$$\mathrm{VSS} = \frac{1}{5}\sum_{i=1}^{5} y_{i00}^2 - \mathrm{CF} = \frac{1}{5}\left[104.4^2 + 177.4^2 + \cdots + 108.9^2\right] - 20689.9456 = 947.146;$$

$$\mathrm{ErSS} = \mathrm{TSS} - \mathrm{RSS} - \mathrm{CSS} - \mathrm{VSS} = 1385.914 - 74.498 - 44.178 - 947.146 = 320.091,$$

where TSS, RSS, CSS, VSS, and ErSS are the total, row, column, variety, and error sum of squares, respectively.

Construct the ANOVA table as given below.

ANOVA				
SOV	d.f.	SS	MS	F
Row	4	74.498	18.625	0.698
Column	4	44.178	11.045	0.414
Variety	4	947.146	236.787	8.877
Error	12	320.091	26.674	
Total	24	1385.914		

The table value of $F_{0.05;4,12} = 3.26$. Thus, we find that the calculated values of F are less, except for variety, than the corresponding table value. So the tests for row and columns are non-significant. We conclude that neither the row effects nor the column effects are significant. But the effects of varieties are significant. So we are to identify the best variety.

Calculate CD(0.05) using the following formula:

$$\mathrm{CD}_{0.05}(\mathrm{Variety}) = \sqrt{\frac{2 \times \mathrm{MSE}}{t}} \times t_{0.025,\ \mathrm{err.\ d.f.}}$$
$$= \sqrt{\frac{2 \times 26.674}{5}} \times t_{0.025,12}$$
$$= \sqrt{\frac{2 \times 26.674}{5}} \times 2.179$$
$$= 7.118.$$

Arrange the varietal mean values in descending order and compare the difference between any two treatment mean differences with that of the critical difference value. If the critical difference value is greater than the difference of two varietal means, then the treatments are statistically at par; there exists no significant difference among the means under comparison.

Variety	Avg. yield (q/ha)
B	35.48
C	32.86
D	32.84
E	21.78
A	20.88

Variety B is the best variety having the highest yield but variety C and variety D are also at par with that of variety B. Variety A and E are significantly different from the other three varieties and are also statistically at par with variety A as the lowest yielders among the varieties.

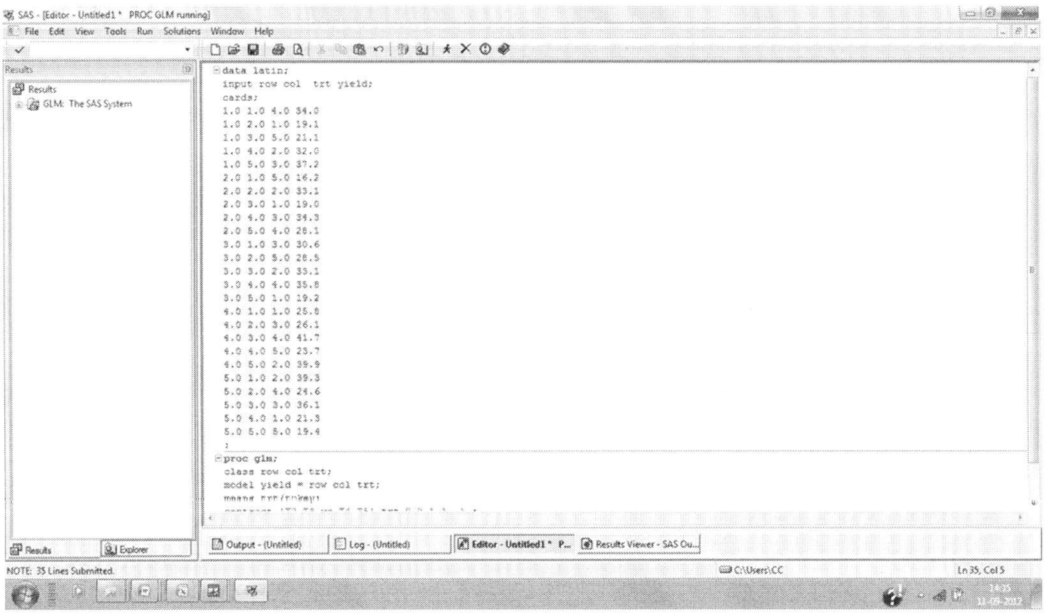

Slide 10.24: Step 1 showing data entry and portion of commands for LSD analysis using SAS

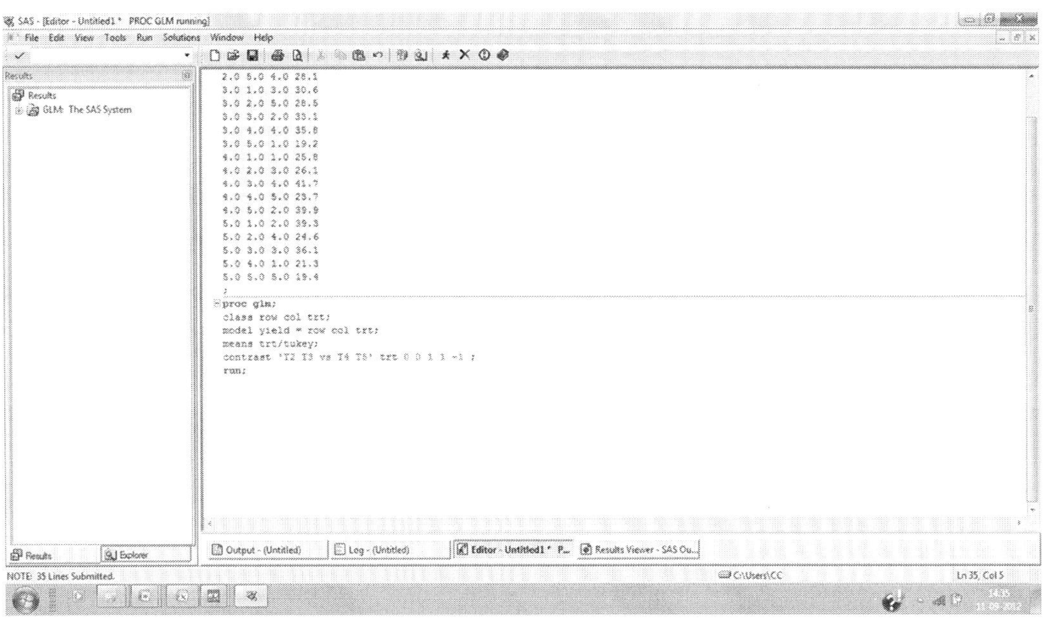

Slide 10.25: Step 2 showing portion of data entry and commands for LSD analysis using SAS

The above analysis can also be done using SAS statistical software. The following few slides present the data entry, command syntax, and the output for the same.

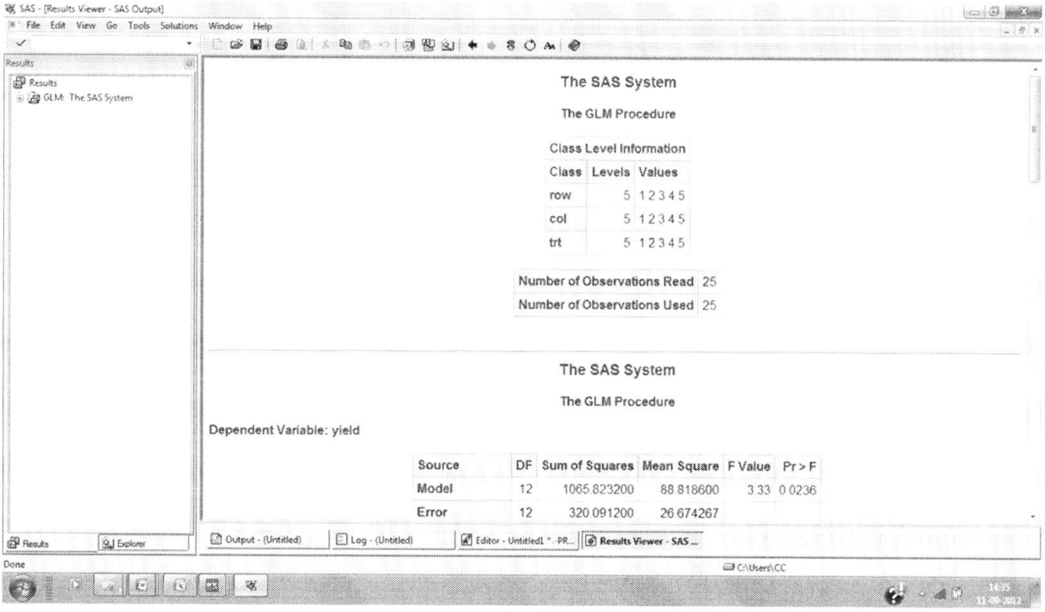

Slide 10.26: Step 3 showing portion of output for LSD analysis using SAS

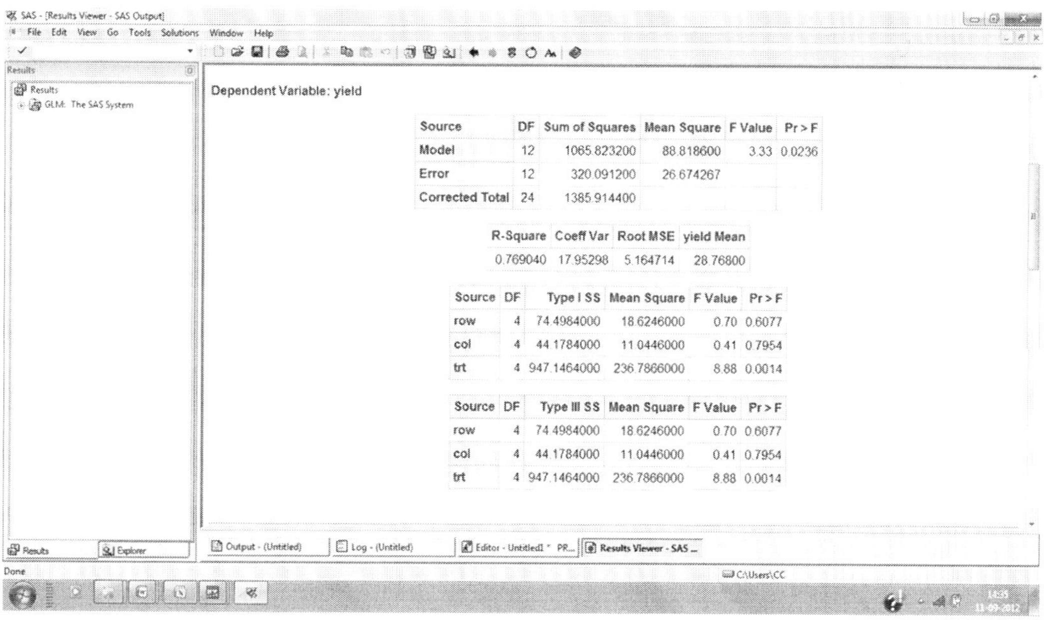

Slide 10.27: Step 3 showing portion of output for LSD analysis using SAS

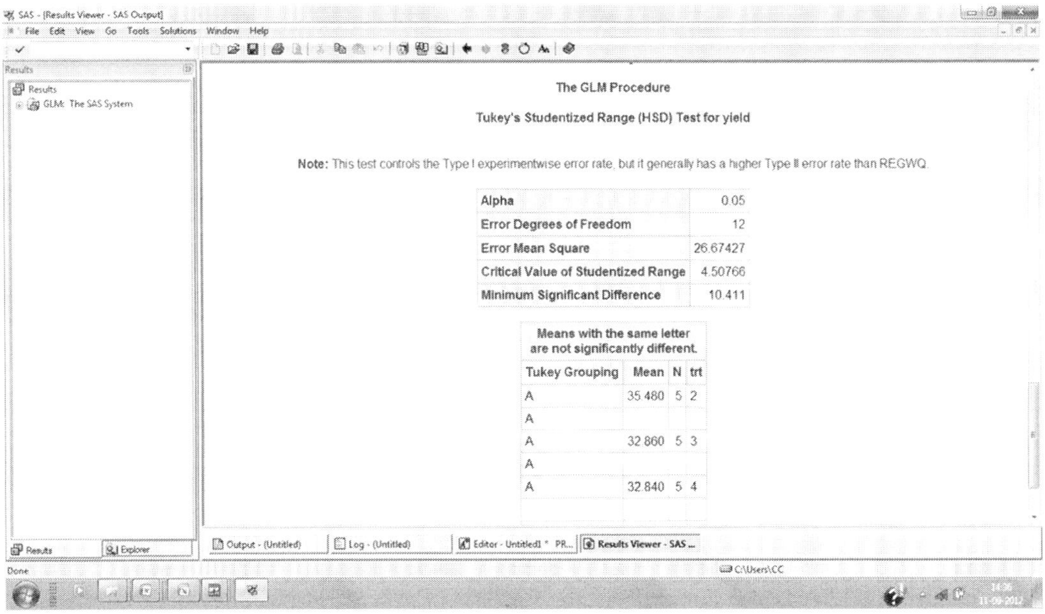

Slide 10.28: Step 3 showing portion of output for LSD analysis using SAS

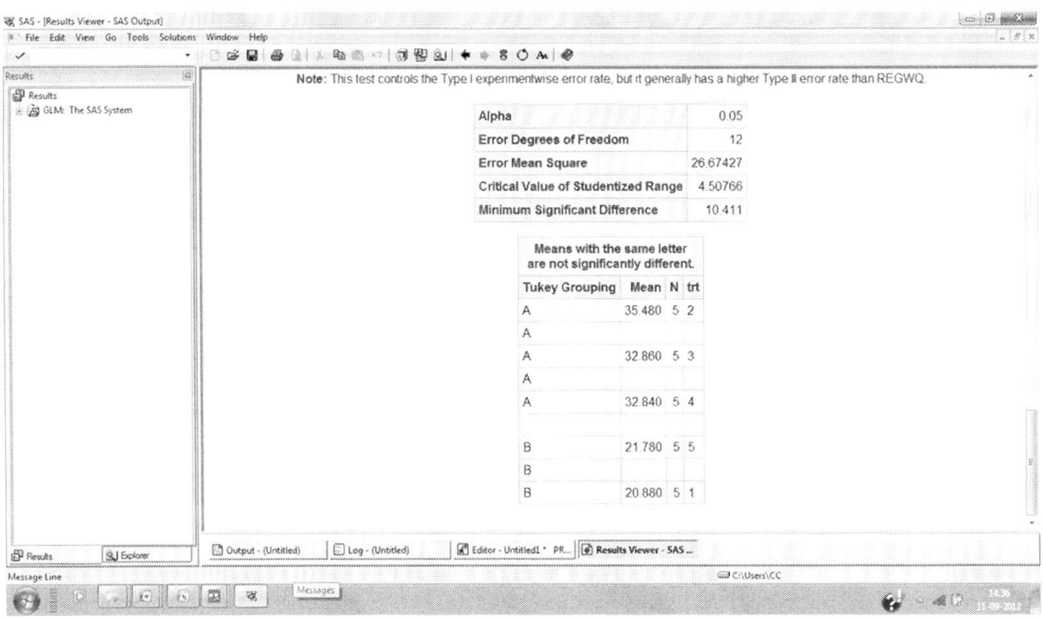

Slide 10.29: Step 3 showing portion of output for LSD analysis using SAS

10.10 Missing Plot Technique

In many of the experiments, it is found that the information from an experimental unit is missing because of some reasons or otherwise. Responses from a particular experimental unit may be lost. Crops of a particular experimental unit may be destroyed, animals under a particular treatment may die because of some reason, fruits/flowers from a particular experimental unit may be stolen, errors on the part of the data recorder during recording time, etc., may result in missing data. *In this connection, the difference between the missing observation and the zero observation should be clearly understood.* For example, if because of an insecticide treatment, the number of insects count per plant is zero or if because of a treatment in studying the pest control measure the yield of any experimental units is zero, then these should be entered zero, not to be treated as missing values. If the information from the whole experiments is to be discarded because of one or two missing values, it will be a great loss of time, resources, and other factors. In order to avoid and overcome the situations, missing plot technique has been developed. Missing observation can, however, be estimated following the least square technique, and application of analysis of variance with some modification can be used for practical purposes to provide reasonably correct result.

In CRD, the above technique is of little use because of the fact that in CRD, analysis of variance is possible with variable number of replications for different treatments. Thus, if one observation from a treatment is missing, then analysis of variance is to be taken up with $(r - 1)$ replication for the corresponding treatment and total $(n - 1)$ number of observations instead of n observation for the whole experiment. But the effect of missing observations on the surrounding experimental units should be noted carefully.

10.10.1 Missing Plot Technique in RBD

Let the following table provide the observations of an RBD with a missing observation y_{ij} (Table 10.13):

With reference to the above table, the observation y_{ij} is missing. This missing value $y_{ij} = y$ (say) is estimated by minimizing error sum of squares, and the corresponding estimate would be

$$y = \frac{ty'_{i0} + ry'_{0j} - y'_{00}}{(r-1)(t-1)}$$

where

y'_{i0} is the total of known observations in the ith treatment

y'_{0j} is the total of known observations in jth replication (block)

y'_{00} is the total of all known observations

Once the estimated value for the missing observation is worked out, the usual analysis of variance is taken up with the estimated value of the missing observation. The treatment sum of squares is corrected by subtracting the upward bias

Table 10.13 Observations from RBD with one missing value

	Replications (blocks)						
Treatments	1	2	....	j	....	r	Total
1	y_{11}	y_{12}	....	y_{1j}	....	y_{1r}	y_{10}
2	y_{21}	y_{22}	....	y_{2j}	....	y_{2r}	y_{20}
:	:	:	:	:	:	:	:
i	y_{i1}	y_{i2}	....	–	....	y_{ir}	y'_{i0}
:	:	:	:	:	:	:	:
t	y_{t1}	y_{t2}	....	y_{tj}	....	y_{tr}	y_{t0}
Total	y_{01}	y_{02}	....	y'_{0j}	....	y_{0r}	y'_{00}

$$B = \frac{\left[y'_{0j} - (t-1)y\right]^2}{t(t-1)}.$$

The degrees of freedom for both total and error sum of squares are reduced by one in each case. The treatment means are compared with the mean having missing value and no missing value using the formula for standard error of difference as

$$\mathrm{SE}_d = \sqrt{\frac{\mathrm{ErMS}}{r}\left[2 + \frac{t}{(r-1)(t-1)}\right]}.$$

Example 10.13. An experiment with six varieties of onion was conducted using RBD with four replications. The following table gives the layout and data pertaining to yield (t/ha) from the experiment. Analyze the data and find out the best variety of onion.

Rep-1	Rep-2	Rep-3	Rep-4
V1 17.8	V2 23.8	V3 27.9	V4 15.5
V5 19.2	V3 26.7	V5 19.7	V6 23.5
V2 23.6	V5	V2 23.5	V3 26.8
V4 16.2	V6 22.5	V4 15.0	V2 24.0
V6 22.3	V4 16	V1 18.5	V5 19.4
V3 27.7	V1 17.3	V6 22.8	V1 18.5

As per the given information, the appropriate model is given by $y_{ij} = \mu + \alpha_i + \beta_j + e_{ij}$

where $i = 6, j = 4$

y_{ij} = effect due to the ith variety in jth replicates

μ = general effect

α_i = additional effect due to the ith variety

β_j = additional effect due to the jth replicate

e_{ij} = errors that are associated with the ith variety in the jth replicate and are i.i.d. $N(0, \sigma^2)$

The hypotheses to be tested are

$$H_0: \alpha_1 = \alpha_2 = \alpha_3 = \alpha_4 = \alpha_5 = \alpha_6,$$
$$\beta_1 = \beta_2 = \beta_3 = \beta_4$$

against

$$H_1: \alpha_i\text{'s are not all equal,}$$
$$\beta_j\text{'s are not all equal.}$$

Let the level of significance be $\alpha = 0.05$.

We shall analyze the data in the following steps:

Make the following table from the given information.

Variety	Rep-1	Rep-2	Rep-3	Rep-4	Total
V1	17.8	17.3	18.5	18.5	72.10
V2	23.6	23.8	23.5	24	94.90
V3	27.7	26.7	27.9	26.8	109.10
V4	16.2	16	15	15.5	62.70
V5	19.2	*X*	19.7	19.4	58.30
V6	22.3	22.5	22.8	23.5	91.10
Total	126.80	106.30	127.40	127.70	**488.20**

The estimate of the missing value is given by

$$\hat{y}_{ij} = X = \frac{rR' + tT' - G'}{(r-1)(t-1)}$$

where

X = estimate of the missing value

R' = total of available entries of the replication having the missing observation

T' = total of available entries of the treatment having the missing observation

G' = total of available entries in the whole design

For this problem $X = \frac{rR' + tT' - G'}{(r-1)(t-1)} = \frac{4 \times 106.3 + 6 \times 58.3 - 488.2}{(4-1)(6-1)} = 19.12.$

Now, $G = G' + X = 488.2 + 19.12 = 507.32$

Total of variety five $= 58.30 + X = 77.42$

Total of replication four $= 106.30 + X = 125.42$

Now we proceed for usual analysis of variance is taken up with the estimated values of the missing observation.

$$\mathrm{CF} = \frac{\mathrm{GT}^2}{n} = \frac{507.32^2}{6 \times 4} = 10723.90,$$

$$\mathrm{TSS} = \sum \mathrm{Obs.}^2 - \mathrm{CF} = 18.5^2 + 19.7^2 + \cdots + 19.12^2 + 17.3^2 - 10723.90 = 363.20,$$

$$\mathrm{RSS} = \frac{1}{6}\sum_{j=1}^{4} R_j^2 - \mathrm{CF} = \frac{1}{6}\left[127.4^2 + 127.7^2 + 126.8^2 + 106.3^2\right] - 10723.90 = 0.51,$$

$$\mathrm{TrSS} = \frac{1}{4}\sum_{i=1}^{6} V_i^2 - \mathrm{CF} = \frac{1}{4}\left[72.1^2 + 94.9^2 + 109.1^2 + 62.7^2 + 77.42^2 + 91.1^2\right] - 10723.90 = 359.00.$$

The TrSS is an overestimate and has to be corrected by subtracting from a quantity (bias)

$$B = \frac{[R' - (t-1)X]^2}{t(t-1)}$$

$$= \frac{[106.30 - (6-1)19.12]^2}{6(6-1)} = 3.816$$

$$\text{Corrected TrSS} = \text{TrSS} - B = 359.00 - 3.816 = 355.18;$$

$$\text{ErSS} = \text{TSS} - \text{RSS} - \text{TrSS(corrected)} = 363.20 - 0.51 - 355.18 = 7.50,$$

where TSS, RSS, TrSS, and ErSS are the total, replication, treatment, and error sum of squares, respectively.

Construct the ANOVA table as given below with the help of the above sum of squares values.

ANOVA				
SOV	d.f.	SS	MS	F
Replication	3	0.51	0.17	0.31733
Variety	5	359.00	71.80	272.38
Error	14	3.69	0.2636	
Total	22	363.20		
Variety (corrected)	5	355.18	71.036	132.51
Error	14	7.506	0.5361	

Let the level of significance be $\alpha = 0.05$.

Thus, one can find that the varietal effects differ significantly among themselves. So we are to calculate the CD values for comparing the treatment means.

The treatment means having no missing value are compared by usual CD value given as

$$\text{CD}_{0.05} = \sqrt{\frac{2\text{ErMS}}{r}} \times t_{0.025,\ \text{error df}}$$

$$= \sqrt{\frac{2 \times 0.5361}{4}} \times t_{0.025,14}$$

$$= \sqrt{\frac{2 \times 0.5361}{4}} \times 2.145 = 1.110.$$

The treatment means are compared with the mean having missing values using the formula for standard error of difference as

$$\text{SE}_d = \sqrt{\frac{\text{ErMS}}{r}\left[2 + \frac{t}{(r-1)(t-1)}\right]}$$

$$= \sqrt{\frac{0.2636}{4}\left[2 + \frac{6}{(4-1)(6-1)}\right]} = 0.398.$$

Thus, to compare the variety having missing value (V5) with the other varieties having no missing value, $\text{SE}_d \times t_{0.025,14} = 0.398 \times 2.145 = 0.854$.

Variety	Average yield
V3	27.28
V2	23.73
V6	22.78
V5	19.36
V1	18.03
V4	15.68

From the table of means it is clear that variety V3 is the highest yielder followed by V2, V6, and so on.

10.10.2 Missing Plot Technique in LSD

For a missing observation in $t \times t$ Latin square, let it be denoted by y_{ijk} and let T', R', C', and G' be the total of available observations (excluding the missing value) of ith treatment, jth row, kth column, and all available observations, respectively; then the least square estimate for the missing value is given by

$$\hat{y} = \frac{t(T' + R' + C') - 2G'}{(t-1)(t-2)}.$$

Once the estimated value for the missing observation is worked out, the usual analysis of variance is taken up with the estimated value of the missing observation. The treatment sum of squares is to be corrected by subtracting the upward biased

$$B = \frac{[(t-1)T' + R' + C' - G']^2}{[(t-1)(t-2)]^2}.$$

The degrees of freedom for both total and error sum of squares is reduced by one in each case. The treatment means are compared with the mean having missing value using the formula for standard error of difference as

$$\mathrm{SE}_d = \sqrt{\frac{\mathrm{ErMS}}{t}\left[2 + \frac{t}{(t-1)(t-2)}\right]}.$$

Example 10.14. Five varieties of paddy were tested in LSD for yield. The following is the layout and responses due to different treatments with a missing value in variety 3. Estimate the missing yield and find out which variety is the best yielder among the varieties of paddy.

V1 10.5	V4 9.5	V5 13.0	V3 14.5	V2 17.5
V4 10.0	V2 16.5	V1 11.0	V5 12.5	V3 13.5
V5 13.5	V3	V4 9.0	V2 17.0	V1 12.0
V3 14.0	V5 12.75	V2 17.6	V1 10.0	V4 10.0
V2 17.9	V1 11.5	V3 14.5	V4 10.5	V5 12.25

Solution. The model for LSD is
$y_{ijk} = \mu + \alpha_i + \beta_j + \gamma_k + e_{ijk}$
where $i = j = k = 5$
y_{ijk} = effect due to the ith variety in the jth row and the kth column
μ = general effect
α_i = additional effect due to the ith variety $\sum_i \alpha_i = 0$
β_j = additional effect due to the jth row $\sum_j \beta_j = 0$
γ_k = additional effect due to the kth column $\sum_k \gamma_k = 0$
e_{ijk} = errors that are associated with the ith variety in the jth row and the kth column and are i.i.d. $N(0, \sigma^2)$.

The hypotheses to be tested are

$$H_0: \alpha_1 = \alpha_2 = \alpha_3 = \alpha_4 = \alpha_5,$$
$$\beta_1 = \beta_2 = \beta_3 = \beta_4 = \beta_5,$$
$$\gamma_1 = \gamma_2 = \gamma_3 = \gamma_4 = \gamma_5$$

against

H_1 : α_i's are not all equal,
β_j's are not all equal,
γ_k's are not all equal.

Let the level of significance be $\alpha = 0.05$.

We shall analyze the data in the following steps:

Make the following two tables from the given information.

	C1	C2	C3	C4	C5	Total (y_{0j0})	Average ($\bar{y}_{0j0}$)
R1	10.5	9.5	13.0	14.5	17.5	65.0	13.0
R2	10.0	16.5	11.0	12.5	13.5	63.5	12.7
R3	13.5	V3	9.0	17.0	12.0	51.5	12.9
R4	14.0	12.75	17.6	10.0	10.0	64.35	12.9
R5	17.9	11.5	14.5	10.5	12.25	66.65	13.3
Total (y_{00k})	65.9	50.25	65.1	64.5	65.25	311	
Average ($\bar{y}_{00k}$)	13.18	12.563	13.02	12.9	13.05		

	Varieties				
	V1	V2	V3	V4	V5
	10.5	17.5	14.5	9.5	13
	11	16.5	13.5	10	12.5
	12	17	y	9	13.5
	10	17.6	14	10	12.75
	11.5	17.9	14.5	10.5	12.25
Total (y_{i00})	55	86.5	56.5	49	64
Average ($\bar{y}_{i00}$)	11	17.3	14.125	9.8	12.8

$$\hat{y} = X = \frac{t(R' + C' + T') - 2G'}{(t-1)(t-2)}$$
$$= \frac{5(51.5 + 50.25 + 56.5) - 2 \times 311}{(5-1)(5-2)} = 14.104$$

where

X = estimate of the missing value

R' = total of available entries of the row having the missing observation

T' = total of available entries of the variety V3 having the missing observation

C' = total of available entries of the column having the missing observation

G' = total of available entries in the whole design

Now, $G = G' + X = 311 + 14.104 = 325.104$

Total of variety three $= 56.5 + X = 70.604$

Total of row three $= 51.5 + X = 65.604$

Total of column two $= 50.25 + X = 64.354$

Calculate the following quantities:

$$\text{CF} = \frac{G^2}{t \times t} = \frac{325.104^2}{5 \times 5} = 4227.704,$$
$$\text{TSS} = \sum \text{Obs.}^2 - \text{CF} = 10.5^2 + 9.5^2 + \cdots + 10.5^2 + 12.25^2 - 4227.704 = 176.7634,$$
$$\text{RSS} = \frac{1}{5}\sum_{j=1}^{5} y_{0j0}^2 - \text{CF} = \frac{1}{5}\left[65^2 + 63.5^2 + \cdots + 66.65^2\right] - 4227.704 = 1.151531,$$
$$\text{CSS} = \frac{1}{5}\sum_{k=1}^{5} y_{00k}^2 - \text{CF} = \frac{1}{5}\left[65.9^2 + 64.354^2 + \cdots + 65.25^2\right] - 4227.704 = 0.309531,$$
$$\text{VSS} = \frac{1}{5}\sum_{i=1}^{5} y_{i00}^2 - \text{CF} = \frac{1}{5}\left[55^2 + 86.5^2 + 70.604^2 + 49^2 + 64^2\right] - 4227.704 = 170.1305,$$
$$\text{ErSS} = \text{TSS} - \text{RSS} - \text{CSS} - \text{VSS} = 176.7634 - 1.151531 - 0.309531 - 170.1305 = 5.171792$$

where TSS, RSS, CSS, VSS, and ErSS are the total, row, column, variety, and error sum of squares, respectively.

The upward bias is calculated as follows:

$$B = \frac{(G' - R' - C' - (t-1)T')^2}{[(t-1)(t-2)]^2}$$
$$= \frac{(311 - 51.5 - 50.25 - (5-1)56.3)^2}{[(5-1)(5-2)]^2} = 1.948.$$

Construct the ANOVA table as given below using the above values:

ANOVA				
SOV	d.f.	SS	MS	F
Row	4	1.152	0.288	<1
Column	4	0.309	0.077	<1
Variety	4	170.131		
Error	11	5.172		
Total	23	176.763		
Variety (corrected)		168.182	42.046	64.957
Error		7.120	0.647	

The table value of $F_{0.05;4,11} = 3.36$. Thus, we find that the calculated values of F are less than the corresponding table value, except for variety. So the tests for row and columns are nonsignificant. We conclude that neither the row effects nor the column effects are significant. But the effects of varieties are significant. So we are to identify the best variety.

To compare the varietal means involving no missing value, the CD(0.05) is calculated using the following formula:

$$\text{CD}_{0.05}(\text{Variety}) = \sqrt{\frac{2 \times \text{MSE}}{t}} \times t_{0.025,\,\text{err.df.}}$$
$$= \sqrt{\frac{2 \times MSE}{5}} \times t_{0.025,11}$$
$$= \sqrt{\frac{2 \times 0.647286}{5}} \times 2.201 = 1.12,$$

and

to compare the varietal means with variety 3, involving missing value, the CD(0.05) is calculated using the following formula:

$$\mathrm{CD}_{0.05}(\text{Variety}) = \sqrt{\frac{\mathrm{MSE}}{t}\left(2+\frac{t}{(t-1)(t-2)}\right)} \times t_{0.025,\,\text{err.df.}} = \sqrt{\frac{\mathrm{MSE}}{t}\left(2+\frac{t}{(t-1)(t-2)}\right)} \times t_{0.025,11}$$

$$= \sqrt{\frac{0.647286}{5}\left(2+\frac{5}{4\times 3}\right)} \times 2.201 = 1.231.$$

Variety	Mean yield
V2	17.30
V3	14.12
V5	12.80
V1	11.00
V4	9.80

Comparing the varietal differences with appropriate CD values, it can be inferred that all the varieties are significantly different from each other. Variety V2 is the best yielder, while variety V4 is the lowest yielder.

10.11 Factorial Experiment

Instead of conducting many single-factor experiments to fulfill the objectives, a researcher is always in search of experimental method in which more than one set of treatments could be tested or compared. Moreover, if the experimenter not only wants to know the level or the doses of individual factor giving better result but also wants to know the combination/interaction of the levels of different factors which is producing the best result, factorial experiment is the answer. For example, an experimenter may be interested to know not only the best dose of nitrogen, phosphorus, and potassium but also the best dose combination of these three essential plant nutrients to get the best result in guava. This can be done by designing his experiment in such a way that the best levels of all these three factors, N, P, and K, can be identified along with the combination of N, P, and K to get the best result. Factorial experiments are the methods for inclusions of more than one factor to be compared for their individual effects as well as interaction effects in a single experiment. The essential analysis of factorial experiments is accomplished through two-way analysis of variance, as already discussed and other methods.

Factors

A factor is a group of treatments. In a factorial experiment different varieties may form a factor and different doses of nitrogen or different irrigation methods may form other factors in the same experiment. Here variety, nitrogen (irrespective dose), and irrigation schedule (irrespective of types) are the factors of the experiment.

Levels of Factors

Different components of a factor are known as the levels of the factor. Different varieties under the factor variety form the level of the factor variety; similarly doses of nitrogen, types of irrigation, etc., are the levels of the factors nitrogen, irrigation, etc., respectively.

10.11.1 $m \times n$ Factorial Experiment

The dimension of a factorial experiment is indicated by the number of factors and the number of levels of each factor. Thus, a 3×4 factorial experiment means an experiment with 2 factors, one with 3 levels and another with 4 levels.

Let us assume that an $m \times n$ experiment is conducted in a randomized block design with r replication. So there would be two factors: the first factor is having m levels and the second factor is having n levels and, all together, $m \times n$ treatment combinations in each replication. Thus, the model for the design can be presented as $y_{ijk} = \mu + \alpha_i + \beta_j + (\alpha\beta)_{ij} + \gamma_k + e_{ijk}$

where $i = 1, 2, \ldots, m; j = 1, 2, \ldots, n; k = 1, 2, \ldots, r$

y_{ijk} = response in the kth observation due to the ith level of the first factor A and the jth level of the second factor B

μ = general effect

α_i = additional effect due to the ith level of first factor A, $\sum \alpha_i = 0$

Table 10.14 Plot-wise observations from factorial RBD experiment

B	Replication 1				Replication 2					Replication r			
A	b_1	b_2	$\ldots b_j \ldots$	b_n	b_1	b_2	$\ldots b_j \ldots$	b_n	$\ldots$	b_1	b_2	$\ldots b_j \ldots$	b_n
a_1	y_{111}	y_{121}	y_{1j1}	y_{1n1}	y_{112}	y_{122}	y_{1j2}	y_{1n2}	$\ldots$	y_{11r}	y_{12r}	y_{1jr}	y_{1nr}
a_2	y_{211}	y_{221}	y_{2j1}	y_{2n1}	y_{212}	y_{222}	y_{2j2}	y_{2n2}	$\ldots$	y_{21r}	y_{22r}	y_{2jr}	y_{2nr}
.	.	.	……										
:	:	:	……										
:	:	:	……										
a_i	y_{i11}	y_{i21}	y_{ij1}	y_{in1}	y_{i12}	y_{i22}	y_{ij2}	y_{in2}		y_{i1r}	y_{i2r}	y_{ijr}	y_{inr}
:	:	:	:										
:	:	:	:										
a_m	y_{m11}	y_{m21}	y_{mj1}	y_{mn1}	y_{m12}	y_{m22}	y_{mj2}	y_{mn2}		y_{m1r}	y_{m2r}	y_{mjr}	y_{mnr}

β_j = additional effect due to the jth level of B, $\sum \beta_j = 0$

$(\alpha\beta)_{ij}$ = interaction effect of the ith level of the first factor and the jth level of B, $\sum_i (\alpha\beta)_{ij} = \sum_j (\alpha\beta)_{ij} = 0$

γ_k = additional effect due to the kth replicate, $\sum \gamma_k = 0$

e_{ijk} = error component associated with the ith level of the first factor and the jth level of B in kth replicate and $e_{ijk} \sim$ i.i.d. $N(0, \sigma^2)$

We have the following hypotheses for $m \times n$ factorial experiment in RBD to be tested:

$$H_{01} : \alpha_1 = \alpha_2 = \ldots = \alpha_i = \ldots = \alpha_m = 0,$$
$$H_{02} : \beta_1 = \beta_2 = \ldots = \beta_j = \ldots = \beta_n = 0,$$
$$H_{03} : \gamma_1 = \gamma_2 = \ldots = \gamma_k = \ldots = \gamma_r = 0,$$
$$H_{04} : \text{All } (\alpha\beta)_{ij}\text{s are equal}$$

against the alternative hypotheses

$$H_{11} : \text{All } \alpha\text{'s are not equal},$$
$$H_{12} : \text{All } \beta\text{'s are not equal},$$
$$H_{13} : \text{All } \gamma\text{'s are not equal},$$
$$H_{14} : \text{All } (\alpha\beta)\text{'s are not equal}.$$

Let the level of significance be 0.05.

We calculate the following quantities from the table:

$$\text{Grand total } (G) = \sum_{i=1}^{m}\sum_{j=1}^{n}\sum_{k=1}^{r} y_{ijk}$$

$$\text{Correction factor(CF)} = \frac{G^2}{mnr}$$

$$\text{TSS} = \sum_{i=1}^{m}\sum_{j=1}^{n}\sum_{k=1}^{r} y_{ijk}^2 - \text{CF}$$

$$\text{RSS} = \frac{1}{mn}\sum_{k=1}^{r} y_{ook}^2 - \text{CF},$$

$$\text{where } y_{00k} = \sum_{i}^{m}\sum_{j}^{n} y_{ijk} = k\text{th replication total.}$$

Let us make the following table for calculating other sums of squares (Tables 10.15 and 10.16).

$$\text{TrSS} = \frac{1}{r}\sum_{i=1}^{m}\sum_{j=1}^{n} y_{ij0}^2 - \text{CF},$$

$$\text{where } y_{ij0} = \sum_{k=1}^{r} y_{ijk}$$

$$\text{ErSS} = \text{TSS} - \text{RSS} - \text{TrSS};$$

$$\text{SS(A)} = \frac{1}{nr}\sum_{i=1}^{m} y_{i00}^2 - \text{CF}$$

$$\text{SS(B)} = \frac{1}{mr}\sum_{j=1}^{n} y_{0j0}^2 - \text{CF};$$

$$\text{SS(AB)} = \text{TrSS} - \text{SS(A)} - \text{SS(B)}$$

If the cal. F (for A) $> F_{0.05;(m-1),\ (r-1)(mn-1)}$ then H_{01} is rejected. This means the main effects

Table 10.15 Table of A × B treatment totals over the replications

	b_1	b_2	........b_j.......	b_n	Total
a_1	$\sum_{k=1}^{r} y_{11k}$	$\sum_{k=1}^{r} y_{12k}$	$\sum_{k=1}^{r} y_{1jk}$	$\sum_{k=1}^{r} y_{1nk}$	y_{100}
a_2	$\sum_{k=1}^{r} y_{21k}$	$\sum_{k=1}^{r} y_{22k}$	$\sum_{k=1}^{r} y_{2jk}$	$\sum_{k=1}^{r} y_{2nk}$	y_{200}
.	.	.	....		
:	:	:	::::::::::::::::::::		
:	:	:	::::::::::::::::::::		
a_i	$\sum_{k=1}^{r} y_{i1k}$	$\sum_{k=1}^{r} y_{i2k}$	$\sum_{k=1}^{r} y_{2jk}$	$\sum_{k=1}^{r} y_{2nk}$	y_{i00}
:	:	:	:	:	
:	:	:	:	:	
a_m	$\sum_{k=1}^{r} y_{m1k}$	$\sum_{k=1}^{r} y_{m2k}$	$\sum_{k=1}^{r} y_{mjk}$	$\sum_{k=1}^{r} y_{mnk}$	y_{m00}
Total	y_{010}	y_{020}	y_{0j0}	y_{0n0}	y_{000}

Table 10.16 Analysis of variance table for FRBD with two factorsThe structure of analysis of variance table is as follows:

SOV	d.f.	SS	MS	*F*
Replication	$r-1$	RSS	RMS = RSS/$(r-1)$	RMS/ErMS
Factor A	$m-1$	ASS	AMS = ASS/$(m-1)$	AMS/ErMS
Factor B	$n-1$	BSS	BMS = BSS/$(n-1)$	BMS/ErMS
Interaction (A × B)	$(m-1)(n-1)$	ABSS	ABMS = ABSS/$(m-1)(n-1)$	ABMS/ErMS
Error	$(r-1)(mn-1)$	ErSS	ErMS = ErSS/$(r-1)(mn-1)$	
Total	$mnr-1$	TSS		

due to factor A are not all equal to zero. Similarly H_{02}, H_{03}, and H_{04} are rejected if the calculated value of $F >$ table value of $F_{0.05}$ with corresponding d.f. If the main effects and/or the interaction effects are significant, then we calculate the critical difference (CD)/least significant difference (LSD):

$$\text{LSD(A)} = \sqrt{\frac{2\text{ErMS}}{r \times n}} \times t_{\alpha/2,\ \text{err.d.f}}$$

$$\text{LSD(B)} = \sqrt{\frac{2\text{ErMS}}{r \times m}} \times t_{\alpha/2,\ \text{err.d.f}}$$

$$\text{LSD(AB)} = \sqrt{\frac{\text{ErMS}}{r}} \times t_{\alpha/2,\ \text{err.d.f}}.$$

Note: The $m \times n$ factorial experiment can also be conducted in a basic CRD or LSD if situation permits.

Example 10.15. The following table gives the yield (q/ha) of rice from an experiment conducted with four doses of nitrogen under three spacing treatments using RBD with three replications. Find out the best dose of nitrogen, spacing, and interaction of these two factors to yield maximum.

	R1			R2			R3		
Nitrogen	S1	S2	S3	S1	S2	S3	S1	S2	S3
N1	13.50	16.00	15.20	14.00	16.00	15.00	14.50	15.50	14.80
N2	14.80	16.50	15.80	15.20	16.80	15.20	16.00	16.20	15.50
N3	14.00	15.20	14.80	13.80	15.50	14.30	13.50	14.80	14.50
N4	13.70	15.80	15.00	14.50	15.70	14.80	15.00	15.20	14.70

Solution. From the given information it is clear that the experiment is an asymmetrical (4 × 3) factorial experiment conducted in randomized block design, so the appropriate statistical model for the analysis will be

$y_{ijk} = \mu + \alpha_i + \beta_j + \gamma_k + (\alpha\beta)_{ij} + e_{ijk}$, where $i = 4; j = 3; k = 3$

y_{ijk} = response in the kth replicate due to the ith level of the first factor (variety) and the jth level of the second factor (seed rate)

μ = general effect

α_i = additional effect due to the ith level of the first factor (nitrogen), $\sum \alpha_i = 0$

β_j = additional effect due to the jth level of the second factor (spacing), $\sum \beta_j = 0$

γ_k = additional effect due to the kth replicate, $\sum \gamma_k = 0$

$(\alpha\beta)_{ij}$ = interaction effect of the ith level of the first factor (nitrogen) and the jth level of the second factor (spacing), $\sum_i (\alpha\beta)_{ij} = \sum_j (\alpha\beta)_{ij} = 0$

e_{ijk} = error component associated with the ith level of the first factor (nitrogen), the jth level of the second factor (spacing), and the kth replicates and $e_{ijk} \sim N(0, \sigma^2)$

Hypothesis to Be Tested

$$H_0 : \alpha_1 = \alpha_2 = \alpha_3 = \alpha_4 = 0,$$
$$\beta_1 = \beta_2 = \beta_3 = 0,$$
$$\gamma_1 = \gamma_2 = \gamma_3 = 0,$$
$$\text{All}(\alpha\beta)_{ij}\text{'s are equal}$$

against

$$\alpha\text{'s are not all equal},$$
$$\beta_j\text{'s are not all equal},$$
$$\gamma\text{'s are not all equal},$$
$$All(\alpha\beta)_{ij}\text{'s are not equal}.$$

Let the level of significance be 0.05.

From the given data table, let us calculate the following quantities:

$$\text{Grand total (GT)} = \sum_{i=1}^{4}\sum_{j=1}^{3}\sum_{k=1}^{3} y_{ijk} = 541.300$$

$$\text{Correction factor (CF)} = \frac{GT^2}{4.3.3} = \frac{541.3^2}{36} = 8139.047$$

$$TSS = \sum_{i=1}^{4}\sum_{j=1}^{3}\sum_{k=1}^{3} y_{ijk}^2 - CF$$
$$= 8162.810 - 8139.047 = 23.763$$

$$RSS = \frac{1}{v.s}\sum_{k=1}^{3} y_{ijk}^2 - CF$$
$$= \frac{180.3^2 + 180.8^2 + 180.2^2}{12} - 8139.047$$
$$= 0.017$$

From the above table first let us form the following table and from the table get the following quantities:

$$TrSS = \frac{1}{3}\sum_{i=1}^{4}\sum_{j=1}^{3} y_{ij0}^2 - CF$$
$$= \frac{42^2 + 47.5^2 + 45.0^2 + \cdots + 45.5^2 + 43.6^2}{3}$$
$$- 8139.047$$
$$= 20.296$$

Table of totals

	Nitrogen					
Spacing	N1	N2	N3	N4	Total	Average
S1	42.0	46.0	41.3	43.2	172.5	14.4
S2	47.5	49.5	45.5	46.7	189.2	15.8
S3	45.0	46.5	43.6	44.5	179.6	15.0
Total	134.5	142.0	130.4	134.4		
Average	14.9	15.8	14.5	14.9		

$$\text{ErSS} = \text{TSS} - \text{TrSS} - \text{RSS} = 23.763 - 20.296 - 0.017 = 3.449$$

$$\text{SS (Nitrogen)} = \frac{1}{3.3}\sum_{i=1}^{4} y_{i00}^2 - \text{CF} = \frac{134.5^2 + 142.0^2 + 130.4^2 + 134.4^2}{9} - 8139.047 = 7.816$$

$$\text{SS (Spacing)} = \frac{1}{4.3}\sum_{j=1}^{3} y_{0j0}^2 - \text{CF} = \frac{172.5^2 + 189.2^2 + 179.6^2}{12} - 8139.047 = 11.70$$

$$\text{SS (NS)} = \text{TrSS} - \text{SS (Nitrogen)} - \text{SS (Spacing)} = 20.296 - 7.816 - 11.706 = 0.773$$

Now we make the following analysis of variance table with the help of the above quantities:

ANOVA						
SOV	d.f.	MS	MS	F-ratio	Table value of F	
Replication	$3 - 1 = 2$	0.017	0.009	0.055	$F_{0.05;2,22} = 3.44$	$F_{0.01;2,22} = 5.72$
Treatment	$12 - 1 = 11$	20.296	1.845	11.768	$F_{0.05;11,22} = 2.27$	$F_{0.01;11,22} = 3.19$
Nitrogen	$4 - 1 = 3$	7.816	2.605	16.617	$F_{0.05;3,22} = 3.05$	$F_{0.01;3,22} = 4.82$
Spacing	$3 - 1 = 2$	11.707	5.854	37.333	$F_{0.05;2,22} = 3.44$	$F_{0.01;2,22} = 5.72$
NS	$3 \times 2 = 6$	0.773	0.129	0.821	$F_{0.05;6,22} = 2.55$	$F_{0.01;6,22} = 3.76$
Error	$35 - 2 - 11 = 22$	3.449	0.157			
Total	$36 - 1 = 35$	23.763				

It is clear from the above table that all the effects of variety as well as the spacing are significant at 1% level of significance. But the replication interaction effects of nitrogen and spacing are not significant even at 5% (desired level of significance) level of significance.

To find out which dose of nitrogen and which spacing have maximum yield potentiality, we are to calculate the critical difference values for nitrogen and spacing effects separately using the following formulae:

$$\text{CD}_{0.01}(\text{Nitrogen}) = \sqrt{\frac{2\text{ErMS}}{r.s}}t_{0.005,\,\text{err.d.f}} = \sqrt{\frac{2 \times 0.157}{3 \times 3}}2.819 = 0.527,$$

$$\text{CD}_{0.01}(\text{Spacing}) = \sqrt{\frac{2\text{ErMS}}{r.n}}t_{0.005,\,\text{err.d.f}} = \sqrt{\frac{2 \times 0.157}{3 \times 4}}2.819 = 0.456.$$

Table of averages

	Yield (q/ha)
Variety	
N2	15.78
N1	14.94
N4	14.93
N3	14.49
Seed rate	
S2	15.77
S3	14.97
S1	14.38

Nitrogen N2 has recorded significantly the highest yield compared to other doses of nitrogen. The difference of means between any two varieties for the rest three varieties (N1, N3, N4) is not greater than the critical difference value; they are statistically at par. There exists no significant difference among the nitrogen doses N1, N3, and N4. On the other hand, spacing S2 is the best plant density for getting maximum yield of paddy followed by S3 and S1. So far about the interaction effect of nitrogen and spacing is concerned, no significant difference is recorded among the different treatment combinations.

The above analysis can also be done using SAS statistical software. The following few slides present the data entry, the command syntax, and the output for the same:

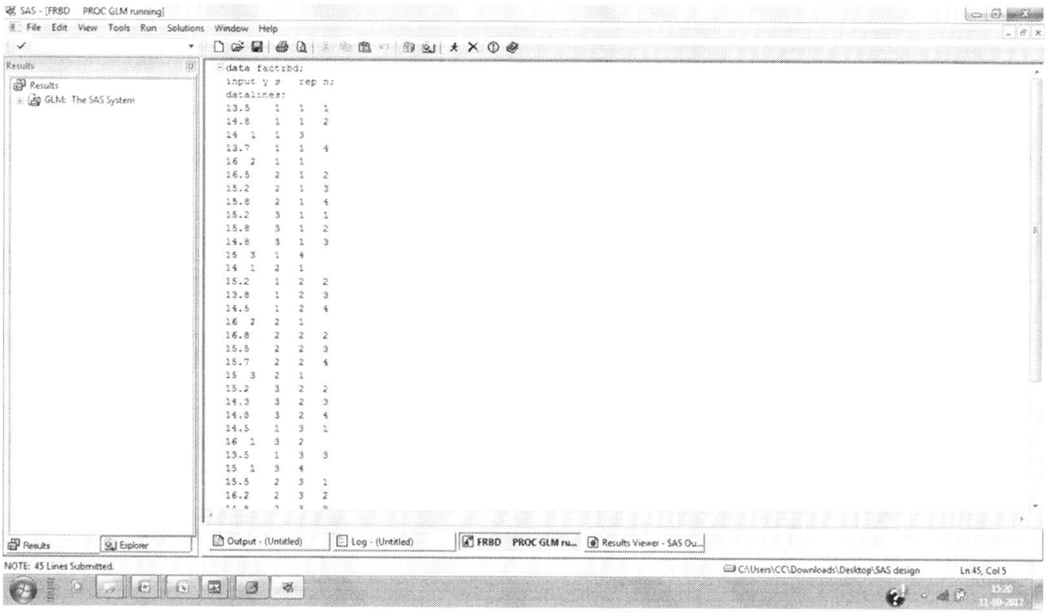

Slide 10.30: Step 1 showing data entry for two-factor factorial RBD analysis using SAS

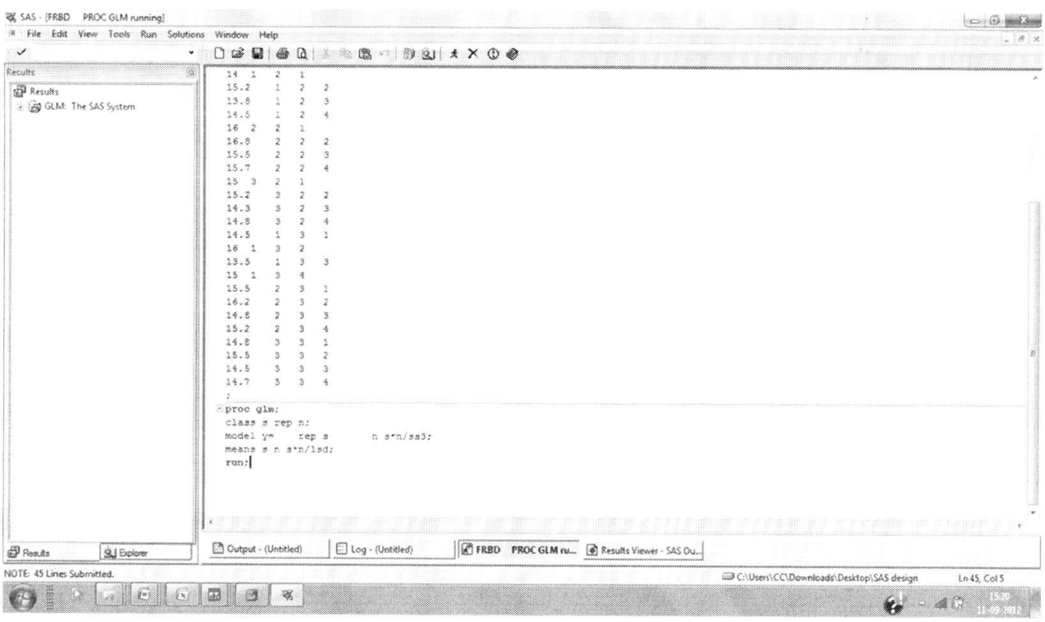

Slide 10.31: Step 2 showing portion of data entry and commands for two-factor factorial RBD analysis using SAS

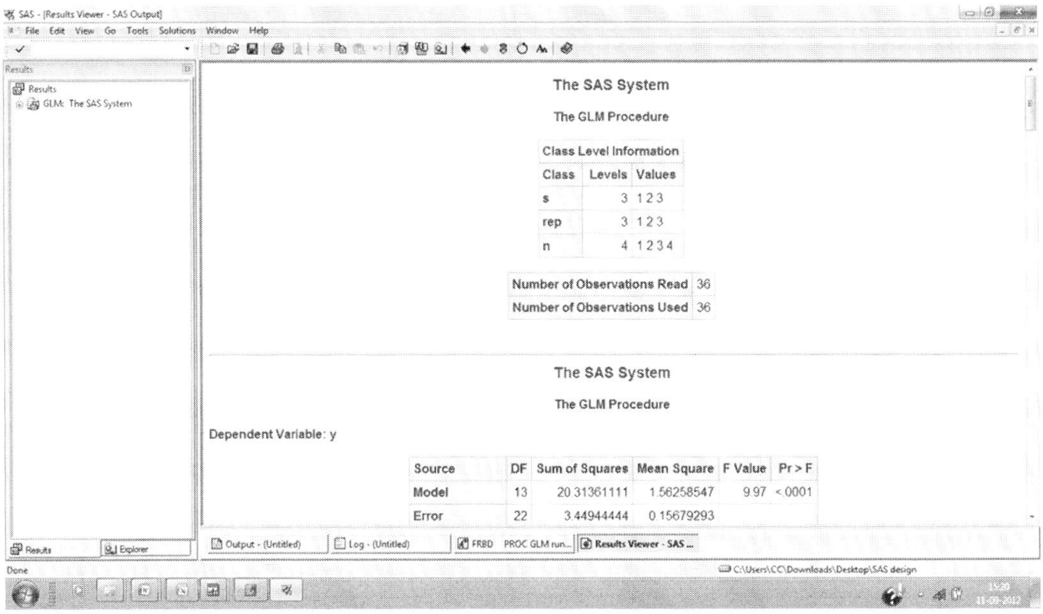

Slide 10.32: Step 3 showing portion of output for two-factor factorial RBD analysis using SAS

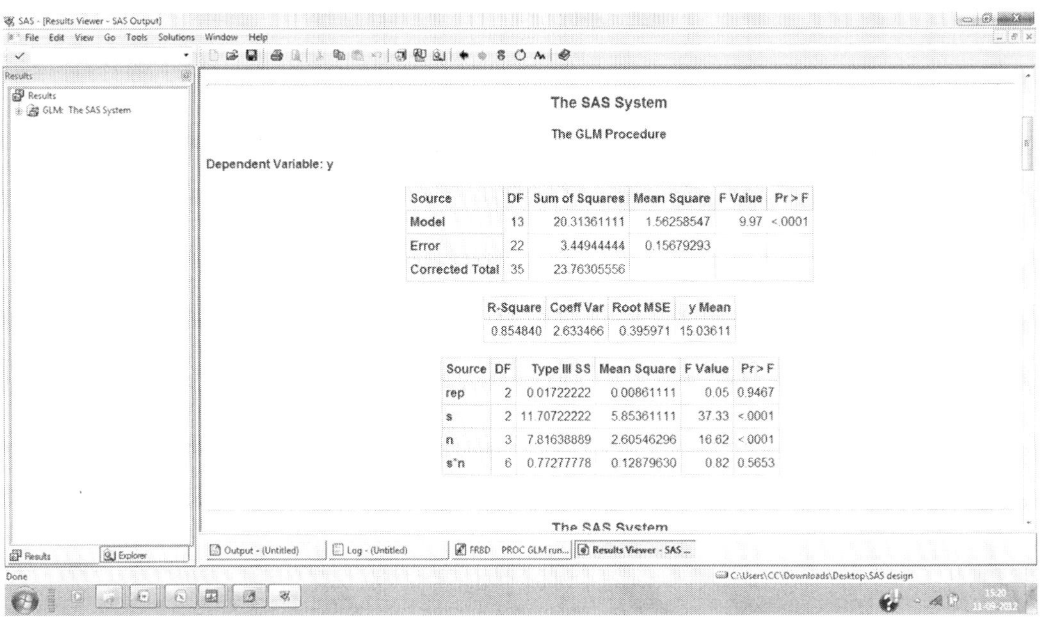

Slide 10.33: Step 3 showing portion of output for two-factor factorial RBD analysis using SAS

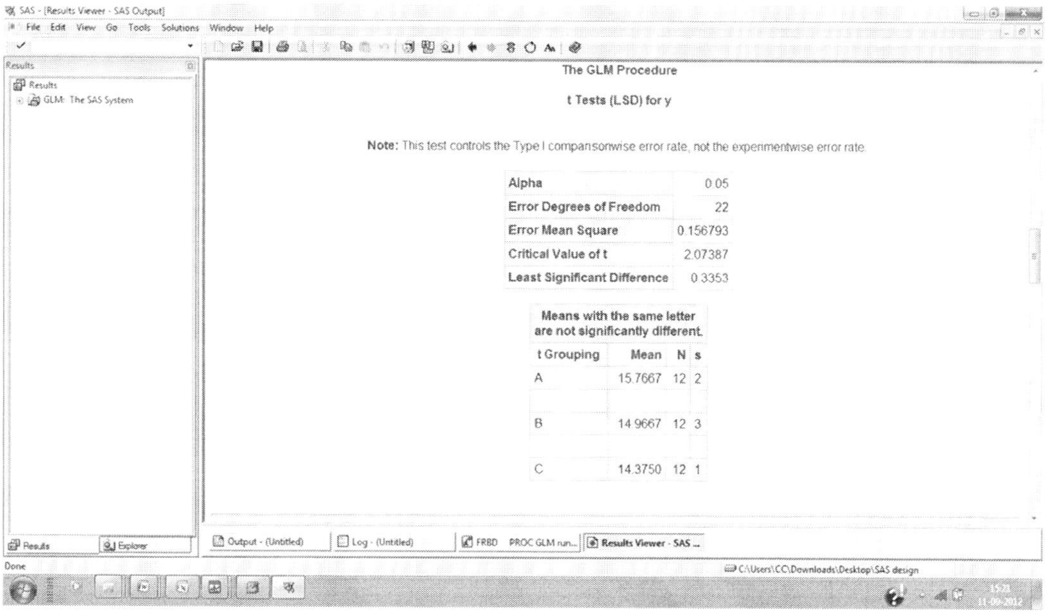

Slide 10.34: Step 3 showing portion of output for two-factor Factorial RBD analysis using SAS

10.11.2 *m* × *n* × *p* Factorial Experiment

In an $m \times n \times p$ three-factor factorial experiment, three factors are tested of which the first, second, and third factors are at m, n, and p levels, respectively; the efficacy of levels of different factors separately and the interactions of the levels of different factors are worked out. A three-factor factorial experiment can be designed in a CRD, RBD, or LSD fashion. But the problem with LSD is that even at the lowest level of three-factor factorial experiment, that is, 2^3 factorial experiment, we require to get a plot of 8×8 experimental area, and any factor having a level more than two in a three-factor factorial experiment requires at least a plot of 12×12 number of experimental units. In reality most of the times, it becomes difficult to get such a plot and impossible when the levels of any factor increase beyond this three-level limit. Thus, three-factor and more-than-three-factor factorial experiments are generally conducted in CRD or RBD design. Let us take the following examples of $2 \times 3 \times 4$ FRBD.

Example 10.16. The following table gives the yield (q/ha) from a field experiment of paddy with three doses of phosphate fertilizer and four doses of nitrogen in two types of irrigation in a randomized block design. Analyze the data to find out the best dose of phosphate, the best dose of nitrogen, the best irrigation, and the best interaction effects among the factors.

Solution. The experiment is an asymmetrical $(2 \times 3 \times 4)$ factorial experiment conducted in randomized block design, so the appropriate statistical model for the analysis will be

$$y_{ijkl} = \mu + \alpha_i + \beta_j + (\alpha\beta)_{ij} + \gamma_k + (\alpha\gamma)_{ik} + (\beta\gamma)_{jk} + (\alpha\beta\gamma)_{ijk} + \delta_l + e_{ijkl}, \quad \text{where} \quad i = 2; j = 3; k = 4; l = 3$$

y_{ijkl} = response in the lth replicate due to the ith level of irrigation, the jth level of phosphate, and the kth level of nitrogen

μ = general effect

Irrigation	I1												I2											
Phosphate	P1				P2				P3				P1				P2				P3			
Nitrogen	N0	N1	N2	N3	N0	N1	N2	N3	N0	N1	N2	N3	N0	N1	N2	N3	N0	N1	N2	N3	N0	N1	N2	N3
R1	19.5	21.0	22.7	16.8	15.5	16.7	17.5	16.0	17.2	13.2	17.0	16.5	21.5	23.0	24.7	20.2	17.5	18.7	19.0	18.0	19.7	15.2	19.0	20.5
R2	20.0	20.5	22.5	16.5	16.2	16.0	17.0	15.5	16.7	14.0	16.5	16.2	22.0	22.5	24.5	19.2	18.2	18.9	18.8	17.5	19.2	16.0	18.7	19.5
R3	19.2	20.8	21.9	17.0	16.0	16.5	16.9	15.2	16.5	15.0	16.7	16.2	21.4	22.8	24.2	19.5	18.0	18.5	18.2	17.2	19.5	17.0	18.8	19.2

α_i = additional effect due to the ith level of irrigation, $\sum \alpha_i = 0$

β_j = additional effect due to the jth level of phosphate, $\sum \beta_j = 0$

γ_k = additional effect due to the kth level of nitrogen, $\sum \gamma_k = 0$

$(\alpha\beta)_{ij}$ = interaction effect of the ith level of irrigation and the jth level of phosphate, $\sum_i (\alpha\beta)_{ij} = \sum_j (\alpha\beta)_{ij} = 0$

$(\alpha\gamma)_{ik}$ = interaction effect of the ith level of irrigation and the kth level of nitrogen, $\sum_i (\alpha\gamma)_{ik} = \sum_k (\alpha\gamma)_{ik} = 0$

$(\beta\gamma)_{jk}$ = interaction effect of the jth level of phosphate and the lth level of nitrogen, $\sum_j (\beta\gamma)_{jk} = \sum_k (\beta\gamma)_{jk} = 0$;

$(\alpha\beta\lambda)_{ijk}$ = interaction effect of the ith level of irrigation, the jth level of phosphate, and the kth level of nitrogen, $\sum_i (\alpha\beta\lambda)_{ijk} = \sum_j (\alpha\beta\lambda)_{ijk} = \sum_k (\alpha\beta\lambda)_{ijk} = 0$

δ_l = additional effect due to the lth replication, $\sum \delta_l = 0$

e_{ijkl} = error component associated with the lth replicate due to the ith level of irrigation and the jth level of phosphate and the kth level of nitrogen and $e_{ijkl} \sim$ i.i.d. $N(0, \sigma^2)$

Hypothesis to Be Tested

$$H_0: \alpha_1 = \alpha_2 = 0,$$
$$\beta_1 = \beta_2 = \beta_3 = 0,$$
$$\gamma_1 = \gamma_2 = \gamma_3 = \gamma_4 = 0,$$
$$\delta_1 = \delta_2 = \delta_3 = 0,$$

All interaction effects = 0 against

$$H_1: \text{All } \alpha\text{'s are not equal,}$$
$$\text{All } \beta\text{'s are not equal,}$$
$$\text{All } \gamma\text{'s are not equal,}$$
$$\text{All } \delta\text{'s are not equal,}$$

All interaction effects are not equal.

Let the level of significance be 0.05.

From the given data table, let us calculate the following quantities:

$$\text{Grand total (GT)} = \sum_{i=1}^{2}\sum_{j=1}^{3}\sum_{k=1}^{4}\sum_{l=1}^{3} y_{ijkl} = 1331.4$$

$$\text{Correction factor (CF)} = \frac{\text{GT}^2}{2.3.4.3} = \frac{1331.4^2}{72} = 24619.805$$

$$\text{TSS} = \sum_{i=1}^{2}\sum_{j=1}^{3}\sum_{k=1}^{4}\sum_{l=1}^{3} y^2_{ijkl} - \text{CF} = 461.767$$

$$\text{RSS} = \frac{1}{m.n.p}\sum_{l=1}^{3} y^2_{000l} - \text{CF}$$
$$= \frac{446.6^2 + 442.2^2 + 442.2^2}{24} - 24619.805$$
$$= 0.500.$$

From the above table first let us form the following tables and from the table get the following quantities:

Table of totals for irrigation × nitrogen

	Irrigation			
Nitrogen	I1	I2	Total	Average
N0	156.800	177.000	333.800	18.544
N1	153.650	172.600	326.250	18.125
N2	168.700	185.900	354.600	19.700
N3	145.900	170.800	316.700	17.594
Total	625.050	706.300		
Average	17.363	19.619		

Table of totals for irrigation × phosphate

	Irrigation			
Phosphate	I1	I2	Total	Average
P1	238.350	265.500	503.850	20.994
P2	195.000	218.500	413.500	17.229
P3	191.700	222.300	414.000	17.250
Total	625.050	706.300		
Average	17.363	19.619		

$$\text{SS (Irrigation)} = \frac{1}{n.p.r}\sum_{i=1}^{2} y^2_{i000} - \text{CF}$$
$$= \frac{625.05^2 + 706.3^2}{3.4.3} - 24619.805$$
$$= 91.688$$

$$\text{SS (Phosphate)} = \frac{1}{m.p.r}\sum_{j=1}^{3} y_{ojoo}^2 - \text{CF} = \frac{503.85^2 + 413.5^2 + 414.0^2}{2.4.3} - 24619.805 = 225.505$$

$$\text{SS (I} \times \text{P)} = \frac{1}{n.r}\sum\sum y_{ij00}^2 - \text{CF} - \text{SS (V)} - \text{SS (S)}$$
$$= \frac{238.35^2 + 265.5^2 + 195.0^2 + 218.5^2 + 191.7^2 + 222.3^2}{4.3} - 24619.805 - 225.505 - 91.688 = 1.050.$$

Table of totals for phosphate × nitrogen

	Phosphate				
Nitrogen	P1	P2	P3	Total	Average
N0	123.600	101.400	108.800	333.800	18.544
N1	130.550	105.300	90.400	326.250	18.125
N2	140.500	107.400	106.700	354.600	19.700
N3	109.200	99.400	108.100	316.700	17.594
Total	503.850	413.500	414.000		
Average	20.994	17.229	17.250		

Table totals for irrigation × phosphate × nitrogen (treatments)

	I1			I2				
	P1	P2	P3	P1	P2	P3	Total	Average
N0	58.700	47.700	50.400	64.900	53.700	58.400	333.800	18.544
N1	62.250	49.200	42.200	68.300	56.100	48.200	326.250	18.125
N2	67.100	51.400	50.200	73.400	56.000	56.500	354.600	19.700
N3	50.300	46.700	48.900	58.900	52.700	59.200	316.700	17.594
Total	238.350	195.000	191.700	265.500	218.500	222.300		
Average	19.863	16.250	15.975	22.125	18.208	18.525		

$$\text{SS (Nitrogen)} = \frac{1}{m.n.r}\sum_{k=1}^{4} y_{00k0}^2 - \text{CF} = \frac{333.80^2 + 326.25^2 + 354.60^2 + 316.70^2}{2.3.3} - 24619.805 = 43.241$$

$$\text{SS (I} \times \text{N)} = \frac{1}{n.r}\sum_{i=1}^{2}\sum_{k=1}^{4} y_{i0k0}^2 - \text{CF} - \text{SS (I)} - \text{SS (N)}$$
$$= \frac{156.8^2 + 177.0^2 + 153.65^2 + \cdots + 145.9^2 + 170.8^2}{3.3} - 24619.805 - 91.688 - 43.241 = 1.811.$$

$$\text{SS(P} \times \text{N)} = \frac{1}{m.r}\sum_{j=1}^{3}\sum_{k=1}^{4} y_{0jk0}^2 - \text{CF} - \text{SS (P)} - \text{SS (N)}$$
$$= \frac{123.6^2 + 101.4^2 + 108.8^2 + \cdots + 99.4^2 + 108.1^2}{2.3} - 24619.805 - 225.505 - 43.241 = 88.368.$$

$$\text{SS (I} \times \text{P} \times \text{N)} = \frac{1}{r}\sum_{i=1}^{2}\sum_{j=1}^{3}\sum_{k=1}^{4} y_{0jk0}^2 - \text{CF} - \text{SS (I)} - \text{SS (P)} - \text{SS (N)} - \text{SS (IP)} - \text{SS (IN)} - \text{SS (PN)}$$
$$= \frac{58.7^2 + 47.7^2 + 50.4^2 + \cdots + 52.7^2 + 59.2^2}{3} - 24619.805 - 91.688 - 225.505 - 43.241$$
$$- 1.05 - 1.81 - 88.368 = 1.324,$$

$$\text{SS (Error)}: \text{TSS} - \text{RSS} - \text{SS(I)} - \text{SS(P)} - \text{SS(N)} - \text{SS(IP)} - \text{SS(IN)} - \text{SS(PN)} - \text{SS(IPN)}$$
$$= 461.767 - 0.500 - 91.688 - 225.505 - 43.241 - 1.050 - 1.811 - 88.368 - 1.324 = 8.278.$$

SOV	d.f.	SS	MS	F-ratio	Table value of F at $p = 0.05$	$p = 0.01$
Replication	2	0.500	0.250	1.389	3.21	5.12
Irrigation	1	91.688	91.688	509.487	4.07	7.25
Phosphate	2	225.505	112.753	626.536	3.21	5.12
I × P	2	1.050	0.525	2.919	3.21	5.12
Nitrogen	3	43.241	14.414	80.094	2.83	4.26
I × N	3	1.811	0.604	3.355	2.83	4.26
P × N	6	88.368	14.728	81.839	2.22	3.23
I × P × N	6	1.324	0.221	1.226	2.22	3.23
Error	46	8.278	0.180			
Total	71	461.767				

Now using the above values, we frame the following ANOVA table:

From the above table we have all the effects except for replication, irrigation × phosphate, and irrigation × phosphate × nitrogen interaction significant at 5% level of significance. Moreover, the main effects of all the factors and the interaction effect of phosphate and nitrogen are significant even at 1% level of significance.

Now, we are to find out the levels of different factors which are significantly different from others and also the best level of each factor. For the purpose we are to calculate the critical difference values for irrigation, phosphate, nitrogen, and interaction effects of irrigation × nitrogen and phosphate × nitrogen using the following formulae:

$$\text{LSD}_{(0.05)}\text{for irrigation} = \sqrt{\frac{2 \times \text{ErMS}}{npr}} \times t_{\frac{\alpha}{2};\text{error d.f.}} = \sqrt{\frac{2 \times 0.180}{3.4.3}} \times t_{0.025;46.}$$

$$= \sqrt{\frac{2 \times 0.180}{3.4.3}} 2.016 = 0.2016$$

$$\text{LSD}_{(0.05)}\text{for phosphate} = \sqrt{\frac{2 \times \text{ErMS}}{mpr}} \times t_{\frac{\alpha}{2};\text{error d.f.}} = \sqrt{\frac{2 \times 0.180}{2.4.3}} \times t_{0.025;46.}$$

$$= \sqrt{\frac{2 \times 0.180}{2.4.3}} 2.016 = 0.247$$

$$\text{LSD}_{(0.05)}\text{for nitrogen} = \sqrt{\frac{2 \times \text{ErMS}}{mnr}} \times t_{\frac{\alpha}{2};\text{error d.f.}} = \sqrt{\frac{2 \times 0.180}{2.3.3}} \times t_{0.025;46.}$$

$$= \sqrt{\frac{2 \times 0.180}{2.3.3}} 2.016 = 0.2851$$

$$\text{LSD}_{(0.05)}\text{for irrigation} \times \text{nitrogen} = \sqrt{\frac{2 \times \text{ErMS}}{nr}} \times t_{\frac{\alpha}{2};\text{error d.f.}} = \sqrt{\frac{2 \times 0.180}{3.3}} \times t_{0.025;46.}$$

$$= \sqrt{\frac{2 \times 0.180}{3.3}} 2.016 = 0.403$$

$$\text{LSD}_{(0.05)}\text{for phosphate} \times \text{nitrogen} = \sqrt{\frac{2 \times \text{ErMS}}{mr}} \times t_{\frac{\alpha}{2};\text{error d.f.}} = \sqrt{\frac{2 \times 0.180}{2.3}} \times t_{0.025;46.}$$

$$= \sqrt{\frac{2 \times 0.180}{2.3}} 2.016 = 0.4938$$

Table of means

	Yield (q/ha)	LSD(0.05)
Irrigation		
I2	19.619	0.2016
I1	17.363	
Phosphate		
P1	20.994	0.247
P3	17.250	
P2	17.229	
Nitrogen		
N2	19.700	0.2851
N0	18.544	
N1	18.125	
N3	17.594	
Irrigation × nitrogen(I × N)		
I2N2	20.656	0.403
I2N0	19.667	
I2N1	19.178	
I2N3	18.978	
I1N2	18.744	
I1N0	17.422	
I1N1	17.072	
I1N3	16.211	
Phosphate × nitrogen (P × N)		
P1N2	23.417	0.494
P1N1	21.758	
P1N0	20.600	
P1N3	18.200	
P3N0	18.133	
P3N3	18.017	
P2N2	17.900	
P3N2	17.783	
P2N1	17.550	
P2N0	16.900	
P2N3	16.567	
P3N1	15.067	

From the above table we have the following conclusions:

1. I2 irrigation is significantly better than the I1 irrigation.
2. Phosphate P1 is the best one and P2 and P3 are statistically at par.
3. All the doses of nitrogen are significantly different from each other and the best dose of nitrogen is N2.
4. Among the interaction effects of irrigation and nitrogen, I2N2 is the best followed by I2N0 and so on.
5. Among the interaction effects of variety and nitrogen, P1N2 is the best followed by P1N1, P1N0, and so on.

10.12 Incomplete Block Design

As the number of factors or levels of factors or both increase, the number of treatment combination increases, and it becomes very difficult to accommodate all the treatments in a single block. As a result the idea of incomplete blocks comes under consideration. Each replication is now being constituted of a number of blocks which can accommodate a part of the total treatments. Thus, the blocks are incomplete in the sense that they do not contain/accommodate all the treatments in each of them. In the process a complete block equivalent to a complete replication changes to a replication of number of incomplete blocks. Thus, in an incomplete block design, blocks are no longer equivalent to replication.

10.12.1 Split Plot Design

The simplest of all incomplete block designs involving two factors is the split plot design. In field experiment certain factors like type of irrigation, tillage, drainage, and weed management require comparatively larger size of plots for convenience of the treatments compared to the other factors. Thus, factorial experiments involving these factors along with other factors require different sizes of experimental plots in the same experiment. In split plot design, each and every replication is divided into the number of blocks as the number of levels of the factors requiring higher plot size and lesser precision, the main plot factor. Each and every block is constituted of as many numbers of experimental units as the levels of the other factors which require lesser plot size compared to the main plot factor and higher precision. The subplot effects and the interaction effects of subplot factors are more precisely estimated in split plot design compared to the main plot effects.

Layout and Randomization

The step-by-step procedure for layout of an $m \times n$ factorial experiment conducted in split plot design with r replication is given as follows:

Step 1: Divide the whole experimental area into r replications considering the field condition and local situation.

Step 2: Divide each and every replication into m number of blocks (main plot) of equal size and homogeneity.

Step 3: Divide each and every main plot into n number of experimental units of equal size.

Step 4: Randomly allocate the m levels of main plot factor into m main plots of each and every replication separately.

Step 5: The n levels of subplot factor is randomly allocated to n subplots of each and every m main plot factor.

Step 6: Continue the procedure in step 2 to step 5 for r replications separately.

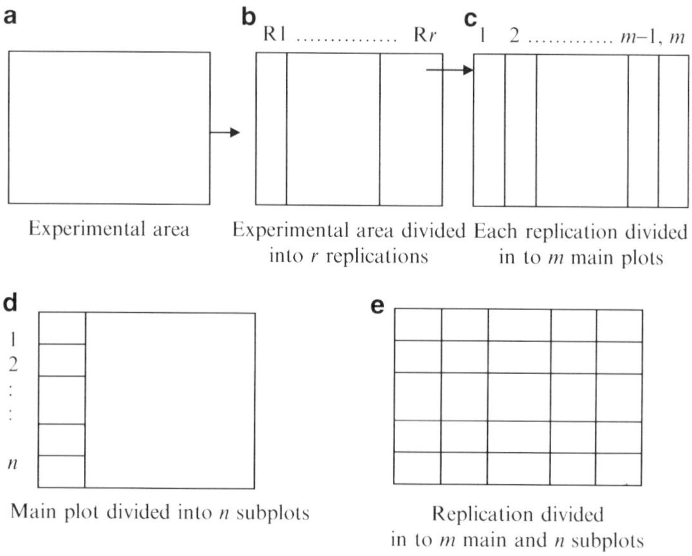

Analysis

The appropriate model split plot experiment with m levels of main plot factors and n levels of subplot factors in r replications is $y_{ijk} = \mu + \gamma_i + \alpha_j + e_{ij} + \beta_k + \upsilon_{jk} + e'_{ijk}$

where $i = 1, 2, \ldots, r$; $j = 1, 2, \ldots, m$; $k = 1, 2, \ldots, n$

μ = general effect

γ_i = additional effect due to the ith replication

α_j = additional effect due to the jth level of main plot factor A and $\sum_{j=1}^{m} \alpha_j = 0$

β_k = additional effect due to the kth level of subplot factor B and $\sum_{k=1}^{n} \beta_k = 0$

υ_{jk} = interaction effect due to the jth level of main plot factor A and the kth level of subplot factor B and $\sum_j \upsilon_{jk} = \sum_k \upsilon_{jk} = 0$ for all k for all j

e_{ij} (error I) = associated with the ith replication and the jth level of main plot factor and $e_{ij} \sim$ i.i.d. $N(0, \sigma_m^2)$

e'_{ijk} (error II) = error associated with the ith replication, the jth level of main plot factor, and the kth level of subplot factor and $e'_{ijk} \sim$ i.i.d. $N(0, \sigma_s^2)$

Hypothesis to Be Tested

$$
\begin{aligned}
H_0 : \gamma_1 &= \gamma_2 = \cdots = \gamma_i = \cdots = \gamma_r = 0, \\
\alpha_1 &= \alpha_2 = \cdots = \alpha_j = \cdots = \alpha_m = 0, \\
\beta_1 &= \beta_2 = \cdots = \beta_k = \cdots = \beta_n = 0, \\
\upsilon_{11} &= \upsilon_{12} = \cdots = \upsilon_{jk} = \cdots = \upsilon_{mn} = 0.
\end{aligned}
$$

Table 10.17 ANOVA for split plot design

SOV	d.f.	MS	MS	F-ratio	Table value of F
Replication	$r-1$	RMS	RMS	RMS/Er.MS I	
Main plot factor(A)	$m-1$	MS(A)	MS(A)	MS(A)/Er.MS I	
Error I	$(r-1)(m-1)$	Er.MS(I)	Er.MS I		
Subplot factor(B)	$(n-1)$	MS(B)	MS(B)	MS(B)/Er.MS II	
Interaction (A × B)	$(m-1)(n-1)$	MS(AB)	MS(AB)	MS(AB)/Er.MS II	
Error II	$m(n-1)(r-1)$	Er.MS II	Er.MS II		
Total	$mnr-1$				

Table 10.18 Table totals for main plot × replication

	a_1	a_2	……	a_j	……	a_m	Total	Mean
R_1	y_{110}	y_{120}	……	y_{1j0}	……	y_{1m0}	y_{100}	$\bar{y}_{100}$
R_2	y_{210}	y_{220}	……	y_{2j0}	……	y_{2m0}	y_{200}	$\bar{y}_{200}$
:	:	:	:	:	:	:	:	
R_i	y_{i10}	y_{i20}	……	y_{ij0}	……	y_{im0}	y_{i00}	$\bar{y}_{i00}$
:	:	:	:	:	:	:	:	
R_r	y_{r10}	y_{r20}	……	y_{rj0}	……	y_{rm0}	y_{r00}	$\bar{y}_{r00}$
Total	y_{010}	y_{020}	……	y_{0j0}	……	y_{0m0}	y_{000}	$\bar{y}_{000}$
Mean	$\bar{y}_{010}$	$\bar{y}_{020}$		$\bar{y}_{0j0}$		$\bar{y}_{0m0}$		

H_1 : γ's are not all equal,
α's are not all equal,
β's are not all equal,
υ's are not all equal.

Let the level of significance be 0.05 (Table 10.17).

The step-by-step procedure for the computation of different sums of squares and mean sum of squares is given as follows:

1. Grand total $= \sum_{i=1}^{r}\sum_{j=1}^{m}\sum_{k=1}^{n} y_{ijk} = \text{GT}$.
2. Correction factor (CF) $= \frac{\text{GT}^2}{mnr}$.
3. Total MS $= \sum_{i=1}^{r}\sum_{j=1}^{m}\sum_{k=1}^{n} y_{ijk}^2 - \text{CF}$.
4. Work out the sum of squares due to the main plot factor and the replication. For the purpose, the following table of totals is required to be framed (Table 10.18):

   where $y_{ij0} = \sum_{k=1}^{n} y_{ijk}$

   TSS (Table 10.18) $= \frac{1}{n}\sum_{i}^{r}\sum_{j}^{m} y_{ij0}^2 - \text{CF}$,

   Replication SS $= \frac{1}{mn}\sum_{i=1}^{r} y_{i00}^2 - \text{CF}$,

   SS(A) $= \frac{1}{nr}\sum_{j=1}^{m} y_{0j0}^2 - \text{CF}$, SS Error I = TSS (Table 10.18) – RSS – SS(A).
5. Work out the sum of squares due to the subplot factor and interaction. For the purpose, the following table of totals is required to the formed (Table 10.19):

   Note: In both tables the totals for main factor A at different levels will be the same.

$$\text{TSS (Table 10.19)} = \frac{1}{r}\sum_{j=1}^{m}\sum_{k=1}^{n} y_{0jk}^2 - \text{CF},$$

$$\text{SS (B)} = \frac{1}{mr}\sum_{k=1}^{n} y_{00k}^2 - \text{CF}$$

$$\begin{aligned}\text{SS(AB)} &= \text{TSS (Table 10.19)} - \text{SS (A)} \\ &\quad - \text{SS (B)},\ \text{SS(Error II)} \\ &= \text{TSS} - \text{RSS} - \text{SS(A)} \\ &\quad - \text{SS(Error I)} - \text{SS(B)} - \text{SS(AB)}.\end{aligned}$$

Mean sum of squares is calculated by dividing the sum of squares by the corresponding degrees of freedom.

Table 10.19 Table totals for main plot × subplot

	b_1	b_2	……	b_k	……	b_n	Total	Mean
a_1	y_{011}	y_{012}	……	y_{01k}	……	y_{01n}	y_{010}	$\bar{y}_{010}$
a_2	y_{021}	y_{022}	……	y_{02k}	……	y_{02n}	y_{020}	$\bar{y}_{020}$
:	:	:	:	:	:	:	:	
a_j	y_{0j1}	y_{0j2}	……	y_{0jk}	……	y_{0jn}	y_{0j0}	$\bar{y}_{0j0}$
:	:	:	:	:	:	:	:	
a_m	y_{0m1}	y_{0m2}	……	y_{0mk}	……	y_{0mn}	y_{0m0}	$\bar{y}_{0m0}$
Total	y_{001}	y_{002}	……	y_{00k}	……	y_{00n}	y_{000}	$\bar{y}_{000}$
Mean	$\bar{y}_{001}$	$\bar{y}_{002}$		$\bar{y}_{00k}$		$\bar{y}_{00n}$		

6. F-ratio for replication and main plot factors are obtained by comparing the respective mean sum of squares against mean sum of squares due to error I. On the other hand, the F-ratios corresponding to subplot factor and the interaction effects are worked out by comparing the respective mean sum of squares against the mean sum of squares due to error II.
7. Calculated F-ratios are compared with tabulated value of F at appropriate level of significance and degrees of freedom.
8. Once the F-test becomes significant, the next task will be to estimate the standard errors (SE) for different types of comparison as given below:
   (a) LSD for difference between two replication means $= \sqrt{\frac{2\text{ErMS-I}}{mn}} \times t_{\frac{\alpha}{2};\text{error-I d.f.}}$.
   (b) LSD for difference between two main plot treatment means $= \sqrt{\frac{2\text{ErMS-I}}{rn}} \times t_{\frac{\alpha}{2};\text{error-I d.f.}}$.
   (c) LSD for difference between two subplot treatment means $= \sqrt{\frac{2\text{ErMS-II}}{rm}} \times t_{\frac{\alpha}{2};\text{error-II d.f.}}$.
   (d) LSD for difference between two subplot treatment means at the same level of main plot treatment $= \sqrt{\frac{2\text{ErMSII}}{r}} \times t_{\frac{\alpha}{2};\text{errorII d.f.}}$.
   (e) SE for difference between two main plot treatment means at the same or different levels of subplot treatment $= \sqrt{\frac{2[(n-1)\text{ErMS-II}+\text{ErMS-I}]}{rn}}$, but the ratio of the treatment mean difference and the above SE does not follow t distribution. An approximate value of t is given by $t = \frac{t_1\text{ErMS-I}+t_2(n-1)\text{ErMS-II}}{\text{ErMS-I}+(n-1)\text{ErMS-II}}$, where t_1 and t_2 are tabulated values at error I and error II degrees of freedom, respectively, at the chosen significance level and the corresponding CD value could be $\text{CD}_\alpha = \text{SE}_d \times t(\text{cal})$.

Example 10.17. Three different methods of tillage and four varieties were tested in a field experiment using split plot design with tillage as main plot factor and variety as subplot factor. Yield (t/ha) are recorded from the individual plots and given below. Analyze the data and draw your conclusion.

Tillage	Tillage 1				Tillage 2				Tillage 3			
Variety	V_1	V_2	V_3	V_4	V_1	V_2	V_3	V_4	V_1	V_2	V_3	V_4
Rep -1	8.7	9.1	7.8	7.2	9.5	12.6	11.2	9.8	7.5	9.5	8.2	7.9
Rep -2	8.6	9.2	7.9	7.3	9.4	12.5	11	9.6	7.6	9.8	8.4	8
Rep -3	8.5	9.3	8.2	7.4	9.6	12.3	10.9	10	7.4	9.7	8.5	8.1

For the above experiment, the appropriate model is

$y_{ijk} = \mu + \gamma_i + \alpha_j + e_{ij} + \beta_k + \upsilon_{jk} + e_{ijk}$, $i = 3$, $j = 3, k = 4$,

where

μ = general effect

γ_i = additional effect due to the ith replication

α_j = additional effect due to the jth level of main plot factor, tillage, and $\sum_{j=1}^{3} \alpha_j = 0$

β_k = additional effect due to the kth level of subplot factor, variety, and $\sum_{k=1}^{4} \beta_k = 0$

υ_{jk} = interaction effect due to the jth level of main plot factor (tillage) and the kth level of subplot factor (variety) and $\sum_{j=1}^{3} \upsilon_{jk} = \sum_{k=1}^{4} \upsilon_{jk} = 0$

e_{ij} (error I) = error associated with the ith replication and the jth level of tillage and $e_{ij} \sim$ i.i.d. $N(0, \sigma_m^2)$

e'_{ijk} (error II) = error associated with the ith replication, the jth level of tillage, and the kth level of subplot factor (variety) and $e_{ijk} \sim$ i.i.d. $N(0, \sigma_s^2)$

Hypothesis to Be Tested

$$H_0: \gamma_1 = \gamma_2 = \gamma_3 = 0,$$
$$\alpha_1 = \alpha_2 = \alpha_3 = 0,$$
$$\beta_1 = \beta_2 = \beta_3 = \beta_4 = 0,$$
$$\upsilon\text{'s are all equal to zero.}$$

$$H_1: \alpha\text{'s are not all equal,}$$
$$\gamma\text{'s are not all equal,}$$
$$\beta\text{'s are not all equal,}$$
$$\upsilon\text{'s are not all equal.}$$

Let the level of significance be 0.05.

$$\text{Grand total} = \sum_{i=1}^{3}\sum_{j=1}^{3}\sum_{k=1}^{4} y_{ijk} = \text{GT} = 8.7 + 9.1 + \cdots + 8.5 + 8.1 = 328.20$$

$$\text{Correction factor (CF)} = \frac{\text{GT}^2}{mnr} = \frac{328.20^2}{3.4.3} = 2992.09$$

$$\text{Total SS} = \sum_{i=1}^{3}\sum_{j=1}^{3}\sum_{k=1}^{4} y_{ijk}^2 - \text{CF} = 8.7^2 + 9.1^2 + + \cdots + 8.5^2 + 8.1^2 - 2992.09 = 75.87$$

Table I Table of totals for tillage × replication

Tillage	Replication R$_1$	R$_2$	R$_3$	Total
T$_1$	32.800	33.000	33.400	99.20
T$_2$	43.100	42.500	42.800	128.40
T$_3$	33.100	33.800	33.700	100.60
Total	109.000	109.300	109.900	328.20
Average	36.3333	36.43333	36.6333	

$$\text{TSS (Table 1)} = \frac{1}{n}\sum_{i=1}^{r}\sum_{j=1}^{m} y_{ij0}^2 - \text{CF}$$
$$= 45.37$$

$$\text{SS (Error I)} = \text{TSS (Table I)} - \text{RSS} - \text{SS (P)}$$
$$= 45.37 - 45.207 - 0.035 = 0.128.$$

$$\text{Replication SS} = \frac{1}{mn}\sum_{i=1}^{r} y_{i00}^2 - \text{CF}$$
$$= \frac{1}{3.4}\sum_{i=1}^{3} y_{i00}^2 - \text{CF}$$
$$= \frac{109^2 + 109.3^2 + 109.9^2}{12} - 2992.09$$
$$= 0.035$$

$$\text{SS (tillage)} = \frac{1}{nr}\sum_{j=1}^{m} y_{0j0}^2 - \text{CF}$$
$$= \frac{1}{4.3}\sum_{j=1}^{3} y_{0j0}^2 - \text{CF}$$
$$= \frac{99.2^2 + 128.4^2 + 100.6^2}{12} - 2992.09$$
$$= 45.207$$

Table II Table of totals for tillage × variety

Tillage	Variety V_1	V_2	V_3	V_4	Total	Average
T_1	25.80	27.60	23.90	21.90	99.20	8.267
T_2	28.50	37.40	33.10	29.40	128.40	10.700
T_3	22.50	29.00	25.10	24.00	100.60	8.383
Total	76.80	94.00	82.10	75.30		
Average	8.533	10.444	9.122	8.367		

$$\text{TSS (Table II)} = \frac{1}{3}\sum_{j=1}^{m}\sum_{k=1}^{4} y_{0jk} - \text{CF} = \frac{1}{3}\left[25.80^2 + \cdots + 24.00^2\right] - \text{CF}$$
$$= 75.397$$

$$\text{SS (variety)} = \frac{1}{mr}\sum_{k=1}^{n} y_{00k}^2 - \text{CF} = \frac{1}{3.3}\sum_{k=1}^{4} y_{00k}^2 - \text{CF}$$
$$= \frac{76.8^2 + 94.0^2 + 82.10^2 + 75.30^2}{3.3} - 2992.09 = 23.992.$$

$$\text{SS (T} \times \text{V)} = \text{TSS (Table II)} - \text{SS(T)} - \text{SS(V)}$$
$$= 75.397 - 45.207 - 23.992 = 6.198,$$

$$\text{ErSS II} = \text{TSS} - \text{RSS} - \text{SS(T)} - \text{Er.SS I} - \text{SS(V)} - \text{SS(T} \times \text{V)}$$
$$= 75.87 - 0.035 - 45.207 - 23.992 - 6.198 - 0.128 = 0.310.$$

F-ratios for replication and tillage are obtained by comparing the respective mean sum of squares against mean sum of squares due to error I. On the other hand, the F-ratios corresponding to subplot factor and the interaction effects are worked out by comparing the respective mean sum of squares against the mean sum of squares due to error II.

It is found that the effects of tillage, variety and their interaction are significant at both 5% and 1% level of significance.

So the next task will be to estimate the SEs for different types of comparison as given below:

1. Standard error for difference between two tillage means $= \sqrt{\frac{2\text{ErMS-I}}{rn}} = \sqrt{\frac{2(0.032)}{3\times4}}$ and the corresponding LSD value could be

$$\text{LSD}_{(0.05)} = \sqrt{\frac{2\text{ErMS-I}}{rn}}\, t_{(0.025);\ \text{error-I d.f.}}$$
$$= \sqrt{\frac{2(0.032)}{3\times 4}} \times 2.776 = 0.2027.$$

2. Standard error for difference between two subplot treatment means $= \sqrt{\frac{2\text{ErMS-II}}{r.m}} = \sqrt{\frac{2(0.017)}{3\times3}}$ and the corresponding LSD value could be

$$\text{LSD}_{(0.05)} = \sqrt{\frac{2\text{ErMS-II}}{r.m}}\, t_{0.025;\ \text{error-II d.f.}}$$
$$= \sqrt{\frac{2(0.017)}{3\text{x}3}} \times 2.101 = 0.129.$$

ANOVA table for 3 × 4 split plot experiment					Table value of F	
SOV	d.f.	MS	MS	F-ratio	$p = 0.05$	$p = 0.01$
Replication	$3 - 1 = 2$	0.035	0.018	0.545	6.94	18.00
Main Plot Factor(P)	$3 - 1 = 2$	45.207	22.603	704.519	6.94	18.00
Error I	$(3 - 1)(3 - 1) = 4$	0.128	0.032			
Subplot factor(V)	$(4 - 1) = 3$	23.992	7.997	464.366	3.16	5.09
Interaction (P × V)	$(3 - 1)(4 - 1) = 6$	6.198	1.033	59.978	2.66	4.01
Error II	$3(4 - 1)(3 - 1) = 18$	0.310	0.017			
Total	$3.3.4 - 1 = 35$	75.87				

3. Standard error for difference between two tillage means at the same or different level of subplot treatment $= \sqrt{\frac{2[(n-1)\,\text{ErMS-II} + \text{ErMS-I}]}{rn}} = \sqrt{\frac{2[(4-1)\,0.017 + 0.032]}{3.4}} = 0.1386$, but the ratio of the treatment mean difference and the above SE does not follow t distribution, and the approximate value of t is given by $t = \frac{t_1\,\text{ErMS-I} + t_2(n-1)\,\text{ErMS-II}}{\text{ErMS-I} + (n-1)\,\text{ErMS-II}} = \frac{(2.776)(0.032) + (2.101)(4-1)\,0.017}{(0.032) + (4-1)(0.017)} = 2.361$, where $t_1 = t_{0.025,4}$ value and $t_2 = t_{0.025,18}$ value and the corresponding CD value could be $\text{LSD}_{(0.05)} = \text{SE}_d \times t(\text{cal}) = 0.1386 \times 2.361 = 0.327$.

Table of mean comparison

	Average	LSD(0.05)
Tillage		
T_2	10.700	0.2027
T_3	8.383	
T_1	8.267	
Variety		
V_2	10.444	0.129
V_3	9.122	
V_1	8.533	
V_4	8.367	
$T \times V$		
T_2V_2	12.467	0.227
T_2V_3	11.033	and 0.324
T_2V_4	9.800	
T_3V_2	9.667	
T_2V_1	9.500	
T_1V_2	9.200	
T_1V_1	8.600	
T_3V_3	8.367	
T_3V_4	8.000	
T_1V_3	7.967	
T_3V_1	7.500	
T_1V_4	7.300	

Thus, second tillage is the best, which is significantly superior to the other two tillages. The two methods of tillage, namely, T_3 and T_1, are statistically at par. So far as the effect of variety is concerned, maximum yield of ginger is obtained from variety V_2 followed by V_3, V_1, and V_4. All the varieties are significantly different from each other with respect to yield of ginger. The interaction of the method of tillage and the varieties has significantly different effects from each other; the interaction effects which are not significantly different have been put under the same lines. Variety in combination with tillage $T2V_2$ has produced significantly higher yield than any other combination, followed by P_2V_3, P_2V_4, and so on.

The above analysis can also be done using SAS statistical software. The following few slides present the data entry, command syntax, and the output for the same:

10.12.2 Strip Plot Design

In split plot design, if both the factors require a large plot like that in the main plot factor, then it may not be possible in split plot design. Strip plot design is the solution. In agricultural field experimentation with factors like different methods of irrigation, different methods of pest control, different methods of mulching, different methods of plowing, etc., require larger plot size for convenience of management. In experiments involving these factors, strip plot design may become very useful. In strip plot design each replication is divided into the number of horizontal rows and the number of vertical columns equals to the number of levels of factor 1 and factor 2, respectively, or vice versa.

The factor assigned to the horizontal rows is called *horizontal factor* and the factor assigned to the vertical column is called the *vertical factor*. Thus, horizontal factor and vertical factor always remain perpendicular to each other. *This does not necessarily mean that the shape and/or size of horizontal strips and vertical strips will be the same.* The smallest plot in strip plot design is the intersection plot and in strip plot design, the interaction effects

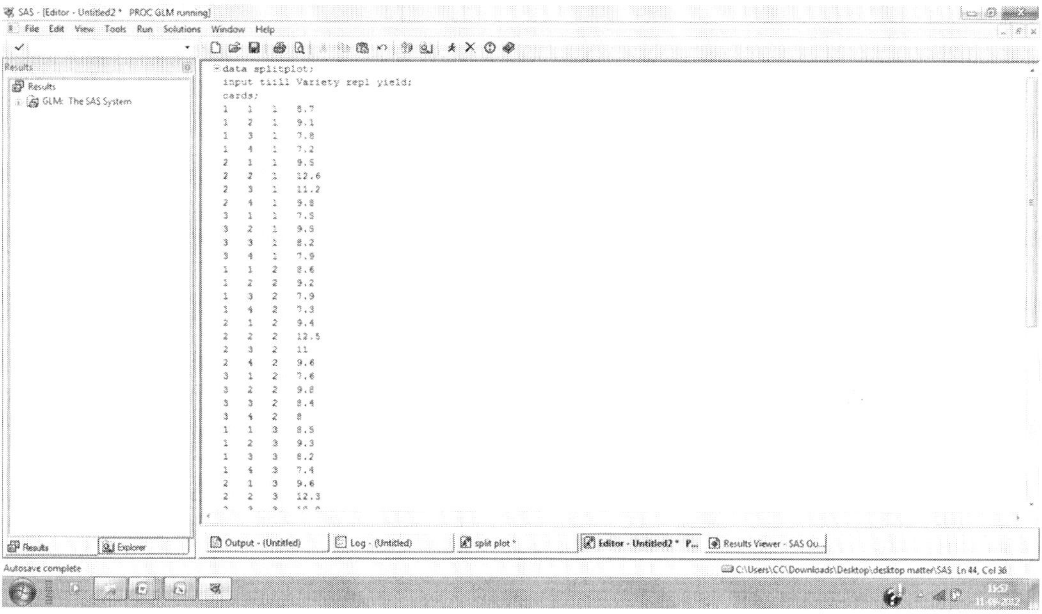

Slide 10.35: Step 1 showing data entry for split plot data analysis using SAS

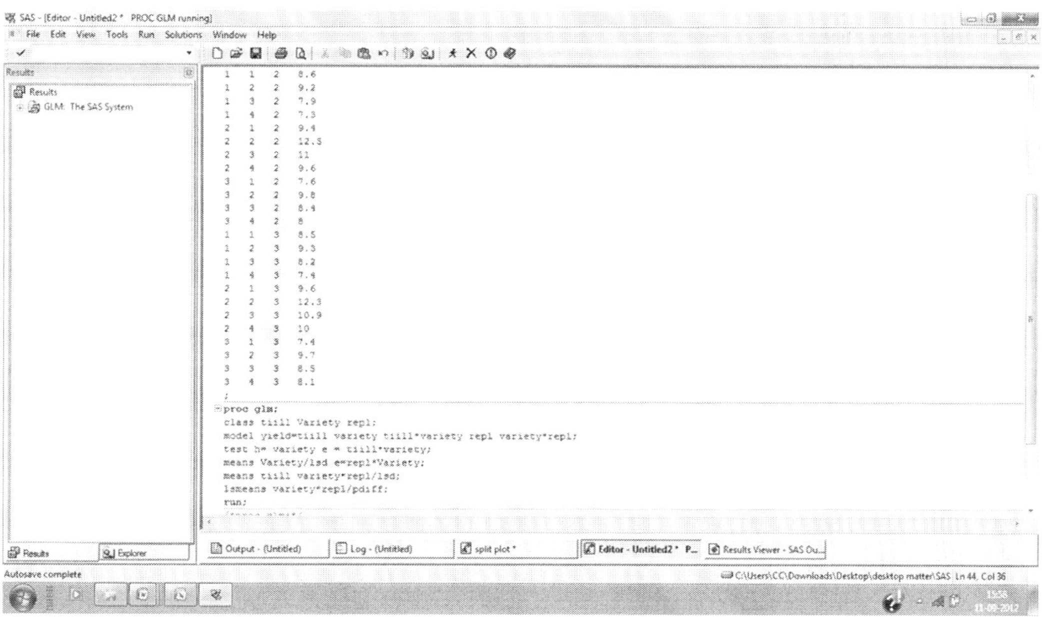

Slide 10.36: Step 2 showing a portion of data entry and commands for split plot data analysis using SAS

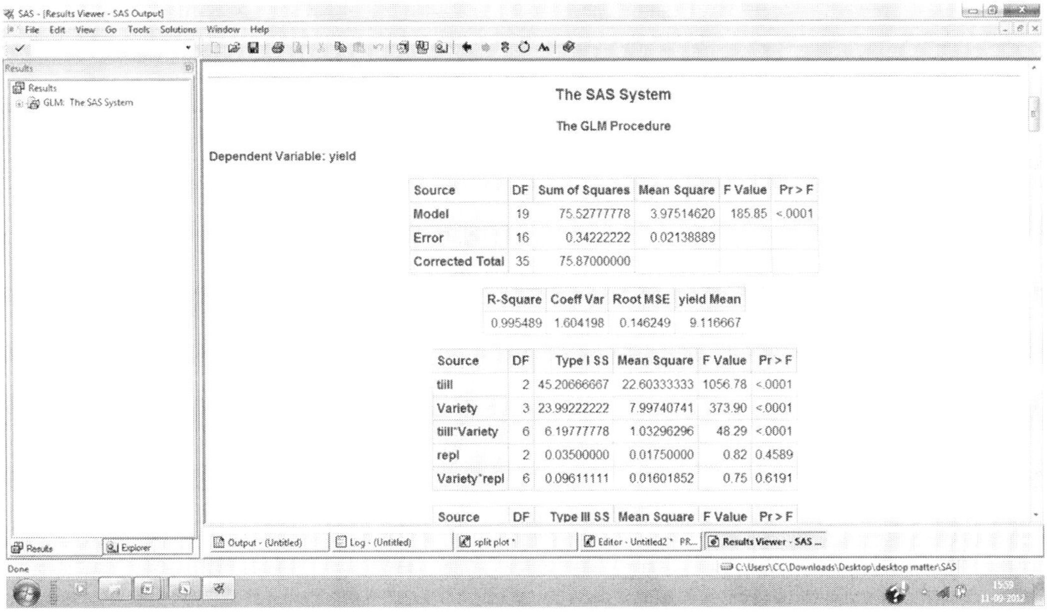
The SAS System

The GLM Procedure

Dependent Variable: yield

Source	DF	Sum of Squares	Mean Square	F Value	Pr > F
Model	19	75.52777778	3.97514620	185.85	<.0001
Error	16	0.34222222	0.02138889		
Corrected Total	35	75.87000000			

R-Square	Coeff Var	Root MSE	yield Mean
0.995489	1.604198	0.146249	9.116667

Source	DF	Type I SS	Mean Square	F Value	Pr > F
tiill	2	45.20666667	22.60333333	1056.78	<.0001
Variety	3	23.99222222	7.99740741	373.90	<.0001
tiill*Variety	6	6.19777778	1.03296296	48.29	<.0001
repl	2	0.03500000	0.01750000	0.82	0.4589
Variety*repl	6	0.09611111	0.01601852	0.75	0.6191

Slide 10.37: Step 3 showing a portion of output for split plot data analysis using SAS

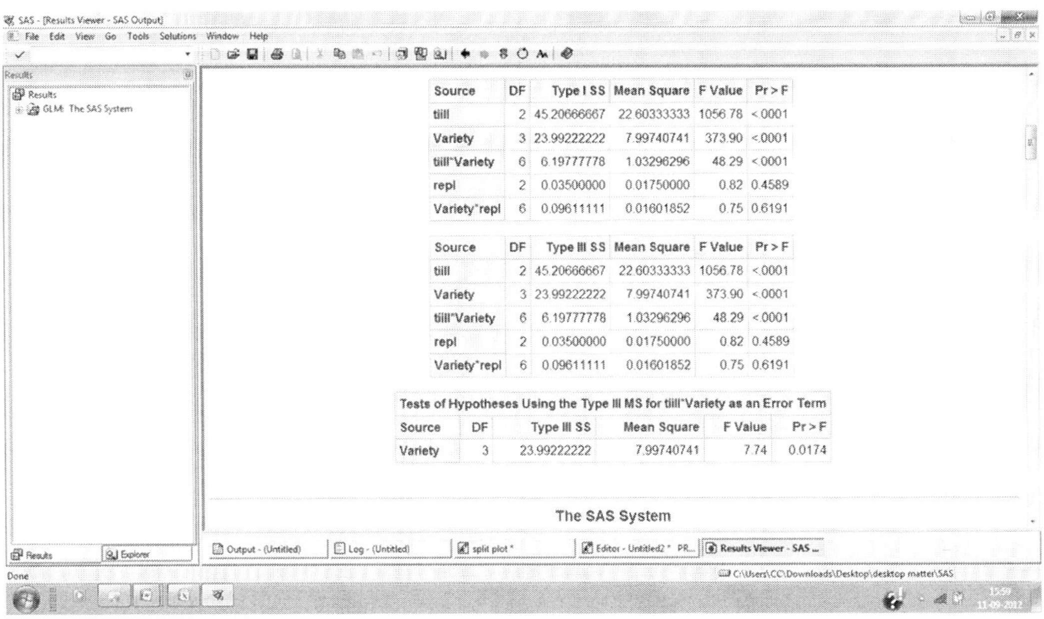
Source	DF	Type I SS	Mean Square	F Value	Pr > F
tiill	2	45.20666667	22.60333333	1056.78	<.0001
Variety	3	23.99222222	7.99740741	373.90	<.0001
tiill*Variety	6	6.19777778	1.03296296	48.29	<.0001
repl	2	0.03500000	0.01750000	0.82	0.4589
Variety*repl	6	0.09611111	0.01601852	0.75	0.6191

Source	DF	Type III SS	Mean Square	F Value	Pr > F
tiill	2	45.20666667	22.60333333	1056.78	<.0001
Variety	3	23.99222222	7.99740741	373.90	<.0001
tiill*Variety	6	6.19777778	1.03296296	48.29	<.0001
repl	2	0.03500000	0.01750000	0.82	0.4589
Variety*repl	6	0.09611111	0.01601852	0.75	0.6191

Tests of Hypotheses Using the Type III MS for tiill*Variety as an Error Term

Source	DF	Type III SS	Mean Square	F Value	Pr > F
Variety	3	23.99222222	7.99740741	7.74	0.0174

The SAS System

Slide 10.38: Step 3 showing a portion of output for split plot data analysis using SAS

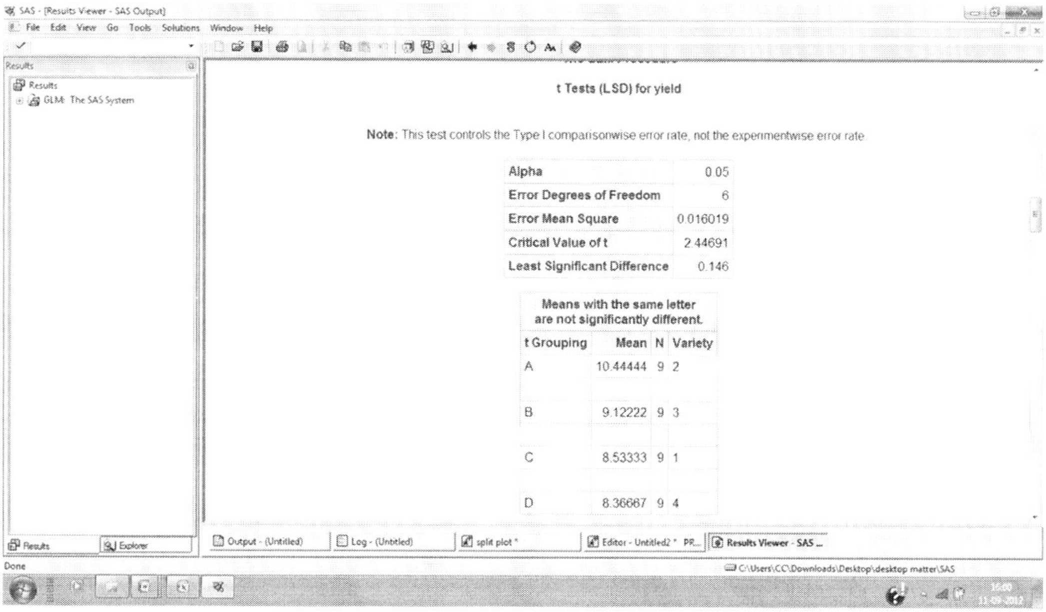

Alpha	0.05
Error Degrees of Freedom	6
Error Mean Square	0.016019
Critical Value of t	2.44691
Least Significant Difference	0.146

Means with the same letter are not significantly different.

t Grouping	Mean	N	Variety
A	10.44444	9	2
B	9.12222	9	3
C	8.53333	9	1
D	8.36667	9	4

Slide 10.39: Step 3 showing a portion of output for split plot data analysis using SAS

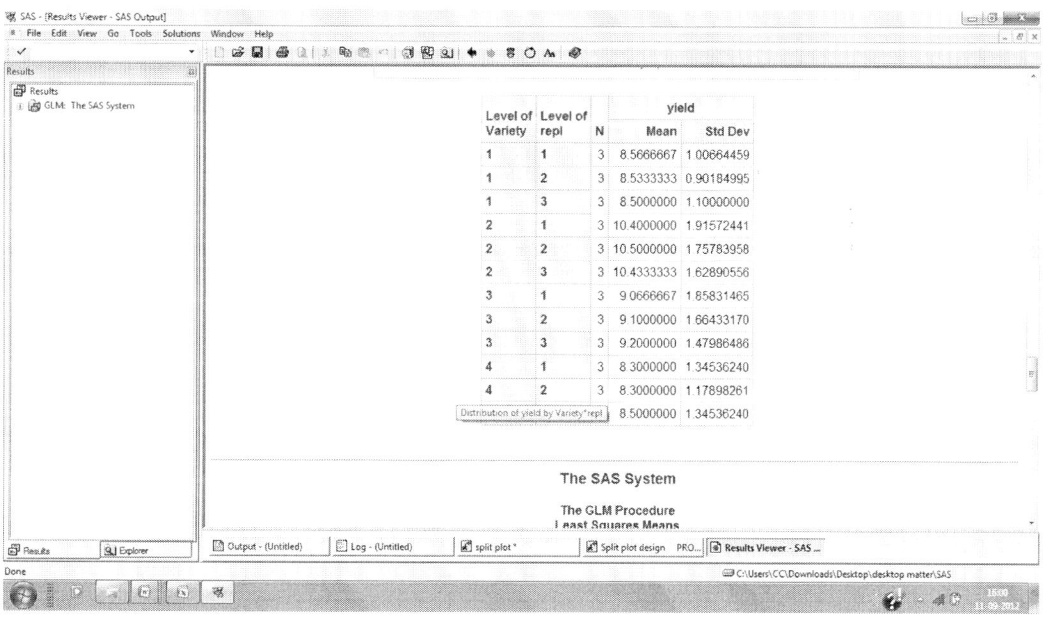

Level of Variety	Level of repl	N	yield Mean	yield Std Dev
1	1	3	8.5666667	1.00664459
1	2	3	8.5333333	0.90184995
1	3	3	8.5000000	1.10000000
2	1	3	10.4000000	1.91572441
2	2	3	10.5000000	1.75783958
2	3	3	10.4333333	1.62890556
3	1	3	9.0666667	1.85831465
3	2	3	9.1000000	1.66433170
3	3	3	9.2000000	1.47986486
4	1	3	8.3000000	1.34536240
4	2	3	8.3000000	1.17898261
			8.5000000	1.34536240

Slide 10.40: Step 3 showing a portion of output for split plot data analysis using SAS

between the two factors are measured with higher precision than either of the two factors.

Randomization and Layout

Like split plot design, the randomization and allocation of the two factors in strip plot design is done in two steps. First, n horizontal factors are randomly allocated to n horizontal strips. Next the m levels of factor B are distributed among the m vertical strips independently and randomly. The procedure of randomization is repeated for each and every replication independently. The ultimate layout of an $m \times n$ strip plot design in r replication is given in (Table 10.20):

Analysis

The analysis of strip plot design is performed in three steps: (a) analysis of horizontal factor effects, (b) analysis of vertical factor effects, and (c) analysis of interaction factor effects.

Statistical Model

$$y_{ijk} = \mu + \gamma_i + \alpha_j + e_{ij} + \beta_k + e'_{ik} + (\alpha\beta)_{jk} + e''_{ijk}$$

where $i = 1, 2, \ldots, r$; $j = 1, 2, \ldots, m$; $k = 1, 2, \ldots, n$

μ = general effect

γ_i = additional effect due to the ith replication

α_j = additional effect due to the jth level of vertical factor A and $\sum_{j=1}^{m} \alpha_j = 0$

β_k = additional effect due to the kth level of horizontal factor B and $\sum_{j=1}^{n} \beta_k = 0$

$(\alpha\beta)_{jk}$ = interaction effect due to the jth level of factor A and the kth level of factor B and $\sum_j (\alpha\beta)_{jk} = \sum_k (\alpha\beta)_{jk} = 0\ r$

e_{ij} (error I) = error associated with the ith replication and the jth level of vertical factor A and $\sim e_{ik} \sim$ i.i.d. $N(0, \sigma_1^2)$

Table 10.20 Layout of $m \times n$ strip plot design

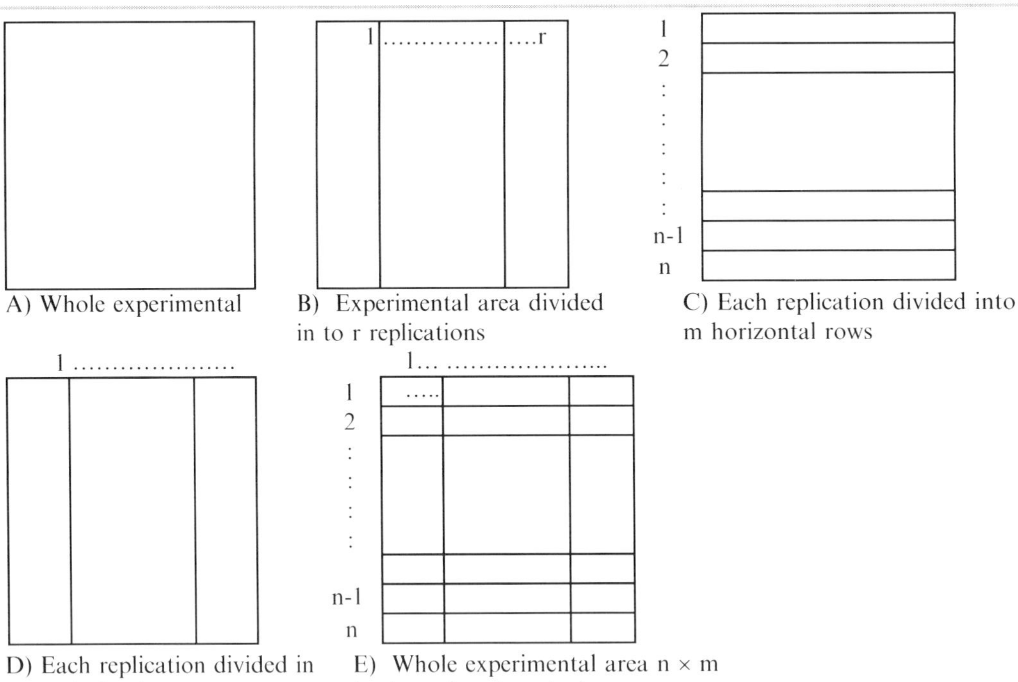

A) Whole experimental

B) Experimental area divided in to r replications

C) Each replication divided into m horizontal rows

D) Each replication divided in m vertical columns

E) Whole experimental area n × m horizontal and vertical rows

Table 10.21 Table of totals for replication × vertical factor A

	a_1	a_2	......	a_j	......	a_m	Total	Mean
R_1	y_{110}	y_{120}	......	y_{1j0}	......	y_{1m0}	y_{100}	$\bar{y}_{100}$
R_2	y_{210}	y_{220}	......	y_{2j0}	......	y_{2m0}	y_{200}	$\bar{y}_{200}$
:	:	:	:	:	:	:	:	
R_i	y_{i10}	y_{i20}	......	y_{ij0}	......	y_{im0}	y_{i00}	$\bar{y}_{i00}$
:	:	:	:	:	:	:	:	
R_r	y_{r10}	y_{r20}	......	y_{rj0}	......	y_{rm0}	y_{r00}	$\bar{y}_{r00}$
Total	y_{010}	y_{020}	......	y_{0j0}	......	y_{0m0}	y_{000}	$\bar{y}_{000}$
Mean	$\bar{y}_{010}$	$\bar{y}_{020}$		$\bar{y}_{0j0}$		$\bar{y}_{0m0}$		

e'_{ik} (error II) = error associated with the ith level of replication and the kth level of horizontal factor B and e'_{ij} ~ i.i.d. $N(0, \sigma_2^2)$

e''_{ijk} (error III) = error associated with the ith replication, the jth level of vertical factor A, and the kth level of horizontal factor B and e_{ijk} ~ i.i.d. $N(0, \sigma_3^2)$

Hypothesis to Be Tested

$$H_0: \gamma_1 = \gamma_2 = \cdots = \gamma_i = \cdots = \gamma_r = 0,$$
$$\alpha_1 = \alpha_2 = \cdots = \alpha_j = \cdots = \alpha_m = 0,$$
$$\beta_1 = \beta_2 = \cdots = \beta_k = \cdots = \beta_n = 0,$$
$$(\alpha\beta)_{11} = (\alpha\beta)_{12} = \cdots = (\alpha\beta)_{jk} = \cdots$$
$$= (\alpha\beta)_{mn} = 0.$$

H_1: γ's are not all equal,
α's are not all equal,
β's are not all equal,
$(\alpha\beta)$'s are not all equal.

Let the level of significance be 0.05.

First we construct the three two-way tables of totals for (a) replication × horizontal factor, (b) replication × vertical factor, and (c) horizontal × vertical factor. The step-by-step procedure for computation of different sums of squares and mean sums of squares is given as follows:

1. Grand total $= \sum_{i=1}^{r}\sum_{j=1}^{m}\sum_{k=1}^{n} y_{ijk} = G$.
2. Correction factor (CF) $= \frac{G^2}{mnr}$.
3. Total MS $= \sum_{i=1}^{r}\sum_{j=1}^{m}\sum_{k=1}^{n} y_{ijk}^2 - \text{CF}$.
4. Work out the sum of squares due to the vertical factor A and the replication. Table 10.21 (table of totals) is required to be framed.
where

$$y_{ij0} = \sum_{k=1}^{n} y_{ijk}$$

$$\text{TSS (Table I)} = \frac{1}{n}\sum_{i=1}^{r}\sum_{j=1}^{m} y_{ij0}^2 - \text{CF}$$

$$\text{Replication SS} = \frac{1}{mn}\sum_{i=1}^{r} y_{i00}^2 - \text{CF}$$

$$\text{SS(A)} = \frac{1}{nr}\sum_{j=1}^{m} y_{0j0}^2 - \text{CF}$$

$$\text{SS(Error I)} = \text{TSS(Table I)} - \text{RSS} - \text{SS(A)}$$

5. Work out the sum of squares due to the horizontal factor B and the replication. Table 10.22 (table of totals) is required to be framed.

$$\text{TSS (Table II)} = \frac{1}{m}\sum_{i=1}^{r}\sum_{k=1}^{n} y_{i0k}^2 - \text{CF}$$

$$\text{SS(B)} = \frac{1}{mr}\sum_{k=1}^{n} y_{00k}^2 - \text{CF}$$

$$\text{SS(Error II)} = \text{TSS(Table II)} - \text{RSS} - \text{SS(B)}$$

Table 10.22 Table of totals for replication × horizontal factor B

	b_1	b_2	……	b_k	……	b_n	Total	Mean
R_1	y_{101}	y_{102}	……	y_{10k}	……	y_{10n}	y_{100}	$\bar{y}_{100}$
R_2	y_{201}	y_{202}	……	y_{20k}	……	y_{20n}	y_{200}	$\bar{y}_{200}$
:	:	:	:	:	:	:	:	
R_i	y_{i01}	y_{i02}	……	y_{i0k}	……	y_{i0n}	y_{i00}	$\bar{y}_{i00}$
:	:	:	:	:	:	:	:	
R_r	y_{r01}	Y_{r02}	……	y_{r0k}	……	y_{r0n}	y_{r00}	$\bar{y}_{r00}$
Total	y_{001}	y_{002}	……	y_{00kj}	……	y_{00nm}	y_{000}	$\bar{y}_{000}$
Mean	$\bar{y}_{001}$	$\bar{y}_{002}$		$\bar{y}_{00k}$		$\bar{y}_{00n}$		

Table 10.23 Table of totals for vertical factor A × horizontal factor B

	b_1	b_2	……	b_k	……	b_n	Total	Mean
a_1	y_{011}	y_{012}	……	y_{01k}	……	y_{01n}	y_{010}	$\bar{y}_{010}$
a_2	y_{021}	y_{022}	……	y_{02k}	……	y_{02n}	y_{020}	$\bar{y}_{020}$
:	:	:	:	:	:	:	:	
a_j	y_{0j1}	y_{0j2}	……	y_{0jk}	……	y_{0jn}	y_{0j0}	$\bar{y}_{0j0}$
:	:	:	:	:	:	:	:	
a_m	y_{0m1}	y_{0m2}	……	y_{0mk}	……	y_{0mn}	y_{0m0}	$\bar{y}_{0m0}$
Total	y_{001}	y_{002}	……	y_{00k}	……	y_{00n}	y_{000}	$\bar{y}_{000}$
Mean	$\bar{y}_{001}$	$\bar{y}_{002}$		$\bar{y}_{00k}$		$\bar{y}_{00n}$		

6. Work out the sum of squares due to vertical factor A and horizontal factor B interaction. Table 10.23 (table of totals) is required to be formed.

$$\text{TSS(Table III)} = \frac{1}{r}\sum_{j=1}^{m}\sum_{k=1}^{n} y_{0jk}^2 - \text{CF}$$

$$\text{SS(AB)} = \text{TSS(Table III)} - \text{SS(A)} - \text{SS(B)}$$

$$\begin{aligned}\text{SS(Error III)} = {} & \text{TSS} - \text{RSS} - \text{SS(A)} \\ & - \text{SS(B)} - \text{SS(AB)} - \text{ErSS I} \\ & - \text{ErSS II}\end{aligned}$$

Mean sum of squares is calculated by dividing the sum of squares by the corresponding degrees of freedom (Table 10.24).

7. F-ratios for replication and vertical factor effects are obtained by comparing the replication mean sum of squares against the mean sum of squares due to error I. On the other hand, the F-ratios corresponding to horizontal factor and interaction between horizontal and vertical factors are worked out by comparing the respective mean sum of squares against the mean sum of squares due to error II and error III, respectively.
8. Calculated F-ratios are compared with tabulated value of F at appropriate level of significance and degrees of freedom.
9. Once the F-test becomes significant, our next task will be to estimate the SEs for different types of comparison as given below:
   (a) Standard error for difference between two replication means $= \sqrt{\frac{2\text{ErMS-I}}{mn}}$ and the corresponding LSD value will be $\text{LSD}_\alpha = \sqrt{\frac{2\text{ErMS-I}}{mn}} \times t_{\frac{\alpha}{2};\text{error-I d.f.}}$.
   (b) Standard error for difference between two vertical plot treatment means $= \sqrt{\frac{2\text{ErMS-I}}{rn}}$ and the corresponding LSD value will be $\text{LSD}_\alpha = \sqrt{\frac{2\text{ErMS-I}}{rn}} \times t_{\frac{\alpha}{2};\text{error-I d.f.}}$.
   (c) Standard error for difference between two horizontal plot treatment

Table 10.24 ANOVA table for $m \times n$ strip plot design in r replication

SOV	d.f.	SS	MS	F-ratio
Replication	$r-1$	RSS	RMS	RMS/ErMSI
Vertical plot factor (A)	$m-1$	SS(A)	MS(A)	MS(A)/ErMSI
Error I	$(r-1)(m-1)$	ErSSI	ErMSI	
Horizontal plot factor (B)	$(n-1)$	SS(B)	MS(B)	MS(B)/ErMSII
Error II	$(r-1)(n-1)$	ErSSII	ErMSII	
Interaction (A × B)	$(m-1)(n-1)$	SS(AB)	MS(AB)	MS(AB)/ErMSII
Error III	$(m-1)(n-1)(r-1)$	ErSSIII	ErMSIII	
Total	$mnr-1$	TSS		

means $= \sqrt{\frac{2\text{ErMS-II}}{rm}}$ and the corresponding LSD value will be $\text{LSD}_\alpha = \sqrt{\frac{2\text{ErMS-II}}{rm}} \times t_{\frac{\alpha}{2};\ \text{error-II d.f.}}$.

(d) Standard error for difference between two vertical plot treatment means at the same level of horizontal plot treatment $= \sqrt{\frac{2[(n-1)\ \text{ErMS-III} + \text{ErMS-I}]}{rn}}$ and the corresponding LSD value will be $\text{LSD}_\alpha = \sqrt{\frac{2[(n-1)\ \text{ErMS-III} + \text{ErMS-I}]}{rn}} \times t_{\frac{\alpha}{2};\text{errorII d.f.}}$.

(e) Standard error for difference between two horizontal plot treatment means at the same level of vertical plot treatment $= \sqrt{\frac{2[(m-1)\ \text{ErMS-III} + \text{ErMS-II}]}{rm}}$.
But the ratio of the treatment difference and the above SE in (d) and (e) does not follow the t distribution, and the approximate weighted t is calculated as follows:

$$\frac{\{(n-1)\ \text{ErMS-III} \times t_{\text{III}} + (\text{ErMS-I} \times t_{\text{I}})\}}{\{(n-1)\ \text{ErMS-III} + \text{ErMS-I}\}} \text{ and}$$

$$\frac{[(m-1)\ \text{ErMS-III} \times t_{\text{III}} + \text{ErMS-II} \times t_{\text{II}}]}{(m-1)\ \text{ErMS-III} + \text{ErM-SII}}, \text{ respectively,}$$

where $t_{\text{I}} = t$ value at error I degrees of freedom, $t_{\text{II}} = t$ value at error II degrees of freedom, and $t_{\text{III}} = t$ value at error III d.f. with specified level of significance. Corresponding LSD values will be $\text{LSD}_\alpha = \text{SE}_d \times t(\text{cal})$.

Example 10.18. To find the efficacy of three different sources of organic manure and four irrigation schedules, an experiment was conducted in strip plot design with three replications. Given below are the yield (t/ha) data for different treatments. Analyze the information to identify the best organic manure and irrigation schedule along with their combination.

Statistical Model

$$y_{ijk} = \mu + \gamma_i + \alpha_j + e_{ij} + \beta_k + e'_{jk} + (\alpha\beta)_{jk} + e''_{ijk}$$

where $i = 3;\ j = 3;\ k = 4;$

μ = general effect

r_i = additional effect due to the ith replication

α_j = additional effect due to the jth level of vertical factor, manure, and $\sum_{j=1}^{3} \alpha_j = 0$

β_k = additional effect due to the kth level of horizontal factor, irrigation, and $\sum_{k=1}^{4} \beta_k = 0$

$(\alpha\beta)_{jk}$ = interaction effect due to the jth level of manure and the kth level of irrigation and $\sum_j (\alpha\beta)_{jk} = \sum_k (\alpha\beta)_{jk} = 0$

e_{ij} (error I) = error associated with the ith replication and the jth level of vertical factor A and $e_{ij} \sim$ i.i.d. $N(0, \sigma_1^2)$

e^2_{jk} (error II) = error associated with the jth level of vertical factor A (manure) and the kth level of horizontal factor B (irrigation) and $e_{jk} \sim$ i.i.d. $N(0, \sigma_2^2)$

e''_{ijk} (error III) = error associated with the ith replication, the jth level of vertical factor A and the kth level of horizontal factor B and $e_{ijk} \sim$ i.i.d. $N(0, \sigma_3^2)$

Horizontal plot (irrigation)	Vertical plot (manure)											
	M1				M2				M3			
	I1	I2	I3	I4	I1	I2	I3	I4	I1	I2	I3	I4
R1	11.2	10.2	14.5	12.3	11.8	10.9	16.2	13.5	11	9.5	12.5	10.6
R2	11.3	10.5	14.6	12.5	11.9	10.8	16.5	13.8	10.9	9.8	12	10.8
R3	11.5	10.4	14.2	12.8	11.6	10.7	16.6	13.9	10.5	9.7	12.4	10.7

Hypothesis to Be Tested

$$H_0: \gamma_1 = \gamma_2 = \gamma_3 = 0,$$
$$\alpha_1 = \alpha_2 = \alpha_3 = 0,$$
$$\beta_1 = \beta_2 = \beta_3 = \beta_4 = 0,$$
$$(\alpha\beta)_{11} = (\alpha\beta)_{12} = \cdots = (\alpha\beta)_{jk} = \cdots = (\alpha\beta)_{34} = 0.$$

H_1 : γ's are not all equal,
α's are not all equal,
β's are not all equa,
$(\alpha\beta)$'s are not all equal.

Let the level of significance be 0.05.

First we construct three two-way tables of totals for replication × manure, replication × irrigation, and manure × irrigation. The step-by-step procedure for computation of different sums of squares and mean sum of squares is given as follows:

1. Grand total $= \sum_{i=1}^{3}\sum_{j=1}^{3}\sum_{k=1}^{4} y_{ijk} = \text{GT} = 11.2 + 10.2 + \cdots + 12.4 + 10.7 = 434.6.$
2. Correction factor (CF) $= \frac{\text{GT}^2}{3.4.3} = \frac{434.6^2}{36} = 5246.488.$
3. Work out the sum of squares due to vertical factor A and replication. The following table of totals is required to be framed:

$$\text{TSS(Table I)} = \frac{1}{4}\sum_{i=1}^{3}\sum_{j=1}^{3} y_{ij0}^2 - \text{CF}$$
$$= 5279.09 - 5246.588 = 32.502$$

$$\text{Replication SS} = \frac{1}{3.4}\sum_{i=1}^{3} y_{i00}^2 - \text{CF}$$
$$= \frac{144.2^2 + 145.4^2 + 145^2}{12} - 5246.588$$
$$= 0.062$$

$$\text{SS(A)} = \frac{1}{4.3}\sum_{j=1}^{3} y_{0j0}^2 - \text{CF}$$
$$= \frac{146^2 + 158.2^2 + 130.4^2}{12} - 5246.588$$
$$= 32.362$$

$$\text{SS(Error I)} = \text{TSS (Table I)} - \text{RSS} - \text{SS (A)}$$
$$= 32.502 - 0.062 - 32.362$$
$$= 0.078$$

4. Work out the sum of squares due to horizontal factor B and replication. The following table of totals is required to be framed:

Table I Table of totals for replication × manure

	M1	M2	M3	Total	Average
R1	48.200	52.400	43.600	144.200	12.017
R2	48.900	53.000	43.500	145.400	12.117
R3	48.900	52.800	43.300	145.000	12.083
Total	146.000	158.200	130.400		
Average	12.167	13.183	10.867		

Table 2 Table of totals for replication × irrigation

	I1	I2	I3	I4	Total	Average
R1	34.000	30.600	43.200	36.400	144.200	12.017
R2	34.100	31.100	43.100	37.100	145.400	12.117
R3	33.600	30.800	43.200	37.400	145.000	12.083
Total	101.700	92.500	129.500	110.900		
Average	11.300	10.278	14.389	12.322		

$$\text{(Table II) ErSS II} = \tfrac{1}{3}\sum_{i=1}^{3}\sum_{k=1}^{4} y_{i0k}^2 - \text{CF}$$

$$= \frac{34^2 + 30.6^2 + \cdots + 43.2^2 + 37.4^2}{3} - 5246.588$$

$$= 83.416$$

$$\text{SS(B)} = \frac{1}{3.3}\sum_{k=1}^{4} y_{00k}^2 - \text{CF}$$

$$= \frac{101.7^2 + 92.5^2 + 129.5^2 + 110.9^2}{9}$$

$$- 5246.588$$

$$= 83.212$$

$$\text{ErSS II} = \frac{1}{3}\sum_{i=1}^{3}\sum_{k=1}^{4} y_{i0k}^2 - \text{CF} - \text{SS (Irrg.)}$$

$$= \frac{34^2 + 30.6^2 + \cdots + 43.2^2 + 37.4^2}{3}$$

$$-5246.588 - 83.212 = 0.204.$$

5. Work out the sum of squares due to vertical factor and horizontal factor interaction. The following table of totals is required to be framed:

Table 3 Table of totals for irrigation × manure

	M1	M2	M3	Total	Average
I1	34.0	35.3	32.4	101.700	11.300
I2	31.1	32.4	29.0	92.500	10.280
I3	43.3	49.3	36.9	129.500	14.389
I4	37.6	41.2	32.1	110.900	12.322
Total	146.000	158.200	130.400		
Average	12.167	13.183	10.867		

$$\text{SS(M} \times \text{I)} = \frac{1}{r}\sum_{j=1}^{m}\sum_{k=1}^{n} y_{0jk}^2 - \text{CF} - \text{SS(A)} - \text{SS(B)}$$

$$= \frac{34^2 + 35.3^2 + 32.4^2 + \cdots + 41.2^2 + 32.1^2}{3} - 5246.588 - 32.362 - 83.212 = 10.644$$

$$\begin{aligned}\text{Error III} &= \text{TMS} - \text{RMS} - \text{MS(A)} - \text{MS(B)} - \text{MS(AB)} - \text{ErMS-I} - \text{ErMS-II}\\ &= 127.112 - 0.062 - 32.362 - 83.212 - 10.644 - 0.078 - 0.204\\ &= 0.549.\end{aligned}$$

Mean sum of squares is calculated by dividing the sum of squares by the corresponding degrees of freedom.

ANOVA table for 3 × 4 strip plot design with 3 replications

6. F-ratios for replication and vertical factor effects are obtained by comparing the respective mean sum of squares against the mean sum of squares due to error I. On the other hand, the F-ratios corresponding to horizontal factor and interaction between horizontal and vertical factors are worked out by comparing the respective mean sum of squares against the mean sum of squares due to error II and error III, respectively.
7. Calculated F-ratios are compared with tabulated value of F at appropriate level of significance and degrees of freedom. It is found from the above table that except for replication effect, all other effects are significant at both 5 and 1% level of significance.
8. Once the F-test becomes significant, our next task is to estimate the SEs for different types of comparison as given below:
   (a) Standard error for difference between two vertical plot treatment means $=\sqrt{\frac{2\text{ErMS-I}}{rn}}$ and the corresponding CD value could be

$$\text{CD}_{\alpha}=\sqrt{\frac{2\text{ErMS-I}}{rn}}\times t_{\frac{\alpha}{2};\,\text{error-I d.f.}}$$
$$=\sqrt{\frac{2\times 0.019}{3.4}}\times 2.776=0.1562.$$

   (b) Standard error for difference between two horizontal plot treatment means $=\sqrt{\frac{2\text{ErMS-II}}{rm}}$ and the corresponding CD value could be

$$\text{CD}_{(0.05)}=\sqrt{\frac{2\text{ErMS-II}}{3.3}}\times t_{\frac{\alpha}{2};\text{error-II d.f.}}$$
$$=\sqrt{\frac{2\times 0.034}{3.3}}\times 2.447=0.213.$$

   (c) Standard error for difference between two vertical plot treatment means at the same level of horizontal plot treatment

$$=\sqrt{\frac{2[(n-1)\text{ErMS-III}+\text{ErMS-I}]}{rn}}$$
$$=\sqrt{\frac{2(4-1)0.046+0.019}{3.4}}=0.162.$$

   (d) Standard error for difference between two horizontal plot treatment means at the same level of vertical plot treatment

$$=\sqrt{\frac{2[(m-1)\text{ErMSIII}+\text{ErMSII}]}{rm}}$$
$$=\sqrt{\frac{2[(3-1)0.046+0.034]}{3\times 3}}=0.167.$$

But the ratio of the treatment difference and the above SE in (c) and (d) does not follow the t distribution, and the approximate weighted t is calculated as follows:

$$\frac{\{(n-1)\text{ErMS-III}\times t_{\text{III}}\}+(\text{ErMS-I}\times t_{\text{I}})}{\{(n-1)\text{ErMS-III}+\text{ErMS-I}\}}$$
$$=\frac{\{(4-1)0.046\times 2.179\}+(0.019\times 2.776)}{\{(4-1)\times 0.046+0.019\}}=2.248$$

SOV	d.f.	SS	MS	F-value	Table value of F ($p=0.05$)	($p=0.01$)
Replication	2	0.062	0.031	1.6	6.94	18.00
Vertical factor (manure)	2	32.362	16.181	832.1702	6.94	18.00
Error I	4	0.078	0.019			
Horizontal factor (irrig.)	3	83.212	27.737	814.0336	4.76	9.78
Error II	6	0.204	0.034			
M × I	6	10.644	1.774	38.7855	3.00	4.82
Error III	12	0.549	0.046			
Total	35	127.112				

and

$$\frac{[(m-1)\text{ErMS-III} \times t_{\text{III}} + \text{ErMS-II} \times t_{\text{II}}]}{(m-1)\text{ErMS-III} + \text{ErMS-II}}$$
$$= \frac{[(3-1) \times 0.046 \times 2.179 + 0.034 \times 2.447]}{(3-1) \times 0.046 + 0.034}$$
$$= 2.251,$$

where $t_{\text{I}} = t_{0.025,4} = 2.776$, $t_{\text{II}} = t_{0.025,6} = 2.447$, and $t_{\text{III}} = t_{0.025,12} = 2.179$. Corresponding CD values could be CD $(0.05) = \text{SE}_d \times t_{\text{cal}}$.

Thus, the critical difference value to compare two vertical plot treatment means at the same level of horizontal plot treatment is $\text{CD}(0.05) = \text{SE}_d \times t_{\text{cal}} = 0.162 \times 2.248 = 0.364$, and the critical difference value to compare two horizontal plot treatment means at the same level of vertical plot treatment is $\text{CD}(0.05) = \text{SE}_{\text{d}} \times t_{\text{cal}} = 0.167 \times 2.251 = 0.376$.

Manure	Yield (t/ha)	Irrigation	Yield (t/ha)
M2	13.183	I3	14.389
M1	12.167	I4	12.322
M3	10,867	I1	11.300
		I2	10.280

Comparing the treatment means it can be concluded that manure 2 is the best among the manures and irrigation schedule 3 is the best irrigation schedule. Similarly by using appropriate critical difference values as mentioned above, one can find out the best manure at a particular irrigation level and vice versa (Table 10.25).

Table 10.25 Table for transformed values of percentages as arcsine $\sqrt{\text{percentage}}$

Percent (%)	Transformed	Percent (%)	Transformed	Percent (%)	Transformed	Percent (%)	Transformed	Percent (%)	Transformed
0.0	0.00	4.1	11.68	8.2	16.64	12.3	20.53	16.4	23.89
0.1	1.81	4.2	11.83	8.3	16.74	12.4	20.62	16.5	23.97
0.2	2.56	4.3	11.97	8.4	16.85	12.5	20.70	16.6	24.04
0.3	3.14	4.4	12.11	8.5	16.95	12.6	20.79	16.7	24.12
0.4	3.63	4.5	12.25	8.6	17.05	12.7	20.88	16.8	24.20
0.5	4.05	4.6	12.38	8.7	17.15	12.8	20.96	16.9	24.27
0.6	4.44	4.7	12.52	8.8	17.26	12.9	21.05	17.0	24.35
0.7	4.80	4.8	12.66	8.9	17.36	13	21.13	17.1	24.43
0.8	5.13	4.9	12.79	9.0	17.46	13.1	21.22	17.2	24.50
0.9	5.44	5.0	12.92	9.1	17.56	13.2	21.30	17.3	24.58
1.0	5.74	5.1	13.05	9.2	17.66	13.3	21.39	17.4	24.65
1.1	6.02	5.2	13.18	9.3	17.76	13.4	21.47	17.5	24.73
1.2	6.29	5.3	13.31	9.4	17.85	13.5	21.56	17.6	24.80
1.3	6.55	5.4	13.44	9.5	17.95	13.6	21.64	17.7	24.88
1.4	6.80	5.5	13.56	9.6	18.05	13.7	21.72	17.8	24.95
1.5	7.03	5.6	13.69	9.7	18.15	13.8	21.81	17.9	25.03
1.6	7.27	5.7	13.81	9.8	18.24	13.9	21.89	18	25.10
1.7	7.49	5.8	13.94	9.9	18.34	14	21.97	18.1	25.18
1.8	7.71	5.9	14.06	10.0	18.43	14.1	22.06	18.2	25.25
1.9	7.92	6.0	14.18	10.1	18.53	14.2	22.14	18.3	25.33
2.0	8.13	6.1	14.30	10.2	18.63	14.3	22.22	18.4	25.40
2.1	8.33	6.2	14.42	10.3	18.72	14.4	22.30	18.5	25.47
2.2	8.53	6.3	14.54	10.4	18.81	14.5	22.38	18.6	25.55
2.3	8.72	6.4	14.65	10.5	18.91	14.6	22.46	18.7	25.62
2.4	8.91	6.5	14.77	10.6	19.00	14.7	22.54	18.8	25.70
2.5	9.10	6.6	14.89	10.7	19.09	14.8	22.63	18.9	25.77
2.6	9.28	6.7	15.00	10.8	19.19	14.9	22.71	19.0	25.84
2.7	9.46	6.8	15.12	10.9	19.28	15.0	22.79	19.1	25.91
2.8	9.63	6.9	15.23	11.0	19.37	15.1	22.87	19.2	25.99
2.9	9.80	7.0	15.34	11.1	19.46	15.2	22.95	19.3	26.06
3.0	9.97	7.1	15.45	11.2	19.55	15.3	23.03	19.4	26.13

3.1	10.14	7.2	15.56	11.3	19.64	15.4	23.11	19.5	26.21
3.2	10.30	7.3	15.68	11.4	19.73	15.5	23.18	19.6	26.28
3.3	10.47	7.4	15.79	11.5	19.82	15.6	23.26	19.7	26.35
3.4	10.63	7.5	15.89	11.6	19.91	15.7	23.34	19.8	26.42
3.5	10.78	7.6	16.00	11.7	20.00	15.8	23.42	19.9	26.49
3.6	10.94	7.7	16.11	11.8	20.09	15.9	23.50	20.0	26.57
3.7	11.09	7.8	16.22	11.9	20.18	16.0	23.58	20.1	26.64
3.8	11.24	7.9	16.32	12.0	20.27	16.1	23.66	20.2	26.71
3.9	11.39	8.0	16.43	12.1	20.36	16.2	23.73	20.3	26.78
4.0	11.54	8.1	16.54	12.2	20.44	16.3	23.81	20.4	26.85
20.5	26.92	24.6	29.73	28.7	32.39	32.8	34.94	36.9	37.41
20.6	26.99	24.7	29.80	28.8	32.46	32.9	35.00	37.0	37.46
20.7	27.06	24.8	29.87	28.9	32.52	33.0	35.06	37.1	37.52
20.8	27.13	24.9	29.93	29.0	32.58	33.1	35.12	37.2	37.58
20.9	27.20	25.0	30.00	29.1	32.65	33.2	35.18	37.3	37.64
21.0	27.27	25.1	30.07	29.2	32.71	33.3	35.24	37.4	37.70
21.1	27.35	25.2	30.13	29.3	32.77	33.4	35.30	37.5	37.76
21.2	27.42	25.3	30.20	29.4	32.83	33.5	35.37	37.6	37.82
21.3	27.49	25.4	30.26	29.5	32.90	33.6	35.43	37.7	37.88
21.4	27.56	25.5	30.33	29.6	32.96	33.7	35.49	37.8	37.94
21.5	27.62	25.6	30.40	29.7	33.02	33.8	35.55	37.9	38.00
22.0	27.69	25.7	30.46	29.8	33.09	33.9	35.61	38.0	38.06
21.7	27.76	25.8	30.53	29.9	33.15	34.0	35.67	38.1	38.12
21.8	27.83	25.9	30.59	30.0	33.21	34.1	35.73	38.2	38.17
21.9	27.90	26.0	30.66	30.1	33.27	34.2	35.79	38.3	38.23
22.0	27.97	26.1	30.72	30.2	33.34	34.3	35.85	38.4	38.29
22.1	28.04	26.2	30.79	30.3	33.40	34.4	35.91	38.5	38.35
22.2	28.11	26.3	30.85	30.4	33.46	34.5	35.97	38.6	38.41
22.3	28.18	26.4	30.92	30.5	33.52	34.6	36.03	38.7	38.47
22.4	28.25	26.5	30.98	30.6	33.58	34.7	36.09	38.8	38.53
22.5	28.32	26.6	31.05	30.7	33.65	34.8	36.15	38.9	38.59
22.6	28.39	26.7	31.11	30.8	33.71	34.9	36.21	39.0	38.65
22.7	28.45	26.8	31.18	30.9	33.77	35.0	36.27	39.1	38.70

(continued)

Table 10.25 (continued)

Percent (%)	Transformed	Percent (%)	Transformed	Percent (%)	Transformed	Percent (%)	Transformed	Percent (%)	Transformed
22.8	28.52	26.9	31.24	31.0	33.83	35.1	36.33	39.2	38.76
22.9	28.59	27.0	31.31	31.1	33.90	35.2	36.39	39.3	38.82
23.0	28.66	27.1	31.37	31.2	33.96	35.3	36.45	39.4	38.88
23.1	28.73	27.2	31.44	31.3	34.02	35.4	36.51	39.5	38.94
23.2	28.79	27.3	31.50	31.4	34.08	35.5	36.57	39.6	39.00
23.3	28.86	27.4	31.56	31.5	34.14	35.6	36.63	39.7	39.06
23.4	28.93	27.5	31.63	31.6	34.20	35.7	36.69	39.8	39.11
23.5	29.00	27.6	31.69	31.7	34.27	35.8	36.75	39.9	39.17
23.6	29.06	27.7	31.76	31.8	34.33	35.9	36.81	40.0	39.23
23.7	29.13	27.8	31.82	31.9	34.39	36.0	36.87	40.1	39.29
23.8	29.20	27.9	31.88	32.0	34.45	36.1	36.93	40.2	39.35
23.9	29.27	28.0	31.95	32.1	34.51	36.2	36.99	40.3	39.41
24.0	29.33	28.1	32.01	32.2	34.57	36.3	37.05	40.4	39.47
24.1	29.40	28.2	32.08	32.3	34.63	36.4	37.11	40.5	39.52
24.2	29.47	28.3	32.14	32.4	34.70	36.5	37.17	40.6	39.58
24.3	29.53	28.4	32.20	32.5	34.76	36.6	37.23	40.7	39.64
24.4	29.60	28.5	32.27	32.6	34.82	36.7	37.29	40.8	39.70
24.5	29.67	28.6	32.33	32.7	34.88	36.8	37.35	40.9	39.76
41.0	39.82	45.1	42.19	49.2	44.54	53.3	46.89	57.4	49.26
41.1	39.87	45.2	42.25	49.3	44.60	53.4	46.95	57.5	49.31
41.2	39.93	45.3	42.30	49.4	44.66	53.5	47.01	57.6	49.37
41.3	39.99	45.4	42.36	49.5	44.71	53.6	47.06	57.7	49.43
41.4	40.05	45.5	42.42	49.6	44.77	53.7	47.12	57.8	49.49
41.5	40.11	45.6	42.48	49.7	44.83	53.8	47.18	57.9	49.55
41.6	40.16	45.7	42.53	49.8	44.89	53.9	47.24	58.0	49.60
41.7	40.22	45.8	42.59	49.9	44.94	54.0	47.29	58.1	49.66
41.8	40.28	45.9	42.65	50.0	45.00	54.1	47.35	58.2	49.72
41.9	40.34	46.0	42.71	50.1	45.06	54.2	47.41	58.3	49.78
42.0	40.40	46.1	42.76	50.2	45.11	54.3	47.47	58.4	49.84
42.1	40.45	46.2	42.82	50.3	45.17	54.4	47.52	58.5	49.89
42.2	40.51	46.3	42.88	50.4	45.23	54.5	47.58	58.6	49.95

42.3	40.57	46.4	42.94	50.5	45.29	54.6	47.64	58.7	50.01
42.4	40.63	46.5	42.99	50.6	45.34	54.7	47.70	58.8	50.07
42.5	40.69	46.6	43.05	50.7	45.40	54.8	47.75	58.9	50.13
42.6	40.74	46.7	43.11	50.8	45.46	54.9	47.81	59.0	50.18
42.7	40.80	46.8	43.17	50.9	45.52	55.0	47.87	59.1	50.24
42.8	40.86	46.9	43.22	51.0	45.57	55.1	47.93	59.2	50.30
42.9	40.92	47.0	43.28	51.1	45.63	55.2	47.98	59.3	50.36
43.0	40.98	47.1	43.34	51.2	45.69	55.3	48.04	59.4	50.42
43.1	41.03	47.2	43.39	51.3	45.74	55.4	48.10	59.5	50.48
43.2	41.09	47.3	43.45	51.4	45.80	55.5	48.16	59.6	50.53
43.3	41.15	47.4	43.51	51.5	45.86	55.6	48.22	59.7	50.59
43.4	41.21	47.5	43.57	51.6	45.92	55.7	48.27	59.8	50.65
43.5	41.27	47.6	43.62	51.7	45.97	55.8	48.33	59.9	50.71
43.6	41.32	47.7	43.68	51.8	46.03	55.9	48.39	60.0	50.77
43.7	41.38	47.8	43.74	51.9	46.09	56.0	48.45	60.1	50.83
43.8	41.44	47.9	43.80	52.0	46.15	56.1	48.50	60.2	50.89
43.9	41.50	48.0	43.85	52.1	46.20	56.2	48.56	60.3	50.94
44.0	41.55	48.1	43.91	52.2	46.26	56.3	48.62	60.4	51.00
44.1	41.61	48.2	43.97	52.3	46.32	56.4	48.68	60.5	51.06
44.2	41.67	48.3	44.03	52.4	46.38	56.5	48.73	60.6	51.12
44.3	41.73	48.4	44.08	52.5	46.43	56.6	48.79	60.7	51.18
44.4	41.78	48.5	44.14	52.6	46.49	56.7	48.85	60.8	51.24
44.5	41.84	48.6	44.20	52.7	46.55	56.8	48.91	60.9	51.30
44.6	41.90	48.7	44.26	52.8	46.61	56.9	48.97	61.0	51.35
44.7	41.96	48.8	44.31	52.9	46.66	57.0	49.02	61.1	51.41
44.8	42.02	48.9	44.37	53.0	46.72	57.1	49.08	61.2	51.47
44.9	42.07	49.0	44.43	53.1	46.78	57.2	49.14	61.3	51.53
45.0	42.13	49.1	44.48	53.2	46.83	57.3	49.20	61.4	51.59
61.5	51.65	65.6	54.09	69.7	56.60	73.8	59.21	77.9	61.96
61.6	51.71	65.7	54.15	69.8	56.66	73.9	59.28	78.0	62.03
61.7	51.77	65.8	54.21	69.9	56.73	74.0	59.34	78.1	62.10
61.8	51.83	65.9	54.27	70.0	56.79	74.1	59.41	78.2	62.17
61.9	51.88	66.0	54.33	70.1	56.85	74.2	59.47	78.3	62.24

(continued)

Table 10.25 (continued)

Percent (%)	Transformed	Percent (%)	Transformed	Percent (%)	Transformed	Percent (%)	Transformed	Percent (%)	Transformed
62.0	51.94	66.1	54.39	70.2	56.91	74.3	59.54	78.4	62.31
62.1	52.00	66.2	54.45	70.3	56.98	74.4	59.60	78.5	62.38
62.2	52.06	66.3	54.51	70.4	57.04	74.5	59.67	78.6	62.44
62.3	52.12	66.4	54.57	70.5	57.10	74.6	59.74	78.7	62.51
62.4	52.18	66.5	54.63	70.6	57.17	74.7	59.80	78.8	62.58
62.5	52.24	66.6	54.70	70.7	57.23	74.8	59.87	78.9	62.65
62.6	52.30	66.7	54.76	70.8	57.29	74.9	59.93	79.0	62.73
62.7	52.36	66.8	54.82	70.9	57.35	75.0	60.00	79.1	62.80
62.8	52.42	66.9	54.88	71.0	57.42	75.1	60.07	79.2	62.87
62.9	52.48	67.0	54.94	71.1	57.48	75.2	60.13	79.3	62.94
63.0	52.54	67.1	55.00	71.2	57.54	75.3	60.20	79.4	63.01
63.1	52.59	67.2	55.06	71.3	57.61	75.4	60.27	79.5	63.08
63.2	52.65	67.3	55.12	71.4	57.67	75.5	60.33	79.6	63.15
63.3	52.71	67.4	55.18	71.5	57.73	75.6	60.40	79.7	63.22
63.4	52.77	67.5	55.24	71.6	57.80	75.7	60.47	79.8	63.29
63.5	52.83	67.6	55.30	71.7	57.86	75.8	60.53	79.9	63.36
63.6	52.89	67.7	55.37	71.8	57.92	75.9	60.60	80.0	63.43
63.7	52.95	67.8	55.43	71.9	57.99	76.0	60.67	80.1	63.51
63.8	53.01	67.9	55.49	72.0	58.05	76.1	60.73	80.2	63.58
63.9	53.07	68.0	55.55	72.1	58.12	76.2	60.80	80.3	63.65
64.0	53.13	68.1	55.61	72.2	58.18	76.3	60.87	80.4	63.72
64.1	53.19	68.2	55.67	72.3	58.24	76.4	60.94	80.5	63.79
64.2	53.25	68.3	55.73	72.4	58.31	76.5	61.00	80.6	63.87
64.3	53.31	68.4	55.80	72.5	58.37	76.6	61.07	80.7	63.94
64.4	53.37	68.5	55.86	72.6	58.44	76.7	61.14	80.8	64.01
64.5	53.43	68.6	55.92	72.7	58.50	76.8	61.21	80.9	64.09
64.6	53.49	68.7	55.98	72.8	58.56	76.9	61.27	81.0	64.16
64.7	53.55	68.8	56.04	72.9	58.63	77.0	61.34	81.1	64.23
64.8	53.61	68.9	56.10	73.0	58.69	77.1	61.41	81.2	64.30
64.9	53.67	69.0	56.17	73.1	58.76	77.2	61.48	81.3	64.38
65.0	53.73	69.1	56.23	73.2	58.82	77.3	61.55	81.4	64.45

65.1	53.79	69.2	56.29	73.3	58.89	77.4	61.61	81.5	64.53
65.2	53.85	69.3	56.35	73.4	58.95	77.5	61.68	81.6	64.60
65.3	53.91	69.4	56.42	73.5	59.02	77.6	61.75	81.7	64.67
65.4	53.97	69.5	56.48	73.6	59.08	77.7	61.82	81.8	64.75
65.5	54.03	69.6	56.54	73.7	59.15	77.8	61.89	81.9	64.82
82.0	64.90	86.1	68.11	90.2	71.76	94.3	76.19	98.4	82.73
82.1	64.97	86.2	68.19	90.3	71.85	94.4	76.31	98.5	82.97
82.2	65.05	86.3	68.28	90.4	71.95	94.5	76.44	98.6	83.20
82.3	65.12	86.4	68.36	90.5			76.56	98.7	83.45
82.4	65.20	86.5	68.44	90.6	72.15	94.7	72.05	94.6	83.71
82.5	65.27	86.6	68.53	90.7	72.24	94.8	76.82	98.9	83.98
82.6	65.35	86.7	68.61	90.8	72.34	94.9	76.95	99	84.26
82.7	65.42	86.8	68.70	90.9	72.44	95.0	77.08	99.1	84.56
82.8	65.50	86.9	68.78	91.0	72.54	95.1	77.21	99.2	84.87
82.9	65.57	87.0	68.87	91.1	72.64	95.2	77.34	99.3	85.20
83.0	65.65	87.1	68.95	91.2	72.74	95.3	77.48	99.4	85.56
83.1	65.73	87.2	69.04	91.3	72.85	95.4	77.62	99.5	85.95
83.2	65.80	87.3	69.12	91.4	72.95	95.5	77.75	99.6	86.37
83.3	65.88	87.4	69.21	91.5	73.05	95.6	77.89	99.7	86.86
83.4	65.96	87.5	69.30	91.6	73.15	95.7	78.03	99.8	87.44
83.5	66.03	87.6	69.38	91.7	73.26	95.8	78.17	99.9	88.19
83.6	66.11	87.7	69.47	91.8	73.36	95.9	78.32	100	90.00
83.7	66.19	87.8	69.56	91.9	73.46	96.0	78.46		
83.8	66.27	87.9	69.64	92.0	73.57	96.1	78.61		
83.9	66.34	88.0	69.73	92.1	73.68	96.2	78.76		
84.0	66.42	88.1	69.82	92.2	73.78	96.3	78.91		
84.1	66.50	88.2	69.91	92.3	73.89	96.4	79.06		
84.2	66.58	88.3	70.00	92.4	74.00	96.5	79.22		
84.3	66.66	88.4	70.09	92.5	74.11	96.6	79.37		
84.4	66.74	88.5	70.18	92.6	74.21	96.7	79.53		
84.5	66.82	88.6	70.27	92.7	74.32	96.8	79.70		
84.6	66.89	88.7	70.36	92.8	74.44	96.9	79.86		
84.7	66.97	88.8	70.45	92.9	74.55	97.0	80.03		

(continued)

Table 10.25 (continued)

Percent (%)	Transformed	Percent (%)	Transformed	Percent (%)	Transformed	Percent (%)	Transformed	Percent (%)	Transformed
84.8	67.05	88.9	70.54	93.0	74.66	97.1	80.20		
84.9	67.13	89.0	70.63	93.1	74.77	97.2	80.37		
85.0	67.21	89.1	70.72	93.2	74.88	97.3	80.54		
85.1	67.29	89.2	70.81	93.3	75.00	97.4	80.72		
85.2	67.37	89.3	70.91	93.4	75.[illegible]1	97.5	80.90		
85.3	67.46	89.4	71.00	93.5	75.23	97.6	81.09		
85.4	67.54	89.5	71.09	93.6	75.35	97.7	81.28		
85.5	67.62	89.6	71.19	93.7	75.46	97.8	81.47		
85.6	67.70	89.7	71.28	93.8	75.58	97.9	81.67		
85.7	67.78	89.8	71.37	93.9	75.70	98.0	81.87		
85.8	67.86	89.9	71.47	94.0	75.82	98.1	82.08		
85.9	67.94	90.0	71.57	94.1	75.94	98.2	82.29		
86.0	68.03	90.1	71.66	94.2	76.06	98.3	82.51		

11 Analysis Related to Breeding Researches

Variability is one of the major concerns of any research. In a real-life situation, we are to deal with a number of variables at a time, so co-variability plays an important role. In social, agricultural, and other fields, the biophysical features in any experiment rarely behave independently; rather these are found to be functionally related to each other. There are several examples where the analysis of covariance can be used effectively in augmenting the precession of the experimental results. For example, in yield component analysis of paddy, the yield components, namely, the number of hills per unit area, the number of effective tillers per hill, and the number of grains per panicles, can be used as covariates or concomitant variables. In a study of health drinks on the growth and physique of school-going children, initial body weight, height, age, physical agility, etc., can be taken as concomitant variables during the analysis of covariance. In the analysis of covariance, there are two types of variables: the characteristic of the main interest and the information on the secondary or auxiliary interest or the covariates. In the analysis of covariance, the expected (true) value of the response is the resultant of two components, one because of the linear combination of the values of the concomitant variables which are functionally related with the response and another one already obtained in the analysis of variance. Thus, the analysis of covariance is the synthesis of the analysis of variance and the regression. Similar to that of the partitioning of variances into different components, one can also partition the covariance among the variables into different components like genotypic and environmental.

11.1 Analysis of Covariance

The analysis of covariance can be taken up for one-way and two-way layouts and other specific types of experimental design. In the analysis of covariance, there is one dependent variable (y) and one or more concomitant variables. The basic difference between the analysis of variance and the analysis of covariance models is that in the former, each response (y) is partitioned into two components, one because of its true value and the error part. The model in its simplest form may be written as

$$y_i = \sum_j \alpha_{ij}\tau_j + \sum_k \beta_k x_{ik} + e_i, \text{ where } \alpha_{ij} \text{ are known,}$$

x_{ik} is the value of the kth concomitant variable corresponding to y_i,

β_k's are the regression coefficients of y on the covariates x_k's,

τ_j's are the effects (main, interaction, blocks, etc.),

e_i is the random component in the model.

P.K. Sahu, *Research Methodology: A Guide for Researchers in Agricultural Science, Social Science and Other Related Fields*, DOI 10.1007/978-81-322-1020-7_11, © Springer India 2013

11.1.1 Analysis of Covariance for One-Way Classified Data with One Covariate

Let the criterion variate be denoted by y and the covariate by x. Thus, for each experimental unit, there is a pair of observations (x,y).

Let y_{ij} and x_{ij} denote the jth observation of the variate and covariate, respectively, for the ith treatment in a one-way classified data. Then the linear model is given by

$$y_{ij} = \mu + \alpha_i + \beta x_{ij} + e_{ij}; \quad \begin{array}{l} i = 1,2,3 \ldots . t \\ j = 1,2,3, \ldots . n_i \end{array}$$

where

$\mu =$ general effect,
$\alpha_i =$ effect due to ith treatment,
$\beta =$ regression coefficient of y on x,
$e_{ij} =$ independent normal variate with mean zero(0) and variance, σ^2.

The above model is based on the following assumptions:

1. The regression of y on x is linear and independent of the treatment so that the treatment and regression effects are additive.
2. The residuals are independently and normally distributed with mean 0 and the common variance.
3. The covariate x is fixed and is measured without error.

The model is often written in the form

$$y_{ij} = \mu^* + \alpha_i + \beta\left(x_{ij} - \bar{x}\right) + e_{ij}; \quad \begin{array}{l} i = 1,2,3 \ldots . t \\ j = 1,2,3, \ldots . n_i \end{array}$$

where

$$\mu^* = \mu + \beta\bar{x}$$

The least square estimates of μ, α_i, and β are obtained by minimizing $L = \sum_i \sum_j \left(y_{ij} - \mu - \alpha_i - \beta x_{ij}\right)^2$. The normal equations are

$$\sum\sum y_{ij} = n\mu + \sum_i n_i \alpha_i + \beta \sum\sum x_{ij},$$

$$\sum_j y_{ij} = n_i\mu + n_i\alpha_i + \beta \sum_j x_{ij},$$

$$\sum\sum x_{ij}y_{ij} - \mu \sum\sum x_{ij} + \sum_i \sum_j \alpha_i x_{ij} + \beta \sum_j x_{ij}^2.$$

Since $\sum_i n_i\alpha_i = 0, \quad \hat{\mu} = \bar{y} - \hat{\beta}\bar{x};$

$$\hat{\alpha}_i = \bar{y}_i - \bar{y} - \hat{\beta}(\bar{x}_i - \bar{x})$$

$$\sum\sum x_{ij}y_{ij} = \left(\bar{y} - \hat{\beta}\bar{x}\right)n\bar{x} + \sum_i \left[\bar{y}_i - \bar{y} - \hat{\beta}(\bar{x}_i - \bar{x})\right]n_i\bar{x}_i + \hat{\beta}\sum\sum x_{ij}^2$$

$$= n\bar{y}\,\bar{x} - n\hat{\beta}\bar{x}^2 + \sum n_i\bar{x}_i\bar{y}_i - n\bar{y}\,\bar{x} - \hat{\beta}\sum(\bar{x}_i - \bar{x})n_i\bar{x}_i + \hat{\beta}\sum\sum x_{ij}^2$$

$$\text{or } \hat{\beta}\left[\sum\sum x_{ij}^2 - \sum n_i\bar{x}_i^2\right] = \sum\sum x_{ij}y_{ij} - \sum n_i\bar{x}_i\bar{y}_i$$

$$\text{or} \quad \hat{\beta} = \frac{\sum\sum\left(x_{ij} - \bar{x}_i\right)\left(y_{ij} - \bar{y}_i\right)}{\sum\sum\left(x_{ij} - \bar{x}_i\right)^2}.$$

From the analysis of variance, we know that $\hat{\alpha}_i = \bar{y}_i - \bar{y}$, whereas from the analysis of covariance, $\hat{\alpha}_i = \bar{y}_i - \bar{y} - \hat{\beta}(\bar{x}_i - \bar{x})$. The factor $\hat{\beta}(\bar{x}_i - \bar{x})$ is called the adjustment factor which is the amount of the linear effect of x on the ith treatment effect. The adjustment is zero for the treatment for which $\bar{x}_i$ is equal to $\bar{x}$.

We want to test
$H_{01} : \beta = 0,$
$H_{02} : \alpha_1 = \alpha_2 = \alpha_3 = \ldots\ldots = \alpha_t = 0.$

To test the above, we are to find out the residual sum of squares for the full model. The residual or the minimum error sum of squares under the full model $y_{ij} = \mu + \alpha_i + \beta x_{ij} + e_{ij}$ is given by

$$
\begin{aligned}
R &= \sum_i \sum_j \left(y_{ij} - \mu - \alpha_i - \beta x_{ij}\right)^2 \qquad \text{minimum for } \mu, \alpha_i, \beta \\
&= \sum_i \sum_j \left[y_{ij} - \bar{y} + \hat{\beta}\bar{x} - \bar{y}_i + \bar{y} + \hat{\beta}(\bar{x}_i - \bar{x}) - \hat{\beta}x_{ij}\right]^2 \\
&= \sum_i \sum_j \left[y_{ij} - \bar{y}_i - \hat{\beta}(x_{ij} - \bar{x}_i)\right]^2 = \sum_i \sum_j \left(y_{ij} - \bar{y}_i\right)^2 - \hat{\beta}^2 \sum \sum \left(x_{ij} - \bar{x}_i\right)^2 \\
&= E_{yy} - \frac{E_{xy}^2}{E_{xx}}, \text{ since } \hat{\beta} = \frac{E_{xy}}{E_{xx}} \text{ with } (n - t - 1) \text{ d.f.}
\end{aligned}
$$

The residual sum of squares under the restricted model when H_{01} holds good, that is, $y_{ij} = \mu + \alpha_i + e_{ij}$, is

$$
\begin{aligned}
R_2 &= \sum_i \sum_j \left(y_{ij} - \mu - \alpha_i\right)^2 \quad \text{minimum } \mu, \alpha_i \\
&= \sum_i \sum_j \left(y_{ij} - \bar{y} - \bar{y}_i + \bar{y}\right)^2 = \sum_i \sum_j \left(y_{ij} - \bar{y}_i\right)^2 \\
&= E_{yy} \text{with } (n - t) \text{ d.f.}
\end{aligned}
$$

Hence, the regression sum of squares will be $R_0 = R_2 - R = \frac{E_{xy}^2}{E_{xx}}$ with 1 d.f.

For testing $H_{01}\colon \beta = 0$, we have the test statistic $F = \dfrac{R_0/1}{R/(n-t-1)}$ with $(1, n - t - 1)$ d.f.

If the hypothesis is not rejected, we may not go for working out the adjusted treatment sum of squares because in that case, the assumed model will turn out to be $y_{ij} = \mu + \alpha_i + e_{ij}$, the model of one-way classified data. If the regression coefficient proves to be significant, we will proceed to test H_{02}.

The residual sum of squares under the restricted model when H_{02} holds good, that is, $y_{ij} = \mu + \beta x_{ij} + e_{ij}$, is

$$
\begin{aligned}
R_1 &= \sum_i \sum_j \left(y_{ij} - \mu - \beta x_{ij}\right)^2 \quad \text{minimum } \mu, \beta \\
&= \sum \sum \left(y_{ij} - \bar{y} + \hat{\hat{\beta}}\bar{x} - \hat{\hat{\beta}}x_{ij}\right)^2 \\
&= \sum \sum \left[y_{ij} - \bar{y} - \hat{\hat{\beta}}\left(x_{ij} - \bar{x}\right)\right]^2, \text{where } \hat{\hat{\beta}} = \frac{\sum \sum \left(x_{ij} - \bar{x}\right)\left(y_{ij} - \bar{y}\right)}{\sum \sum \left(x_{ij} - \bar{x}\right)^2} = \frac{S_{xy}}{S_{xx}} \\
&= \sum \sum \left[y_{ij} - \bar{y}\right]^2 - \hat{\hat{\beta}}^2 \sum \sum \left(x_{ij} - \bar{x}\right)^2 \\
&= S_{yy} - \frac{S_{xy}^2}{S_{xx}} \text{ with } (n - 2)\text{d.f.}
\end{aligned}
$$

Hence, the adjusted treatment sum of squares will be $R_{00} = R_1 - R = \left(S_{yy} - \frac{S_{xy}^2}{S_{xx}}\right) - \left(E_{yy} - \frac{E_{xy}^2}{E_{xy}}\right)$ with $t - 1$d.f.

For testing H_{02}, we have $F = \dfrac{R_{00}/(t-1)}{R/(n-t-1)}$ with $(t - 1, n - t - 1)$ d.f.

In case F is significant, we conclude that the treatment means after they are adjusted for x differ significantly.

The analysis of covariance table is as follows (Table 11.1):

Let $\bar{y}'_i$ be the mean of the ith treatment adjusted for the linear regression of y on x. Then,

Table 11.1 Analysis of covariance table

SOV	d.f.	SS(x)	SP(xy)	SS(y)	SS(Reg.)	SS(y)adj.	d.f.(adj.)
Treatment	$t-1$	T_{xx}	T_{xy}	T_{yy}			
Error	$n-t$	E_{xx}	E_{xy}	E_{yy}	$\frac{E_{xy}^2}{E_{xx}}=R_0$	$E_{yy}-\frac{E_{xy}^2}{E_{xx}}=R$	$n-t-1$
Treatment + error	$n-1$	$T_{xx}+E_{xx}=S_{xx}$	$T_{xy}+E_{xy}=S_{xy}$	$T_{yy}+E_{yy}=S_{yy}$	$\frac{S_{xy}^2}{S_{xx}}$	$S_{yy}-\frac{S_{xy}^2}{S_{xx}}=R_1$	$n-2$
Treatment (adjusted)						$R_1-R=R_{00}$	$t-1$

$$y'_i=\bar{y}_i-\hat{\beta}(\bar{x}_i-\bar{x}),$$

$$V(\bar{y}'_i)=V(\bar{y}_i)+(\bar{x}_i-\bar{x})^2V(\hat{\beta})=\sigma^2\left[\frac{1}{n_i}+\frac{(\bar{x}_i-\bar{x})^2}{E_{xx}}\right],$$

$$V\left(\bar{y}'_i-\bar{y}'_j\right)=\sigma^2\left[\frac{1}{n_i}+\frac{1}{n_j}+\frac{(\bar{x}_i-\bar{x}_j)^2}{E_{xx}}\right].$$

This requires a separate calculation of $(\bar{x}_i-\bar{x}_j)^2$ for each i and j. D. J. Finney (1946) has, however, suggested to use $\frac{2\upsilon^2}{n'}\left[1+\frac{T_{xx}/(t-1)}{E_{xx}}\right]$ as an average value of $V\left(\bar{y}'_i-\bar{y}'_j\right)$ averaged over all $t(t-1)$ pairs when $n_i=n_j=n'$. The estimated variances are given by the corresponding expressions with σ^2 substituted by

$$s_{y.x}^2=\frac{\left(E_{yy}-\frac{E_{xy}^2}{E_{xx}}\right)}{(n-t-1)}.$$

For testing the hypothesis $H_0:\alpha_i=\alpha_j$, we have the test statistic:

$$F=\frac{\left(\bar{y}'_i-\bar{y}'_j\right)^2}{\frac{2s_{y.x}^2}{n'}\left[1+\frac{T_{xx}/(t-1)}{E_{xx}}\right]}\quad\text{with }(1,n-t-1)\text{ d.f.}$$

It may be noted that the denominator remains unchanged for any i and j. Below, we present an example by Sahu and Das (2010).

Example 11.1. Four different feeds are given to four calves each; initial age in months (x) and average weekly gain (y) in body weight of the calves over a period of 8 weeks are recorded. Analyze the data of gain in body weight using the information on character x. Give the standard error of the difference between average effects of two feeds after adjusting the means for variation in x.

Feed							
F1		F2		F3		F4	
x	y	x	y	x	y	x	y
11	120	12	145	9	123	10	132
10	135	11	152	11	110	12	130
13	121	9	160	10	120	9	140
10	130	13	142	12	112	10	128

Solution. From the above analysis, it is clear that the response variable here is the average weekly gain in body weight (y) and the covariate is the initial body weight (x). The appropriate covariance model is

$$y_{ij}=\mu+\alpha_i+\beta x_{ij}+e_{ij};\quad i=1,2,3,4;\ j=1,2,3,4,$$

where

y_{ij} = gain in body weight for jth calves fed with ith feed,

μ = general effect,

α_i = effect due to ith feed,

β = regression coefficient of y on x,

e_{ij} = independent normal variate with mean zero(0) and variance, σ^2.

We want to test

$$H_{01}:\beta=0,$$
$$H_{02}:\alpha_1=\alpha_2=\alpha_3=\alpha_4=0.$$

From the given information, we first calculate the following quantities:

Feed							
F1		F2		F3		F4	
x	y	x	y	x	y	x	y
11	120	12	145	9	123	10	132
10	135	11	152	11	110	12	130
13	121	9	160	10	120	9	140
10	130	13	142	12	112	10	128
Total: 44	506	45	599	42	465	41	530

$$\sum x = 172 = G_x; \sum y = 2,100 = G_y; \sum x^2 = 1,876; \sum y^2 = 278,540; \sum xy = 22,522$$

$$\mathrm{CF}(x) = \frac{G_x^2}{4 \times 4} = 1,849;\ \mathrm{CF}(y) = \frac{G_y^2}{4 \times 4} = 275,625; \mathrm{CF}(x,y) = \frac{G_x G_y}{4 \times 4} = 22,575.$$

$$\mathrm{TSS}(x) = \sum x^2 - \mathrm{CF}(x) = 11^2 + 10^2 + \cdots\cdots + 9^2 + 10^2 - 1,849 = 27,$$

$$\mathrm{TrSS}(x) = \frac{44^2 + 45^2 + 42^2 + 41^2}{4} - \mathrm{CF}(x) = 2.50,$$

$$\mathrm{ErSS}(x) = \mathrm{TSS}(x) - \mathrm{TrSS}(X) = 27 - 2.5 = 24.5.$$

Similarly,

$$\mathrm{TSS}(y) = \sum y^2 - \mathrm{CF}(y) = 120^2 + 135^2 + \cdots\cdots + 140^2 + 128^2 - 275,625 = 2,915,$$

$$\mathrm{TrSS}(y) = \frac{506^2 + 599^2 + 465^2 + 530^2}{4} - 275,625 = 2365.50,$$

$$\mathrm{ErSS}(y) = \mathrm{TSS}(y) - \mathrm{TrSS}(y) = 2,915 - 2365.50 = 549.50,$$

$$\mathrm{TSP}(xy) = \sum xy - \mathrm{CF}(x,y) = 11 \times 120 + 10 \times 135 + \cdots\cdots + 9 \times 140 + 10 \times 128 - 22,575 = -53.00,$$

$$\mathrm{TrSP}(xy) = \frac{44 \times 506 + 45 \times 599 + 42 \times 465 + 41 \times 530}{4} - 22,575 = 44.75.$$

$$\mathrm{ErSP}(xy) = \mathrm{TSS}(xy) - \mathrm{TrSS}(xy) = -53.00 - 44.75 = -97.75.$$

Thus, the analysis of covariance table can be written as follows:

SOV	d.f.	SS(x)	SS(y)	SP(x,y)	$\hat{\beta}$	SS(Reg.)	SS(y)$_{adj.}$	d.f.$_{adj.}$	
Treatment	3	2.50	2365.50	44.75					
Error	12	24.54	549.50	−97.75	−3.9898	390.0026	159.4974	11	$s^2_{y.x} = 14.499$
Treatment + Error	15	27.00	2915.00	−53.00	−1.9630	104.0370	2810.963	14	
Treatment (adjusted)							2651.4656	3	

For testing H_{01}: $\beta = 0$, we have test statistic

$$F = \frac{\text{SS(Reg.)}/1}{\text{SS}(y)_{\text{adj.}}/11} = \frac{390.0026/1}{159.4974/11} = 26.40 \text{ with } (1,11) \text{ d.f.}$$

The table value of $F_{0.05;1,11} = 4.84$; hence, the test is significant, and the null hypothesis is rejected.

For testing H_{02}: $\alpha_1 = \alpha_2 = \alpha_3 = \alpha_4 = 0$,

we have the test statistic $F = \frac{\text{Tr.SS}(y)_{adj.}/3}{\text{SS}(y)_{\text{adj.}}/11} = \frac{2651.4656/3}{159.4974/11} = 60.95$ with (3,11) d.f.

The table value of $F_{0.05;3,11} = 3.59$; hence, the test is significant, and the null hypothesis is rejected. Thus, we can infer that the gain in body weight due to different feeds varies and also depends on the initial body weight of the calves.

As both null hypotheses are rejected, we are to work out the adjusted treatment means and also the estimated adjusted variances for different pairs of adjusted treatment means. We have $\hat{\beta} = -3.9898, \bar{x} = 10.75, E_{xx} = 24.5$, and $s^2_{y.x} = 14.4998$.

We make the following table:

$\bar{x}_i$	$\bar{y}_i$	$\bar{x}_i - \bar{x}$	$\bar{y}'_i = \bar{y}_i - \hat{\beta}(\bar{x}_i - \bar{x})$	$\hat{V}\left(\bar{y}'_i - \bar{y}'_j\right) = s^2_{y.x}\left[\frac{1}{n_i} + \frac{1}{n_j} + \frac{(x_i - x_j)^2}{E_{xx}}\right]$	
				$\left(\bar{y}'_i - \bar{y}'_j\right)$	$s^2_{y.x}\left[\frac{1}{n_i} + \frac{1}{n_j} + \frac{(x_i - x_j)^2}{E_{xx}}\right]$
11.00	126.50	0.25	127.4975	$\bar{y}'_1 - \bar{y}'_2$	7.2869
11.25	149.75	0.50	151.7449	$\bar{y}'_1 - \bar{y}'_3$	7.3979
10.50	116.25	−0.25	115.2526	$\bar{y}'_1 - \bar{y}'_4$	7.5828
10.25	132.50	−0.50	130.5051	$\bar{y}'_2 - \bar{y}'_3$	7.5828
				$\bar{y}'_2 - \bar{y}'_4$	7.8417
				$\bar{y}'_3 - \bar{y}'_4$	7.2869

On the other hand, if we go for the method given by Finney, then we shall have

$$\hat{V}\left(\bar{y}'_i - \bar{y}'_j\right) = \frac{2s^2_{y.x}}{4}\left[1 + \frac{T_{xx}/(t-1)}{E_{xx}}\right]$$
$$= 7.4965.$$

$$H_0 : \alpha_1 = \alpha_2$$
$$F = \frac{(127.49745 - 151.7449)^2}{7.4965}$$
$$= 78.43 \text{ with } (1,11) \text{ d.f.}$$

$$H_0 : \alpha_2 = \alpha_4$$
$$F = \frac{(151.7449 - 130.5051)^2}{7.4965}$$
$$= 60.18 \text{ with } (1,11) \text{ d.f.}$$

$$\text{CD}_{0.05} = \sqrt{7.4965} \times t_{0.025,11} = 2.7379 \times 2.201$$
$$= 6.0263.$$

If we arrange the adjusted treatment means, we have the following:

Treatment means: $\bar{y}'_i$	Value
$\bar{y}'_2$	151.74
$\bar{y}'_4$	130.51
$\bar{y}'_1$	127.5
$\bar{y}'_3$	115.25

Feed two is significantly producing higher gain in body weight compared to the other feeds, and feed three provides the lowest gain in body weight among all feeds. On the other hand, feed four and feed one are the same; they are statistically at par.

11.2 Partitioning of Variance and Covariance

Phenotypic characters which we measure on individuals are the manifestation of the interaction between the genetic architecture and the environment around them. The environmental factors mainly include all the nonheritable components. The variation in phenotypic values is therefore governed by the variances attributed to genotypic values and environmental deviations. Thus, the partitioning of phenotypic variations into different components is of utmost importance. Relative magnitude or proportions of these variations actually govern the genetic properties of the population. Again, covariability of different characters is measured in terms of covariance. Covariability measured between two characters is the phenotypic value. Similar to that of phenotypic variability, phenotypic covariability can also be partitioned into genotypic and environmental (nonheritable) components. The partitioning of variances and covariances will facilitate the estimation of various genetic/breeding parameters like heritability, coheritability, and genotypic correlations which can help breeders tremendously.

11.2.1 Components of Variance

During the analysis of variance, the mean sum of squares obtained due to the treatments consists of two components: (1) genotypic differences among the treatments and (2) environmental variations. Thus, $E(\text{MS}_\text{G}) = \sigma^2_\text{e} + r\sigma^2_\text{g}$, where $E(\text{MSE}) = \sigma^2_\text{e}$, and MS_G and MSE are the mean sum of square due to genotype and error, respectively, in ANOVA table. It can be seen that $\sigma^2_\text{g} = \frac{E(\text{MS}_\text{G}) - E(\text{MSE})}{r}$.

Thus, σ^2_g is estimated by $\hat{\sigma}^2_\text{g} = \frac{\text{MS}_\text{G} - \text{MSE}}{r}$, and σ^2_e is estimated by $\hat{\sigma}^2_\text{e} = \text{MSE}$.

Hence, $\sigma^2_\text{p} = \sigma^2_\text{g} + \sigma^2_\text{e}$ where $\sigma^2_\text{p}, \sigma^2_\text{g}$, and σ^2_e are the phenotypic, genotypic, and environmental variances, respectively. From the given data, one can estimate $\hat{\sigma}^2_\text{p} = \hat{\sigma}^2_\text{g} + \hat{\sigma}^2_\text{e}$.

To test H_0: $\sigma^2_\text{g} = 0$ means the genotypes do not differ significantly among themselves. The test statistic under H_0 is $F = \frac{\text{MSg}}{\text{MSE}}$ with $(t-1), (r-1)(t-1)$ d.f. If F is significant, we can say that the genotypes differ among themselves.

11.2.2 Components of Covariance

The covariances can also be partitioned due to environmental and genotypic factors in a similar way as that of the partitioning of treatment variances from the analysis of variance table. Thus, the phenotypic covariance between any

Table 11.2 Table of variance, covariance, and correlation

Level	Variance	Covariance	Correlation coefficients
Phenotypic	$\hat{\sigma}^2_{\mathrm{p}i}$ and $\hat{\sigma}^2_{\mathrm{p}i'}$	$\hat{\sigma}_{\mathrm{p}ii'}$	$r_{\mathrm{p}_{ii'}}$
Genotypic	$\hat{\sigma}^2_{\mathrm{g}i}$ and $\hat{\sigma}^2_{\mathrm{g}i'}$	$\hat{\sigma}_{\mathrm{g}_{ii'}}$	$r_{\mathrm{g}_{ii'}}$
Environmental	$\hat{\sigma}^2_{\mathrm{e}i}$ and $\hat{\sigma}^2_{\mathrm{e}i'}$	$\hat{\sigma}_{\mathrm{e}_{ii'}}$	$r_{\mathrm{e}_{ii'}}$

two characters can be written as a combination of genotypic and environmental covariances, that is, $\sigma_{\mathrm{p}_{12}} = \sigma_{\mathrm{g}_{12}} + \sigma_{\mathrm{e}_{12}}$. One can have the correlation coefficients at the genotypic, phenotypic, and environmental levels by using the variance and covariance of appropriate level in the formula for the correlation coefficient. The significance of the correlation coefficients can usually be tested using *t*-test described already in Chap. 9 with $(n - 2)$ degrees of freedom where n is the number of genotypes under consideration.

We can write for any two characters i and i' (Table 11.2):

From the above variances and covariances at the genotypic, phenotypic, and environmental levels, one can work out different genetic parameters like (a) *phenotypic coefficient of variation (PCV),* (b) *genotypic coefficient of variation (GCV),* (c) *broad sense heritability, and* (d) *co-heritability*, which are of paramount importance in the selection process of breeding experiments:

(a) *Phenotypic coefficient of variation (PCV%)*: $= \frac{\sigma_{\mathrm{p}}}{\bar{X}} \times 100$.

(b) *Genotypic coefficient of variation (GCV)*: $= \frac{\sigma_{\mathrm{g}}}{\bar{X}} \times 100$.

(c) *Heritability (broad sense)*: Another parameter, heritability (broad sense), for any character, is expressed as the ratio of the genotypic variance to the phenotypic variance $h^2 = \frac{\hat{\sigma}^2_{\mathrm{g}}}{\hat{\sigma}^2_{\mathrm{p}}}$. This gives an abstract idea about the amount of variability for a particular character due to its genotypic feature.

(d) *Co-heritability*: Co-heritability between two characters is defined as the ratio of the genotypic and the phenotypic covariances between the two characters under consideration and is given as $\frac{\sigma_{\mathrm{g}_{ii'}}}{\sigma_{\mathrm{p}_{ii'}}}$, where $\sigma_{\mathrm{g}_{12}}$ and $\sigma_{\mathrm{p}_{12}}$ are the genotypic and phenotypic covariances between characters 1 and 2 as obtained from the partitioning of the mean sum of products in the analysis of covariance.

(e) *Genetic advance/gain*: The difference between the mean of the selected individuals $(\bar{X}_{\mathrm{s}})$ and the mean of the population from which the selection has been made $(\bar{X}_{\mathrm{p}})$ is known as genetic gain or advance. But for all practical purposes, the genetic gain or advance is worked out as $\mathrm{GA} = k.h^2.\hat{\sigma}_{\mathrm{p}}$, where k is the standardized selection differential having constant values at different selection intensities (e.g., the values of standardized selection differentials at 1, 5, 10, and 20 % selection intensities are, respectively, 2.64, 2.06, 1.40, and 1.16) and h^2 and σ_{p} have their usual meanings.

(f) *Genotypic and environmental correlation*: The correlation coefficient between two characters at the genotypic and environmental levels is worked out using the variances and covariances due to the genotype and nonheritable factors.

The association between two characters is measured in terms of the correlation coefficient which is based on the phenotypic values of the variances and the covariances between the characters. Since the phenotypic values are determined by the genotype and the environmental interactions, the phenotypic correlation is also governed by these factors.

Example 11.2. The following table gives the information on four yield components and the yield of ten genotypes of mulberry. Analyze the data with respect to the variability and covariabilities to work out the correlation coefficients and the path of correlation coefficients of yield with other characters at the genotypic and phenotypic levels. Also find out the genetic parameters, phenotypic coefficient of variation (PCV), genotypic coefficient of variation (GCV), broad sense heritability, and co-heritability of the characters under study.

Genotype	Rep	Yield	Shoot length	100 leaf wt	Leaf area	Leaf wt/m^2
1	1	158.89	344.00	94.11	57.96	148.91
1	2	164.57	352.20	95.65	57.00	150.20
1	3	152.54	345.00	93.36	55.45	148.70
2	1	192.59	629.66	188.22	106.73	162.89
2	2	193.24	638.25	190.00	107.00	160.40
2	3	191.45	647.00	192.20	106.00	165.40
3	1	195.18	453.22	87.44	61.25	131.59
3	2	190.24	455.32	85.25	62.32	132.20
3	3	201.45	465.32	86.56	60.74	130.60
4	1	211.11	363.11	225.56	106.13	360.94
4	2	209.12	385.00	229.00	107.58	364.50
4	3	213.45	345.32	226.32	105.45	358.40
5	1	202.22	557.87	169.22	95.15	160.55
5	2	200.12	552.45	172.25	99.69	162.50
5	3	204.52	563.00	168.58	94.47	160.30
6	1	205.56	431.22	211.22	109.23	174.96
6	2	207.42	452.21	215.30	110.23	175.40
6	3	203.28	414.56	223.00	108.78	176.40
7	1	217.78	687.78	159.33	93.73	157.30
7	2	215.65	693.56	162.30	94.25	160.20
7	3	219.45	705.45	156.20	90.21	158.70
8	1	170.00	253.89	158.44	93.93	154.60
8	2	165.25	258.32	165.30	95.56	155.60
8	3	175.59	264.47	160.00	92.32	153.70
9	1	117.93	383.56	49.56	35.96	116.64
9	2	116.39	389.21	50.23	35.65	116.40
9	3	118.42	375.41	47.58	34.47	115.40
10	1	175.33	414.78	109.00	71.29	134.12
10	2	177.45	425.36	110.20	72.45	135.40
10	3	175.21	458.33	105.25	73.65	140.20

So as to get the path coefficients at the genotypic and phenotypic levels, one needs to have the correlation coefficients at the genotypic and phenotypic levels. The genotypic and phenotypic correlations could be obtained by partitioning the sum of squares and products due to the treatments into genotypic and phenotypic components.

We demonstrate the whole process in a stepwise manner as provided in Sahu and Das (2009)

Step 1: Analysis of Variance for the Characters Under Study

The analysis of variance for all the characters is taken up one by one. In the following steps, we demonstrate the analysis of variance procedure for a single character, that is, leaf yield per plant in details, and only the results of the analysis of variance for the rest of the characters are presented.

Analysis of Variance for Leaf Yield

The appropriate statistical model for RBD will be

$$y_{ij} = \mu + g_i + b_j + e_{ij}$$

where

y_{ij} = response corresponding to the jth replication of the ith genotype

μ = general effect

g_i = additional effect due to the ith genotype

b_j = additional effect due to the jth replication

e_{ij} = error associated with jth replication of the ith genotype

Let the level of significance be $\alpha = 0.05$.

Let us make the following table:

	Leaf yield/plant(Y)			
Genotype	R1	R2	R3	Total
G1	158.89	164.57	152.54	476.00
G2	192.59	193.24	191.45	577.28
G3	195.18	190.24	201.45	586.87
G4	211.11	209.12	213.45	633.68
G5	202.22	200.12	204.52	606.86
G6	205.56	207.42	203.28	616.26
G7	217.78	215.65	219.45	652.88
G8	170.00	165.25	175.59	510.84
G9	117.93	116.39	118.42	352.74
G10	175.33	177.45	175.21	527.99
Total	1846.59	1839.45	1855.36	5541.4

From the above table, the following quantities are calculated:

$$\text{Grand total} = G = \sum_{i=1}^{10}\sum_{j=1}^{3} y_{ij}$$
$$= 158.89 + 164.57 + \cdots\cdots\cdots + 175.21$$
$$= 5541.4$$

$$\text{Correction factor} = \text{CF} = \frac{G^2}{r \times t} = \frac{5541.4^2}{3 \times 10}$$
$$= 1{,}023{,}570$$

Total sum of squares = TSS

$$= \sum_{i=1}^{10}\sum_{j=1}^{3} {y_{ij}}^2 - \text{CF}$$
$$= 158.89^2 + 164.57^2 + \cdots\cdots\cdots + 175.21^2 - \text{CF} = 24747.19$$

Replication sum of squares

$$= \text{RSS} = \frac{1}{t}\sum_{j=1}^{3} y_{0j}^2 - \text{CF}$$
$$= \frac{1}{10}\left[1846.59^2 + 1839.45^2 + 1855.36^2\right] - \text{CF} = 12.75$$

Treatment sum of squares

$$= \text{TrSS} = \frac{1}{r}\sum_{i=1}^{10} y_{i0}^2 - \text{CF}$$
$$= \frac{1}{3}\left[476^2 + 577.28^2 + \cdots\cdots\cdots + 527.99^2\right] - \text{CF} = 24515.98$$

Error sum of squares = TSS − RSS − TrSS = 24747.19 − 12.75 − 24515.98 = 218.46

With the help of the above quantities, we now make the following ANOVA table for leaf yield per plant:

ANOVA table for yield/plant (y)				
SOV	d.f.	S.S	M.S	F-ratio
REP	2	12.75	6.38	0.53
GEN.	9	24515.98	2724.00	224.44
EROR	18	218.459	12.14	
Total	29	24747.19		

Similarly, following the same procedures, we can also perform the analysis of variance for other characters and make the following ANOVA tables:

ANOVA table for total shoot length (x_1)				
SOV	d.f.	S.S	M.S	F-ratio
REP	2	379.00	189.50	1.22
GEN.	9	499621.8	55513.54	356.98
EROR	18	2799.156	155.51	
Total	29	502800		

ANOVA table for fresh wt of 100 leaves (x_2)

SOV	d.f.	S.S	M.S	F-ratio
REP	2	28.8125	14.41	1.97
GEN.	9	94476.82	10497.42	1433.50
EROR	18	131.8125	7.32	
Total	29	94637.44		

ANOVA table for unit leaf area (x_3)

SOV	d.f.	S.S	M.S	F-ratio
REP	2	20.39062	10.20	7.91
GEN.	9	17180.62	1908.96	1480.51
EROR	18	23.20898	1.29	
Total	29	17224.22		

ANOVA table for leaf weight/m^2 (x_4)

SOV	d.f.	S.S	M.S	F-ratio
REP	2	5.25	2.63	0.79
GEN.	9	129114.3	14346.04	4292.08
EROR	18	60.16406	3.34	
Total	29	129179.7		

The results of the analysis of variance show that genotypes have different effects from one another for all characters y, x_1, x_2, x_3, and x_4.

From the above analysis of variance tables using the formula for genotypic, phenotypic, and environmental variances discussed in Sect. (), one can work out the estimated variances of the different characters due to genotypic, phenotypic, and environmental effects. Thus, the genotypic, phenotypic, and environmental variances for leaf yield per plant can be estimated as follows:

$$\hat{\sigma}_g^2 = \frac{(MS_G) - (MSE)}{r} = \frac{2724.00 - 12.14}{3}$$
$$= 903.95$$

and $\hat{\sigma}_p^2 = \hat{\sigma}_g^2 + \hat{\sigma}_e^2 = 903.95 + 12.14 = 916.09$

and $\hat{\sigma}_e^2 = MSE = 12.14.$

Similarly, these components for other characters can also be worked out. The following table gives the genotypic, phenotypic, and environmental variances of different characters under study.

Variance component	Yield/plant	Shoot length	Fresh weight of leaf	Leaf area	Leaf weight/m^2
Genotypic $\left(\hat{\sigma}_g^2\right)$	903.95	18452.68	3496.70	635.89	4780.90
Phenotypic $\left(\hat{\sigma}_p^2\right)$	916.09	18608.19	3504.02	637.18	4784.24
Environmental $(\hat{\sigma}_e^2)$	12.14	155.51	7.32	1.29	3.34

Step 2: Analysis of Covariance Among the Characters Under Study

We first take up the analysis of covariance of leaf yield per plant with the total shoot length.

To facilitate the analysis, we make the following table:

	R1		R2		R3		Total	
Genotype	Y	X_1	Y	X_1	Y	X_1	Y	X_1
1	158.89	344.00	164.57	352.20	152.54	345.00	476.00	1041.20
2	192.59	629.66	193.24	638.25	191.45	647.00	577.28	1914.91
3	195.18	453.22	190.24	455.32	201.45	465.32	586.87	1373.86
4	211.11	363.11	209.12	385.00	213.45	345.32	633.68	1093.43
5	202.22	557.87	200.12	552.45	204.52	563.00	606.86	1673.32
6	205.56	431.22	207.42	452.21	203.28	414.56	616.26	1297.99
7	217.78	687.78	215.65	693.56	219.45	705.45	652.88	2086.79
8	170.00	253.89	165.25	258.32	175.59	264.47	510.84	776.68
9	117.93	383.56	116.39	389.21	118.42	375.41	352.74	1148.18
10	175.33	414.78	177.45	425.36	175.21	458.33	527.99	1298.47
Total	1846.59	4519.09	1839.45	4601.88	1855.36	4583.86	5541.40	13704.83

From the above table, the following quantities are worked out:

Correction factor (YX_1)

$$= \mathrm{CF}_{YX_1} = \frac{G_{X_1} \times G_Y}{r \times t} = \frac{13704.83 \times 5541.40}{3 \times 10} = 2531464.83$$

Total sum of products (YX_1)

$$= \mathrm{TSP}_{YX_1} = \sum_{i=1}^{10}\sum_{j=1}^{3} y_{ij}x1_{ij} - \mathrm{CF}_{YX_1}$$
$$= (158.89 \times 344) + (164.57 \times 352.2) + \cdots\cdots\cdots + (175.21 \times 458.33) - \mathrm{CF}_{YX_1} = 57115.75,$$

Replication sum of products (YX_1)

$$= \mathrm{RSP}_{YX_1} = \frac{1}{t}\sum_{j=1}^{3} y_{0j}x1_{0j} - \mathrm{CF}_{YX_1}$$
$$= \frac{1}{10}[(1846.59 \times 4519.09) + (1839.45 \times 4601.88) + (1855.36 \times 4583.86)] - \mathrm{CF}_{YX_1} = -10.25,$$

Treatment sum of products (YX_1)

$$= \mathrm{TrSP}_{YX_1} = \frac{1}{r}\sum_{i=1}^{10} y_{i0}x1_{i0} - \mathrm{CF}_{YX_1}$$
$$= \frac{1}{3}[(476 \times 1041.20) + (577.28 \times 1914.91) + \cdots\cdots + (527.99 \times 1298.47)] - \mathrm{CF}_{YX_1} = 56986.42,$$

Error sum of products (YX_1)

$$= \mathrm{TSP}_{YX_1} - \mathrm{RSP}_{YX_1} - \mathrm{TrSP}_{YX_1} = 57115.75 + 10.25 - 56986.42 = 139.58$$

A. Analysis of Covariance of Leaf Yield with Other Characters

With the help of the above quantities, we make the following analysis of covariance table (1 × 2):

SOV	d.f.	S.P	M.S.P
REP	2	−10.25	−5.13
GEN.	9	56986.42	6331.82
EROR	18	139.58	7.75
Total	29	57115.75	

Similarly using the same method, one can frame the other analysis of covariance tables as given below:

1. Analysis of covariance table of yield/plant with fresh weight of leaf (1 × 3)

SOV	d.f.	S.P	M.S.P
REP	2	−12.19	−6.09
GEN.	9	36342.36	4038.04
EROR	18	−36.04	−2.00
Total	29	36294.13	

2. Analysis of covariance table of yield/plant with leaf area (1 × 4)

SOV	d.f.	S.P	M.S.P
REP	2	−16.06	−8.03
GEN.	9	16453.56	1828.17
EROR	18	−20.22	−1.12
Total	29	16417.28	

3. Analysis of covariance table of yield/plant with leaf weight per square meter (1 × 5)

SOV	d.f.	S.P	M.S.P
REP	2	−3.63	−1.81
GEN.	9	26560.38	2951.15
EROR	18	−37.88	−2.10
Total	29	26518.88	

B. Analysis of Covariance of Total Shoot Length with Other Characters

1. Analysis of covariance table of total shoot length with fresh weight of leaf (2 × 3)

SOV	d.f.	S.P	M.S.P
REP	2	89.25	44.63
GEN.	9	45279.50	5031.06
EROR	18	−260.88	−14.49
Total	29	45107.88	

2. Analysis of covariance table of total shoot length with leaf area (2 × 4)

SOV	d.f.	S.P	M.S.P
REP	2	19.50	9.75
GEN.	9	28888.83	3209.87
EROR	18	25.29	1.41
Total	29	28933.62	

3. Analysis of covariance table of total shoot length with leaf weight per square meter (2 × 5)

SOV	d.f.	S.P	M.S.P
REP	2	42.75	21.38
GEN.	9	−37827.92	−4203.10
EROR	18	224.92	12.50
Total	29	−37560.25	

C. Analysis of Covariance of Fresh Leaf Weight with Other Characters

1. Analysis of covariance table of fresh leaf weight with leaf area (3 × 4)

SOV	d.f.	S.P	M.S.P
REP	2	16.88	8.44
GEN.	9	39197.84	4355.32
EROR	18	10.53	0.59
Total	29	39225.25	

2. Analysis of covariance table of fresh leaf weight with leaf weight per square meter (3 × 5)

SOV	d.f.	S.P	M.S.P
REP	2	11.94	5.97
GEN.	9	74448.35	8272.04
EROR	18	13.40	0.74
Total	29	74473.69	

D. Analysis of Covariance of Leaf Area with Leaf Weight per Square Meter (4 × 5)

SOV	d.f.	S.P	M.S.P
REP	2	5.16	2.58
GEN.	9	25041.16	2782.35
EROR	18	19.19	1.07
Total	29	25065.50	

Using the same procedure of partitioning variances into genotypic, phenotypic, and environmental components, the covariances between the characters can also be partitioned. Thus,

$$\sigma_{g_{ij}} = \frac{E(\mathrm{MSP}_{G_{ij}}) - E(\mathrm{MSP}_{e_{ij}})}{r}$$

where $\sigma^2_{g_{ij}}$ = genotypic covariance between the ith and jth character,

$E(\mathrm{MSP}_{G_{ij}})$ = the expected treatment (genotype) mean sum of product value of ith and jth character,

$E(\mathrm{MSP}_{e_{ij}})$ = the expected error mean sum of product value of ith and jth character,

r = number of replications.

Thus, the genotypic, phenotypic, and environmental covariances between leaf yield per plant and total shoot length are

$$\hat{\sigma}_{g_{yx_1}} = \frac{\mathrm{MSP}_{G_{yx_1}} - \mathrm{MSP}_{e_{yx_1}}}{r} = \frac{6331.82 - 7.75}{3} = 2108.02, \text{ and}$$

$$\hat{\sigma}_{p_{yx_1}} = \hat{\sigma}_{G_{yx_1}} + \hat{\sigma}_{e_{yx_1}} = 2108.02 + 7.75 = 2115.77,$$

where $\hat{\sigma}_{e_{yx_1}} = 7.75$ is the environmental covariance between y and x_1.

Similarly, the covariances among the characters under study at the genotypic, phenotypic, and environmental levels can also be worked out from the respective analysis of covariance tables as given below:

Step 3: Calculation of Correlation Coefficients

	Covariance component	Yield/plant (Y)	Shoot length (X_1)	Fresh wt of leaf (X_2)	Leaf area (X_3)
Shoot length (X_1)	Genotypic $\left(\hat{\sigma}_{g_{ij}}\right)$	2108.02			
	Phenotypic $\left(\hat{\sigma}_{p_{ij}}\right)$	2115.78			
	Environmental $\left(\hat{\sigma}_{e_{ij}}\right)$	7.75			
Fresh weight of leaf (X_2)	Genotypic $\left(\hat{\sigma}_{g_{ij}}\right)$	1346.68	1681.85		
	Phenotypic $\left(\hat{\sigma}_{p_{ij}}\right)$	1344.68	1667.36		
	Environmental $\left(\hat{\sigma}_{e_{ij}}\right)$	−2.00	−14.49		
Leaf area (X_3)	Genotypic $\left(\hat{\sigma}_{g_{ij}}\right)$	609.77	1069.49	1451.58	
	Phenotypic $\left(\hat{\sigma}_{p_{ij}}\right)$	608.64	1070.89	1452.16	
	Environmental $\left(\hat{\sigma}_{e_{ij}}\right)$	−1.12	1.41	0.59	
Leaf weight/m^2 (X_4)	Genotypic $\left(\hat{\sigma}_{g_{ij}}\right)$	984.42	−1405.20	2757.10	927.09
	Phenotypic $\left(\hat{\sigma}_{p_{ij}}\right)$	982.31	−1392.70	2757.84	928.16
	Environmental $\left(\hat{\sigma}_{e_{ij}}\right)$	−2.10	12.50	0.74	1.07

After partitioning the variances and the covariances into genotypic, phenotypic, and environmental components, the next task is to calculate the correlation coefficients among the variables at the genotypic, phenotypic, and environmental levels, which will be used in path analysis. The usual formula for the correlation coefficients as described in Chap. 9, Sect. 9.1.1 of Sahu (2007) is also used for obtaining the correlation coefficients at different levels:

$$r_{yx_1} = \frac{\mathrm{Cov}(Y, X_1)_{\mathrm{gen}}}{\sqrt{V(Y)_{\mathrm{gen}} \times V(X_1)_{\mathrm{gen}}}},$$

where $\mathrm{Cov}(Y, X_1)_{\mathrm{gen}}$ = covariance between Y and X_1 at genotypic level,

$V(Y)_{\mathrm{gen}}$ = variance of the character Y at genotypic level,

$V(X_1)_{\mathrm{gen}}$ = variance of the character X_1 at genotypic level.

Similarly, the correlation coefficients at the phenotypic and environmental levels can be worked out by using the covariance and variances at the phenotypic and environmental levels, respectively. Thus, the correlation coefficient at the genotypic, phenotypic, and environmental levels for the characters leaf yield and total shoot length can be worked out as follows:

Correlation coefficients between leaf yield and total shoot length

$$\text{Genotypic level: } r_{yx_1} = \frac{\mathrm{Cov}(Y, X_1)_{\mathrm{gen}}}{\sqrt{V(Y)_{\mathrm{gen}} \times V(X_1)_{\mathrm{gen}}}} = \frac{2108.02}{\sqrt{903.95 \times 18452.68}} = 0.516,$$

Phenotypic level:

$$r_{yx_1} = \frac{\text{Cov}(Y, X_1)_{\text{phen}}}{\sqrt{V(Y)_{\text{phen}} \times V(X_1)_{\text{phen}}}} = \frac{2115.78}{\sqrt{916.09 \times 18608.19}} = 0.512,$$

Environmental level:

$$r_{yx_1} = \frac{\text{Cov}(Y, X_1)_{\text{env}}}{\sqrt{V(Y)_{\text{env}} \times V(X_1)_{\text{env}}}} = \frac{7.75}{\sqrt{12.14 \times 155.51}} = 0.178.$$

Similarly, other correlation coefficients can also be worked out as per the above formula and are presented below:

Genotypic correlation

	Y	X_1	X_2	X_3	X_4
Y	1.000	0.516	0.757	0.804	0.474
X_1	0.516	1.000	0.209	0.312	−0.15
X_2	0.757	0.209	1.000	0.973	0.674
X_3	0.804	0.312	0.973	1.000	0.532
X_4	0.474	−0.15	0.674	0.532	1.000

Phenotypic correlation

	Y	X_1	X_2	X_3	X_4
Y	1.000	0.512	0.751	0.797	0.469
X_1	0.512	1.000	0.206	0.311	−0.148
X_2	0.751	0.206	1.000	0.972	0.674
X_3	0.797	0.311	0.972	1.000	0.532
X_4	0.469	−0.148	0.674	0.532	1.000

Environmental correlation

	Y	X_1	X_2	X_3	X_4
Y	1.000	0.178	−0.212	−0.284	−0.330
X_1	0.178	1.000	−0.429	0.099	0.548
X_2	−0.212	−0.429	1.000	0.190	0.150
X_3	−0.284	0.099	0.190	1.000	0.513
X_4	−0.330	0.548	0.150	0.513	1.000

Step 4: Path Coefficients Analysis

Using the above correlation matrices, path coefficients at the genotypic, phenotypic, and environmental levels can be worked out as per the method described in Chap. 10. We present below the tables for the direct and indirect effects of path coefficient analyses at the genotypic, phenotypic, and environmental levels (details are provided in chapter path analysis):

Direct and indirect effects of the yield components on leaf yield of mulberry

Genotypic path				
Shoot length (X_1)	0.323	−0.314	0.587	−0.080
Fresh weight of leaf (X_2)	0.068	−1.501	1.831	0.360
Leaf area (X_3)	0.101	−1.462	1.881	0.284
Leaf weight/m^2 (X_4)	−0.048	−1.012	1	0.534
Residual = 0.2048				
Phenotypic path				
Shoot length (X_1)	0.330	−0.241	0.492	−0.068
Fresh weight of leaf (X_2)	0.068	−1.166	1.537	0.312
Leaf area (X_3)	0.103	−1.133	1.581	0.246
Leaf weight/ m^2 (X_4)	0.049	0.785	0.841	0.463
Residual = 0.2293				

The direct and indirect effects are the diagonal and off-diagonal elements, respectively.

Step 5: Calculation of Genetic Parameters

With the help of the variance–covariance values of the characters under considerations at the genotypic and phenotypic levels, we can work out the genetic parameters like PCV, GCV, broad sense heritability, and co-heritability. From the above calculation of analyses of variance and covariances, we have the following:

A. Variance components of different characters of mulberry

Variance component	Yield/plant	Shoot length	Fresh weight of leaf	Leaf area	Leaf weight/m^2
Genotypic $\left(\hat{\sigma}_g^2\right)$	903.95	18452.68	3496.70	635.89	4780.90
Phenotypic $\left(\hat{\sigma}_p^2\right)$	916.09	18608.19	3504.02	637.18	4784.24

B. Covariance components among the different characters of mulberry

	Covariance component	Yield/plant (Y)	Shoot length (X_1)	Fresh weight of leaf (X_2)	Leaf area (X_3)
Shoot length (X_1)	Genotypic $\left(\hat{\sigma}_{g_{ij}}\right)$	2108.02			
	Phenotypic $\left(\hat{\sigma}_{p_{ij}}\right)$	2115.78			
Fresh weight of leaf (X_2)	Genotypic $\left(\hat{\sigma}_{g_{ij}}\right)$	1346.68	1681.85		
	Phenotypic $\left(\hat{\sigma}_{p_{ij}}\right)$	1344.68	1667.36		
Leaf area (X_3)	Genotypic $\left(\hat{\sigma}_{g_{ij}}\right)$	609.77	1069.49	1451.58	
	Phenotypic $\left(\hat{\sigma}_{p_{ij}}\right)$	608.64	1070.89	1452.16	
Leaf weight/m^2 (X_4)	Genotypic $\left(\hat{\sigma}_{g_{ij}}\right)$	984.42	−1405.20	2757.10	927.09
	Phenotypic $\left(\hat{\sigma}_{p_{ij}}\right)$	982.31	−1392.70	2757.84	928.16

C. Average performance with respect to different characters in ten genotypes of mulberry

Genotype	Leaf yield/plant	Total shoot length	Fresh wt of 100 leaves	Unit leaf area	Leaf weight/m^2
1	158.67	347.07	94.37	56.80	149.27
2	192.43	638.30	190.14	106.58	162.90
3	195.62	457.95	86.42	61.44	131.46
4	211.23	364.48	226.96	106.39	361.28
5	202.29	557.77	170.02	96.44	161.12
6	205.42	432.66	216.51	109.41	175.59
7	217.63	695.60	159.28	92.73	158.73
8	170.28	258.89	161.25	93.94	154.63
9	117.58	382.73	49.12	35.36	116.15
10	176.00	432.82	108.15	72.46	136.57
Mean	184.71	456.83	146.22	83.15	170.77

(a) *Phenotypic Coefficient of Variation (PCV)*
Phenotypic coefficient of variation of any character is given by $PCV = \frac{\hat{\sigma}_p}{\bar{x}} \times 100$, where $\hat{\sigma}_p^2$ is the estimated phenotypic variance and $\bar{x}$ is the arithmetic mean of the character (X) under consideration. Thus, the PCV for leaf yield per plant is $PCV = \frac{\hat{\sigma}_p}{\bar{x}} \times 100 = \frac{\sqrt{916.09}}{184.71} \times 100 = 16.39$; similarly the phenotypic coefficient of variation for other characters also worked out and is presented in the table of genetic parameters.

(b) *Genotypic Coefficient of Variation (GCV)*
Genotypic coefficient of variation is given by $GCV = \frac{\hat{\sigma}_g}{\bar{x}} \times 100$, where $\hat{\sigma}_g^2$ is the genotypic variance and $\bar{x}$ is the arithmetic mean of the character (X) under consideration. Thus, the GCV for leaf yield per plant is $GCV = \frac{\hat{\sigma}_g}{\bar{x}} \times 100 = \frac{\sqrt{903.95}}{184.71} \times 100 = 16.28$; similarly the genotypic coefficient of variation for other characters also worked out.

(c) *Broad Sense Heritability*
Heritability in broad sense is given as $h^2 = \frac{\hat{\sigma}_g^2}{\hat{\sigma}_p^2}$, where the estimate of genotypic variance is $\hat{\sigma}_g^2$ and that of phenotypic variance is

$\hat{\sigma}_p^2$. Thus, the broad sense heritability for leaf yield is $h^2 = \frac{\hat{\sigma}_g^2}{\hat{\sigma}_p^2} = \frac{903.95}{916.09} = 0.987$; similarly the broad sense heritability values for other characters also worked out.

(d) *Co-heritability*

Co-heritability between two characters 1 and 2 is given as $\frac{\hat{\sigma}_{g12}}{\hat{\sigma}_{p12}}$, where $\hat{\sigma}_{g_{12}}$ and $\hat{\sigma}_{p_{12}}$ are the estimates of genotypic and phenotypic covariances between the characters 1 and 2, respectively. Thus, the co-heritability between the leaf yield and the total shoot length is $\frac{\hat{\sigma}_{g12}}{\hat{\sigma}_{p12}} = \frac{2108.02}{2115.78} = 0.996$; similarly the co-heritability values among the other characters also worked out.

(e) *Genetic Advance/Gain*

Genetic gain or advance is $GA = k.h^2.\hat{\sigma}_p$, where k is the standardized selection differential, h^2 is the broad sense heritability, and $\hat{\sigma}_p$ is the estimated phenotypic standard deviation of the character concerned. Thus, at 5% (say) selection intensity, the genetic advance for the character leaf yield per plant is $GA = k.h^2.\hat{\sigma}_p = 2.06 \times 0.987 \times 30.267 = 61.54$; similarly the genetic advance orgain values for the other characters also worked out.

Sometimes it is customary to present the genetic gain or advance values as percentage over mean. As such the genetic advance of leaf yield per plant at percentage over mean is $\frac{61.54}{184.71} \times 100 = 33.32$.

All the above genetic parameters, namely, phenotypic and genotypic coefficients of variation, broad sense heritability, and genetic advance, for the characters are presented in the following table.

Table of genetic parameters

Genetic parameter	Yield/plant	Shoot length	Fresh weight of leaf	Leaf area	Leaf weight/m^2
Phenotypic coefficient of variation	16.39	29.86	40.48	30.36	40.50
Genotypic coefficient of variation	16.28	29.74	40.44	30.33	40.49
Broad sense heritability	0.99	0.99	1.00	1.00	1.00
G. A./gain	61.54	278.66	121.69	51.89	142.39
G.A. as percent over mean	33.32	61.00	83.22	62.40	83.38

Co-heritability	Yield/plant	Shoot length	Fresh weight of leaf	Leaf area
Shoot length	0.999			
Fresh weight of leaf	0.999	0.991		
Leaf area	0.998	1.00	0.999	
Leaf weight/m^2	0.998	0.991	0.999	0.998

The above analysis can be made using the SPAR1 software, and the output is as follows:

ANALYSIS OF VAR COV
Source D.F. MEAN SUM OF SQ F-VALUE

CHARACTERS 1 1
REP 2. 0.63750000E+01 0.52527020E+00
TRET 9. 0.27239976E+04 0.22444470E+03
EROR 18. 0.12136610E+02 0.10000000E+01
CHAR 1 TREATMENT MEANS
158.67 192.43 195.62 211.23 202.29 205.42 217.63 170.28 117.58 176.00
S.E. OF DIFF BET TWO MEANS= 0.28444810E+01
GRAND MEAN= 184.7133 COEFF. OF VARIATION= 1.8860

(continued)

(continued)

```
CHARACTERS 1 2
REP    2.  -0.51250000E+01   -0.66090170E+00
TRET  9.  0.63318242E+04    0.81652940E+03
EROR 18. 0.77545573E+01    0.10000000E+01
CHARACTERS 1 3
REP    2.   -0.60937500E+01
TRET  9.   0.40380395E+04   -0.20166130E+04
EROR 18. -0.20023872E+01    0.10000000E+01
CHARACTERS 1 4
REP    2.   -0.80312500E+01    0.71499230E+01
TRET  9.   0.18281736E+04   -0.16275550E+04
EROR 18. -0.11232639E+01    0.10000000E+01
CHARACTERS 1 5
REP    2.   -0.18125000E+01    0.86138610E+00
TRET  9.   0.29511528E+04   -0.14025280E+04
EROR 18. -0.21041667E+01    0.10000000E+01

CHARACTERS 2 2
REP    2.   0.18950000E+03 0.12185810E+01
TRET  9.   0.55513538E+05 0.35698030E+03
EROR 18. 0.15550868E+03 0.10000000E+01
CHAR 2 TREATMENT MEANS
347.07 638.30 457.95 364.48 557.77 432.66 695.60 258.89 382.73 432.82
S.E. OF DIFF BET TWO MEANS= 0.10181970E+02
GRAND MEAN=   456.8277    COEFF. OF VARIATION=    2.7298

CHARACTERS 2 3
REP    2.    0.44625000E+02  -0.30790610E+01
TRET  9.    0.50310556E+04  -0.34713560E+03
EROR 18. -0.14493056E+02   0.10000000E+01
CHARACTERS 2 4
REP    2.   0.97500000E+01    0.69392230E+01
TRET  9.   0.32098704E+04    0.22845140E+04
EROR 18. 0.14050564E+01    0.10000000E+01
CHARACTERS 2 5
REP    2.   0.21375000E+02    0.17106240E+01
TRET  9.  -0.42031020E+04   -0.33637080E+03
EROR 18. 0.12495443E+02    0.10000000E+01

CHARACTERS 3 3
REP    2.   0.14406250E+02    0.19672830E+01
RET    9.   0.10497424E+05    0.14335030E+04
EROR 18. 0.73229167E+01    0.10000000E+01
CHAR 3 TREATMENT MEANS
94.37 190.14 86.42 226.96 170.02 216.51 159.28 161.25 49.12 108.15
S.E. OF DIFF BET TWO MEANS= 0.22095120E+01
GRAND MEAN=    146.2210    COEFF. OF VARIATION=    1.8507

CHARACTERS 3 4
REP    2.   0.84375000E+01    0.14421360E+02
TRET  9.   0.43553160E+04    0.74441010E+04
EROR 18. 0.58506944E+00    0.10000000E+01

CHARACTERS 3 5
REP    2.   0.59687500E+01    0.80186590E+01
TRET  9.   0.82720391E+04    0.11112990E+05
EROR 18. 0.74435764E+00    0.10000000E+01
```

(continued)

(continued)

```
CHARACTERS 4 4
REP    2.  0.10195312E+02  0.79070940E+01
TRET   9.  0.19089577E+04  0.14805140E+04
EROR  18.  0.12893880E+01  0.10000000E+01
CHAR 4 TREATMENT MEANS
56.80 106.58 61.44 106.39 96.44 109.41 92.73 93.94 35.36 72.46
S.E. OF DIFF BET TWO MEANS= 0.9271418
GRAND MEAN=   83.1543   COEFF. OF VARIATION=   1.3655

CHARACTERS 4 5
REP    2.  0.25781250E+01  0.24185670E+01
TRET   9.  0.27823507E+04  0.26101530E+04
EROR  18.  0.10659722E+01  0.10000000E+01

CHARACTERS 5 5
REP    2.  0.26250000E+01  0.78535250E+00
TRET   9.  0.14346037E+05  0.42920750E+04
EROR  18.  0.33424479E+01  0.10000000E+01
CHAR 5 TREATMENT MEANS
149.27 162.90 131.46 361.28 161.12 175.59 158.73 154.63 116.15 136.57
S.E. OF DIFF BET TWO MEANS= 0.14927490E+01
GRAND MEAN=   170.7700   COEFF. OF VARIATION=   1.0706

ENVIRONMENTAL CORR
1   1.000   0.178  -0.212  -0.284  -0.330
2   0.178   1.000  -0.429   0.099   0.548
3  -0.212  -0.429   1.000   0.190   0.150
4  -0.284   0.099   0.190   1.000   0.513
5  -0.330   0.548   0.150   0.513   1.000

GENOTYPIC CORRELATIONS
1  1.000   0.516  0.757  0.804   0.474
2  0.516   1.000  0.209  0.312  -0.150
3  0.757   0.209  1.000  0.973   0.674
4  0.804   0.312  0.973  1.000   0.532
5  0.474  -0.150  0.674  0.532   1.000

PHENOTYPIC CORRELATIONS
1  1.000   0.512  0.751  0.797   0.469
2  0.512   1.000  0.206  0.311  -0.148
3  0.751   0.206  1.000  0.972   0.674
4  0.797   0.311  0.972  1.000   0.532
5  0.469  -0.148  0.674  0.532   1.000

HERITABILITY VALUES
0.987 0.992 0.998 0.998 0.999

GENETIC ADVANCE VALUES (K=2.06)
61.52 278.66 121.69  51.89 142.39

COEFF. OF VARIATION (GENO)
16.28  29.74  40.44  30.33  40.49

COEFF. OF VARIATION (PHENO)
16.39  29.86  40.48  30.36  40.50
```

(continued)

(continued)

PATH ANALYSIS
VARIABLES FOR PATH ANALYSIS (FIRST DEPENDENT)
1 2 3 4 5
GENETIC ADVANCE= 0.54798203E+02

GENOTYPIC PATH
MATRIX OF DIRECT AND INDIRECT EFFECTS
DIRECT EFFECTS ON MAIN DIAGONAL

2	0.323	−0.314	0.587	−0.080
3	0.068	−1.501	1.831	0.360
4	0.101	−1.462	1.881	0.284
5	−0.048	−1.012	1.000	0.534

RESIDUAL= 0.2048

PHENOTYPIC PATH
MATRIX OF DIRECT AND INDIRECT EFFECTS
DIRECT EFFECTS ON MAIN DIAGONAL

2	0.330	−0.241	0.492	−0.068
3	0.068	−1.166	1.537	0.312
4	0.103	−1.133	1.581	0.246
5	−0.049	−0.785	0.841	0.463

RESIDUAL= 0.2293

11.3 Path Analysis

In agriculture and allied fields, including social sciences, the researchers come across with a number of variables at a time, in which different variables are not only related with response variables but also related among themselves. As such, the degree of association which we measured through the correlation coefficient may not provide the actual intensity of relationship because other variables might have influenced both variables under consideration. For example, the paddy yield is governed by different yield components like the number of hills per square meter, the number of tillers per hill, the number of panicle bearing tillers per hill, the length of panicle, and the number of grains per panicle. All these components are influencing the yield. The yield components are not only individually correlated with yield but also correlated among themselves. Similarly, while studying the physical fitness of human being, one considers the body characteristics like height and body weight of body, leg length, and BMR. In socioeconomic studies, the social status of a person is influenced by his/her age, education, family size, income, expenditure, involvement in social program, etc.

As the variables/characters are correlated not only with the response variable but also among themselves, the question arises on how to work out the direct influence of a particular variable/character on the response variable/character and the effects of the variable/character through other variables/characters because of their interrelationship. Path coefficient analysis is one of the possible answers. In a path coefficient analysis, there are at least two groups of variables: the independent or exogenous variables and the dependent or endogenous variables. Path coefficients indicate the direct and indirect effects of the variables on another variable. A path diagram is a flowchart to represent the direction of effects of different variables on the dependent variables (Fig. 11.1).

In the above diagram, X_1, X_2, and X_3 are the exogenous independent variables, and X_4 and X_5 are the endogenous dependent variables. Here the variable X_5 is dependent not only on X_4 but also on the independent variables X_1, X_2, and X_3. Similarly, the dependent variable X_4 is dependent on X_1, X_2, and X_3. The direction of arrows indicates the response variable by the causal variables. It may be noted that there may be both-way arrows among the exogenous independent variables, but there are one-way arrows from the exogenous independent variables to the endogenous variables. Thus, for X_5, the variable X_4 is independent but it is not exogenous. The both-way arrows mean the variables are related among themselves.

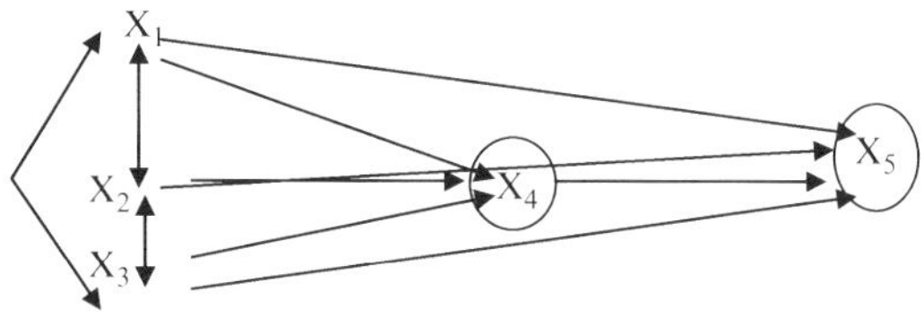

Fig. 11.1 Path diagram with two dependent and three independent variables

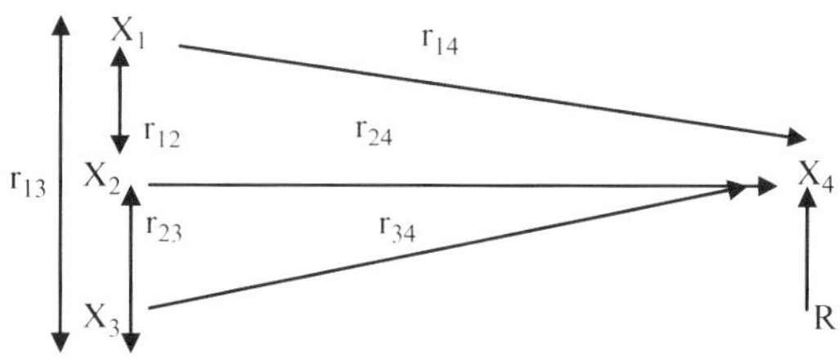

Fig. 11.2 Path diagram of one dependent and three correlated causal variables along with the residual part

The correlation between the response variable and any component variable (causal variable) can be looked upon as the additive effect of the direct and indirect effects of the causal variables. If a_i and $b_{i'}$ $(i \neq i' = 1, 2, \ldots\ldots, k)$ are the direct and indirect effects of the ith variable in a system of $(k + 1)$ variables, then $a_i + \sum_{i \neq i' = 1}^{k} b_{i'} = r_{x_i y}$, where a_i is the direct effect of the ith causal variable on the response variable y and $b_{i'}$s are the indirect effects of the ith causal variable on the response variable via the i'th causal variable (other than the ith variable).

In this fashion, each of the k number of individual correlation coefficients of causal variables with the response variable can be partitioned into the additive effect of the direct effect of the respective variables and the sum of the indirect effects of the other variables:

$$\begin{aligned}
r_{x_1 y} &= a_1 + b_2 + b_3 + b_4 + \cdots\cdots\cdots + b_k \\
r_{x_2 y} &= b_1 + a_2 + b_3 + b_4 + \cdots\cdots\cdots + b_k \\
r_{x_3 y} &= b_1 + b_2 + a_3 + b_4 + \cdots\cdots\cdots + b_k \\
&\vdots \\
&\vdots \\
r_{x_k y} &= b_1 + b_2 + b_3 + \cdots\cdots\cdots + a_k
\end{aligned}$$

Path analysis is the method of studying the direct and indirect effects of the variables on the response variables. But it is not directed towards discovering the causes of a response.

11.3.1 Calculation of Path Coefficient

Generally the path coefficients are denoted by p_{ij} (a, in our earlier notation), where i denotes for the dependent/response variable and j for the independent/causal variable and is defined as the ratio of the standard deviations due to a given cause to the total standard deviation of the response, that is, $p_{ij} = \frac{\sigma_j}{\sigma_i}$.

Let us have four variables situation in which X_1, X_2, and X_3 are the independent variables and X_4 is the dependent variable. So the variation X_4 is supposed to be explained by X_1, X_2, and X_3 and an unexplained portion R (Fig. 11.2).

Thus, we have $X_4 = X_1 + X_2 + X_3 + R$. So the correlation coefficients of X_4 with that of X_1, X_2, X_3, and R are given as

$$\begin{aligned}
r_{14} &= \frac{\text{Cov}(x_1,x_4)}{\sqrt{V(x_1)V(x_4)}} = \frac{\text{Cov}(x_1,x_1+x_2+x_3+R)}{\sqrt{V(x_1)V(x_4)}} \\
&= \frac{\text{Cov}(x_1,x_1)}{\sqrt{V(x_1)V(x_4)}} + \frac{\text{Cov}(x_1,x_2)}{\sqrt{V(x_1)V(x_4)}} + \frac{\text{Cov}(x_1,x_3)}{\sqrt{V(x_1)V(x_4)}} + \frac{\text{Cov}(x_1,R)}{\sqrt{V(x_1)V(x_4)}} \\
&= \frac{V(x_1)}{\sqrt{V(x_1)V(x_4)}} + \frac{r_{12}\sqrt{V(x_1)V(x_2)}}{\sqrt{V(x_1)V(x_4)}} + \frac{r_{13}\sqrt{V(x_1)V(x_3)}}{\sqrt{V(x_1)V(x_4)}} + 0 \\
&= \frac{\sigma_1}{\sigma_4} + \frac{r_{12}\sigma_2}{\sigma_4} + \frac{r_{13}\sigma_3}{\sigma_4} = p_{41} + r_{12}p_{42} + r_{13}p_{43}.
\end{aligned}$$

Note: $\text{Cov}(x_1, R)$ is assumed to be zero, and p_{41}, p_{42}, and p_{43} are the path coefficients of X_1, X_2, and X_3, respectively.

Following the same procedure, we have

$$\begin{aligned}
r_{24} &= \frac{\text{Cov}(x_2,x_4)}{\sqrt{V(x_2)V(x_4)}} = \frac{\text{Cov}(x_2,x_1+x_2+x_3+R)}{\sqrt{V(x_2)V(x_4)}} = p_{41}r_{12} + p_{42} + p_{43}r_{23}, \\
r_{34} &= \frac{\text{Cov}(x_3,x_4)}{\sqrt{V(x_3)V(x_4)}} = \frac{\text{Cov}(x_3,x_1+x_2+x_3+R)}{\sqrt{V(x_3)V(x_4)}} = p_{41}r_{13} + p_{42}r_{23} + p_{43}, \\
\text{and } r_{R4} &= \frac{\text{Cov}(R,x_4)}{\sqrt{V(R)V(x_4)}} = \frac{\text{Cov}(R,x_1+x_2+x_3+R)}{\sqrt{V(R)V(x_4)}} \\
&= \frac{\text{Cov}(R,x_1)}{\sqrt{V(x_1)V(R)}} + \frac{\text{Cov}(R,x_2)}{\sqrt{V(R)V(x_4)}} + \frac{\text{Cov}(R,x_3)}{\sqrt{V(R)V(x_4)}} + \frac{\text{Cov}(R,R)}{\sqrt{V(R)V(x_4)}} \\
&= \frac{0}{\sqrt{V(R)V(x_4)}} + \frac{0}{\sqrt{V(R)V(x_4)}} + \frac{0}{\sqrt{V(x_1)V(x_4)}} + \frac{V(R)}{\sqrt{V(R)V(x_4)}} \\
&= 0+0+0+C \\
&= C(\text{say}) \text{ (since covariance between the variables and the residual } R \text{ is assumed to be zero).}
\end{aligned}$$

Thus, the three correlation coefficients of the three independent variables X_1, X_2, and X_3 with X_4 are

$$r_{14} = p_{41} + r_{12}p_{42} + r_{13}p_{43},$$

$$r_{24} = p_{41}r_{12} + p_{42} + p_{43}r_{23},$$

$$r_{34} = p_{41}r_{13} + p_{42}r_{23} + p_{43}.$$

In matrix notation, the above three simultaneous equations can be written as

$$\underset{\sim}{A}\begin{pmatrix} r_{14} \\ r_{24} \\ r_{34} \end{pmatrix} = \underset{\sim}{B}\begin{pmatrix} r_{11} & r_{12} & r_{13} \\ r_{21} & r_{22} & r_{23} \\ r_{31} & r_{32} & r_{33} \end{pmatrix} \underset{\sim}{P}\text{ (say)}\begin{pmatrix} p_{41} \\ p_{42} \\ p_{43} \end{pmatrix}$$

Thus, $\underset{\sim}{A} = \underset{\sim}{B} \times \underset{\sim}{P}$

or $\underset{\sim}{B}^{-1} \times \underset{\sim}{A} = \underset{\sim}{B}^{-1} \times \underset{\sim}{B} \times \underset{\sim}{P}$

or $\underset{\sim}{P} = \underset{\sim}{B}^{-1} \times \underset{\sim}{A}$.

Given a set of observations for a number of independent variables and a dependent variable, the matrices A and B can be framed with the correlation coefficients among the variables.

The inverse matrix B^{-1} can be worked out following the method of matrix inversion. The inversion of matrix is available in the commonly used MS Excel subprogram of the MS Office software also. In Example 1, we shall demonstrate how, with the help of the MS Excel software, path analysis can be done.

11.3.2 Calculation of Residual

Once the path coefficient values corresponding to different independent variables are worked out, the next problem lies in estimating the residual (R), that is, the unexplained part of the model. We have assumed that $X_4 = X_1 + X_2 + X_3 + R$

or $\sigma_4^2 = \sigma^2(X_1 + X_2 + X_3 + R)$

$= \sigma_1^2 + \sigma_2^2 + \sigma_3^2 + \sigma_R^2 + 2\sigma_{12} + 2\sigma_{13} + 2\sigma_{23}$, where, σ_1^2, σ_2^2, σ_3^2, and σ_R^2 are the variance of the X_1, X_2, X_3 variables and residual and $\sigma_{ij}(i \neq j = 1,2,3)$ is the covariance between X_i and X_j (covariances between the variables and the residual are assumed to be zero)

$$= \sigma_1^2 + \sigma_2^2 + \sigma_3^2 + \sigma_R^2 + 2r_{12}\sigma_1\sigma_2 + 2r_{13}\sigma_1\sigma_3 + 2r_{23}\sigma_2\sigma_3;$$

$$\text{or } \sigma_4^2 = \sigma_1^2 + \sigma_2^2 + \sigma_3^2 + \sigma_R^2 + 2r_{12}\sigma_1\sigma_2 + 2r_{13}\sigma_1\sigma_3 + 2r_{23}\sigma_2\sigma_3,$$

$$\frac{\sigma_4^2}{\sigma_4^2} = \frac{\sigma_1^2 + \sigma_2^2 + \sigma_3^2 + \sigma_R^2 + 2r_{12}\sigma_1\sigma_2 + 2r_{13}\sigma_1\sigma_3 + 2r_{23}\sigma_2\sigma_3}{\sigma_4^2},$$

$$1 = \frac{\sigma_1^2}{\sigma_4^2} + \frac{\sigma_2^2}{\sigma_4^2} + \frac{\sigma_3^2}{\sigma_4^2} + \frac{\sigma_R^2}{\sigma_4^2} + \frac{2r_{12}\sigma_1\sigma_2}{\sigma_4^2} + \frac{2r_{13}\sigma_1\sigma_3}{\sigma_4^2} + \frac{2r_{23}\sigma_2\sigma_3}{\sigma_4^2}$$

$$= p_{41}^2 + p_{42}^2 + p_{43}^2 + p_{4R}^2 + 2r_{12}p_{41}p_{42} + 2r_{13}p_{41}p_{43} + 2r_{23}p_{24}p_{34}$$

$$\therefore p_{4R}^2 = 1 - \left[p_{41}^2 + p_{42}^2 + p_{43}^2 + 2r_{12}p_{41}p_{42} + 2r_{13}p_{41}p_{43} + 2r_{23}p_{42}p_{43}\right].$$

For the k number of independent variables, the generalized formula for getting residual will be

$$p_{yR}^2 = 1 - \sum_{j=1}^{k} p_{yj}^2 - 2\sum_{j<j'}^{k}\sum^{k} r_{jj'}p_{yj}p_{yj'},$$

where $p_{yj}, p_{yj'}$ are the path coefficients of jth and j'th variables, respectively, and $r_{jj'}$'s are the correlation coefficients among the independent variables.

11.3.3 Types of Path Coefficients

Using the variances and covariances at the phenotypic, genotypic, and environmental levels, the correlation coefficients among the variables at the respective levels can also be worked out. From the correlation coefficients, one can work out the path coefficients at the phenotypic, genotypic, and environmental levels also. The analysis of correlation and path coefficients at the phenotypic, genotypic, and environmental levels gives an indication about the genetic architecture and their interrelationships among the characters.

Interpretation of Path Coefficients

1. *Value of the residual*: If the value of the residual is substantial, then the question is whether or not the variables/characters under consideration are sufficient to explain the variations in the dependent variable/character. If one

gets a residual of 0.40, then it is quite clear that only 60% variation in the response variable can be explained by the variables/characters under consideration and 40% variation in the response variable/character will remain unexplained. Are we satisfied? It is very difficult to answer such. Actually the decision norm is governed by several factors like the type of experiment and material being handled and the situation of the experimentation. In agronomic field trials, one may be satisfied with 0.30 residual value; in laboratory experiment, one may be satisfied with a residual less than 0.10; on the other hand, in medical research, one may be satisfied with a residual less than 0.001 or so.

2. When the path coefficient value, that is, the direct effect of the causal character/variable, is almost equal to that of the correlation coefficients between them, the correlation coefficient presents the true picture of linear association between the response variable/character and the independent causal character.
3. When the direct effect is negligible or negative but the correlation coefficient between the causal variable/character is positive and significant, then the indirect effects are the cause of such correlation coefficients. In this situation, it is better to consider the other causal variables/characters rather than this variable/character.
4. When the direct effect is high and positive but the correlation coefficient is negative or negligible, then the indirect effects are the cause of such correlation. It is better to restrict the undesirable characters.
5. When both the direct effect and the correlation coefficient are negligible or negative, then one should discard such characters or variables.

Example 11.3. The following data give the yield (Y) corresponding to four yield components (X_1, X_2, X_3, X_4) for a certain agricultural crop. Work out the path of correlations of yield with the yield components.

Y	X_1	X_2	X_3	X_4	Y	X_1	X_2	X_3	X_4
99.56	6.60	4.07	8.20	8.20	121.87	6.40	6.62	7.40	8.60
89.23	7.20	3.82	4.80	5.20	102.06	6.20	6.52	6.60	7.20
98.76	8.00	3.63	7.20	7.40	119.68	6.60	6.01	10.20	6.60
136.33	4.20	5.83	8.80	7.40	133.12	5.80	6.22	9.40	8.00
153.12	5.20	5.42	8.20	7.20	129.70	6.20	6.14	9.60	6.80
135.05	4.40	5.62	10.20	6.20	64.84	3.20	5.73	8.20	5.20
93.86	6.20	6.44	6.20	7.20	65.37	3.20	5.04	7.80	4.40
98.08	5.80	5.40	8.60	5.40	61.28	2.80	5.40	9.20	5.20
102.06	5.80	5.50	6.60	5.20	81.32	2.80	7.08	7.20	5.40
82.31	4.20	6.90	8.20	4.00	73.59	2.60	6.32	10.20	5.00
65.37	4.60	6.69	7.40	6.60	71.89	3.60	6.27	7.40	7.40
69.37	4.40	7.12	9.20	5.00	58.33	2.80	7.08	9.20	8.20
69.17	3.60	5.78	9.40	6.00	60.25	3.00	6.43	8.40	6.20
59.24	3.20	5.35	8.60	6.20	49.72	3.20	6.20	6.80	5.20
63.59	3.20	5.45	8.80	4.40	80.94	3.40	6.51	8.40	4.20
61.11	4.20	5.97	8.20	7.20	78.64	2.60	8.32	9.20	4.00
59.69	3.80	7.32	7.40	6.20	83.21	3.00	7.29	6.60	3.60
53.20	4.00	7.19	6.80	5.80	57.38	2.80	5.88	9.60	4.40
83.53	5.20	5.49	10.20	4.40	60.12	2.80	5.38	8.20	4.20
79.11	5.40	4.72	8.40	5.20	61.09	3.40	5.49	9.20	3.20
70.13	4.60	4.82	7.60	4.20	49.86	3.80	7.04	8.80	5.20
129.21	8.00	3.12	9.20	5.20	57.15	3.20	6.88	7.40	3.40

(continued)

(continued)

Y	X_1	X_2	X_3	X_4	Y	X_1	X_2	X_3	X_4
117.29	8.20	4.16	8.80	6.20	52.98	3.40	6.26	7.20	5.00
121.30	7.80	3.63	10.20	5.40	97.58	4.60	7.42	8.20	4.00
121.38	5.60	6.76	8.20	10.20	93.69	3.80	6.43	7.40	5.00

Solution. From the given data, we calculate the following correlation coefficient and prepare the correlation table.

	Y	X_1	X_2	X_3	X_4
Y	1.0000	0.6861	−0.2977	0.1929	0.4559
X_1	0.6861	1.0000	−0.5945	−0.0580	0.4085
X_2	−0.2977	−0.5945	1.0000	−0.0826	−0.0447
X_3	0.1929	−0.0580	−0.0826	1.0000	−0.0360
X_4	0.4559	0.4085	−0.0447	−0.0360	1.0000

From the above correlation table, we have the following correlation matrices:

$$\underset{\sim}{B}\begin{pmatrix} 1.0000 & -0.5945 & -0.0580 & 0.4085 \\ -0.5945 & 1.0000 & -0.0826 & -0.0447 \\ -0.0580 & -0.0826 & 1.0000 & -0.0360 \\ 0.4085 & -0.0447 & -0.0360 & 1.0000 \end{pmatrix} \quad \underset{\sim}{A}\begin{pmatrix} 0.6861 \\ -0.2977 \\ 0.1929 \\ 0.4559 \end{pmatrix}$$

$$\begin{pmatrix} 2.0328612 & 1.1896280 & 0.1884937 & -0.7704334 \\ 1.1896280 & 1.7053523 & 0.1954023 & -0.4026777 \\ 0.1884937 & 0.1954023 & 1.0259480 & -0.0313676 \\ -0.7704334 & -0.4026777 & -0.0313676 & 1.2955812 \end{pmatrix}$$

The inverse matrix of B, that is, $\underset{\sim}{B}^{-1}$.

So $\underset{\sim}{P} = \underset{\sim}{A}.\underset{\sim}{B}^{-1}$ is given by $\begin{pmatrix} 0.725755792 \\ 0.162659179 \\ 0.254770092 \\ 0.175887089 \end{pmatrix}$.

Now the direct and indirect effects of different yield components are given as follows:

Table of path coefficients

r_{yx}	X_1	X_2	X_3	X_4
r_{y1} =0.6861	0.72575579	−0.09669818	−0.014779644	0.071848
r_{y2} =−0.2977	−0.431449783	0.16265918	−0.021045259	−0.007860
r_{y3} =0.1929	−0.042102321	−0.01343644	0.254770090	−0.006320
r_{y4} =0.4559	0.296462008	−0.00727221	−0.009161656	0.175890

Note: Diagonal elements are the direct effects and the off-diagonal elements are the indirect effects.

The indirect path coefficients are obtained by multiplying each element of the row concerned in matrix B (correlation matrix) by the respective elements of the P matrix. Thus, the indirect effect of X_1 through X_2 on Y is $-0.5945 \times 0.1626 = -0.096698$.

The diagonal elements of the table are the direct effects, whereas the off-diagonal elements are the indirect effects. It may be noted that the sum of the direct and indirect effects of any character is equal to its correlation coefficients with the response variables.

The residual is worked with the help of the formula

$$R^2 = 1 - \Big[p_{y1}^2 + p_{y2}^2 + p_{y3}^2 + p_{y4}^2 + 2r_{12}p_{y1}p_{y2} + 2r_{13}p_{y1}p_{y3} + 2r_{14}p_{y1}p_{y4} + 2r_{23}p_{y2}p_{y3} + 2r_{24}p_{y2}p_{y4} + 2r_{34}p_{y3}p_{y4}\Big] = 0.42113.$$

Thus, from the above analysis, it is clear that the yield-contributing characters considered in the example are efficient enough to explain as much as 58% variation in the response variable. Character one (X_1) has high positive direct effect commensurating with high positive correlation coefficient; hence, this character may be utilized for direct selection. On the other hand, the negative correlation and positive direct effect of the character two X_2 are mainly attributed by the indirect effect of X_2 via X_1. The other characters though have low positive correlation coefficients but are associated with positive direct effects; hence, these cannot be ignored during selection.

With the help of the MS Excel program, how one can work out the path coefficients is presented in the following example.

Example 11.4. Data given in the example are pertaining to the yield and three yield components per plant in six different varieties. We are to find out the path coefficients for yield per plant on other characters.

Var no	Replication	Yield Ch1	Branch no Ch2	Test wt Ch3	Pod/plant Ch4
1	1	14.5	1.8	34	27.2
	2	13.8	2.6	32.8	29.2
	3	14.8	2.8	35.1	18.6
2	1	14.9	2.9	38.3	24.8
	2	15.8	2.3	38.5	23.6
	3	17.3	3.2	40.2	17.4
3	1	15.1	2.8	48.5	25.2
	2	15.2	3.1	45.3	28.8
	3	16.3	3.3	45.2	25.6
4	1	15.7	2	36.2	29.2
	2	16.5	2.5	38.1	18.4
	3	17.5	3.1	44.2	24.8
5	1	17.6	3.3	46.1	24.8
	2	17.1	3.5	55.3	22.2
	3	18.3	3	58.6	20.2
6	1	17.2	3.7	60.8	20.8
	2	17.3	3.4	62	20.2
	3	18.5	4.1	59.4	28

Solution. We are provided with replicated data of four characters, including yield per plant for six varieties. One can use the variance–covariance analysis to get the correlation at the phenotypic, genotypic, and environmental levels. But for this particular example, we shall consider the path coefficient from simple correlations considering that there is $6 \times 3 = 18$ number of observations per character. In a stepwise manner, we shall demonstrate how path analysis can be done using the MS Excel program in the MS Office software.

Step 1: Calculation of Correlation Coefficients

1. Go to Tools menu and click Data Analysis submenu.
2. From Data Analysis submenu, select Correlation.

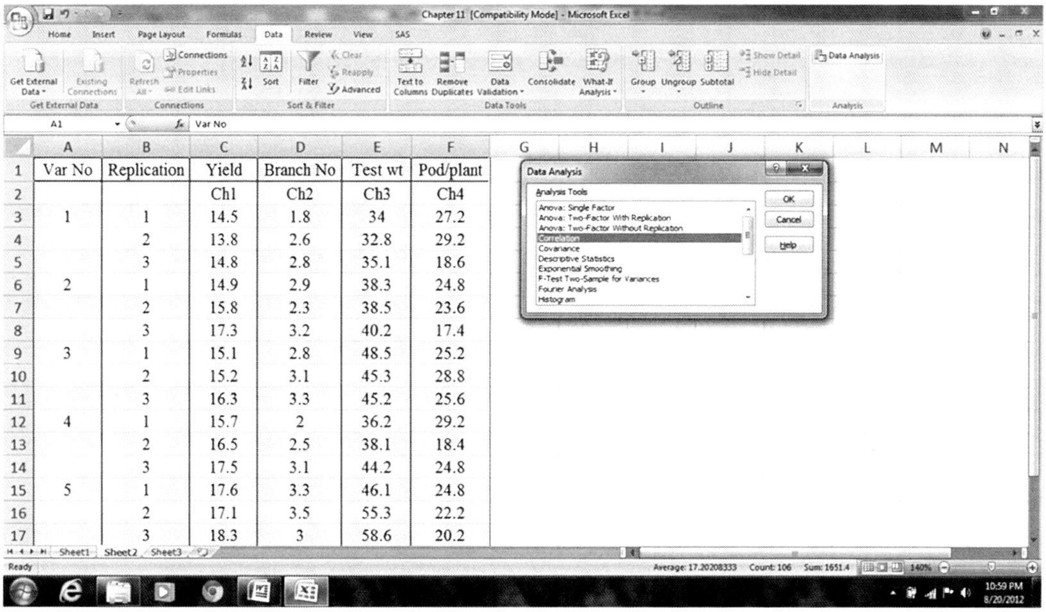

Slide 11.1: Step 1, showing the data and selection of Correlation from Data Analysis submenu using MS Excel

3. Select the input–output ranges, label, etc., as shown in the figure below:

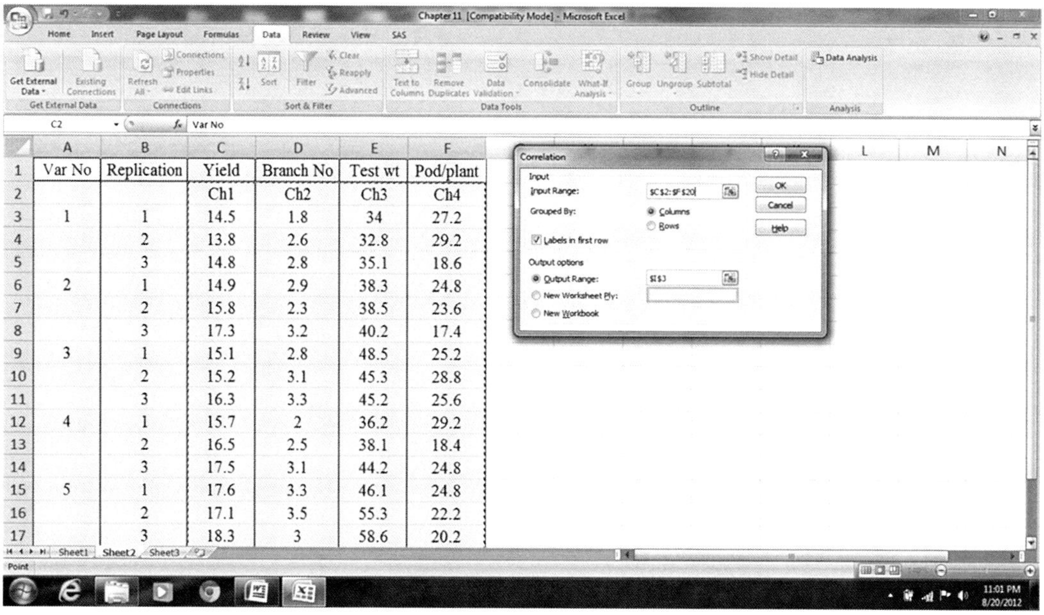

Slide 11.2: Step 2, showing the data and selection of input range and other options from Correlation of Data Analysis submenu using MS Excel

4. Click OK, then correlation table will appear as shown below:

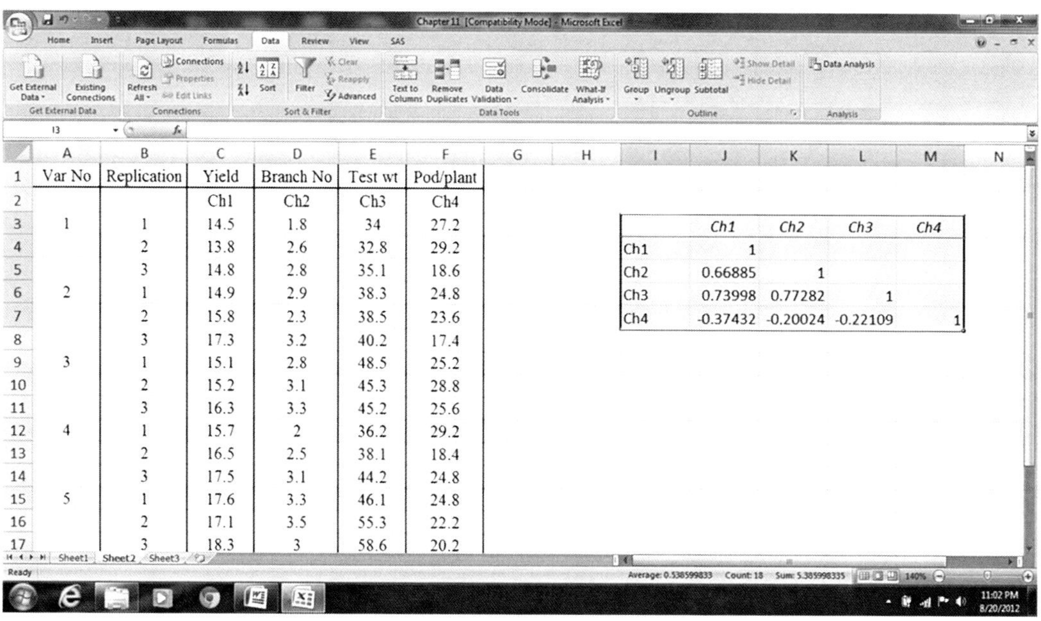

Slide 11.3: Step 3, showing output from Correlation of Data Analysis submenu using MS Excel

Arrange the above correlation table as given below:

Y		Ch1	Ch2	Ch3
0.669		1.000	0.773	−0.200
0.740		0.773	1.000	−0.221
−0.374		−0.200	−0.221	1.000
$\underset{\sim}{A}$			$\underset{\sim}{B}$	

Step 2: Invert the Correlation Coefficients Matrix $\underset{\sim}{B}$

1. **Go to** insert **menu and** click function **button.**

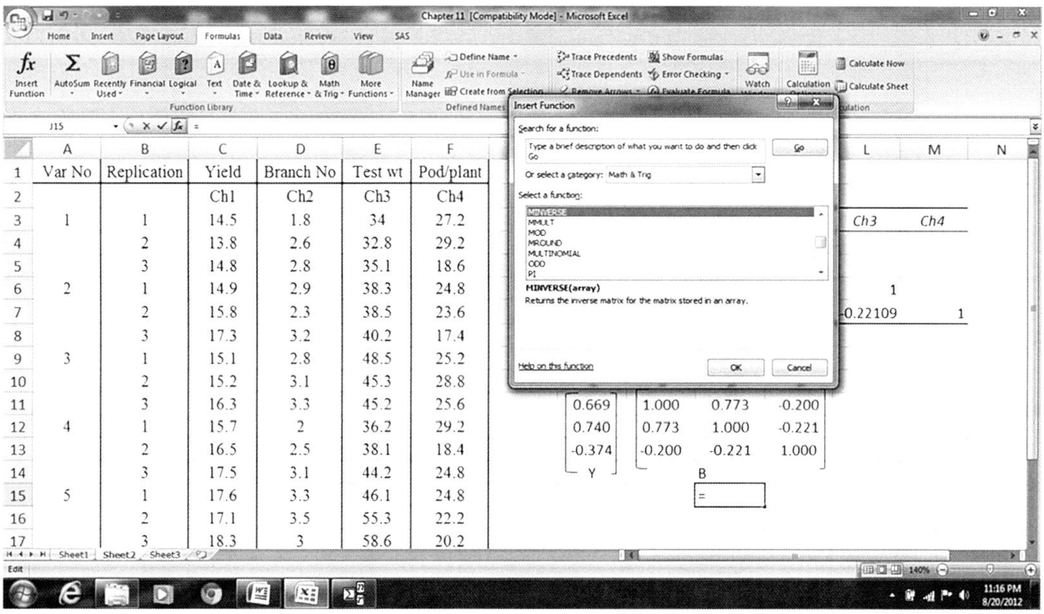

Slide 11.4: Step 4, showing the selection of matrix inverse option from function submenu of MS Excel

2. Select MINVERSE option from the list.
3. Select the range of the matrix to be inverted, that is, $\underset{\sim}{B}$ here.

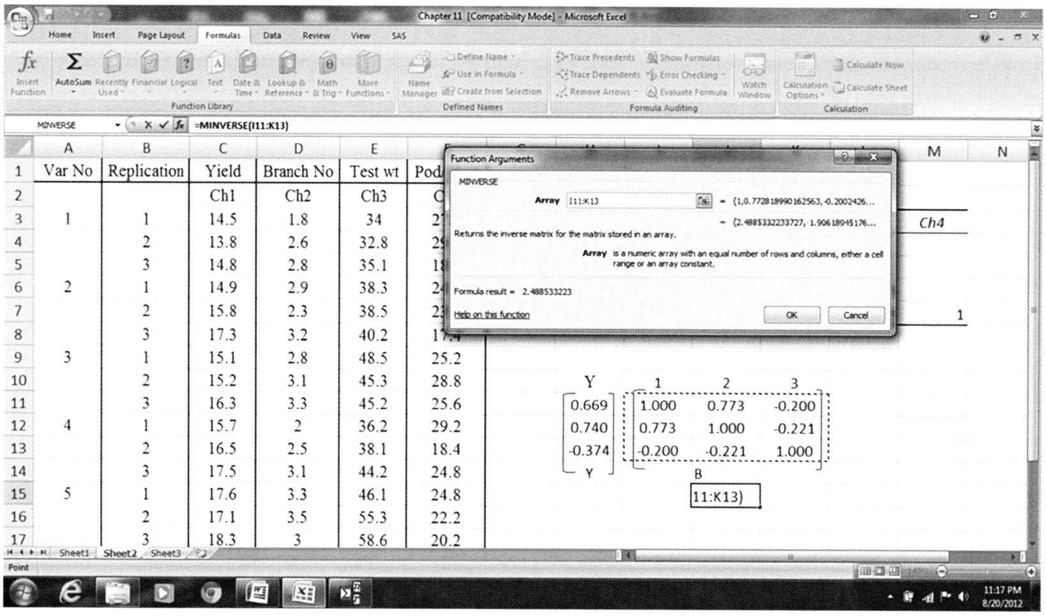

Slide 11.5: Step 4, showing the selection of matrix inverse range option from function submenu of MS Excel

4. Clicking OK **will give us the option given below:**

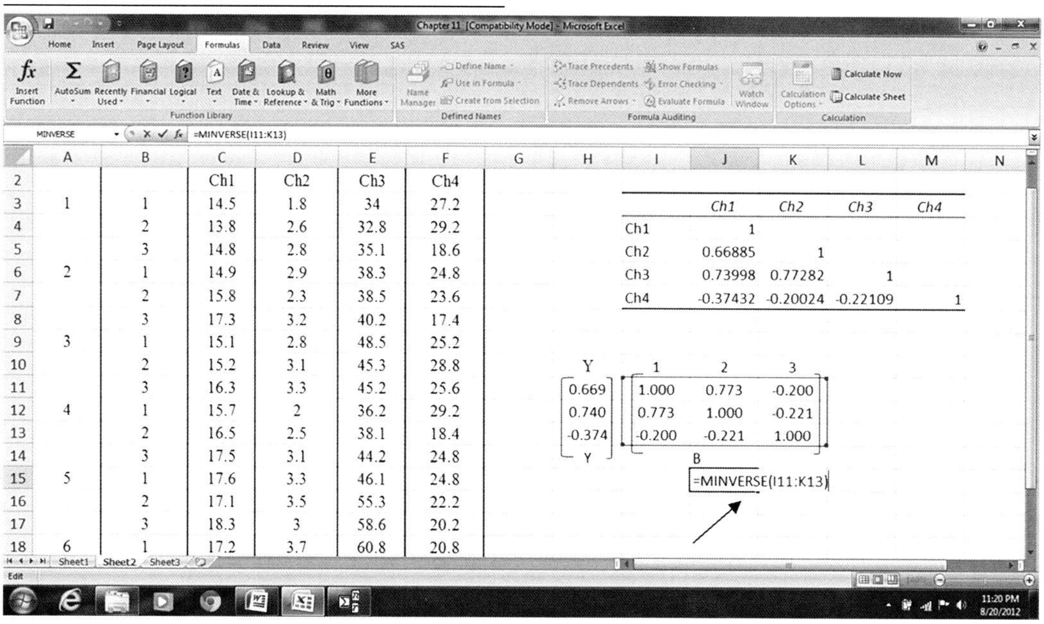

Slide 11.6: Step 4, showing the matrix inverse output option from function submenu of MS Excel

Select the range of cells where the inverse matrix is to be written as shown above.

5. Pressing F2 followed by Ctrl+Shift+Enter will give us the inverted matrix as given below:

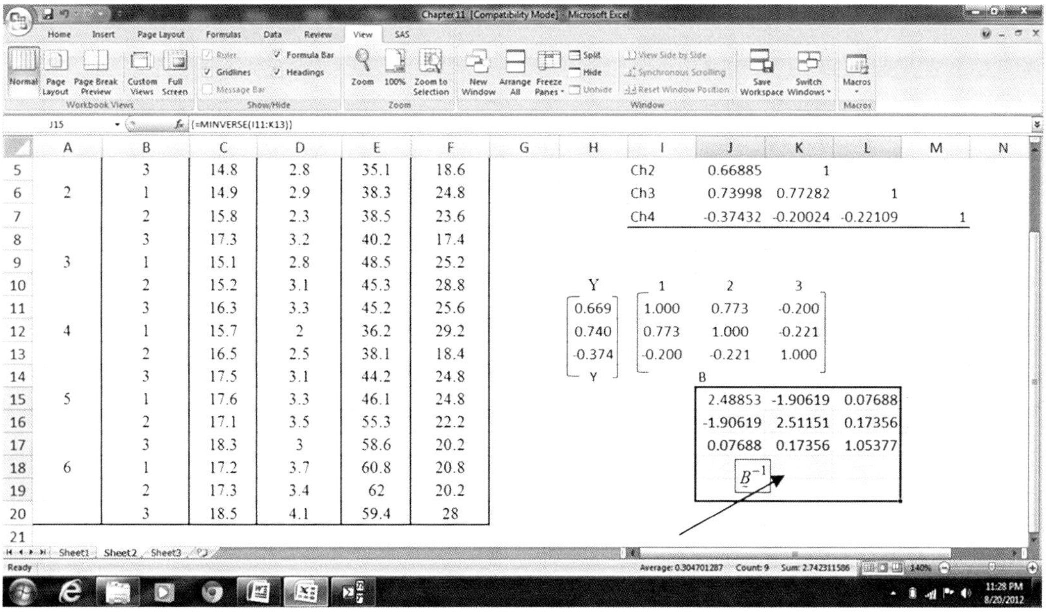

Slide 11.7. Step 4, showing the matrix inverse output total from function submenu of MS Excel

Step 3: Multiplication of $\underset{\sim}{A}$ and $\underset{\sim}{B}^{-1}$ Matrix and Formation Path Coefficient Matrix

1. Go to **Insert** Function submenu and select **MMULT** option as given below:

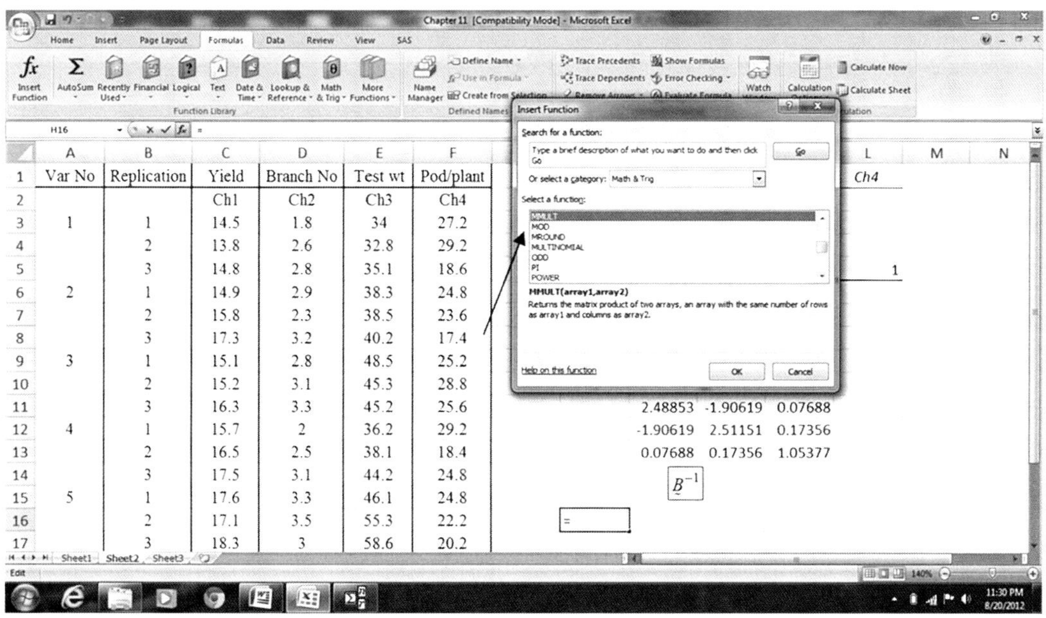

Slide 11.8: Step 5, showing the matrix multiplication option from Insert Function submenu of MS Excel

2. Select the ranges of the matrices in order in the specified boxes as shown below:

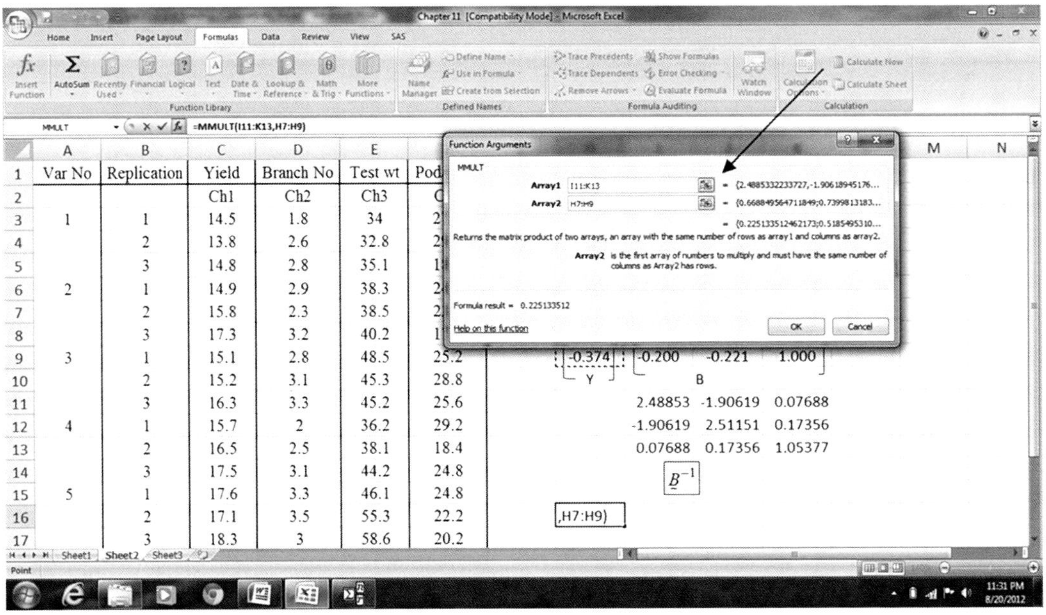

Slide 11.9: Step 5, showing the matrix multiplication range option from Insert Function submenu of MS Excel

3. Select the range where the product matrix is to be written and click OK.
4. Then press F2 followed by Ctrl+Shift+Enter. This will give the product of the matrices, that is, matrix $\underset{\sim}{P}$ of the direct effects as shown below in the shaded cells:

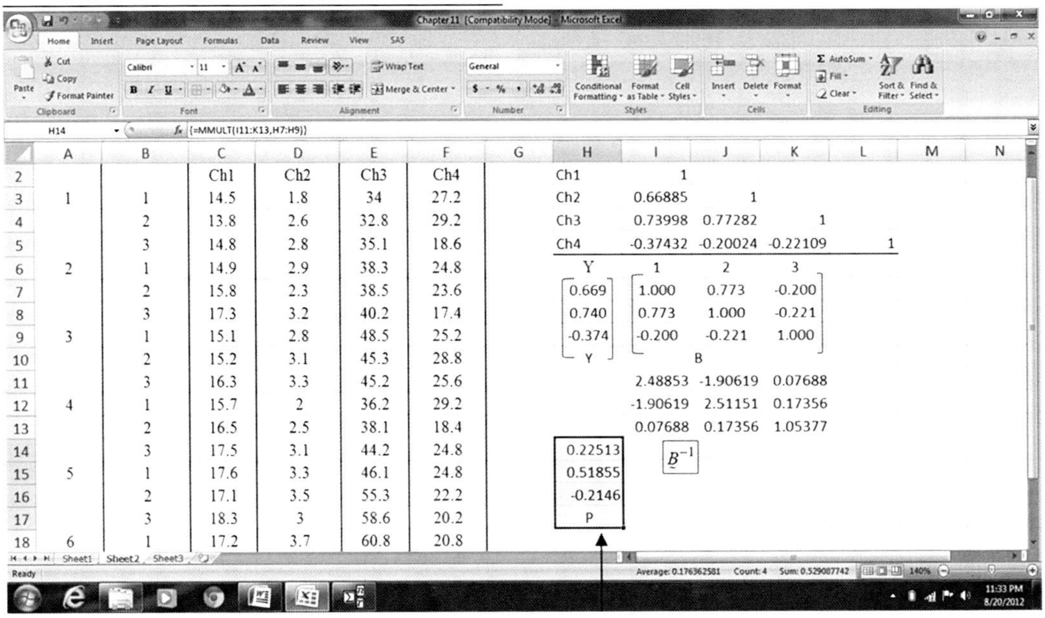

Slide11.10: Step 5, showing the matrix multiplication output from Insert Function submenu of MS Excel

Step 4: Calculation of Indirect Effects

The indirect effects of different factors are calculated with the help of the following formula:

$$
\begin{aligned}
r_{1y} = p_{y1} + p_{y2}r_{12} + p_{y3}r_{13} &= 0.225134 + (0.51855 \times 0.773) + ((-0.2146) \times (-0.2)) \\
&= 0.225134 + 0.4008 + 0.04292, \\
r_{2y} = p_{y1}r_{12} + p_{y2} + p_{y3}r_{23} &= 0.225134 \times 0.773 + 0.51855 + (-0.2146) \times (-0.221) \\
&= 0.174 + 0.51855 + 0.0474, \\
r_{3y} = p_{y1}r_{13} + p_{y2}r_{23} + p_{y3} &= 0.225134 \times (-0.2) + 0.51855 \times (-0.221) - 0.2146 \\
&= -0.045 - 0.11459 - 0.2146. \\
\text{Thus, } r_{1y} &= 0.225134 + 0.4008 + 0.04292,
\end{aligned}
$$

where 0.4008 and 0.04292 are the indirect effects of X_1 via X_2, and X_3, respectively. Similarly, the indirect effects of other variables can also be worked out.

Step 5: Calculation of Residual

Residual value is calculated using the above path and correlation coefficients that we have already calculated using the following relationship:

$$
R^2 = 1 - \left[p_{y1}^2 + p_{y2}^2 + p_{y3}^2 + 2r_{12}p_{y1}p_{y2} + 2r_{13}p_{y1}p_{y3} + 2r_{23}p_{y2}p_{y3}\right] = 0.385.
$$

The procedure for the calculation of path analysis at the phenotypic and genotypic levels using SPAR1 has been presented in Example 2.

11.4 Stability Analysis

One of the major tasks before releasing a genotype, or a variety for cultivation, or a technology at a farmers field is to judge its performance under varied situations. Multi-situational, multilocational, or multi-seasonal or all trials are conducted to understand the adaptability and the performance of the genotypes, the varieties, or the technology under different situations. Multi-situational/multilocational/multi-seasonal (environments) trials are conducted to oversee the stability in the performance of the varieties or the genotypes. When the interaction effects between the treatments and the environment is nonsignificant it implies the consistency of the treatment concerned over varied range of environments. But when this interaction becomes significant, then measuring the stability comes into mind, to find out the environment suitable for specific treatments.

Measuring the stability, mostly related to genotypic stabilities, has been worked out taking the environmental indices as the indicator and using the regression technique. Several measures are in literature; among these, Eberhart and Russell (1966) model is mostly used. But the methods of Yates and Cochran (1938), Comstock and Robinson (1952), Wricke (1962), Perkins and Jinks (1968), Freeman and Perkins (1971), etc., are also found to have been used. For details of these studies, readers may consult *Agriculture and Applied Statistics II* by Sahu and Das (2009). However, here we shall discuss Eberhart and Russell (1966) model.

11.4.1 Eberhart and Russell Model

According to this method,

$Y_{ij} = \mu_i + \beta_i E_j + \delta_{ij}; i = 1,2,\ldots\ldots t$ genotypes and $j = 1,2,\ldots l$ situations or locations,

where Y_{ij} = mean of ith genotype in jth situations,

μ_i = mean of the ith genotype over all the situation,

β_i = regression coefficient ith genotype on the environmental indices,

E_j = Environmental index for jth situation,

δ_{ij} = Deviation of the regression of ith genotype on the environmental indices.

The environmental index is defined as

$$E_j = \frac{\sum_{i=1}^{t} Y_{ij}}{t} - \frac{\sum_{i=1}^{t}\sum_{j=1}^{l} Y_{ij}}{t_l} \quad \text{for} \sum_{j=1}^{l} E_j = 0.$$

The environmental index is an index to measure the capabilities of different environments under the given condition. In Eberhart and Russell model, there are two stability parameters, namely, (1) *the regression coefficient* and (2) *the mean squared deviation from the regression*.

1. The regression coefficient (β_i) is the regression of the performance of each genotype under different environments on environmental average over all the genotypes. This model considers b_i as a measure of responsiveness, and accordingly a genotype having the regression coefficient unity (i.e., $b_i = 1$) is averagely responsive. A regression coefficient less than unity means that the genotype is less responsive to environmental factors like nutrient status, soil status, and plant situation. On the other hand, the regression coefficient greater than unity for a genotype indicates high responsiveness of the genotype towards environmental factors. This type of genotypes generally performs well under rich environmental conditions. On the other hand, the other groups of genotype for which $b < 1$ are suitable for poor environmental conditions.

2. The mean square deviation $\left(\bar{S}^2_{d_i}\right)$ from the regression is given as

$$\bar{S}^2_{d_i} = \frac{\sum_{j=1}^{l} \delta^2_{ij}}{(l-2)} - \frac{S^2_e}{r},$$

where

$$\sum_{j=1}^{l} \delta^2_{ij} = \left[\sum_{j=1}^{l} Y^2_{ij} - \frac{Y^2_{i.}}{t}\right] - \frac{\left(\sum_{j=1}^{l} Y_{ij}E_j\right)^2}{\sum_{j=1}^{l} E^2_j}$$

and S^2_e is the estimate of a pooled error.

A stable genotype is one which has $\bar{S}^2_{d_i} = 0$. Thus, when the null hypothesis $H_0 : S^2_{d_i} = 0$ is accepted, then the genotype is stable, and when the null hypothesis is rejected, then it is less stable. Actually, while testing the above null hypothesis, δ^2_{ij} is compared against S^2_e through F-test.

Thus, using both stability parameters, it can be concluded that for a genotype to be stable, the regression coefficient should be in unity with the mean square deviation $\left(\bar{S}^2_{d_i}\right)$ *not significantly different from zero*. Genotype means are also considered while selecting the best stable suitable genotype (Table 11.3).

Table 11.3 Inference drawn under different situation with respect to stability

Sl No.	Means	Regression coefficient (b_i)	$S^2_{d_i}$	Inference
1	$\bar{g}_i > \bar{y}$	$b_i = 1$	Nonsignificant	Widely adapted stable genotype
2	$\bar{g}_i < \bar{y}$	Do	Do	Stable genotype adapted for poor environment
3	$\bar{g}_i > \bar{y}$	$b_i > 1$	Do	Above-average stable genotype adapted for rich environment
4	$\bar{g}_i > \bar{y}$	$b_i < 1$	Do	Below-average stable genotypes suitable for poor environment
5	$\bar{g}_i > \bar{y}$	$b_i < 1$	Significant	Unstable genotype

		Location			
Variety	Rep	E1	E2	E3	E4
V1	R1	88.59	21.24	92.84	98.96
	R2	100.84	35.53	94.88	101.00
	R3	94.71	47.78	92.84	101.00
V2	R1	98.80	62.06	96.92	98.96
	R2	84.51	68.18	96.92	101.00
	R3	70.22	94.71	94.88	101.00
V3	R1	82.47	45.73	90.80	101.00
	R2	35.53	11.04	86.71	84.67
	R3	86.55	53.90	90.80	98.96
V4	R1	98.80	88.59	96.92	98.96
	R2	51.86	49.82	90.80	92.84
	R3	80.43	88.59	96.92	101.00
V5	R1	31.45	14.10	88.76	92.84
	R2	25.33	15.12	94.88	68.35
	R3	23.29	13.08	86.71	80.59
V6	R1	21.24	9.50	86.71	86.71
	R2	15.12	11.04	96.92	68.35
	R3	15.12	9.50	82.63	82.63
V7	R1	55.94	15.12	96.92	94.88
	R2	45.73	47.78	80.59	98.96
	R3	47.78	60.02	94.88	76.51
V8	R1	72.27	23.29	96.92	84.67
	R2	62.06	13.08	94.88	92.84
	R3	74.31	31.45	94.88	84.67

Example 11.5. The following table gives the yield (q/ha) for eight varieties of wheat tested in four locations to study the yield stability of the varieties. Analyze the data, and using Eberhart–Russell model, find out the best stable variety.

The following is the analysis as per Eberhart–Russell model using the statistical software SPAR1: *Environment wise analysis*

ENVIRNOMENT NO 1 ANOVA FOR TREATMENTS

ANOVA TABLE
SOURCE D.F. SUM OF SQUARES MEAN SUM OF SQ F-VALUE
REPL 2. 0.10375558E+04 0.51877791E+03 0.30540350E+01
TREAT 7. 0.15885272E+05 0.22693245E+04 0.13359470E+02
EROR 14. 0.23781296E+04 0.16986640E+03

(continued)

(continued)

TREATMENT MEANS
94.71 84.51 68.18 77.03 26.69 17.16 49.82 69.55

S.E. OF DIFF. OF TWO MEANS 0.10641630E+02

ENVIRNOMENT NO 2 ANOVA FOR TREATMENTS

ANOVA TABLE

SOURCE	D.F.	SUM OF SQUARES	MEAN SUM OF SQ	F-VALUE
REPL	2.	0.15325399E+04	0.76626993E+03	0.39569090E+01
TREAT	7.	0.13179994E+05	0.18828562E+04	0.97228030E+01
EROR	14.	0.27111510E+04	0.19365365E+03	

TREATMENT MEANS
34.85 74.98 36.89 75.67 14.10 10.01 40.97 22.61

S.E. OF DIFF. OF TWO MEANS 0.11362320E+02

ENVIRNOMENT NO 3 ANOVA FOR TREATMENTS

ANOVA TABLE

SOURCE	D.F.	SUM OF SQUARES	MEAN SUM OF SQ	F-VALUE
REPL	2.	0.10769658E+02	0.53848290E+01	0.22402580E+00
TREAT	7.	0.18601248E+03	0.26573212E+02	0.11055290E+01
EROR	14.	0.33651308E+03	0.24036649E+02	

TREATMENT MEANS
93.52 96.24 89.44 94.88 90.12 88.75 90.80 95.56

S.E. OF DIFF. OF TWO MEANS 0.40030530E+01

ENVIRNOMENT NO 4 ANOVA FOR TREATMENTS

ANOVA TABLE

SOURCE	D.F.	SUM OF SQUARES	MEAN SUM OF SQ	F-VALUE
REPL	2.	0.15303976E+03	0.76519880E+02	0.12410260E+01
TREAT	7.	0.14765731E+04	0.21093901E+03	0.34210830E+01
EROR	14.	0.86321969E+03	0.61658549E+02	

TREATMENT MEANS
100.32 100.32 94.88 97.60 80.59 79.23 90.12 87.39

S.E. OF DIFF. OF TWO MEANS 0.64113730E+01

STABILITY ANALYSIS FOLLOWING EBERHART AND RUSSELLS MODEL

ANOVA TABLE FOR STABILITY

SOURCE	D.F.	SUM OF SQUARES	MEAN SUM OF SQ	F-VALUE
VARIETIES	7	0.59732702E+04	0.85332431E+03	0.41973140E+01
ENVIRONMENTS	3	0.16088347E+05	0.53627825E+04	0.26378340E+02
VARIETYXENVIRONMENT	21	0.42693522E+04	0.20330249E+03	0.18102900E+01
TOTAL	31	0.26330970E+05		

(continued)

(continued)

POOLED ERROR	56	0.62890135E+04	0.11230381E+03	
ENV+(VAR.*ENV.)	24	0.20357700E+05	0.84823748E+03	
ENVRON(LINEAR)	1	0.16088356E+05	0.16088356E+05	0.13016560E+03
VARXENVRON(LINEAR)	7	0.22917573E+04	0.32739391E+03	0.26488360E+01
POOLED DEVIATION	16	0.19775868E+04	0.12359918E+03	0.33017360E+01
POOLED ERROR MSS FOR TESTING POOLED DEVIATION MSS= 0.37434604E+02				
ENV(LIN) AND VARXENV(LIN) ARE TESTED AGAINST POOLED DEVIATION				

VAR	MEAN	VAR. OF MEAN	REG COEF (B)	S(DELTA). SQR	MEAN SQ DEV(SD)
1	80.85	2847.80	1.00	837.95	381.54
2	89.01	397.18	0.44	10.45	−32.21
3	72.35	2074.18	1.00	68.48	−3.20
4	86.29	400.31	0.42	41.80	−16.53
5	52.88	4344.40	1.43	238.88	82.01
6	48.79	5027.75	1.51	427.24	176.19
7	67.93	2069.89	0.99	98.41	11.77
8	68.78	3196.19	1.21	254.37	89.75
POP MEAN = 70.859 S.E.(MEAN) = 0.64187010E+01					
MEAN OF B = 1.0000 S.E. OF B = 0.24791170E+00					

Upon the validity of the null hypotheses w.r.t. b_i and $\bar{S}^2_{d_i}$, we can have the following conclusion:

SL no.	$V_i - \bar{V}$	Regression coefficient (b_i)	Inference
V1	9.991	1.00	Widely adapted stable genotype
V2	18.151	0.44	Stable genotype less sensitive to environmental variation
V3	1.491	1.00	Widely adapted stable genotype
V4	15.431	0.42	Below-average stable genotype suitable for poor environment
V5	−17.979	1.43	Below-average stable genotype suitable for rich environment
V6	−22.069	1.51	Below-average stable genotype suitable for rich environment
V7	−2.929	0.99	Below-average stable genotype
V8	−2.079	1.21	Below-average stable genotype suitable for rich environment

11.5 Sustainability

Sustainability is an idea which has been defined by various authors with varied perspectives and objectivities. In this section, by sustainability, we mean persistence and the capacity of something to continue over a long period of time without damaging or degrading the natural resources. By sustainable agriculture, we mean integration of three main objectives: (1) *maintaining environmental health*, (2) *ensuring economic profitability, and* (3) *social and economic equity*. Sustenance of crop husbandry and cropping system, and farm management are essential criteria for food security in the long run. The measurement of sustainability of the production of a crop or a particular cropping or management system is essential for future food security. Quite a few studies have been attempted to measure the sustainability and thereby compare the different competitive methods or processes in agriculture. Some of the works to mention, though not exhaustive, are of Soni et al. (1988), Narayan et al. (1990), Singh (1990), Katyal et al. (1998, 2000), Gangwar et al. (2003), Sahu et al. (2005), and Pal and Sahu (2007), etc. To frame a measure for sustainability, well-planned long-term experiments of different crops, cropping systems, or farming systems under a given situation are very useful. But such planned experiments for a long period of time are very rare in literature except a few like the long-term experiments conceived and continued in Rothamsted Experimental Station, UK. In India, experiments have been started during the late 1980s of the twentieth century to study the long-term behavior of crop, cropping system, nutrient management system, etc. Let us discuss some of the indices of yield sustainability under the above long-term setup.

Let us suppose that we have an experiment conducted with i treatments for j consecutive years on the same crop/cropping system. Then we have the yield sustainability indices:

Sustainability index SI(1) $= \dfrac{\bar{y} - s}{y_{\max}}$ (Singh et al. 1990),

where

$\bar{y}$, s, $y_{\max}$ are the average, standard deviation and maximum yield respective of a particular crop/cropping sequence or nutrient treatment over a period of time.

The higher the value of the index, the higher the sustainability status but with no definite range:

Sustainability index SI(2) $= \left|\dfrac{1}{b_i}\right|$, where b_i is the regression coefficient in $y_{ij} = a + b_i\bar{y}_j$, (ICARDA 1994)

---

where y_{ij} is the yield corresponding to the ith treatment in the jth time period, and $\left(\bar{y}_j\right)$ is the overall mean yield for the jth time period. According to this method, the higher the b_i value, the lower the sustainability and vice versa. In this method also, the sustainability index does not have any limit.

In an attempt to improve the above measure of ICARDA (1994) and Katyal et al. (2000) introduced a time coefficient in the above regression. Thus, the regression takes the shape of $y_{ij} = a + b_it + c_i\bar{y}_j$ and is the time period (Katyal et al. 2000).

According to this method sustainability index is given as

Sustainability index SI(3) $= \left|\dfrac{1}{c_i}\right|$, where c_i is the regression coefficient in $\bar{y}_j$

The decision rule is similar to that of index 2.

The following two sustainability indices are by Pal and Sahu (2007):

Sustainability index SI(4) $= \left|\frac{1}{b'_i}\right|$ where b'_i is the regression coefficient in $y_{ij} = a + b'_i\bar{y}'_j$ and where $\bar{y}'_j$ is the average yield for jth year excluding the yield for ith treatment in the particular year,

Sustainability index SI(5) $= \left|\frac{1}{c'_i}\right|$ where c'_i is the regression coefficient in $y_{ij} = a + b'_i t + c'_i\bar{y}'_j$ and where $\bar{y}'_j$ is the average yield for jth year excluding the yield for ith treatment in the particular year.

These two measures also do not have limits for sustainability. Another serious objection to all these four measures is the assumption of linearity of the regression. If the linearity of the regression is not valid, then the above measures will be put under question.

Sahu et al. (2005) proposed the sustainability index based on the average performance and the highest ever performance during the period of investigation with the help of the following formula:

Sustainability index SI(6) $= \frac{Y_{\max} - \bar{Y}}{\bar{Y}}$.

In this measure, sustainability has been visualized as the minimum deviation of the average performance over the highest ever achieved performance during the period of investigation. As such, the lower the value of the index, the higher the sustainability.

In addition to the above, Pal and Sahu (2007) proposed the following indices to measure sustainability which do not require any assumption of linearity:

Sustainability index SI(7) $= \frac{s_i}{\bar{y}_i \times s_{\max}}$,

where

s_i is the standard deviation of ith treatment over the entire period,

$\bar{y}_i$ is the average of ith treatment over the entire period,

$s_{\max}$ is the maximum value of the standard deviation of all the treatments;

Sustainability index SI(8) $= \frac{1}{n}\sum_j\left[\frac{|y_{ij} - y_{\max}|}{\bar{y}_i}\right]$

where

$\bar{y}_i$ is the average of ith treatment over the entire time period,

$y_{\max}$ is the maximum value of the ith treatment over the time period;

$$\textbf{Sustainability index SI(9)} = \frac{1}{n}\sum_{j}\left[\frac{|y_{ij} - y_{med}|}{\bar{y}_i}\right]$$

where

$\bar{y}_i$ is the average of ith treatment over the entire time period,

y_{med} is the median value of the ith treatment over the time period;

$$\textbf{Sustainability index SI(10)} = \frac{1}{{}^{n}C_2}\sum_{j}\sum_{j',j<j'}\left[\frac{|y_{ij} - y'_{ij}|}{y_{i\max}}\right]$$

where

n is the number of time periods,

y_{ij} and y'_{ij} are the value of the ith treatment in jth and j'th year, respectively,

$y_{i\max}$ is the maximum value of the ith treatment over the time period.

In all the last five measures, the lower the value of the sustainability index, the higher the sustainability.

Example 11.6. The following table gives the yield of wheat (q/ha) in response to five different nutrient treatments conducted over 17 years. Using the measures of sustainability, find out the most sustainable treatment in terms of yield. Also find out whether or not the most sustainable treatment is the best treatment.

Obs.	T1	T2	T3	T4	T5
1	65.00	63.10	68.50	72.10	82.60
2	66.20	65.80	67.60	75.60	83.60
3	65.95	67.60	71.20	74.30	84.90
4	64.00	68.10	68.90	75.50	84.10
5	64.60	72.10	69.00	74.80	83.60
6	63.50	69.93	69.70	78.20	87.53
7	61.20	65.10	67.50	76.30	87.80
8	60.20	60.70	69.50	73.70	86.90
9	60.10	59.30	68.80	72.20	89.50
10	63.00	62.70	70.50	74.10	87.20
11	63.00	61.90	71.20	73.90	88.30
12	65.10	66.30	74.60	77.80	85.20
13	66.20	69.20	75.40	79.30	85.30
14	67.40	71.80	76.20	80.60	86.40
15	66.20	73.40	85.40	82.30	83.70
16	65.90	72.60	75.90	81.30	87.10
17	68.37	69.59	73.37	78.92	85.45

Solution. First, the average, the maximum, the median, and the standard deviation from the given data set for each treatment are worked out separately. Using the method of regression discussed earlier in Chap. 8 of this book, the regression coefficients for measures 2, 3, 4, and 5 are worked out. The regression coefficients according to measures 2, 3, 4, and 5 are b_i, c_i, b'_i, and c'_i, respectively.

Obs.	T1	T2	T3	T4	T5
1	65.00	63.10	68.50	72.10	82.60
2	66.20	65.80	67.60	75.60	83.60
3	65.95	67.60	71.20	74.30	84.90
4	64.00	68.10	68.90	75.50	84.10
5	64.60	72.10	69.00	74.80	83.60
6	63.50	69.93	69.70	78.20	87.53
7	61.20	65.10	67.50	76.30	87.80
8	60.20	60.70	69.50	73.70	86.90
9	60.10	59.30	68.80	72.20	89.50
10	63.00	62.70	70.50	74.10	87.20
11	63.00	61.90	71.20	73.90	88.30
12	65.10	66.30	74.60	77.80	85.20
13	66.20	69.20	75.40	79.30	85.30
14	67.40	71.80	76.20	80.60	86.40
15	66.20	73.40	85.40	82.30	83.70
16	65.90	72.60	75.90	81.30	87.10
17	68.37	69.59	73.37	78.92	85.45
Average	64.47	67.01	71.96	76.52	85.83
SD	2.39	4.38	4.53	3.16	1.95
Y_{max}	68.37	73.40	85.40	82.30	89.50
Median	65.00	67.60	70.50	75.60	85.45
b_i	0.703	1.574	1.625	1.247	−0.150
b_i'	0.539	1.477	1.52	1.28	−0.205
c	0.953	2.209	1.316	1.185	−0.664
c'	0.714	2.576	0.906	1.153	−0.543

Using the above quantities, that is, average, median, standard deviation, maximum, and regression coefficients, sustainability indices for different treatments as per the formula discussed above are worked out and presented in the following table:

From the above table, it is found that except for SI 7, treatment five is by far the most sustainable treatment. On examination, it is also revealed that treatment five has produced the highest average yield (85.45 q/ha) over the period of experimentation. Hence, treatment five is not only the most sustainable treatment but also the best treatment.

Sustainability index	Inference	Treatment T_1	T_2	T_3	T_4	T_5
SI(1)	The higher the SI value, the higher the sustainability	0.908	0.853	0.790	0.891	**0.937**
SI(2)	Do	1.422	0.635	0.615	0.802	**6.667**
SI(3)	Do	1.049	0.453	0.760	0.844	**1.506**
SI(4)	Do	1.855	0.677	0.658	0.781	**4.878**
SI(5)	Do	1.401	0.388	1.104	0.867	**1.842**
SI(6)	The lower the SI value, the higher the sustainability	0.061	0.095	0.187	0.075	**0.043**
SI(7)	Do	**0.008**	0.014	0.014	0.013	0.012
SI(8)	Do	0.061	0.095	0.187	0.075	**0.043**
SI(9)	Do	0.029	0.054	0.044	0.034	**0.019**
SI(10)	Do	0.040	0.071	0.055	0.045	**0.026**

Bold faces indicate the most sustainable treatment according to the particular measure

12 Multivariate Analysis

Analysis of information, resulting from different research programs, particularly the statistical procedures, may broadly be classified into *univariate analysis and multivariate analysis*. In univariate analyses, we consider one variable at a time contrary to the varied number of variables in multivariate analyses. The simplest case of multivariate analysis is the bivariate analysis, in which two variables are considered together. The variables that we consider in agriculture, economics, anthropology, sociology, psychology, management, etc., tend to move together, and as such multivariate analysis is more useful. Univariate analysis throws light on one character only, but to explain the relationship, interdependence, and relative importance of different variables, multivariate analyses would be more appropriate.

Let us take the example of analysis of innovation index which is associated with a number of parameters like age, gender, education, economic background, and society. These components are not independent of each other; rather these are correlated, interdependent to each other; these have varied importance towards ultimate innovation index. Univariate analysis can throw light separately on each of the character, but to analyze the system as a whole taking due consideration of their interdependence, relationships, importance, etc., multivariate analysis is the more acceptable option. Several examples in other disciplines like agriculture, business, economics, management, and medical science can also be put forward where multivariate analysis can effectively be used.

Multivariate analyses are more complicated, as they take care of the system as a whole, having a number of variables at a time, and consider their interdependence, relationships, importance, etc. But with the advent of computer technology and statistical software, the use of multivariate technique has become user-friendly and is gaining momentum day by day. The area and coverage of multivariate analysis is a huge one, and it is not possible to include all these here. In this book, an attempt will be made to provide an outline of some of the multivariate techniques which can be used in different agriculture and allied fields. The details of analytical/statistical steps for calculation are also avoided; emphasis will be provided to place different multivariate statistical tools using different statistical software to solve different problems. Useful references are provided for interested readers.

12.1 Classification of Multivariate Analysis

According to different authors, multivariate analysis may be classified into two broad groups: (a) *dependence method* and (b) *interdependence method*. In the dependence method of analysis, relationships of some dependent variables are worked out with the independent variables, but in the second method, interrelations among themselves are considered. The examples of first group of analysis are regression analysis, multiple discriminant analysis, multivariate

P.K. Sahu, *Research Methodology: A Guide for Researchers in Agricultural Science, Social Science and Other Related Fields*, DOI 10.1007/978-81-322-1020-7_12,

analysis of variance, canonical analysis, etc., while the other group consists of factor analysis, cluster analysis, etc.

12.2 Regression Analysis

The main task of the researcher is to work out the actual relationship between or among the variables under study. In agricultural and other experiments, mainly three types of variables are recoded: (a) the treatments or factors such as variety, insecticide, doses or type of fertilizers, different chemical treatments, and different management practices; (b) environmental parameters like rainfall, temperature, humidity, sunshine hours, and wind speed; and (c) various responses in the form of different growth and yield parameters, qualitative changes, etc. Regression analysis is one of the most important statistical tools used in this endeavor. Regression analysis is a technique by virtue of which one can study the relationship of the ultimate variables, say, adoption index, awareness, empowerment status, etc., within different fields of studies (different demographic, social, economical, educational, and other parameters).

In regression analysis, the dependent variable (Y) is the function of one or more independent variables (X's) and the error term (u's), which can be represented in the form of $Y = f(X_i, u_i)$. Let us suppose that Y depends on k other variates which are denoted by $X_1, X_2, \ldots X_k$. These need not be independent. The usual problem is to find the best linear predicting equation for Y of the form $Y_c = b_0 + b_1X_1 + b_2X_2 + b_3X_3 + \cdots + b_kX_k$ for the true regression equation in the population $Y = \beta_0 + \beta_1 x_1 + \beta_2 x_2 + \cdots + \beta_k x_k$. We present here three methods of getting such relationships below. Readers are free to select any one of these as per the convenience:

1. Now, using the procedure adopted (Chap. 8) for calculation of simple correlation coefficients among the variables, one can have the following correlation matrix for the variables involved in the regression analysis:

$$\Re = \begin{bmatrix} r_{00} & r_{01} & r_{02} & r_{0j} & \cdots & r_{0k} \\ r_{10} & r_{11} & r_{12} & r_{1j} & \cdots & r_{1k} \\ \cdots & \cdots & \cdots & \cdots & \cdots & \cdots \\ r_{k0} & r_{k1} & r_{k2} & r_{kj} & \cdots & r_{kk} \end{bmatrix},$$

where $r_{01}, r_{02}, \ldots, r_{kj}$'s are the correlation between Y and X_1, Y and $X_2 \ldots X_K$ and X_J and the diagonal elements are 1. If we denote R for the corresponding determinant and R_{ij} for the cofactor of r_{ij} element in $\Re$, then one can have

$$b_j = \frac{s_0}{s_j}(-1)^{j-1} \times (-1)^{j+2} . \frac{R_{0j}}{R_{00}}$$
$$= -\frac{s_0}{s_j} . \frac{R_{0j}}{R_{00}}; \; j = 1, 2, \ldots k \text{ and } S_0 \text{ and } S_j \text{ are}$$

the standard deviations of Y and X_j, respectively,

and

$$b_0 = \bar{Y} - \sum_{j=1}^{n} b_j \bar{X}_j = \bar{Y} + \sum_{j=1}^{k} \frac{s_0}{s_j} . \frac{R_{0j}}{R_{00}} \bar{X}_j.$$

Thus, multiple regression equation of y on $X_1, X_2, \ldots X_k$ becomes

$$y_c = \bar{y} - \frac{s_0}{s_1} . \frac{R_{01}}{R_{00}} (X_1 - \bar{X}_1) - \frac{s_0}{s_2} . \frac{R_{02}}{R_{00}} (X_2 - \bar{X}_2)$$
$$- \cdots - \frac{s_0}{s_k} . \frac{R_{0k}}{R_{00}} (X_k - \bar{X}_k).$$

The coefficient b_j is known as partial regression coefficient of Y on X_j for fixed $X_1, X_2, \ldots X_{j-1}, X_{j+1}, \ldots X_k$ and is written in the form $b_{yj.12\ldots(j-1)(j+1)\ldots k} = b_{0j.12\ldots(j-1)(j+1)\ldots k} = -\frac{s_0}{s_j} \frac{R_{0j}}{R_{00}}$.

2. The objective is to find out the relationship of the form $Y_i = \beta_1 X_{1i} + \beta_2 X_{2i} + \beta_1 X_{3i} + \cdots + \beta_k X_{ki} + u_i$.

   For n number of observations, the regression equation can be written as:

1 $Y_1 = \beta_1 X_{11} + \beta_2 X_{21} + \beta_3 X_{31} + \beta_4 X_{41} + \cdots + \beta_k Xk_1 + u_1$
2 $Y_2 = \beta_1 X_{12} + \beta_2 X_{22} + \beta_3 X_{32} + \beta_4 X_{42} + \cdots + \beta_k Xk_2 + u_2$
3 $Y_3 = \beta_1 X_{13} + \beta_2 X_{23} + \beta_3 X_{33} + \beta_4 X_{43} + \cdots + \beta_k Xk_3 + u_3$
.
.
.
n $Y_n = \beta_1 X_{1n} + \beta_2 X_{2n} + \beta_3 X_{3n} + \beta_4 X_{4n} + \cdots + \beta_k X_{kn} + u_n$

$$\Rightarrow \underset{n \times 1}{\begin{pmatrix} Y_1 \\ Y_2 \\ \\ Y_n \end{pmatrix}} = \underset{n \times k}{\begin{pmatrix} X_{11} & X_{21} & X_{31} & X_{41} \ldots & X_{k1} \\ X_{12} & X_{22} & X_{32} & X_{42} \ldots & X_{k2} \\ \\ X_{1n} & X_{2n} & X_{3n} & X_{4n} \ldots & X_{kn} \end{pmatrix}} \underset{k \times 1}{\begin{pmatrix} \beta_1 \\ \beta_2 \\ \\ \beta_k \end{pmatrix}} + \underset{n \times 1}{\begin{pmatrix} u_1 \\ u_2 \\ \\ u_n \end{pmatrix}}$$

In matrix notation, the above equations can be written as

$\underset{\sim}{Y} = \underset{\sim}{X}\underset{\sim}{\beta} + \underset{\sim}{u}$ with $E(\underset{\sim}{Y}) = \underset{\sim}{X}\underset{\sim}{\beta}$, $E(\underset{\sim}{u}) = \underset{\sim}{0}$ and $E(\underset{\sim}{u}\underset{\sim}{u}') = \sigma^2 \underset{\sim}{I}$, where $\underset{\sim}{I}$ is an $n \times n$ unit matrix.

Our objective will be to minimize $L = \underset{\sim}{u}'\underset{\sim}{u}$.

$$\text{Let } L = \underset{\sim}{u}'\underset{\sim}{u} = \left(\underset{\sim}{Y} - \underset{\sim}{X}\underset{\sim}{\beta}\right)' \left(\underset{\sim}{Y} - \underset{\sim}{X}\underset{\sim}{\beta}\right)'$$
$$= \underset{\sim}{Y}'\underset{\sim}{Y} - 2\,\underset{\sim}{\beta}'\underset{\sim}{X}'\underset{\sim}{Y} + \underset{\sim}{\beta}'\underset{\sim}{X}'\underset{\sim}{X}\underset{\sim}{\beta}.$$

Let $\underset{\sim}{b}$ be the least square estimators of β. Then using the procedure of ordinary least square,

$\frac{\partial L}{\partial \beta} = 0$, and writing $\underset{\sim}{b}$ for β, we have

$$2\underset{\sim}{X}'\underset{\sim}{Y} - 2\underset{\sim}{X}'\underset{\sim}{X}\underset{\sim}{b}$$
$$\text{or } \underset{\sim}{X}'\underset{\sim}{X}\underset{\sim}{b} = \underset{\sim}{X}'\underset{\sim}{Y},$$
$$\text{or } \underset{\sim}{b} = \left(\underset{\sim}{X}'\underset{\sim}{X}\right)^{-1}\underset{\sim}{X}'\underset{\sim}{Y},$$

provided $\left(\underset{\sim}{X}'\underset{\sim}{X}\right)^{-1}$ exists.

Once after getting the b for a given set of observations, one can have the multiple regression equation $y = b_1 x_1 + b_2 x_2 + \cdots + b_k x_k$, the sample regression equation.

3. Most probably the simplest way of getting the above multiple regression equation is the use of data analysis module of MS Excel software of MS Office mostly available in computers. In this method, one need not to go into details about the steps of calculations; rather, one can have the relationship for the research data using the above package. In example one, the same has been demonstrated using MS Excel as well as SPSS software.

12.3 Multiple Correlation

Multiple correlation of y with $x_1, x_2, \ldots x_k$ is nothing but the simple correlation coefficient between the observed y and the predicted y_c. Thus, multiple correlation coefficient is given as

$$R_{y.12\ldots k} = \frac{\text{cov}(y, y_c)}{\sqrt{V(y)}\sqrt{V(y_c)}} \therefore R_{0.12\ldots k}$$
$$= \sqrt{\frac{V(y_c)}{V(y)}};\ R_{0.12\ldots k} = \left(1 - \frac{R}{R_{00}}\right)^{\frac{1}{2}}.$$

We have

$$\text{TSS}(y) = \sum_{i=1}^{n}(y_i - \bar{y})^2 = \sum_{i=1}^{n}(y_i - y_{ic} + y_{ic} - \bar{y})^2,$$
$$= \sum_{i=1}^{n}(y_i - y_{ic})^2 + \sum_{i=1}^{n}(y_{ic} - \bar{y})^2$$
$$= \text{RSS} + R_g\text{SS},$$

where y_i and y_{ic} are the observed and expected values of Y for ith observation, respectively,

$$\therefore \text{TSS} = \text{RSS} + R_g\text{SS} \Rightarrow \frac{\text{TSS}}{\text{TSS}} = \frac{\text{RSS}}{\text{TSS}} + \frac{R_g\text{SS}}{\text{TSS}}$$
$$\Rightarrow 1 = \frac{\text{RSS}}{\text{TSS}} + R^2 \Rightarrow 1 - \frac{\text{RSS}}{\text{TSS}} = R^2.$$

Clearly, from the above, $0 \le R^2 \le 1$

R^2 indicates the proportion of variation of the dependent variable explained by the line of multiple regression.

We shall demonstrate the process of getting regression equation using MS Excel software of MS Office taking the following example:

Example 12.1. The following table gives the information on ten varieties of mulberry. Find out the regression equation of leaf yield on other yield attributing characters.

Leaf area (cm^2): X_1	94.61	42.3	98.15	72.61	174.45	115.62	78.41	133.74	112.94	148.28
Total shoot length (cm): X_2	525.56	345	711.11	307.56	529.44	631.67	526.67	442.22	482.33	422.78
Leaf moisture (%): X_3	79.35	58.41	73.67	74.6	81.21	82.83	78.66	80.76	80.66	78.26
Leaf yield/plant (g): Y	255.56	97.78	237.78	161.11	263.33	237.78	217.78	224.44	228.89	190.00

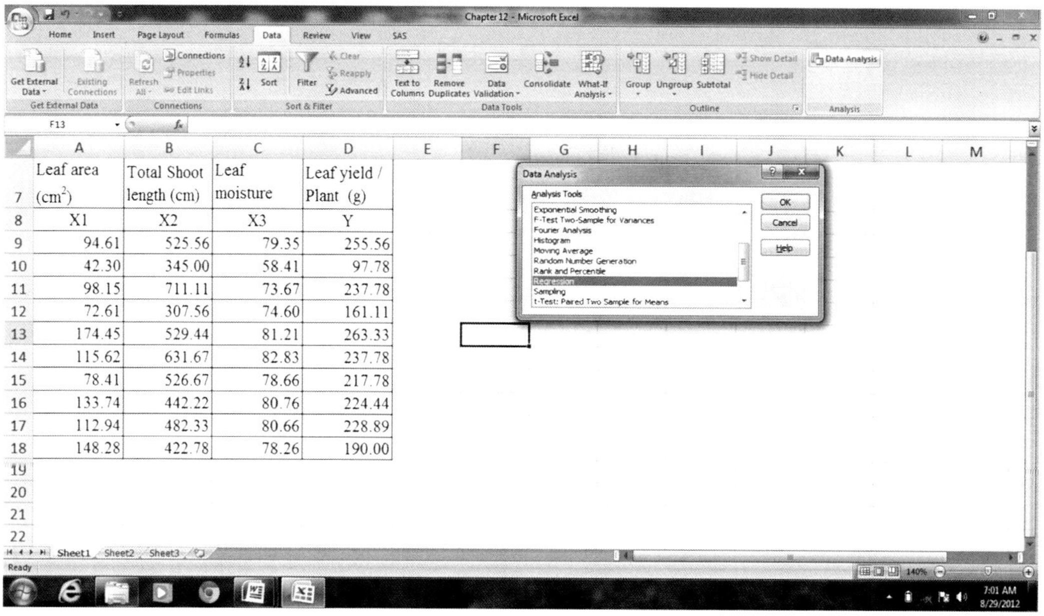

Slide 12.1: Step 1, showing data structure and selection of Regression submenu from Data Analysis menu in MS Excel

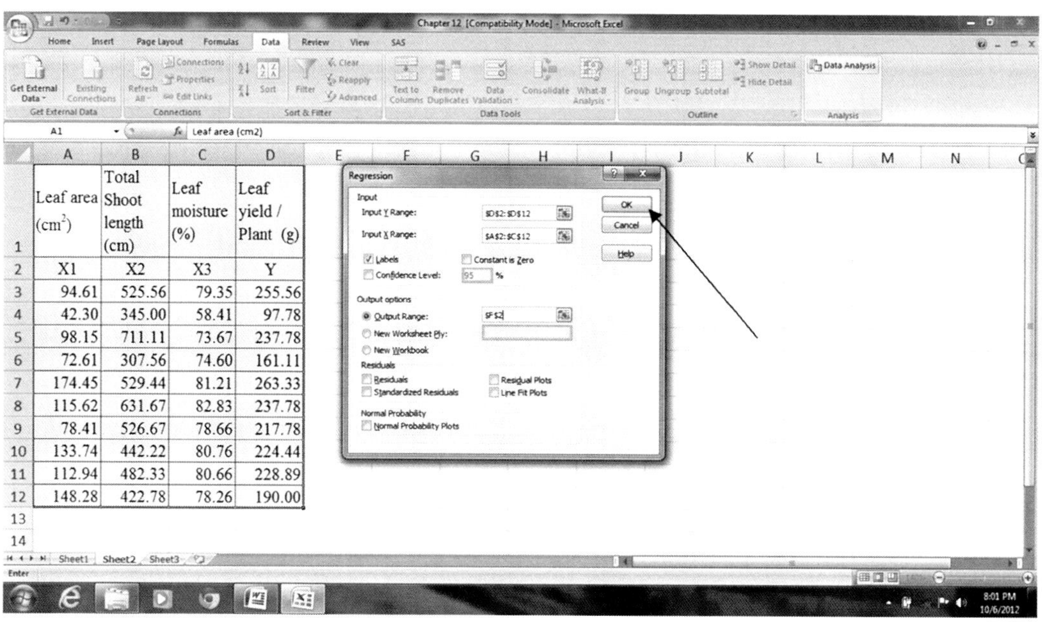

Slide 12.2: Step 2, showing data structure and selection of data range and other options in Regression submenu from Data Analysis menu of MS Excel

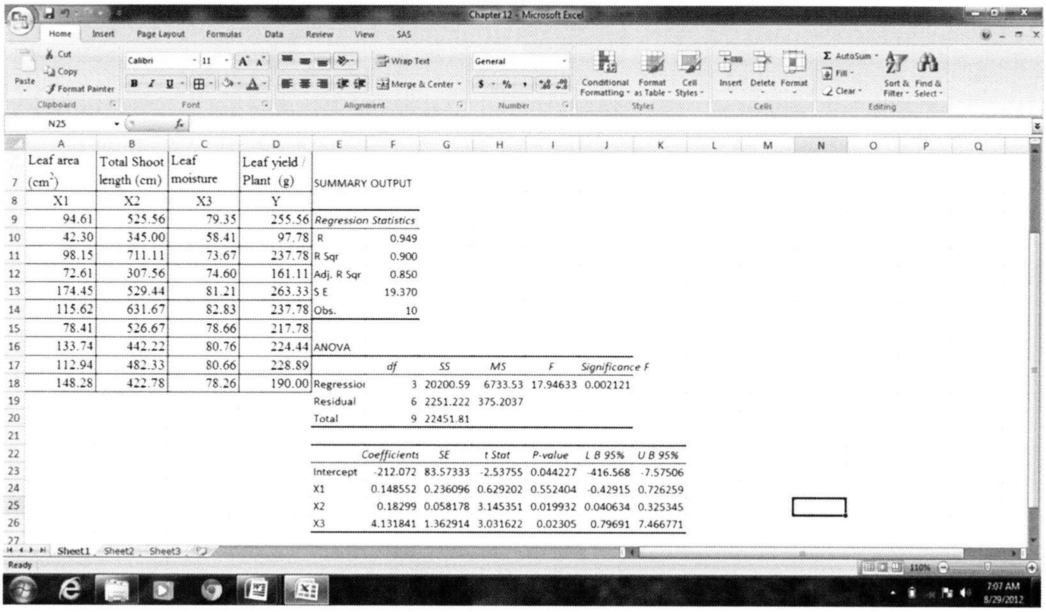

Slide 12.3: Step 3, showing data structure and output of regression analysis using MS Excel

Thus, the above regression equation is found to be $y = -212.072 + 0.149x_1 + 0.183x_2 + 4.132x_3$.

The above calculation can very well be performed using SPSS software as follows:

Step 1:

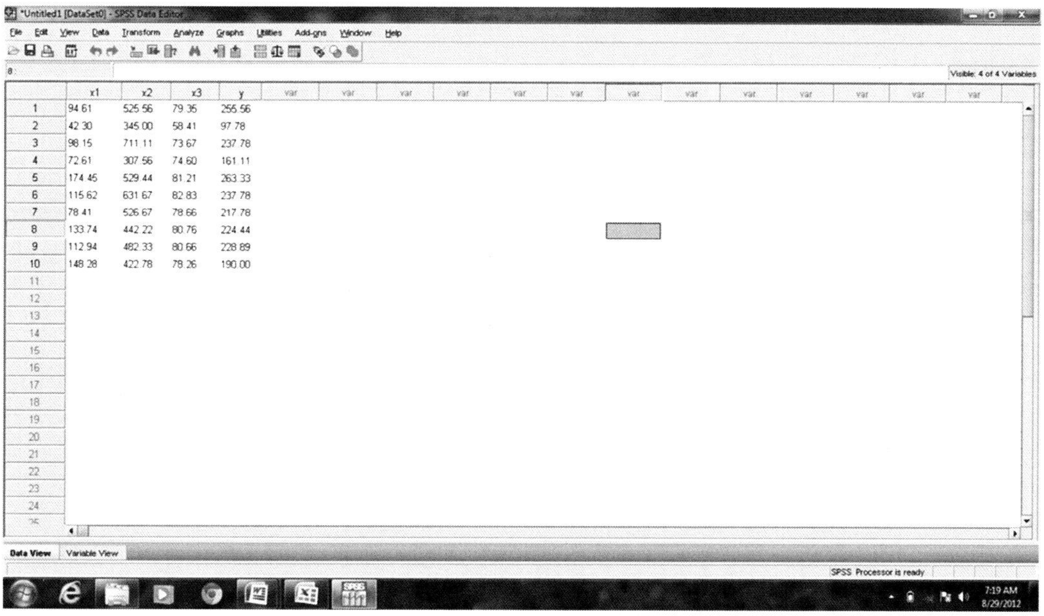

Slide 12.4: Step 1, showing data structure in data editor menu of SPSS

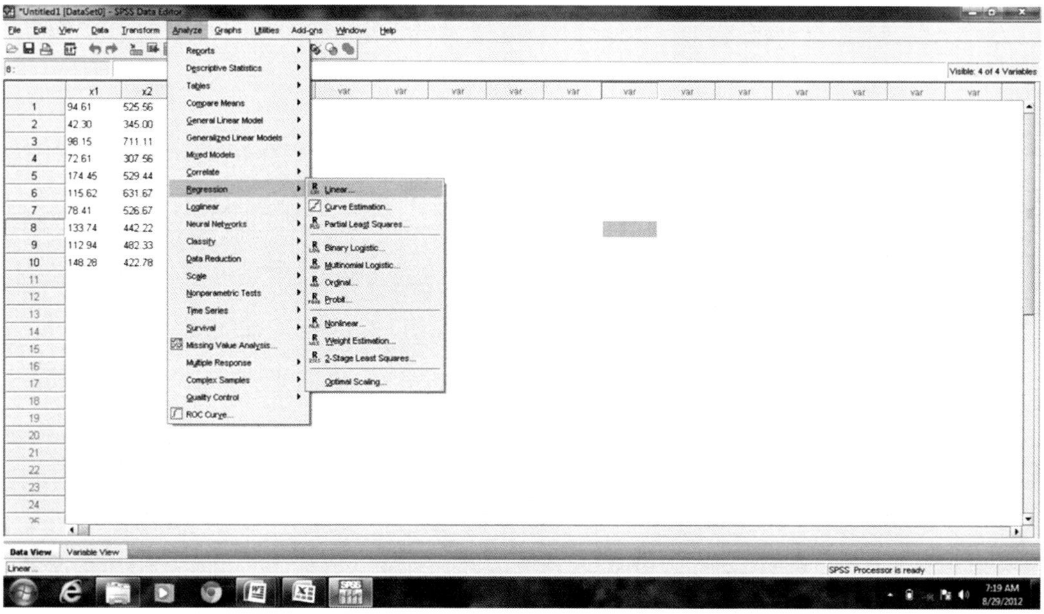

Slide 12.5: Step 2, showing data structure and selection of appropriate options in analysis menu of SPSS

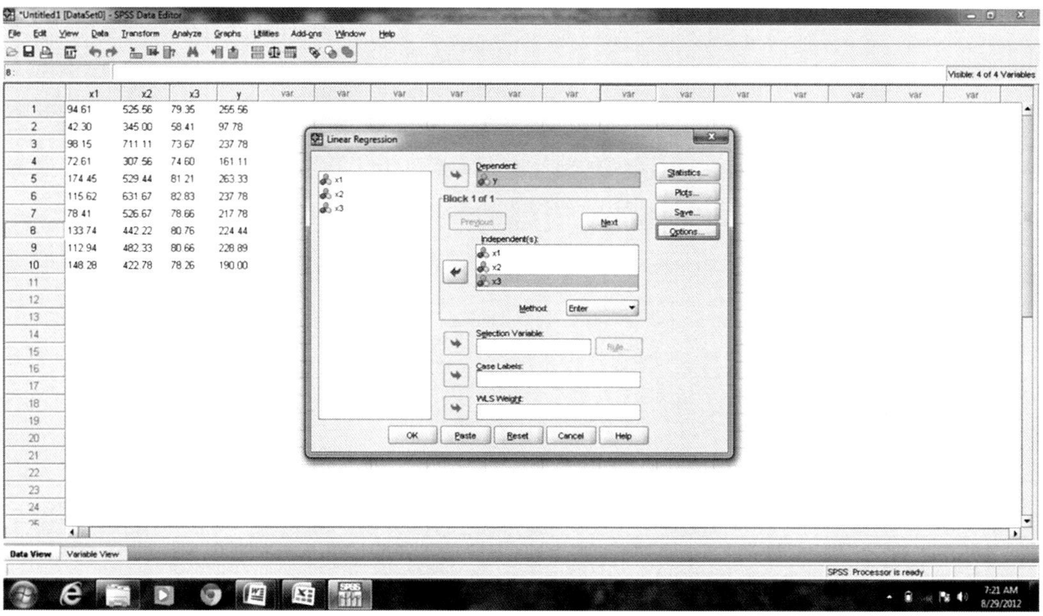

Slide 12.6: Step 3, showing data structure and variable selection of regression analysis using SPSS (Table 12.1)

Thus, both the procedures provide same result; additionally, in SPSS one can get the regression coefficients for standardized variables in the form of beta coefficients. This regression equation can be framed using other standard statistical packages also.

Table 12.1 Regression output through SPSS

Variables Entered/Removed[b]			
Model	Variables Entered	Variables Removed	Method
1	x3, x2, x1[a]		Enter
a. All requested variables entered.			
b. Dependent Variable: y			

Model Summary				
Model	R	R Square	Adjusted R Square	Std. Error of the Estimate
1	.949[a]	.900	.850	19.37018
a. Predictors: (Constant), x3, x2, x1				

ANOVA[b]						
Model		Sum of Squares	df	Mean Square	F	Sig.
1	Regression	20200.591	3	6733.530	17.946	.002[a]
	Residual	2251.222	6	375.204		
	Total	22451.814	9			
a. Predictors: (Constant), x3, x2, x1						
b. Dependent Variable: y						

Coefficients[a]						
Model		Unstandardized Coefficients		Standardized Coefficients		
		B	Std. Error	Beta	t	Sig.
1	(Constant)	-212.072	83.573		-2.538	.044
	x1	.149	.236	.115	.629	.552
	x2	.183	.058	.447	3.145	.020
	x3	4.132	1.363	.586	3.032	.023
a. Dependent Variable: y						

12.4 Stepwise Regression

From the above example, it is clear that the relationship of yield with three yield components having $R^2 = 0.900$ is sufficient to explain more than 90 of the variation in yield. The ANOVA table also shows the significance of the overall relationship as depicted by the significance value of 0.002 that means the R^2 vis-a-vis the relationship is significant at 2 probability level ($P = 0.02$). Now, while analyzing the significance of the individual coefficients, it is found that t-test declares the coefficients of X_2 and X_3 significant as depicted by respective significance levels of 0.02 and 0.023. But the coefficient of X_1 is not found to be significant. So the question is whether to retain X_1 in the model or not. At this juncture, the readers should note that:

1. The one-to-one relationship vis-a-vis the nature of the coefficients in linear relationship may change under multiple variable condition.
2. The coefficient of determination, that is, R^2, is a nondecreasing function of the number of variables in the regression equation. Thus, as we go on introducing more and more number of variables in the regression equation, the R^2 is supposed to increase or at least remain constant. So one may be tempted to include more and more number of variables in the regression model in order to maximize R^2 and thereby increase the possibility of significance of the overall regression equation. This phenomenon is sometimes known as *game of maximizing* R^2. It must be clearly understood that the variables to be included in the regression model should be guided by the knowledge about the variables in relation to the response variable, never should be guided towards maximizing R^2 value. The experimenter should include or try to incorporate only those variables which have got significant coefficients and instead of concentrating on the value of R^2 rather concentrate on adjusted R^2, which is not a nondecreasing

function of the number of variables in the model. The adjusted R^2 is defined as

$$\bar{R}^2 = R^2_{\text{adj}} = \frac{\text{R}_\text{g}\text{SS/d.f}}{\text{TSS/d.f}}$$
$$= \frac{\sum_{i=1}^{n} (y_{ic} - \bar{y})^2/(n-k-1)}{\sum_{i=1}^{n} (y_i - \bar{y})^2/(n-1)}.$$

We have

$$R^2 = \frac{\text{R}_\text{g}\text{SS}}{\text{TSS}} = \frac{\text{TSS} - \text{RSS}}{\text{TSS}} = 1 - \frac{\sum_{i=1}^{n} u_i^2}{\sum_{i=1}^{n} (Y_i - \bar{Y})^2}.$$

On the other hand,

$$\bar{R} = 1 - \frac{\text{RMS}}{\text{TMS}} = 1 - \frac{\text{RSS}/(n-k)}{\text{TSS}/(n-1)}$$
$$= 1 - \frac{n-1}{n-k}\frac{\text{TSS} - \text{R}_\text{g}\text{SS}}{\text{TSS}}$$
$$= 1 - \frac{n-1}{n-k}\left(1 - R^2\right).$$

In regression model, we have:

1. $K > 2$ thereby indicating that $\bar{R}^2 < R^2$; that means as the number of independent variables increases, $\bar{R}^2$ increases less than R^2.
2. $\bar{R}^2 = 1 - \frac{n-1}{n-k}(1-R^2)$ when $R^2 = 1$, $\bar{R}^2 = 1$.
3. When $R^2 = 0 \Rightarrow \bar{R}^2 = 1 - \frac{n-1}{n-k} = \frac{n-k-n+1}{n-k} = \frac{1-k}{n-k}$; if $k \geq 2$ the $\bar{R}^2$ is –ve.

Thus, adjusted R^2 can be negative if $k > 2$ and $R^2 = 0$; when the value of the $\bar{R}^2$ becomes negative, then its value is taken as zero.

While dealing with multiple variable regression equation, particularly with respect to the variable to be retained in the multiple regression equation, we generally follow two procedures—(a) stepwise forward and (b) stepwise backward regression technique—to get the actual relationship. In forward regression technique, the variable to be included first in the model is guided by the theoretical and logical idea about the variables under consideration and is further supported by the higher values of the correlation coefficient with the dependent variable. Similar decisions in subsequent steps are taken to include the other variables in stepwise manner. If the inclusion of a new variable in the model increases the explanatory power of the model, that is, increases the value of R^2, to a great extent without hampering the nature of the coefficient (s) of the previously included variable(s) in the model coupled with a significant coefficient, then the variable is retained in the model and it is useful. If the inclusion of a new variable in the model does not increase the explanatory power of the model, that is, the value of R^2, to a great extent, then the variable is redundant. If with the inclusion of a new variable in the model does not increase the explaining power of the model but rather changes the nature of the coefficient(s) of the variable(s) already in the model, then the variable is detrimental.

In the backward regression technique, all the variables under consideration are included in the model at the first instance to get the multiple regression equation; R^2 value and the nature of the coefficients of the individual variables are noted. The variable having most nonsignificant coefficient (at a preassigned probability level) is dropped at the first instance, and the multiple regression equation is again framed by dropping the variable. In the next subsequent steps, the same procedure is followed to discard the unuseful variables in stepwise manner. The process continued till one gets a regression equation with all the variables having significant coefficient at preassigned level of significance.

Example 12.2. The following table in the next page gives the yield attributing characters along with yield for 37 varieties. We are to work out the linear relationship of yield with other yield components having significant coefficients.

From the given data, one should first make the following correlation table as per the method suggested in Chap. 8 (Table 12.2).

From the correlations of yield with other variables, it is found that the order of correlation coefficients is X11 > X12 > X7 > X10 > X13 and so on. So if one wants to have stepwise forward regression, then he or she should start with X11 first and then in subsequent step as per the order of the correlation coefficient shown above (Table 12.3).

Data Table for Example 12.2

Sl no.	X1	X2	X3	X4	X5	X6	X7	X8	X9	X10	X11	X12	X13	X14	X15	X16	X17	X18	X19	Y
1	32.34	151.81	140.47	21.23	162.67	107.88	23.20	31.33	9.47	149.37	7.55	2.01	195.77	1663.67	313.38	1.93	1.93	9.96	9.49	345.85
2	29.85	140.71	140.95	20.33	150.00	104.49	23.20	33.17	8.87	144.04	7.82	2.11	222.68	1675.33	298.72	1.93	2.45	14.50	9.45	354.41
3	32.38	138.77	144.30	18.63	146.00	93.17	22.23	33.33	9.00	145.20	7.90	2.14	219.74	1672.33	302.74	1.90	2.44	15.45	9.33	358.12
4	38.33	138.31	140.06	17.67	156.33	92.60	22.93	34.50	8.73	141.57	7.67	2.15	228.32	1587.33	263.92	1.57	2.59	15.63	7.93	341.39
5	37.81	140.23	135.62	18.70	153.67	103.54	23.07	33.50	8.77	133.96	7.72	2.15	207.93	1621.00	277.86	1.53	2.40	13.75	7.73	323.82
6	35.02	132.90	128.79	18.47	143.00	108.60	23.33	29.00	9.03	145.32	7.97	2.27	189.59	1629.00	267.74	1.34	2.41	13.37	6.91	369.82
7	32.25	136.10	123.41	19.93	130.00	108.25	23.13	30.33	9.70	149.08	7.80	2.12	161.48	1657.33	264.72	1.72	2.45	13.76	8.44	373.14
8	32.56	137.11	116.18	18.87	143.67	100.95	23.73	28.17	9.47	133.45	7.83	2.08	146.38	1621.67	241.44	1.74	2.25	12.26	7.98	353.28
9	34.10	147.68	125.74	19.67	153.00	98.60	24.13	30.33	9.67	132.12	7.88	2.08	162.86	1634.33	253.30	1.95	2.74	15.22	8.94	350.18
10	37.05	153.77	123.49	19.10	160.00	100.17	24.07	30.00	9.23	137.06	8.58	2.27	184.37	1702.67	263.88	1.77	2.74	15.05	8.18	394.78
11	43.70	161.64	124.89	19.07	166.33	112.08	27.57	25.93	10.30	147.31	10.19	2.73	239.65	1785.67	286.02	1.77	2.73	14.98	8.67	489.41
12	41.96	160.23	114.79	19.60	143.67	107.01	26.27	24.83	10.23	135.48	9.17	2.40	237.48	1693.00	284.65	1.86	2.20	11.29	8.54	416.26
13	38.29	144.75	115.58	20.03	143.00	107.58	26.63	26.17	9.80	129.08	8.97	2.52	201.06	1609.00	276.24	1.74	2.47	12.21	8.01	384.56
14	33.84	155.17	117.81	19.33	157.00	102.07	25.27	24.83	9.17	130.27	7.72	2.16	178.28	1496.00	294.98	1.67	2.29	11.29	7.76	327.73
15	34.56	148.67	119.07	16.67	164.33	116.45	25.87	24.50	9.33	130.48	8.55	2.43	188.79	1511.67	288.41	1.32	2.33	11.52	6.68	363.85
16	32.46	156.00	121.01	20.10	162.67	115.48	25.50	22.50	9.07	138.49	8.38	2.25	202.79	1547.33	307.51	1.44	2.05	10.52	7.22	371.94
17	34.01	156.99	138.86	21.83	144.00	117.92	25.07	27.40	9.10	151.50	10.06	2.54	216.33	1662.67	316.73	1.42	2.66	14.93	7.46	505.86
18	34.58	162.46	145.61	23.90	149.33	104.79	25.17	29.67	9.20	144.95	10.46	2.60	220.73	1685.33	313.12	1.49	2.58	14.54	7.77	502.73
19	34.22	166.22	153.17	22.73	155.67	108.22	24.67	33.83	9.07	150.81	10.01	2.56	208.66	1738.67	288.29	1.85	3.05	17.96	9.29	495.12
20	33.13	162.80	143.02	23.10	144.00	109.56	24.03	30.60	8.93	131.19	8.64	2.34	168.35	1668.33	247.84	1.55	2.40	13.12	7.58	359.93
21	34.27	168.36	135.34	21.20	163.33	115.47	23.13	26.50	11.93	132.39	7.50	1.98	131.32	1711.33	239.02	2.07	2.19	12.33	8.40	325.47
22	31.85	163.56	129.03	20.53	154.67	102.78	24.60	29.00	12.87	117.91	7.89	2.02	117.85	1653.00	234.97	1.81	2.02	10.43	7.41	298.67
23	29.36	160.86	129.24	21.17	193.67	99.50	20.80	37.60	12.50	117.78	7.99	2.11	158.37	1701.67	211.97	2.13	2.38	12.65	8.86	299.57
24	35.01	154.62	140.06	21.53	171.00	90.33	22.53	42.50	9.50	129.01	8.64	2.37	218.03	1692.00	244.12	1.66	2.60	13.66	8.31	356.87
25	35.71	161.44	145.13	22.40	181.00	98.85	21.87	36.73	9.67	131.64	8.67	2.62	263.96	1764.00	289.06	1.87	2.88	15.42	9.13	372.53
26	38.24	159.29	144.58	19.73	161.67	93.00	24.57	37.50	10.57	140.35	9.84	2.84	223.82	1744.67	312.47	1.96	2.65	14.10	9.53	453.42
27	35.36	147.69	144.01	21.93	182.33	104.51	24.37	36.50	10.87	135.28	10.06	2.70	216.55	1747.33	289.87	1.71	2.53	13.53	8.67	450.74
28	33.81	139.24	145.32	22.97	178.00	105.69	23.20	34.50	9.90	137.24	10.00	2.45	197.75	1698.67	286.04	1.49	2.02	10.29	7.83	449.22
29	34.44	139.44	146.65	22.63	193.00	112.40	23.90	29.50	8.50	155.76	8.94	2.08	238.45	1624.00	333.34	1.39	2.01	10.28	7.65	457.62
30	36.05	151.53	154.25	23.53	194.00	103.77	24.10	31.17	8.73	149.60	8.58	2.25	236.20	1542.33	306.56	1.79	2.04	10.45	8.28	420.00
31	37.34	157.67	151.84	21.63	195.33	102.57	24.90	33.50	10.90	161.70	8.42	2.20	232.85	1582.33	291.33	1.76	2.05	10.76	8.12	442.93
32	33.13	151.16	143.56	24.53	191.00	98.26	25.00	29.50	11.03	135.25	7.69	2.20	203.45	1588.33	250.95	2.15	1.80	9.63	7.10	331.91
33	30.20	144.68	133.70	21.73	176.00	95.76	22.50	35.83	10.50	140.91	6.96	1.94	154.35	1611.00	277.85	1.89	2.00	10.19	7.25	321.00
34	30.26	140.64	119.93	21.93	178.67	94.23	20.77	37.67	9.30	128.65	7.12	2.13	159.38	1601.00	286.54	1.89	1.99	10.35	7.28	283.70
35	31.27	141.59	123.96	19.37	180.67	105.51	18.00	35.33	8.93	133.87	6.65	1.91	205.68	1648.00	302.31	1.74	2.23	12.42	8.15	276.98
36	33.17	150.82	131.18	18.80	178.67	107.36	17.37	31.17	9.23	143.59	6.68	1.92	206.11	1634.00	306.61	2.08	2.12	11.87	9.09	300.94
37	32.15	151.23	137.13	22.37	179.67	112.17	20.13	30.83	8.90	143.36	6.83	1.81	203.41	1650.33	304.86	2.12	2.15	11.52	9.27	295.61

Table 12.2 Correlation table

	X1	X2	X3	X4	X5	X6	X7	X8	X9	X10	X11	X12	X13	X14	X15	X16	X17	X18	X19	Y
X1	1.000																			
X2	0.217	1.000																		
X3	−0.026	0.178	1.000																	
X4	−0.237	0.336	0.610	1.000																
X5	−0.129	0.110	0.377	0.429	1.000															
X6	0.066	0.192	−0.138	0.012	−0.142	1.000														
X7	0.551	0.282	−0.067	0.016	−0.335	0.239	1.000													
X8	−0.239	−0.137	0.471	0.253	0.390	−0.689	−0.538	1.000												
X9	−0.041	0.449	−0.059	0.134	0.223	−0.114	0.125	0.068	1.000											
X10	0.203	−0.145	0.488	0.151	0.042	0.264	0.109	−0.084	−0.434	1.000										
X11	0.520	0.360	0.303	0.268	−0.131	0.193	0.653	−0.076	0.028	0.215	1.000									
X12	0.577	0.331	0.197	0.122	−0.142	0.025	0.589	0.029	0.018	0.065	0.896	1.000								
X13	0.528	0.015	0.445	0.109	0.236	−0.011	0.125	0.138	−0.436	0.522	0.430	0.478	1.000							
X14	0.226	0.346	0.267	0.193	−0.069	−0.058	−0.048	0.366	0.276	0.003	0.452	0.441	0.180	1.000						
X15	0.117	−0.117	0.274	0.089	0.127	0.280	0.001	−0.110	−0.566	0.647	0.223	0.178	0.630	−0.095	1.000					
X16	−0.201	0.263	0.052	0.179	0.289	−0.341	−0.368	0.267	0.446	−0.173	−0.400	−0.313	−0.169	0.277	−0.239	1.000				
X17	0.356	0.216	0.072	−0.206	−0.404	−0.100	0.187	0.167	−0.186	0.013	0.483	0.582	0.270	0.531	−0.063	−0.116	1.000			
X18	0.258	0.115	0.174	−0.222	−0.442	−0.135	0.059	0.204	−0.238	0.119	0.360	0.427	0.259	0.532	−0.052	−0.053	0.945	1.000		
X19	0.066	0.199	0.288	0.008	0.022	−0.184	−0.263	0.338	0.018	0.186	0.038	0.052	0.293	0.580	0.161	0.616	0.370	0.421	1.000	
y	0.522	0.254	0.403	0.274	−0.099	0.268	0.613	−0.135	−0.126	0.562	0.920	0.764	0.525	0.358	0.415	−0.400	0.415	0.345	0.073	1.000

Table 12.3 Stepwise regression output

Step 1

SUMMARY OUTPUT

Regression Statistics	
R	0.92
R Sqr	0.85
Adj R Sqr	0.84
SE	25.86
Obs	37.00

ANOVA

	df	*SS*	*MS*	*F*	*Significance F*
Regression	1	128364.52	128364.52	191.88	0.00
Residual	35	23413.83	668.97		
Total	36	151778.35			

	Coefficients	*SE*	*t Stat*	*P-value*	*LB95%*	*UB 95%*
Intercept	−100.710	34.690	−2.903	0.006	−171.134	−30.286
X11	57.043	4.118	13.852	0.000	48.683	65.403

Step 2 Include X12

SUMMARY OUTPUT

Regression Statistics	
R	0.93
R Sqr	0.86
Adj R Sqr	0.86
SE	24.64
Obs	37.00

ANOVA

	df	*SS*	*MS*	*F*	*Significance F*
Regression	2	131141.57	65570.79	108.03	0.00
Residual	34	20636.78	606.96		
Total	36	151778.35			

	Coefficients	*SE*	*t Stat*	*P-value*	*LB95%*	*UB 95%*
Intercept	−66.111	36.790	−1.797	0.081	−140.877	8.655
X12	−78.036	36.482	−2.139	0.040	−152.177	−3.895
X11	73.949	8.824	8.381	0.000	56.017	91.881

Step 3: Include X7

SUMMARY OUTPUT

Regression Statistics	
R	0.93
R Sqr	0.86
Adj R Sqr	0.85
SE	24.98
Obs	37.00

ANOVA

	df	*SS*	*MS*	*F*	*Significance F*
Regression	3	131186.37	43728.79	70.08	0.00
Residual	33	20591.98	624.00		
Total	36	151778.35			

	Coefficients	*SE*	*t Stat*	*P-value*	*LB95%*	*UB 95%*
Intercept	−74.654	49.072	−1.521	0.138	−174.492	25.183
X7	0.692	2.582	0.268	0.790	−4.561	5.944
X12	−78.138	36.993	−2.112	0.042	−153.401	−2.876
X11	73.051	9.554	7.646	0.000	53.614	92.489

Step 4: Include X10 and drop X7

SUMMARY OUTPUT

Regression Statistics	
R	0.99
R Sqr	0.99
Adj R Sqr	0.98
SE	8.14
Obs	37.00

ANOVA

	df	*SS*	*MS*	*F*	*Significance F*
Regression	3.00	149591.99	49864.00	752.63	0.00
Residual	33.00	2186.36	66.25		
Total	36.00	151778.35			

	Coefficients	*SE*	*t Stat*	*P-value*	*LB95%*	*UB 95%*
Intercept	−400.194	23.421	−17.087	0.000	−447.843	−352.544
X10	2.510	0.150	16.688	0.000	2.204	2.816
X12	−16.500	12.605	−1.309	0.200	−42.145	9.144
X11	55.650	3.115	17.867	0.000	49.313	61.986

Step 5: Include X13 and drop X12

SUMMARY OUTPUT

Regression Statistics	
R	0.993
R Sqr	0.987
Adj R Sqr	0.986
SE	7.806
Obs	37.00

ANOVA

	df	*SS*	*MS*	*F*	*Significance F*
Regression	3	149767.55	49922.52	819.30	0.00
Residual	33	2010.79	60.93		
Total	36	151778.35			

	Coefficients	*SE*	*t Stat*	*P-value*	*LB95%*	*UB 95%*
Intercept	−425.760	20.502	−20.766	0.000	−467.472	−384.047
X13	−0.109	0.050	−2.178	0.037	−0.210	−0.007
X10	2.735	0.158	17.319	0.000	2.414	3.056
X11	53.104	1.376	38.581	0.000	50.304	55.904

Step 6: Include X1

Regression Statistics	
R	0.994
R Sqr	0.988
Adj R Sqr	0.986
SE	7.67
Obs	37.00

ANOVA

	df	SS	MS	F	Significance F
Regression	4	149898.23	37474.56	637.83	0.00
Residual	32	1880.11	58.75		
Total	36	151778.35			

	Coefficients	SE	t Stat	P-value	LB95%	UB 95%
Intercept	−442.822	23.156	−19.124	0.000	−489.989	−395.655
X1	0.782	0.525	1.491	0.146	−0.286	1.851
X13	−0.140	0.053	−2.626	0.013	−0.249	−0.031
X10	2.759	0.156	17.698	0.000	2.441	3.076
X11	52.267	1.463	35.715	0.000	49.286	55.248

Following the above procedure of inclusion and dropping of variables, one can reach to the following regression equation in which all the variables are having significant coefficients and as such this equation is retained.

Summary output	
Regression statistics	
Multiple R	0.995846
R square	0.991709
Adjusted R square	0.989708
Standard error	6.5873
Observations	37

ANOVA

	d.f.	SS	MS	F	Significance F
Regression	7	150520	21502.85	495.5428	2.07E-28
Residual	29	1258.383	43.39252		
Total	36	151778.3			

	Coefficients	Standard error	t Stat	P-value	LB95%	UB 95
Intercept	−453.939	26.562	−17.090	0.000	−508.264	−399.613
X10	2.844	0.144	19.802	0.000	2.550	3.137
X11	58.218	2.664	21.852	0.000	52.769	63.666
X12	−36.900	11.677	−3.160	0.004	−60.783	−13.017
X16	26.566	8.358	3.179	0.004	9.473	43.660
X17	53.995	15.977	3.380	0.002	21.318	86.672
X18	−5.115	2.033	−2.516	0.018	−9.274	−0.957
X19	−9.297	2.454	−3.788	0.001	−14.316	−4.278

Similarly, following the backward regression starting with the full model multiple regression equation, one can get the result. We shall demonstrate the same using SPSS. In Step 13 one can get the similar multiple linear regression equation as found above.

Step 1: Activate the SPSS data sheet by either transporting or pasting or importing data files.

Step 2: Proceed to Analyze–Regression–Linear menu as shown below.

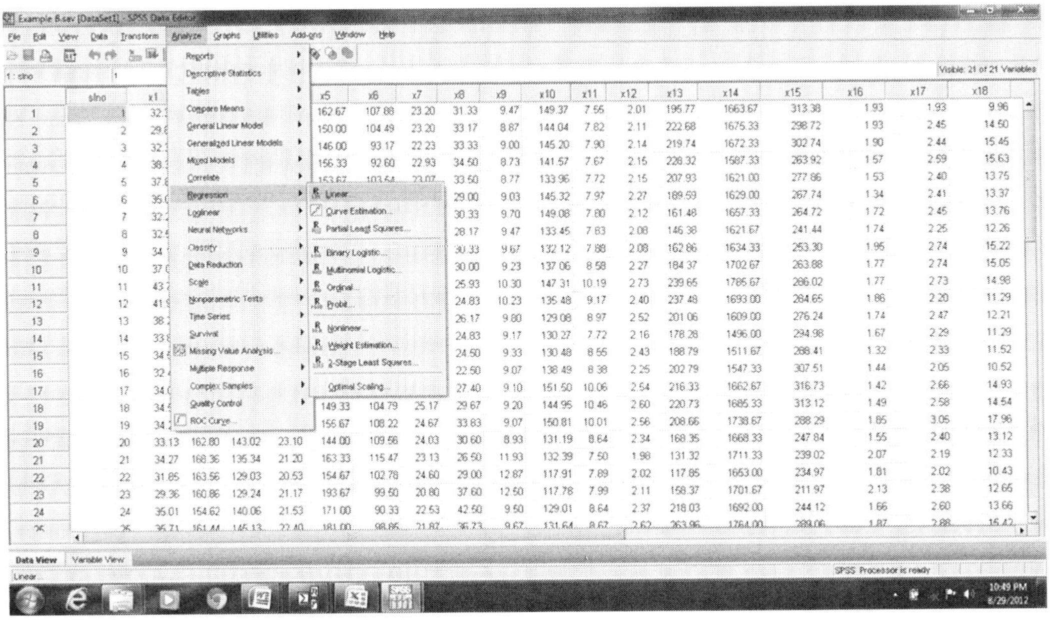

Slide 12.7: Data structure and selection of appropriate regression model from analysis menu of SPSS

Step 3: Select the dependent and the independent variables as shown below and activate Statistics submenu.

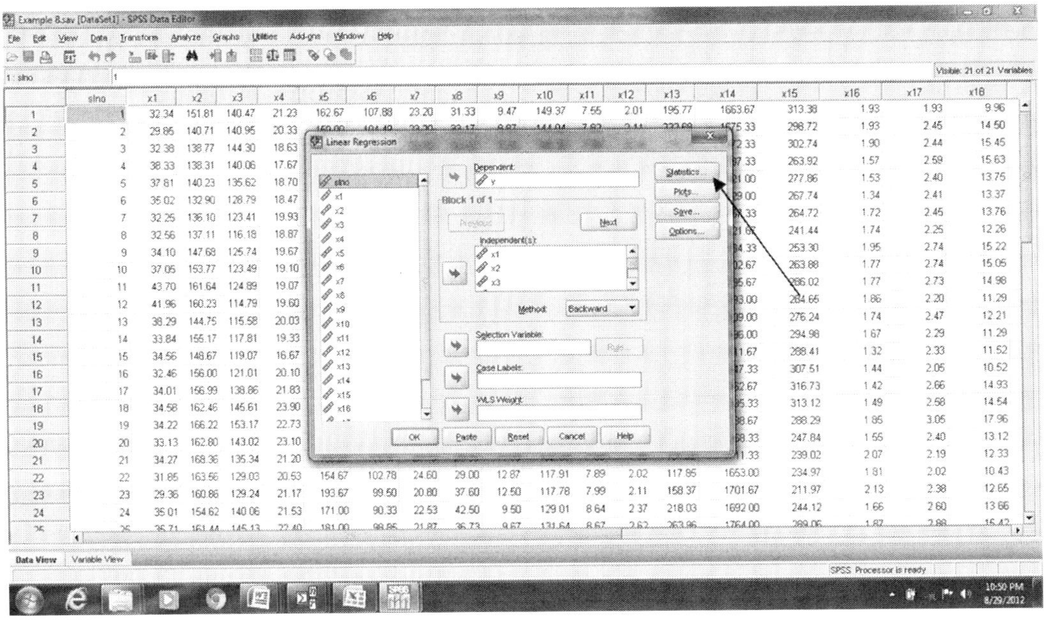

Slide 12.8: Data structure and selection of appropriate variables in Regression submenu from analysis menu of SPSS

Step 4: Get the following window and select the appropriate requirement by ticking the corresponding boxes.

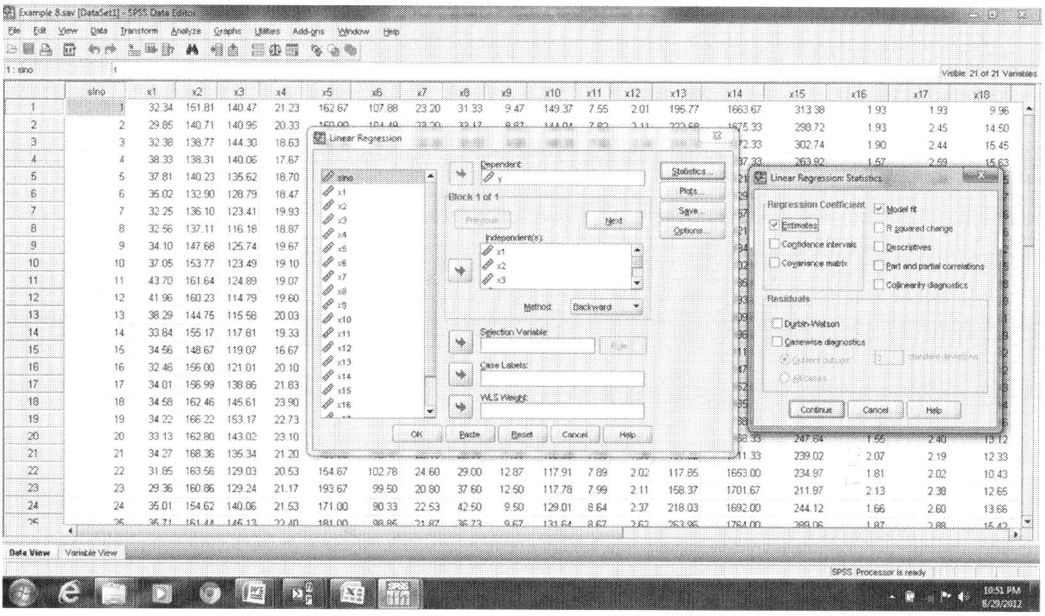

Slide 12.9: Data structure, selection of appropriate variables, and regression statistics in Regression submenu from analysis menu of SPSS

Step 5: Activate the option submenu to get the following window and select the appropriate options.

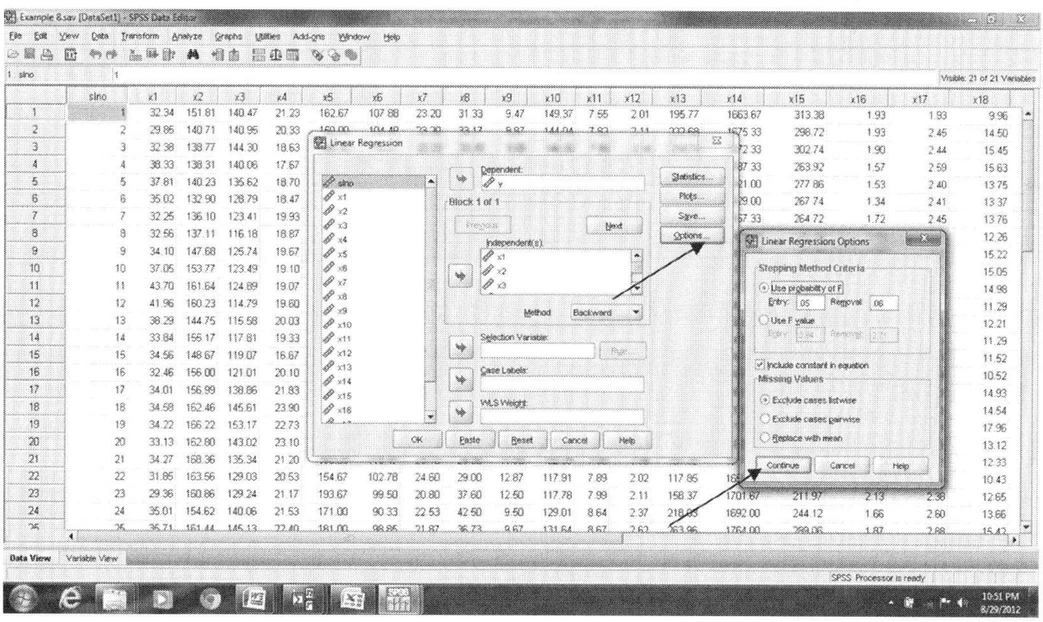

Slide 12.10: Data structure, selection of appropriate variables, and regression options in Regression submenu from analysis menu of SPSS

Step 6: Activate continue followed by OK to get the following results (Table 12.4).

Thus, the ultimate regression equation is the same as was obtained by using MS Excel.

Table 12.4 Backward regression output using SPSS

Variables Entered/Removed[b]

Model	Variables Entered	Variables Removed	Method
1	x19, x4, x12, x9, x6, x5, x10, x2, x1, x18, x7, x3, x15, x14, x13, x16, x8, x11, x17[a]	.	Enter
2	.	x4	Backward (criterion: Probability of F-to-remove >= .060).
3	.	x3	
4	.	x14	
5	.	x1	
6	.	x5	
7	.	x13	
8	.	x15	
9	.	x2	
10	.	x9	
11	.	x7	
12	.	x6	
13	.	x8	

a. All requested variables entered.
b. Dependent Variable: y

Model Summary

Model	R	R Square	Adjusted R Square	Std. Error of the Estimate	Variables in the equation
1	.997[a]	.994	.988	7.09048	a. Predictors: (Constant), x19, x4, x12, x9, x6, x5, x10, x2, x1, x18, x7, x3, x15, x14, x13, x16, x8, x11, x17
2	.997[b]	.994	.989	6.89518	b. Predictors: (Constant), x19, x12, x9, x6, x5, x10, x2, x1, x18, x7, x3, x15, x14, x13, x16, x8, x11, x17
3	.997[c]	.994	.989	6.71770	c. Predictors: (Constant), x19, x12, x9, x6, x5, x10, x2, x1, x18, x7, x15, x14, x13, x16, x8, x11, x17
4	.997[d]	.994	.990	6.55284	d. Predictors: (Constant), x19, x12, x9, x6, x5, x10, x2, x1, x18, x7, x15, x13, x16, x8, x11, x17
5	.997[e]	.994	.990	6.43367	e. Predictors: (Constant), x19, x12, x9, x6, x5, x10, x2, x18, x7, x15, x13, x16, x8, x11, x17
6	.997[f]	.994	.991	6.32285	f. Predictors: (Constant), x19, x12, x9, x6, x10, x2, x18, x7, x15, x13, x16, x8, x11, x17
7	.997[g]	.994	.991	6.23063	g. Predictors: (Constant), x19, x12, x9, x6, x10, x2, x18, x7, x15, x16, x8, x11, x17
8	.997[h]	.994	.991	6.13770	h. Predictors: (Constant), x19, x12, x9, x6, x10, x2, x18, x7, x16, x8, x11, x17
9	.997[i]	.994	.991	6.15038	i. Predictors: (Constant), x19, x12, x9, x6, x10, x18, x7, x16, x8, x11, x17
10	.997[j]	.993	.991	6.16318	j. Predictors: (Constant), x19, x12, x6, x10, x18, x7, x16, x8, x11, x17
11	.996[k]	.993	.991	6.28617	k. Predictors: (Constant), x19, x12, x6, x10, x18, x16, x8, x11, x17
12	.996[l]	.992	.990	6.55662	l. Predictors: (Constant), x19, x12, x10, x18, x16, x8, x11, x17
13	.996[m]	.992	.990	6.60326	m. Predictors: (Constant), x19, x12, x10, x18, x16, x11, x17

ANOVA[n]

Model		Sum of Squares	df	Mean Square	F	Sig.
1	Regression	150924.272	19	7943.383	157.999	.000[a]
	Residual	854.673	17	50.275		
	Total	151778.945	36			
2	Regression	150923.163	18	8384.620	176.357	.000[b]
	Residual	855.782	18	47.543		
	Total	151778.945	36			
3	Regression	150921.522	17	8877.737	196.725	.000[c]
	Residual	857.423	19	45.128		
	Total	151778.945	36			
4	Regression	150920.152	16	9432.509	219.669	.000[d]
	Residual	858.793	20	42.940		
	Total	151778.945	36			
5	Regression	150909.710	15	10060.647	243.057	.000[e]
	Residual	869.235	21	41.392		
	Total	151778.945	36			
6	Regression	150899.419	14	10778.530	269.608	.000[f]
	Residual	879.527	22	39.978		
	Total	151778.945	36			
7	Regression	150886.069	13	11606.621	298.980	.000[g]
	Residual	892.876	23	38.821		
	Total	151778.945	36			
8	Regression	150874.834	12	12572.903	333.753	.000[h]
	Residual	904.111	24	37.671		
	Total	151778.945	36			
9	Regression	150833.266	11	13712.115	362.494	.000[i]
	Residual	945.679	25	37.827		
	Total	151778.945	36			
10	Regression	150791.339	10	15079.134	396.978	.000[j]
	Residual	987.606	26	37.985		
	Total	151778.945	36			
11	Regression	150712.014	9	16745.779	423.773	.000[k]
	Residual	1066.931	27	39.516		
	Total	151778.945	36			
12	Regression	150575.246	8	18821.906	437.828	.000[l]
	Residual	1203.700	28	42.989		
	Total	151778.945	36			
13	Regression	150514.457	7	21502.065	493.132	.000[m]
	Residual	1264.489	29	43.603		
	Total	151778.945	36			

Coefficients[a]

Model		Unstandardized Coefficients		Standardized Coefficients	t	Sig.
		B	Std. Error	Beta		
1	(Constant)	−357.508	75.171		−4.756	.000
	x1	.370	.829	.018	.446	.661
	x2	−.154	.209	−.023	−.735	.472
	x3	−.069	.297	−.012	−.234	.818
	x4	.285	1.922	.008	.149	.884

(continued)

Table 12.4 (continued)

Coefficients[a]						
		Unstandardized Coefficients		Standardized Coefficients		
Model		B	Std. Error	Beta	t	Sig.
	x5	.071	.154	.019	.462	.650
	x6	−.621	.376	−.068	−1.651	.117
	x7	−1.632	1.518	−.054	−1.075	.297
	x8	−1.308	.871	−.090	−1.501	.152
	x9	2.050	2.841	.033	.722	.480
	x10	2.927	.241	.435	12.150	.000
	x11	59.591	5.251	.961	11.349	.000
	x12	−32.486	18.719	−.127	−1.735	.101
	x13	−.074	.093	−.038	−.798	.436
	x14	−.010	.044	−.011	−.232	.820
	x15	.060	.096	.025	.627	.539
	x16	18.444	21.193	.065	.870	.396
	x17	53.480	27.282	.245	1.960	.067
	x18	−4.517	4.045	−.144	−1.117	.280
	x19	−6.919	5.261	−.084	−1.315	.206
2	(Constant)	−356.632	72.876		−4.894	.000
	x1	.304	.682	.015	.446	.661
	x2	−.148	.200	−.022	−.741	.468
	x3	−.042	.224	−.007	−.186	.855
	x5	.064	.143	.017	.450	.658
	x6	−.613	.362	−.067	−1.693	.108
	x7	−1.625	1.475	−.053	−1.102	.285
	x8	−1.292	.840	−.089	−1.537	.142
	x9	1.809	2.265	.029	.798	.435
	x10	2.928	.234	.435	12.497	.000
	x11	60.054	4.108	.968	14.618	.000
	x12	−33.695	16.393	−.131	−2.055	.055
	x13	−.070	.085	−.035	−.816	.425
	x14	−.008	.040	−.008	−.194	.848
	x15	.057	.091	.024	.627	.539
	x16	20.735	14.134	.073	1.467	.160
	x17	55.296	23.718	.253	2.331	.032
	x18	−4.870	3.182	−.155	−1.530	.143
	x19	−7.467	3.649	−.091	−2.046	.056
3	(Constant)	−355.383	70.697		−5.027	.000
	x1	.332	.648	.016	.511	.615
	x2	−.166	.170	−.025	−.976	.341
	x5	.057	.135	.016	.426	.675
	x6	−.623	.350	−.068	−1.782	.091
	x7	−1.696	1.389	−.056	−1.221	.237
	x8	−1.363	.730	−.094	−1.868	.077
	x9	1.869	2.184	.030	.856	.403
	x10	2.914	.216	.433	13.492	.000
	x11	59.960	3.972	.967	15.097	.000
	x12	−33.860	15.948	−.132	−2.123	.047
	x13	−.071	.083	−.036	−.859	.401

(continued)

Table 12.4 (continued)

Coefficients[a]						
		Unstandardized Coefficients		Standardized Coefficients		
Model		B	Std. Error	Beta	t	Sig.
	x14	−.007	.038	−.007	−.174	.864
	x15	.060	.088	.025	.683	.503
	x16	21.082	13.649	.074	1.545	.139
	x17	57.719	19.300	.264	2.991	.008
	x18	−5.182	2.634	−.165	−1.967	.064
	x19	−7.595	3.491	−.093	−2.176	.042
4	(Constant)	−360.069	63.778		−5.646	.000
	x1	.297	.603	.014	.493	.627
	x2	−.164	.166	−.025	−.991	.334
	x5	.065	.125	.018	.520	.609
	x6	−.647	.313	−.071	−2.068	.052
	x7	−1.617	1.281	−.053	−1.263	.221
	x8	−1.409	.663	−.097	−2.124	.046
	x9	1.763	2.045	.029	.862	.399
	x10	2.917	.210	.434	13.918	.000
	x11	59.652	3.470	.962	17.190	.000
	x12	−33.672	15.521	−.131	−2.169	.042
	x13	−.071	.081	−.036	−.884	.387
	x15	.060	.085	.025	.707	.488
	x16	20.393	12.742	.071	1.600	.125
	x17	57.953	18.781	.265	3.086	.006
	x18	−5.246	2.544	−.167	−2.062	.052
	x19	−7.651	3.391	−.093	−2.256	.035
5	(Constant)	−349.457	58.946		−5.928	.000
	x2	−.163	.163	−.024	−1.001	.328
	x5	.061	.122	.017	.499	.623
	x6	−.681	.300	−.074	−2.274	.034
	x7	−1.627	1.257	−.053	−1.294	.210
	x8	−1.519	.613	−.104	−2.476	.022
	x9	1.925	1.981	.031	.972	.342
	x10	2.942	.200	.438	14.739	.000
	x11	59.504	3.394	.959	17.531	.000
	x12	−32.453	15.044	−.127	−2.157	.043
	x13	−.053	.070	−.027	−.753	.460
	x15	.046	.079	.019	.585	.565
	x16	19.683	12.431	.069	1.583	.128
	x17	59.762	18.084	.273	3.305	.003
	x18	−5.506	2.443	−.175	−2.253	.035
	x19	−7.488	3.314	−.091	−2.260	.035
6	(Constant)	−347.982	57.858		−6.014	.000
	x2	−.156	.159	−.023	−.981	.337
	x6	−.654	.289	−.071	−2.259	.034
	x7	−1.750	1.212	−.057	−1.445	.163
	x8	−1.404	.559	−.096	−2.513	.020
	x9	2.148	1.897	.035	1.132	.270
	x10	2.949	.196	.439	15.060	.000
	x11	60.001	3.188	.967	18.818	.000
	x12	−35.185	13.770	−.137	−2.555	.018

(continued)

Table 12.4 (continued)

Coefficients[a]						
		Unstandardized Coefficients		Standardized Coefficients		
Model		B	Std. Error	Beta	t	Sig.
	x13	−.032	.055	−.016	−.578	.569
	x15	.048	.077	.020	.619	.542
	x16	22.238	11.130	.078	1.998	.058
	x17	62.985	16.598	.288	3.795	.001
	x18	−6.078	2.121	−.193	−2.866	.009
	x19	−8.272	2.866	−.101	−2.886	.009
7	(Constant)	−346.763	56.976		−6.086	.000
	x2	−.171	.155	−.026	−1.101	.282
	x6	−.631	.283	−.069	−2.234	.036
	x7	−1.724	1.193	−.057	−1.445	.162
	x8	−1.409	.550	−.097	−2.561	.017
	x9	2.414	1.814	.039	1.331	.196
	x10	2.919	.186	.434	15.674	.000
	x11	60.251	3.113	.971	19.354	.000
	x12	−38.232	12.534	−.149	−3.050	.006
	x15	.040	.075	.017	.538	.596
	x16	22.919	10.906	.080	2.101	.047
	x17	64.111	16.243	.293	3.947	.001
	x18	−6.123	2.088	−.195	−2.932	.007
	x19	−8.651	2.749	−.105	−3.147	.005
8	(Constant)	−335.332	52.078		−6.439	.000
	x2	−.159	.151	−.024	−1.050	.304
	x6	−.644	.277	−.070	−2.321	.029
	x7	−1.931	1.112	−.063	−1.737	.095
	x8	−1.480	.527	−.102	−2.810	.010
	x9	1.984	1.604	.032	1.237	.228
	x10	2.970	.158	.442	18.788	.000
	x11	60.205	3.065	.971	19.640	.000
	x12	−35.667	11.419	−.139	−3.124	.005
	x16	21.941	10.593	.077	2.071	.049
	x17	63.467	15.957	.290	3.977	.001
	x18	−6.324	2.024	−.201	−3.125	.005
	x19	−8.147	2.546	−.099	−3.200	.004
9	(Constant)	−339.343	52.045		−6.520	.000
	x6	−.702	.272	−.077	−2.579	.016
	x7	−1.908	1.114	−.063	−1.712	.099
	x8	−1.457	.527	−.100	−2.764	.011
	x9	1.660	1.577	.027	1.053	.303
	x10	2.977	.158	.443	18.810	.000
	x11	59.433	2.982	.958	19.929	.000
	x12	−34.768	11.410	−.136	−3.047	.005
	x16	17.997	9.926	.063	1.813	.082
	x17	59.673	15.576	.273	3.831	.001
	x18	−5.957	1.997	−.189	−2.982	.006
	x19	−7.821	2.532	−.095	−3.088	.005

(continued)

Table 12.4 (continued)

Coefficients[a]						
		Unstandardized Coefficients		Standardized Coefficients		
Model		B	Std. Error	Beta	t	Sig.
10	(Constant)	−340.110	52.148		−6.522	.000
	x6	−.620	.261	−.068	−2.372	.025
	x7	−1.525	1.055	−.050	−1.445	.160
	x8	−1.272	.498	−.087	−2.555	.017
	x10	2.899	.140	.431	20.665	.000
	x11	60.223	2.892	.971	20.822	.000
	x12	−36.805	11.268	−.144	−3.266	.003
	x16	24.120	8.060	.085	2.993	.006
	x17	57.886	15.515	.265	3.731	.001
	x18	−5.826	1.998	−.185	−2.916	.007
	x19	−8.538	2.444	−.104	−3.494	.002
11	(Constant)	−391.643	38.809		−10.092	.000
	x6	−.424	.228	−.046	−1.860	.074
	x8	−.744	.345	−.051	−2.156	.040
	x10	2.858	.140	.425	20.394	.000
	x11	58.289	2.615	.940	22.289	.000
	x12	−37.182	11.490	−.145	−3.236	.003
	x16	22.678	8.157	.079	2.780	.010
	x17	53.322	15.493	.244	3.442	.002
	x18	−5.176	1.985	−.165	−2.607	.015
	x19	−7.772	2.433	−.095	−3.194	.004
12	(Constant)	−444.651	27.483		−16.179	.000
	x8	−.323	.272	−.022	−1.189	.244
	x10	2.820	.145	.419	19.504	.000
	x11	57.403	2.682	.926	21.403	.000
	x12	−33.111	11.765	−.129	−2.814	.009
	x16	26.011	8.301	.091	3.134	.004
	x17	50.397	16.076	.230	3.135	.004
	x18	−4.647	2.049	−.148	−2.268	.031
	x19	−8.575	2.498	−.105	−3.433	.002
13	(Constant)	−453.177	26.720		−16.960	.000
	x10	2.842	.144	.423	19.687	.000
	x11	58.011	2.652	.935	21.878	.000
	x12	−35.905	11.610	−.140	−3.093	.004
	x16	26.245	8.357	.092	3.140	.004
	x17	53.008	16.039	.242	3.305	.003
	x18	−5.016	2.040	−.160	−2.458	.020
	x19	−9.225	2.455	−.113	−3.758	.001

a. Dependent Variable: y

12.5 Regression vs. Causality

Regression does not necessarily mean cause and effect relationship. Let us consider situations in which two or more variables are related with each other. Can anyone say the variables cause each other? Can anyone assign any direction of causality? *Granger's test for causality* can help in finding which variable is the cause of other one and vice versa. For example, the price of potato can depend not only on its past values but also on the lag values of production area under potato and lag values of area under competing crops grown during the potato season. For the time being, let us restrict to two variables only, that is, price of potato (P_r) and production of potato (P).

$$\begin{aligned} P_{rt} &= \alpha_1 P_1 + \alpha_2 P_2 + \alpha_3 P_3 + \cdots \\ &\quad + \beta_1 P_{r_1} + \beta_2 P_{r_2} + \beta_3 P_{r_3} \cdots + u_{1t} \\ &= \sum_{i=1}^{n} \alpha_i P_{(t-i)} + \sum_{i=1}^{n} \beta_j P_{r(t-j)} + u_{1t} \end{aligned} \quad (12.1)$$

and

$$\begin{aligned} P_t &= \lambda_1 P_{r_1} + \lambda_2 P_{r_2} + \lambda_3 P_{r_3} + \cdots \\ &\quad + \gamma_1 P_1 + \gamma_2 P_2 + \cdots + u_{2t} \\ &= \sum_{i=1}^{m} \lambda_i P_{r(t-i)} + \sum_{j=1}^{m} \gamma_j P_{j(t-j)}. \end{aligned} \quad (12.2)$$

We are to find out whether the (1) price of potato is caused by production of jute, that is, $\sum \alpha_i \neq 0$ and $\sum \gamma_j = 0$; (2) production is caused by the price of potato, that is, $\sum \alpha_i = 0$ and $\sum \lambda_j \neq 0$; (3) two-way causality, that is, both price causes production and production causes price, that is, coefficients are statistically significantly different from zero in both regressions; and (4) existence of no causality, that is, coefficients are not statistically significant from zero.

Solution. The step-by-step procedure is as follows:

1. Regress current price on all lagged price only and get residual sum of square (RSS_1).
2. Regress current price on all lagged price and include lagged productions also. Get RSS_2.
3. Calculate $F = \dfrac{(\text{RSS}_1 - \text{RSS}_2)/m}{\text{RSS}_2/(n-k)}$ with m, $n-k$ d.f., where m is the number of lagged productions and k is the number of parameters estimated in Step 2.
4. If Cal $F >$ Tab F, then H_0: $\sum \alpha_i = 0$ is rejected, that is, production causes price.
5. Repeat the Steps 1–4 with the mode 2, that is, to check whether $P_r \rightarrow P$ and conclude accordingly.

12.6 Partial Correlation

As has already been mentioned, the effect of variables on other variables may not be the same under multiple variable condition than what is found in bivariate condition. The interaction effects among the variables play great role in depicting one-to-one association, measured in terms of simple correlation coefficient also. Thus, the simple correlation coefficient under multiple variable condition may not portray the exact picture. Before calculation of correlation coefficient between any two variables under multiple variable condition, one must eliminate the effects of other variables on both variables under consideration. Precisely, we would like to know what the correlation would be between Y and X_1, say, after eliminating the effects of all other variables such as $X_2, X_3, \ldots, X_k$ on both of these variables. This is called the partial correlation of Y and X_1, and the coefficient is written as $r_{y1.23\ldots k} = r_{01.23\ldots k}$. It is in general differs from the ordinary correlation coefficient r_{01} for Y and X_1. Using the correlation matrix $\Re$, one can work out the partial correlation as follows:

$$r_{01.2} = \frac{-R_{01}}{(R_{00}R_{11})^{\frac{1}{2}}}, \text{ where}$$

$$\Re = \begin{pmatrix} r_{00} & r_{00} & r_{02} \\ r_{10} & r_{11} & r_{12} \\ r_{20} & r_{21} & r_{22} \end{pmatrix}$$

$$= \begin{pmatrix} 1 & r_{01} & r_{02} \\ r_{10} & 1 & r_{12} \\ r_{20} & r_{21} & 1 \end{pmatrix} \text{ for three variable case}$$

and this can be generalized for k variables X_1, $X_2, \ldots, X_k$ as well.

Thus, partial correlation coefficient between ith and jth variables $r_{ij.123\ldots k} = -\frac{R_{ij}}{(R_{ii}R_{jj})^{\frac{1}{2}}}, \quad i \neq j$, where the correlation matrix now has $(k+1)$ rows and columns and R_{ij}, R_{ii}, and R_{jj}, has usual meanings as described earlier. It is to be noted that $-1 \leq r_{ij.123\ldots k} \leq 1$.

Though, using MS Excel package, we may not directly get the partial correlation coefficients, but packages like SPSS and SAS provide partial correlation coefficient.

Let us take the Example 12.1 of this chapter to demonstrate the process of getting partial correlation coefficient using SPSS.

The following table gives the information on ten varieties of mulberry. Find out the partial correlation coefficients of leaf yield on other three yield attributing characters.

Leaf area (cm^2): X_1	94.61	42.3	98.15	72.61	174.45	115.62	78.41	133.74	112.94	148.28
Total shoot length(cm): X_2	525.56	345	711.11	307.56	529.44	631.67	526.67	442.22	482.33	422.78
Leaf moisture(%): X_3	79.35	58.41	73.67	74.6	81.21	82.83	78.66	80.76	80.66	78.26
Leaf yield/plant(g): Y	255.56	97.78	237.78	161.11	263.33	237.78	217.78	224.44	228.89	190

Step 1: Up on transferring the data to SPSS editor, go to Analyze, followed by Correlate and Partial, as shown below:

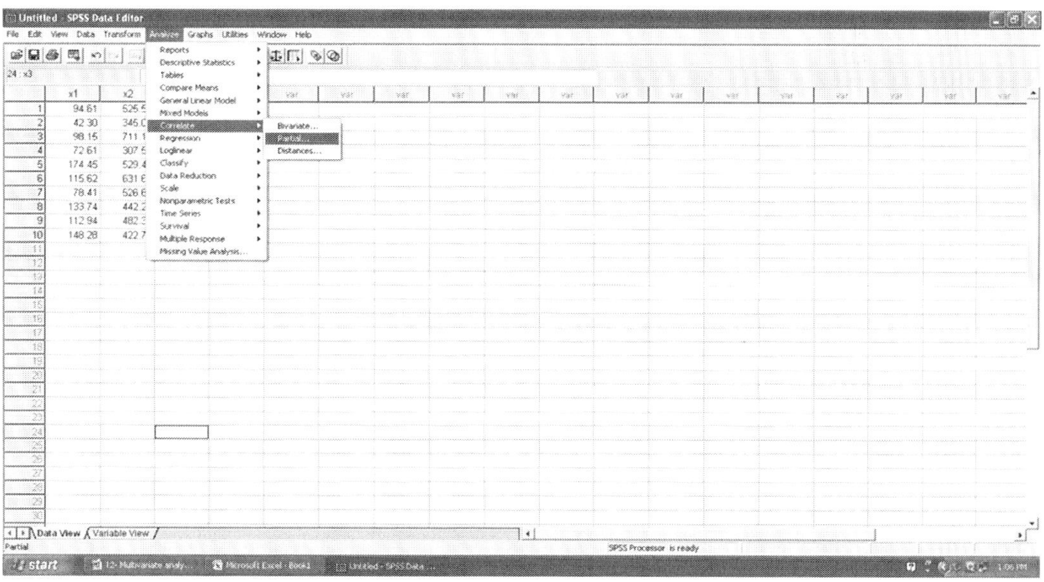

Slide 12.11: Data structure and selection of appropriate correlation submenu in analysis menu of SPSS

Step 2: Select the variables for which partial correlation is to be worked out and also the variables for which effects are to be eliminated, as shown below.

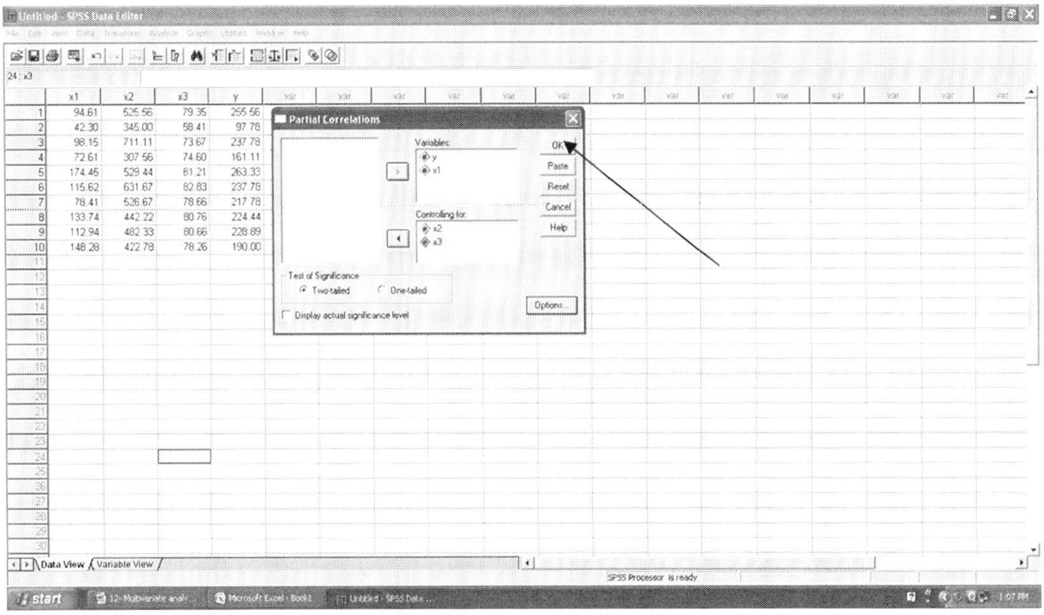

Slide 12.12: Data structure and selection of appropriate variables in correlation submenu in analysis menu of SPSS

Step 3: Click OK to get the following output. In the similar fashion, by changing the variables in Step 2, one can have different combinations of partial correlation coefficients among the variables. Given below are the different partial correlation coefficients of dependent variable *Y* with other variables as worked out following the above process (Table 12.5).

Table 12.5 Output table of partial correlation analysis using SPSS

To get the following partial correlation coefficients:

- - - P A R T I A L C O R R E L A T I O N C O E F F I C I E N T S - - -

Controlling for: X2 X3

	Y	X1
Y	1.0000	.2488
X1	.2488	1.0000

- - - P A R T I A L C O R R E L A T I O N C O E F F I C I E N T S - - -

Controlling for.. X3 X1

	Y	X2
Y	1.0000	.7890*
X2	.7890*	1.0000

- - - P A R T I A L C O R R E L A T I O N C O E F F I C I E N T S - - -

Controlling for: X1 X2

	Y	X3
Y	1.0000	.7778*
X3	.7778*	1.0000

* - Signif. LE .05 ** - Signif. LE .01 (2-tailed)

" . " is printed if a coefficient cannot be computed

Let us consider the simple correlation coefficients among the variables under consideration as given below:

		X1	X2	X3	Y
X1	Pearson correlation	1	.282	.708(*)	.656(*)
	Sig. (2-tailed)	.	.430	.022	.039
	N	10	10	10	10
X2	Pearson correlation	.282	1	.417	.724(*)
	Sig. (2-tailed)	.430	.	.231	.018
	N	10	10	10	10
X3	Pearson correlation	.708(*)	.417	1	.854(**)
	Sig. (2-tailed)	.022	.231	.	.002
	N	10	10	10	10
Y	Pearson correlation	.656(*)	.724(*)	.854(**)	1
	Sig. (2-tailed)	.039	.018	.002	.
	N	10	10	10	10

*Correlation is significant at the 0.05 level (2-tailed)
**Correlation is significant at the 0.01 level (2-tailed)

Thus, from the above, it is clear that in all the three cases, the simple correlation coefficients are different from the partial correlation coefficients of yield and other three variables, individually.

Partial correlation coefficient can also be worked out using SAS.

12.7 Canonical Correlation

During regression analysis, generally one dependent variable is taken at a time to find out its relationship with independent variables. But in many situations, the researchers need to consider a group of dependent and independent variables. The researchers become interested in getting the relationship between a group of dependent variables and a group of independent variables. Canonical correlation analysis facilitates in getting the interrelationship between these two groups of variables. Canonical correlation is a powerful multivariate technique which provides the information of higher quality and in a more interpretable manner. Canonical analysis suggests the number of ways in which two groups (independent and dependent) of variables are related, their strength of linear relationship, and the nature of the relationship which otherwise might be unmanageable by judging huge number of bivariate correlations between sets of variables.

In social or agricultural studies, the innovation index, motivation index, adoption index, etc., of farmers are dependent on age, gender, education, income, family size, etc. Multiple regression equation of innovation index with age, gender, education, income, family size, etc., can predict the innovation index only. Similarly one can predict motivation index and/or adoption index using the independent variables mentioned above. But if the researcher wants to evaluate/identify/index farmers taking all the three indices at a time by assessing the relationship of these groups of dependent variables with the independent variables' group consisting of age, gender, education, income, family size, etc., then canonical correlation analysis could be the possible way out.

Canonical correlation measures the intensity of relationship between the linear combinations of the dependent variables with those of the independent variables. The weighted linear combinations of two or more (either for independent or for dependent) variables are known as *canonical variates* (also known as linear composites, linear compounds). *Thus, the canonical correlation measures the strength of relationship between two canonical variates (two sets of variables).* The canonical correlation analysis searches for optimum structure of each set of variables that maximizes the intensity of relationship between the dependent and independent variable sets. Actually, during the above process, the process develops the number of independent canonical functions that maximizes the correlation between the canonical variates. The characteristic feature of the canonical correlation is that the canonical weights are derived in such a way so as to maximize the correlation between the canonical variates.

In canonical analysis, one comes across with the following terminologies (X_i's are the independent variables and Y_j's are the dependent variables):

(a) *Canonical loadings*: It is the simple correlation between the individual independent variables and the canonical variates made out of the independent variables. That means $r_{x_i \text{ vs. } x_{cv}}$ canonical loading measures the simple linear correlation between an observed variable and the sets of canonical variates. Thus, canonical loading reflects the variance that the variable shares with canonical variate, that is, its relative importance to the canonical variate; the larger the value, the greater the importance.

(b) *Canonical cross loading*: As the name suggests, it is the correlation between the independent/dependent variables and the opposite canonical variates, that is, canonical variates of dependent variables/canonical variates of independent variables, respectively. That means $r_{x_i \text{ vs. } y_{cv}}$ or $r_{y_j \text{ vs. } x_{cv}}$. As such cross loadings provide a more direct measure of the dependent–independent variable relationship.

(c) *Canonical function*: The correlational relationships between two canonical variates are known as canonical function. The two canonical variates are made of independent and dependent variables separately. The canonical functions are derived in such a way that these are mutually independent of each other, that is, orthogonal.

(d) *Canonical root*: The square of the canonical correlation between two canonical variates is known as canonical root. This is also known as eigenvalue. Actually canonical root estimates the amount of variations shared between the two optimally weighted canonical variates of the sets of independent and dependent variables.

(e) *Redundancy index*: It is the amount of variance of a canonical variate explained by the other canonical variate in the canonical function. For example, redundancy index of canonical variate of dependent variables is the amount of variance of dependent canonical variate explained by the canonical variates of independent variates.

Canonical correlation analysis is generally aimed at *three objectives*: (1) assessing the intensity of linear relationship between two sets of variables (dependent and independent) measured from the same elements, (2) working out the weights of the linear combinations, that is, canonical variates in such a way that each set will have maximum correlation, and (3) assessing the contribution of each variable to the canonical function.

Sample size plays a great role in canonical analysis. Small sample size fails to represent the correlation adequately, while very large sample may lead to a very small correlation significant at every instance. As a guideline, the researcher is advised to maintain at least ten observations per variable, though there is no or little importance of variables to be included in the dependent or independent sets, because of the maximization of correlation during productions of canonical variates that affect the entire process. Here lies the importance of selection of variables in the canonical variates. It is always welcome that the researcher should have conceptua understanding in linking the sets for variables before canonical correlation analysis.

Various ways of interpreting the canonical correlations have been propounded. As most of the multivariate analyses require computer software and the output generated may vary among these packages, among these the cross loading approach is mostly preferred, if not available, either to calculate these manually or to use canonical loadings for interpretation.

Example 12.3. Using the same data set of 37 varieties for 19 characters and taking X1–X10 as independent and X11–X19 as dependent variables, let us see the canonical correlation analysis through SAS.

Step 1: Import/transport/copy the data set as given below in two slides.

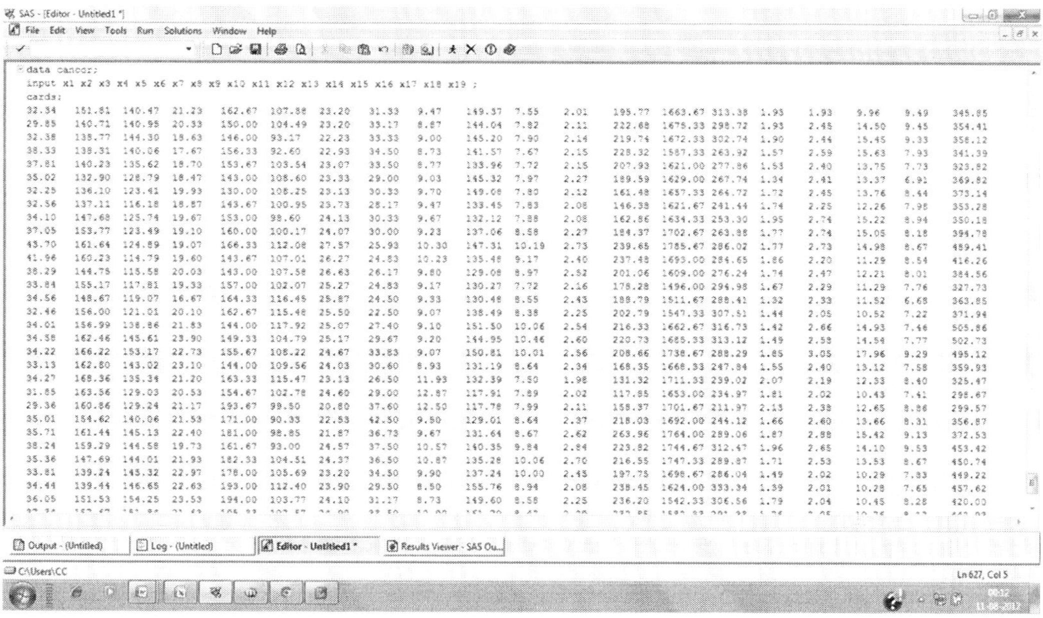

```
data cancor;
input x1 x2 x3 x4 x5 x6 x7 x8 x9 x10 x11 x12 x13 x14 x15 x16 x17 x18 x19 ;
cards;
32.34 151.81 140.47 21.23 162.67 107.88 23.20 31.33 9.47 149.37 7.55 2.01 195.77 1663.67 313.38 1.95 1.93 9.96 9.49 345.85
29.85 140.71 140.95 20.33 150.00 104.49 23.20 33.17 8.87 144.04 7.82 2.11 222.68 1675.33 298.72 1.93 2.45 14.50 9.45 354.41
32.38 138.77 144.30 18.63 146.00 93.17 22.23 33.33 9.00 145.20 7.90 2.14 219.74 1672.33 302.74 1.90 2.44 15.45 9.33 358.12
38.33 138.31 140.06 17.67 156.33 92.60 22.93 34.50 8.73 141.57 7.67 2.15 228.32 1587.33 263.92 1.57 2.59 15.63 7.93 341.39
37.81 140.23 135.62 18.70 153.67 103.54 23.07 33.50 8.77 133.96 7.72 2.15 207.93 1621.00 277.86 1.53 2.40 13.75 7.73 323.82
35.02 132.90 128.79 18.47 143.00 108.60 23.33 29.00 9.03 145.32 7.97 2.27 189.59 1629.00 267.74 1.34 2.41 13.37 6.91 369.82
32.25 136.10 123.41 19.93 130.00 108.25 23.13 30.33 9.70 149.08 7.80 2.12 161.48 1657.33 264.72 1.72 2.45 13.76 8.44 373.14
32.56 137.11 116.18 18.87 143.67 100.95 23.73 28.17 9.47 133.45 7.83 2.08 146.38 1621.67 241.44 1.74 2.25 12.26 7.95 353.28
34.10 147.68 125.74 19.67 153.00 98.60 24.13 30.33 9.67 132.12 7.88 2.08 162.86 1634.33 253.30 1.95 2.74 15.22 8.94 350.18
37.05 153.77 123.49 19.10 160.00 100.17 24.07 30.00 9.23 137.06 8.58 2.27 184.37 1702.67 263.88 1.77 2.74 15.05 8.18 394.78
43.70 161.64 124.89 19.07 166.33 112.08 27.57 25.93 10.30 147.31 10.19 2.73 239.65 1785.67 286.02 1.77 2.73 14.98 8.67 459.41
41.96 160.23 114.79 19.60 143.67 107.01 26.27 24.83 10.23 135.48 9.17 2.40 237.48 1693.00 284.65 1.86 2.20 11.29 8.54 416.26
38.29 144.75 115.58 20.03 143.00 107.58 26.63 26.17 9.80 129.08 8.97 2.52 201.06 1609.00 276.24 1.74 2.47 12.21 8.01 384.56
33.84 155.17 117.81 19.33 157.00 102.07 25.27 24.83 9.17 130.27 7.72 2.16 178.28 1496.00 294.98 1.67 2.29 11.29 7.76 327.73
34.56 148.67 119.07 16.67 164.33 116.45 25.87 24.50 9.33 130.48 8.55 2.43 188.79 1511.67 288.41 1.32 2.33 11.52 6.68 363.85
32.46 156.00 121.01 20.10 162.67 115.48 25.50 22.50 9.07 138.49 8.38 2.25 202.79 1547.33 307.51 1.44 2.05 10.52 7.22 371.94
34.01 156.99 138.86 21.83 144.00 117.92 25.07 27.40 9.10 151.50 10.06 2.54 216.33 1662.67 316.73 1.42 2.66 14.93 7.46 505.86
34.58 162.46 145.61 23.90 149.33 104.79 25.17 29.67 9.20 144.95 10.46 2.60 220.73 1685.33 313.12 1.49 2.58 14.54 7.77 502.73
34.22 166.22 153.17 22.73 155.67 108.22 24.67 33.83 9.07 150.81 10.01 2.56 208.66 1738.67 288.29 1.85 3.05 17.96 9.29 495.12
33.13 162.80 143.02 23.10 144.00 109.56 24.03 30.60 8.93 131.19 8.64 2.34 168.35 1668.33 247.84 1.55 2.40 13.12 7.58 359.93
34.27 168.36 135.34 21.20 163.33 115.47 23.13 26.50 11.93 132.39 7.50 1.98 131.32 1711.33 239.02 2.07 2.19 12.33 8.40 325.47
31.85 163.56 129.03 20.53 154.67 102.78 24.60 29.00 12.87 117.91 7.89 2.02 117.85 1653.00 234.97 1.81 2.02 10.43 7.41 298.67
29.36 160.86 129.24 21.17 193.67 99.50 20.80 37.60 12.50 117.78 7.99 2.11 158.37 1701.67 211.97 2.13 2.38 12.65 8.86 299.57
35.01 154.62 140.06 21.53 171.00 90.33 22.53 42.50 9.50 129.01 8.64 2.37 218.03 1692.00 244.12 1.66 2.60 13.66 8.31 356.87
35.71 161.44 145.13 22.40 181.00 98.85 21.87 36.73 9.67 131.64 8.67 2.62 263.96 1764.00 289.06 1.87 2.88 15.42 9.13 372.53
38.24 159.29 144.58 19.73 161.67 93.00 24.57 37.50 10.57 140.35 9.84 2.84 223.82 1744.67 312.47 1.96 2.65 14.10 9.53 453.42
35.36 147.69 144.01 21.93 182.33 104.51 24.37 36.50 10.87 135.28 10.06 2.70 216.55 1747.33 289.87 1.71 2.53 13.53 8.67 450.74
33.81 139.24 145.32 22.97 178.00 105.69 23.20 34.50 9.90 137.24 10.00 2.45 197.75 1698.67 286.04 1.49 2.02 10.29 7.83 449.22
34.44 139.44 146.65 22.63 193.00 112.40 23.90 29.50 8.50 155.76 8.94 2.08 238.45 1624.00 333.34 1.39 2.01 10.28 7.65 457.62
36.05 151.53 154.25 23.53 194.00 103.77 24.10 31.17 8.73 149.60 8.58 2.25 236.20 1542.33 306.56 1.79 2.04 10.45 8.28 420.00
```

Slide 12.13: Data structure for canonical analysis using SAS

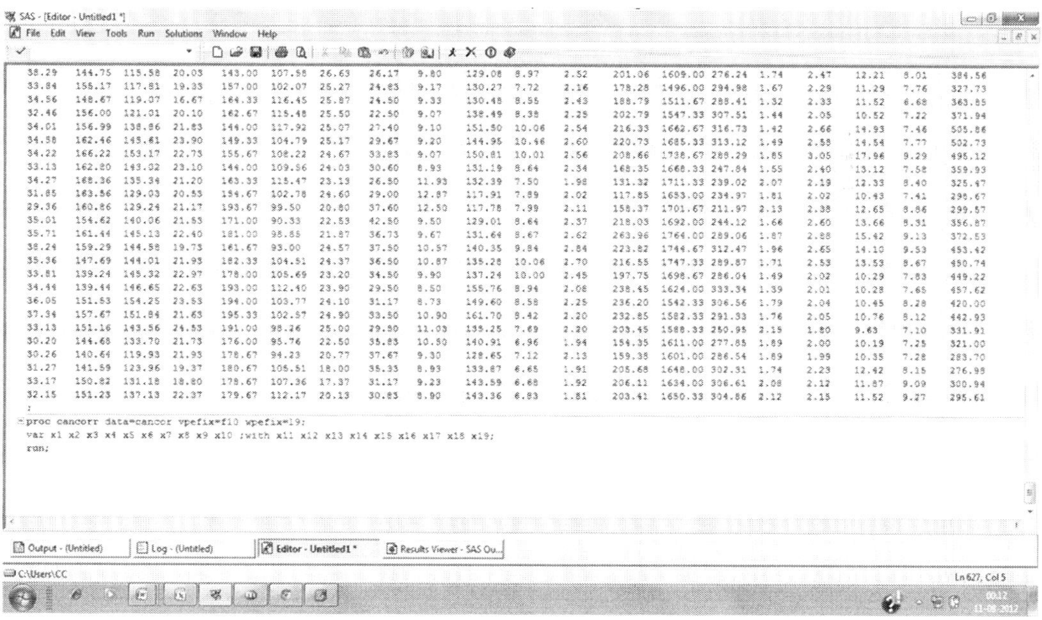

```
38.29 144.75 115.58 20.03 143.00 107.58 26.63 26.17 9.80 129.08 8.97 2.52 201.06 1609.00 276.24 1.74 2.47 12.21 8.01 384.56
33.84 155.17 117.81 19.33 157.00 102.07 25.27 24.83 9.17 130.27 7.72 2.16 178.28 1496.00 294.98 1.67 2.29 11.29 7.76 327.73
34.56 148.67 119.07 16.67 164.33 116.45 25.87 24.50 9.33 130.48 8.55 2.43 188.79 1511.67 288.41 1.32 2.33 11.52 6.68 363.85
32.46 156.00 121.01 20.10 162.67 115.48 25.50 22.50 9.07 138.49 8.38 2.25 202.79 1547.33 307.51 1.44 2.05 10.52 7.22 371.94
34.01 156.99 138.86 21.83 144.00 117.92 25.07 27.40 9.10 151.50 10.06 2.54 216.33 1662.67 316.73 1.42 2.66 14.93 7.46 505.86
34.58 162.46 145.61 23.90 149.33 104.79 25.17 29.67 9.20 144.95 10.46 2.60 220.73 1685.33 313.12 1.49 2.58 14.54 7.77 502.73
34.22 166.22 153.17 22.73 155.67 108.22 24.67 33.83 9.07 150.81 10.01 2.56 208.66 1738.67 288.29 1.85 3.05 17.96 9.29 495.12
33.13 162.80 143.02 23.10 144.00 109.56 24.03 30.60 8.93 131.19 8.64 2.34 168.35 1668.33 247.84 1.55 2.40 13.12 7.58 359.93
34.27 168.36 135.34 21.20 163.33 115.47 23.13 26.50 11.93 132.39 7.50 1.98 131.32 1711.33 239.02 2.07 2.19 12.33 8.40 325.47
31.85 163.56 129.03 20.53 154.67 102.78 24.60 29.00 12.87 117.91 7.89 2.02 117.85 1653.00 234.97 1.81 2.02 10.43 7.41 298.67
29.36 160.86 129.24 21.17 193.67 99.50 20.80 37.60 12.50 117.78 7.99 2.11 158.37 1701.67 211.97 2.13 2.38 12.65 8.86 299.57
35.01 154.62 140.06 21.53 171.00 90.33 22.53 42.50 9.50 129.01 8.64 2.37 218.03 1692.00 244.12 1.66 2.60 13.66 8.31 356.87
35.71 161.44 145.13 22.40 181.00 98.85 21.87 36.73 9.67 131.64 8.67 2.62 263.96 1764.00 289.06 1.87 2.88 15.42 9.13 372.53
38.24 159.29 144.58 19.73 161.67 93.00 24.57 37.50 10.57 140.35 9.84 2.84 223.82 1744.67 312.47 1.96 2.65 14.10 9.53 453.42
35.36 147.69 144.01 21.93 182.33 104.51 24.37 36.50 10.87 135.28 10.06 2.70 216.55 1747.33 289.87 1.71 2.53 13.53 8.67 450.74
33.81 139.24 145.32 22.97 178.00 105.69 23.20 34.50 9.90 137.24 10.00 2.45 197.75 1698.67 286.04 1.49 2.02 10.29 7.83 449.22
34.44 139.44 146.65 22.63 193.00 112.40 23.90 29.50 8.50 155.76 8.94 2.08 238.45 1624.00 333.34 1.39 2.01 10.28 7.65 457.62
36.05 151.53 154.25 23.53 194.00 103.77 24.10 31.17 8.73 149.60 8.58 2.25 236.20 1542.33 306.56 1.79 2.04 10.45 8.28 420.00
37.34 157.67 151.84 21.63 195.33 102.57 24.90 33.50 10.90 161.70 8.42 2.20 232.85 1582.33 291.33 1.76 2.05 10.76 8.12 442.93
33.13 151.16 143.56 24.53 191.00 98.26 25.00 29.50 11.03 135.25 7.69 2.20 203.45 1588.33 250.95 2.15 1.80 9.63 7.10 331.91
30.20 144.68 133.70 21.73 176.00 95.76 22.50 35.83 10.50 140.91 6.96 1.94 154.35 1611.00 277.85 1.89 2.00 10.19 7.25 321.00
30.26 140.64 119.93 21.93 178.67 94.23 20.77 37.67 9.30 128.65 7.12 2.13 159.35 1601.00 286.54 1.89 1.99 10.35 7.28 283.70
31.27 141.59 123.96 19.37 180.67 105.51 18.00 35.33 8.93 133.87 6.65 1.91 205.68 1648.00 302.31 1.74 2.23 12.42 8.15 276.99
33.17 150.82 131.18 18.80 178.67 107.36 17.37 31.17 9.23 143.59 6.68 1.92 206.11 1634.00 306.61 2.08 2.12 11.87 9.09 300.94
32.15 151.23 137.13 22.37 179.67 112.17 20.13 30.83 8.90 143.36 6.83 1.81 203.41 1650.33 304.86 2.12 2.15 11.52 9.27 295.61
;
proc cancorr data=cancor vpefix=f10 wpefix=19;
var x1 x2 x3 x4 x5 x6 x7 x8 x9 x10 ;with x11 x12 x13 x14 x15 x16 x17 x18 x19;
run;
```

Slide 12.14: Data structure and commands for canonical analysis using SAS

Step 2: Write the command along with specification of groups of variables, as shown above.

Step 3: Get the following output.

SAS - [Results Viewer - SAS Output]

cluster analysis using plant data

The CANCORR Procedure

Canonical Correlation Analysis

					Eigenvalues of Inv(E)*H = CanRsq/(1-CanRsq)				Test of H0: The canonical correlations in the current row and all that fol			
	Canonical Correlation	Adjusted Canonical Correlation	Approximate Standard Error	Squared Canonical Correlation	Eigenvalue	Difference	Proportion	Cumulative	Likelihood Ratio	Approximate F Value	Num DF	Den DF
1	0.959219	0.937766	0.013316	0.920102	11.5159	7.8867	0.5161	0.5161	0.00028062	3.44	90	132.34
2	0.885427	.	0.036003	0.783981	3.6292	0.3409	0.1627	0.6788	0.00351222	2.62	72	123.15
3	0.875677	.	0.038865	0.766810	3.2883	1.8531	0.1474	0.8261	0.01625883	2.32	56	113.01
4	0.767704	0.620758	0.068438	0.589369	1.4353	0.2510	0.0643	0.8905	0.06972350	1.86	42	101.95
5	0.736328	.	0.076304	0.542178	1.1843	0.4847	0.0531	0.9435	0.16979601	1.67	30	90
6	0.641573	0.562726	0.098064	0.411616	0.6996	0.1894	0.0314	0.9749	0.37087805	1.35	20	77.232
7	0.581217	.	0.110365	0.337813	0.5101	0.4687	0.0229	0.9978	0.63033282	1.01	12	63.79
8	0.199413	-.001665	0.160039	0.039766	0.0414	0.0327	0.0019	0.9996	0.95189545	0.21	6	50
9	0.093190	.	0.165219	0.008684	0.0088		0.0004	1.0000	0.99131564	0.11	2	26

Output - (Untitled) | Log - (Untitled) | Editor - Untitled1 * | Results Viewer - SAS ...

C:\Users\CC

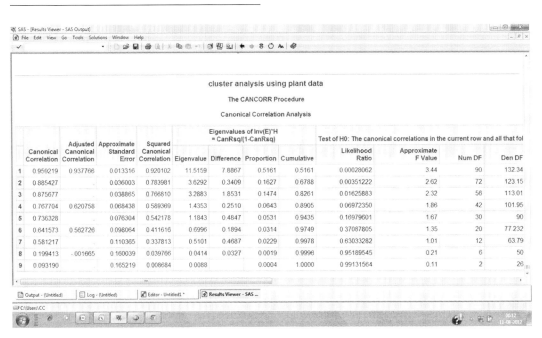

Slide 12.15: Portion of output for canonical analysis using SAS

SAS - [Results Viewer - SAS Output]

cluster analysis using plant data

The CANCORR Procedure

Canonical Correlation Analysis

Raw Canonical Coefficients for the VAR Variables

	f101	f102	f103	f104	f105	f106	f107	f108	f109
x1	0.0922716405	0.1231248396	-0.013298848	0.2662156148	-0.027565205	-0.057408152	-0.169023582	-0.220173537	-0.05006035
x2	-0.002568239	0.0124959285	-0.005799281	0.0120443247	-0.030599373	0.0125605159	0.1301999631	0.0299331986	-0.034194084
x3	-0.044727034	0.0130090617	0.0003017765	-0.002926591	0.0810108228	-0.071273623	-0.052389708	0.0166493818	-0.096199051
x4	0.3905633115	-0.085396782	0.108461776	0.1640227792	0.1028869768	0.2352024805	-0.060761498	-0.636246423	0.1849110668
x5	0.0179238805	0.001804537	0.0429243274	-0.008715978	-0.035539014	-0.012662915	-0.007011087	0.040468309	-0.005308105
x6	-0.008923622	0.0649043005	-0.060424474	0.0112470401	-0.022835848	0.1590498057	-0.09848791	0.0636442884	-0.043459889
x7	0.0929235965	0.4007524738	-0.016208788	-0.465350717	-0.06849391	0.0499995867	0.0427734387	0.3525134943	0.2466957784
x8	-0.004036921	0.1871495786	-0.196298498	0.077000099	-0.128916094	0.2248627327	0.0020358073	0.1603980411	0.1857472256
x9	0.4355253179	-0.480966733	-0.382511278	0.2303285325	0.6646642778	-0.008403049	-0.34019932	0.2617953023	0.2626040519
x10	-0.010943144	-0.024225738	0.0421221247	0.0294962809	0.0266927298	0.0444712638	0.0760326114	0.0348985115	0.1011418681

Raw Canonical Coefficients for the WITH Variables

	l91	l92	l93	l94	l95	l96	l97	l98	l99

Output - (Untitled) | Log - (Untitled) | Editor - Untitled1 * | Results Viewer - SAS ...

C:\Users\CC

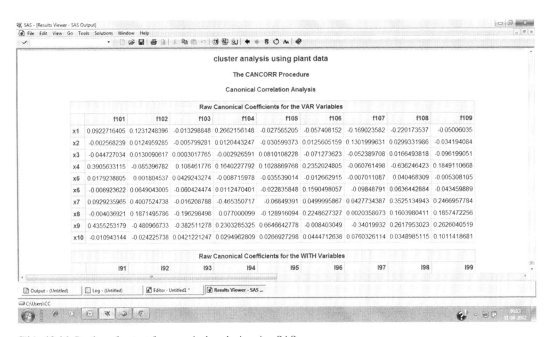

Slide 12.16: Portion of output for canonical analysis using SAS
(Canonical cross loading of first set of variables with opposite canonical variables)

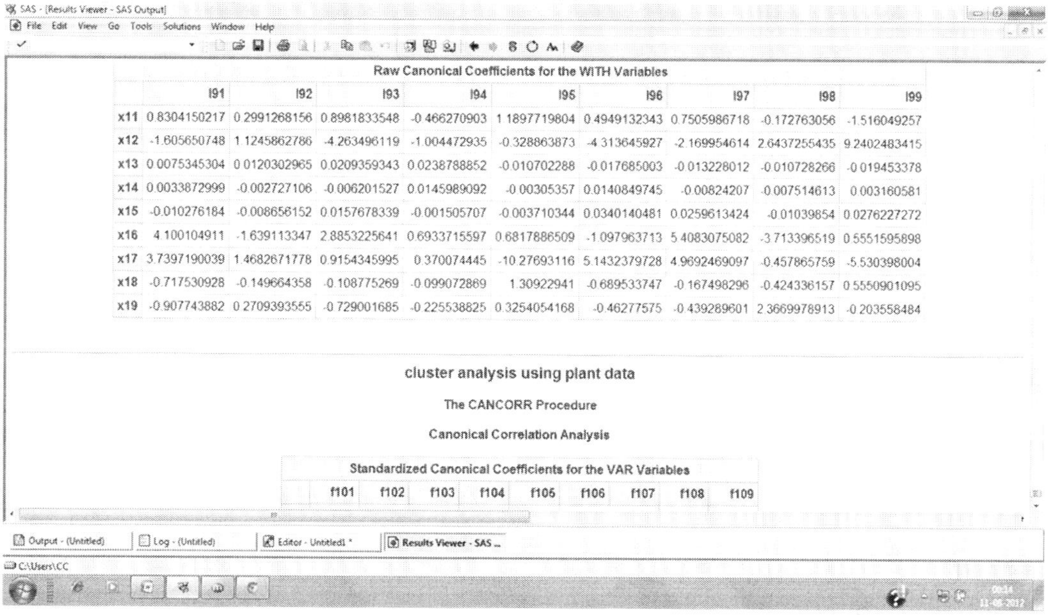

Raw Canonical Coefficients for the WITH Variables

	l91	l92	l93	l94	l95	l96	l97	l98	l99
x11	0.8304150217	0.2991268156	0.8981833548	-0.466270903	1.1897719804	0.4949132343	0.7505986718	-0.172763056	-1.516049257
x12	-1.605650748	1.1245862786	-4.263496119	-1.004472935	-0.328863873	-4.313645927	-2.169954614	2.6437255435	9.2402483415
x13	0.0075345304	0.0120302965	0.0209359343	0.0238788852	-0.010702288	-0.017685003	-0.013228012	-0.010728266	-0.019453378
x14	0.0033872999	-0.002727106	-0.006201527	0.0145989092	-0.00305357	0.0140849745	-0.00824207	-0.007514613	0.003160581
x15	-0.010276184	-0.008656152	0.0157678339	-0.001505707	-0.003710344	0.0340140481	0.0259613424	-0.01039854	0.0276227272
x16	4.100104911	-1.639113347	2.8853225641	0.6933715597	0.6817886509	-1.097963713	5.4083075082	-3.713396519	0.5551595898
x17	3.7397190039	1.4682671778	0.9154345995	0.370074445	-10.27693116	5.1432379728	4.9692469097	-0.457865759	-5.530398004
x18	-0.717530928	-0.149664358	-0.108775269	-0.099072869	1.30922941	-0.689533747	-0.167498296	-0.424336157	0.5550901095
x19	-0.907743882	0.2709393555	-0.729001685	-0.225538825	0.3254054168	-0.46277575	-0.439289601	2.3669978913	-0.203558484

cluster analysis using plant data

The CANCORR Procedure

Canonical Correlation Analysis

Standardized Canonical Coefficients for the VAR Variables

	f101	f102	f103	f104	f106	f106	f107	f108	f109

Slide 12.17: Portion of output for canonical analysis using SAS
(Canonical cross loading of second set of variables with opposite canonical variables)

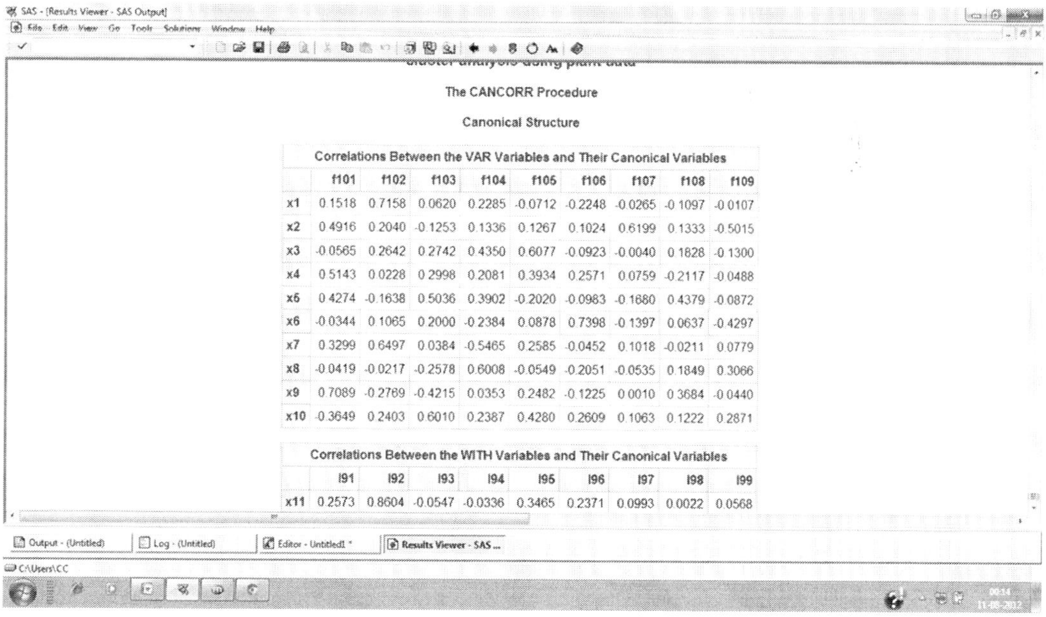

The CANCORR Procedure

Canonical Structure

Correlations Between the VAR Variables and Their Canonical Variables

	f101	f102	f103	f104	f105	f106	f107	f108	f109
x1	0.1518	0.7158	0.0620	0.2285	-0.0712	-0.2248	-0.0265	-0.1097	-0.0107
x2	0.4916	0.2040	-0.1253	0.1336	0.1267	0.1024	0.6199	0.1333	-0.5015
x3	-0.0565	0.2642	0.2742	0.4350	0.6077	-0.0923	-0.0040	0.1828	-0.1300
x4	0.5143	0.0228	0.2998	0.2081	0.3934	0.2571	0.0759	-0.2117	-0.0488
x6	0.4274	-0.1638	0.5036	0.3902	-0.2020	-0.0983	-0.1680	0.4379	-0.0872
x6	-0.0344	0.1065	0.2000	-0.2384	0.0878	0.7398	-0.1397	0.0637	-0.4297
x7	0.3299	0.6497	0.0384	-0.5465	0.2585	-0.0452	0.1018	-0.0211	0.0779
x8	-0.0419	-0.0217	-0.2578	0.6008	-0.0549	-0.2051	-0.0535	0.1849	0.3066
x9	0.7089	-0.2769	-0.4215	0.0353	0.2482	-0.1225	0.0010	0.3684	-0.0440
x10	-0.3649	0.2403	0.6010	0.2387	0.4280	0.2609	0.1063	0.1222	0.2871

Correlations Between the WITH Variables and Their Canonical Variables

	l91	l92	l93	l94	l95	l96	l97	l98	l99
x11	0.2573	0.8604	-0.0547	-0.0336	0.3465	0.2371	0.0993	0.0022	0.0568

Slide 12.18: Portion of output for canonical analysis using SAS
(Canonical loadings)

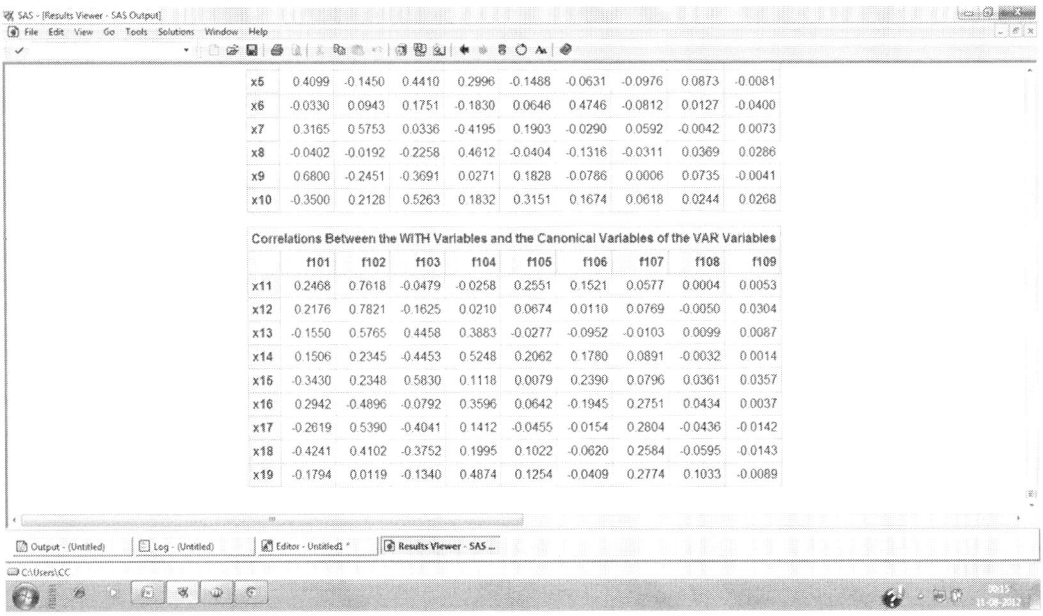

x5	0.4099	-0.1450	0.4410	0.2996	-0.1488	-0.0631	-0.0976	0.0873	-0.0081
x6	-0.0330	0.0943	0.1751	-0.1830	0.0646	0.4746	-0.0812	0.0127	-0.0400
x7	0.3165	0.5753	0.0336	-0.4195	0.1903	-0.0290	0.0592	-0.0042	0.0073
x8	-0.0402	-0.0192	-0.2258	0.4612	-0.0404	-0.1316	-0.0311	0.0369	0.0286
x9	0.6800	-0.2451	-0.3691	0.0271	0.1828	-0.0786	0.0006	0.0735	-0.0041
x10	-0.3500	0.2128	0.5263	0.1832	0.3151	0.1674	0.0618	0.0244	0.0268

Correlations Between the WITH Variables and the Canonical Variables of the VAR Variables

	f101	f102	f103	f104	f105	f106	f107	f108	f109
x11	0.2468	0.7618	-0.0479	-0.0258	0.2551	0.1521	0.0577	0.0004	0.0053
x12	0.2176	0.7821	-0.1625	0.0210	0.0674	0.0110	0.0769	-0.0050	0.0304
x13	-0.1550	0.5765	0.4458	0.3883	-0.0277	-0.0952	-0.0103	0.0099	0.0087
x14	0.1506	0.2345	-0.4453	0.5248	0.2062	0.1780	0.0891	-0.0032	0.0014
x15	-0.3430	0.2348	0.5830	0.1118	0.0079	0.2390	0.0796	0.0361	0.0357
x16	0.2942	-0.4896	-0.0792	0.3596	0.0642	-0.1945	0.2751	0.0434	0.0037
x17	-0.2619	0.5390	-0.4041	0.1412	-0.0455	-0.0154	0.2804	-0.0436	-0.0142
x18	-0.4241	0.4102	-0.3752	0.1995	0.1022	-0.0620	0.2584	-0.0595	-0.0143
x19	-0.1794	0.0119	-0.1340	0.4874	0.1254	-0.0409	0.2774	0.1033	-0.0089

Slide 12.19: Portion of output for canonical analysis using SAS (Canonical loading)

12.8 Multiple Regression Analysis and Multicollinearity

During the regression analysis, we have assumed that the predictor variables are independent not only with each other but also with the disturbance terms. But in real-life situation, hardly one can find any variable which is independent of other variables. Actually, in social, economic, business, biological, and other fields, the variables tend to move together. As a result, the variables will be correlated with each other, thereby violating the assumptions of the regression analysis. This phenomenon of linear association among the explanatory variables during regression analysis is known as multicollinearity. As a result of multicollinearity, the regression estimates sometimes become inestimable, and the standard errors of the estimates become exceptionally high, giving rise to nonsignificance of the regression estimates. Thus, *the relationships worked out, and subsequently the conclusion drawn in the presence of multicollinearity may hamper/effect the quality/validity of the conclusion.* There are different techniques to avoid/overcome the problem of multicollinearity. One of the techniques is to use principal component analysis before conducting regression analysis. The principal component analysis is a type of factor analysis.

12.9 Factor Analysis

One of the most powerful multivariate techniques is the factor analysis aimed at simplifying and analyzing the interrelationship among a set of variables in the form of a relatively few number of hypothetical variables known as factors, which are orthogonal or uncorrelated among themselves. Factor analysis helps in getting an insight into the otherwise hidden structure of the data. The essence of factor analysis lies in explaining the interrelationship among a large number of variables to a small number of factors without losing any essential information or with minimum loss of information.

Types of Factor Analysis: Mainly there are two types of factor analyses:

1. The common factor analysis
2. The principal component analysis

In common factor analysis, estimates of common variance among the original variables are used to generate the factor solution. As such the number of factors will always be less than the number of original variables. It assumes that the variation in variables consists of two parts: a common part and a unique part. On the one hand, the common part of the variable is the part of variable variation shared with other variables; on the other hand, the unique part of the variable variation is specific to the variable concerned. Factor analysis starts with the original variable or standardized variable. In a system of p, variables x_i's ($i = 1,2\ldots p$) are transformed as follows:

$$x_1 = \lambda_{11}F_1 + \lambda_{12}F_2 + \cdots + \lambda_{1k}F_k + \varepsilon_1$$
$$x_2 = \lambda_{21}F_1 + \lambda_{22}F_2 + \cdots + \lambda_{2k}F_k + \varepsilon_2$$
$$x_3 = \lambda_{31}F_1 + \lambda_{32}F_2 + \cdots + \lambda_{3k}F_k + \varepsilon_3$$
$$\cdots\cdots\cdots\cdots\cdots\cdots\cdots\cdots\cdots\cdots$$
$$\cdots\cdots\cdots\cdots\cdots\cdots\cdots\cdots\cdots\cdots$$
$$x_p = \lambda_{p1}F_1 + \lambda_{p2}F_2 + \cdots + \lambda_{pk}F_k + \varepsilon_p,$$
$$\text{where } \operatorname{cov}(\varepsilon_i, \varepsilon_j) = 0, \ \forall \ i \neq j.$$

The regression coefficients λ_{im} ($i = 1,2,3\ldots p$ and $m = 1,2,3\ldots k$) are known as common factor loadings, and the residuals $\varepsilon_1, \varepsilon_2, \varepsilon_3, \ldots \varepsilon_p$ are known as the unique factors. The factor loadings are essentially the correlations between the variable and the factor concerned. The factors F_1, F_2, F_3,… F_k are known as common factors. *Rotation* is a strategy used to clearly obtain the pattern of loadings. In many of the cases, it is found that a single factor results in high factor loadings for all the variables under consideration. Under this situation, the factor analysis fails to provide insight into the interrelationship among the variables. Rotation strategy provides a clear pattern of loadings among the factors. Rotations clearly mark the high factor loadings for some variables in first factor and low loading for the rest of the variables; in subsequent factors, it marks high factor loading for some or all the variables which were not marked in first or subsequent previous factors. Thus, under ideal situation, most of the variables, if not all, will be distributed among the factors with high factor loadings. Various rotational strategies like varimax, varimax normalized, quatrimax, and quatrimax normalized are available in most of the computer packages dealing with factor analysis.

12.9.1 Number of Factors

The problem of retaining the number of factors can be solved through the Kaiser criterion and Cattell scree test. According to the former criterion, one can retain the factors with eigenvalues greater than one. The logic is that unless a factor extracts at least equivalent to one original variable, it is redundant. But according to Cattell scree test, plot the eigenvalues against the number of factors and retain the number of factors beyond which the curve takes almost a smooth safe. The Kaiser method sometimes retains too many factors, while the Cattell scree test sometimes retains too few factors. But both criteria perform well when there are relatively few factors and many cases (Fig. 12.1).

12.9.2 Communality

Communality of a variable is defined as the proportion of the variance of the variable that is shared with other variables, that is, common factor. Thus, the proportion of variance unique to a particular variable is the total variance of the variable less its communality.

12.9.3 Factor Score

One can estimate the actual values of the individual cases for the factors known as factor scores. These factor scores are useful in further analysis involving the factors.

For checking, the users may please note that (1) total of eigenvalues should be equal to the number of variables under consideration and (2) total variance explained by all the factors must be equal to 100.

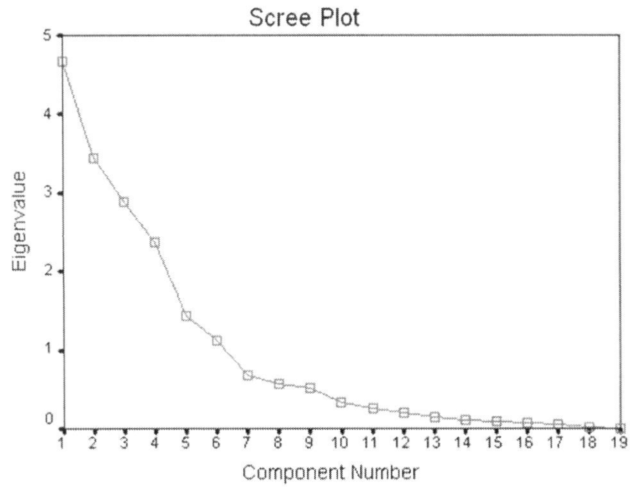

Fig. 12.1 Cattell scree plot

12.9.4 Principal Component Analysis

The principle followed in this method is to construct a set of new variables called "principal components" (P_i's) as linear combination of the given set of variables (mostly correlated) (X_i's) either from the original X's or from their standardized form, that is, $\frac{X_i - \bar{X}}{\sigma_x}$;

$$\text{that is,}\quad \begin{aligned} P_1 &= a_{11}x_1 + a_{12}x_2 + a_{13}x_3 + L + a_{1k}x_k \\ P_2 &= a_{21}x_1 + a_{22}x_2 + a_{23}x_3 + L + a_2kx_k \\ &\vdots \\ P_k &= ak_1x_1 + ak_2x_2 + ak_3x_3 + L + a_{kk}x_k \end{aligned}$$

These a's are called loading and are so chosen that the principal components (P's) are orthogonal, that is, uncorrelated, and the first principal component P_1 absorbs and accounts for the maximum possible proportion of the total variation of the set of all X's, the second P, that is, P_2, absorbs and accounts for maximum of the remaining variation in the X's, and so on.

The step-by-step procedure for PCA is as follows:

1. Construct the correlation matrix of the explanatory variables.
2. Calculate the row total and column total.
3. Obtain the loading for the first principal component P_1 as

$$a_{1j} = \frac{\sum_{j=1}^{k} rxixj}{\sqrt{\sum_{i=1}^{k} \sum_{j=1}^{k} rxixj}} = 1j.$$

4. $$\lambda_1 = \text{Latent root of } P_1 = \sum_{j=1}^{k} l_j^2 = \sum_{j=1}^{k} a_{1j}^2 = a_{11}^2 + a_{12}^2 + a_{13}^2 + \cdots + a_{1k}^2.$$

5. $(\lambda_1/k) \times 100 =$ Percent variation absorbs and accounted for by the P_1 because

$$\sum_{m=1}^{k} \lambda_m = k.$$

6. Regress Y on P_s and substitute X_s in P_s to get the relationship in original X–Y form.

Example 12.4. Using the data for yield components in Example 12.2, let us demonstrate the factor analysis and the PCA through SPSS.

Sl no.	X1	X2	X3	X4	X5	X6	X7	X8	X9	X10	X11	X12	X13	X14	X15	X16	X17	X18	X19	Y
1	32.34	151.81	140.47	21.23	162.67	107.88	23.20	31.33	9.47	149.37	7.55	2.01	195.77	1663.67	313.38	1.93	1.93	9.96	9.49	345.85
2	29.85	140.71	140.95	20.33	150.00	104.49	23.20	33.17	8.87	144.04	7.82	2.11	222.68	1675.33	298.72	1.93	2.45	14.50	9.45	354.41
3	32.38	138.77	144.30	18.63	146.00	93.17	22.23	33.33	9.00	145.20	7.90	2.14	219.74	1672.33	302.74	1.90	2.44	15.45	9.33	358.12
4	38.33	138.31	140.06	17.67	156.33	92.60	22.93	34.50	8.73	141.57	7.67	2.15	228.32	1587.33	263.92	1.57	2.59	15.63	7.93	341.39
5	37.81	140.23	135.62	18.70	153.67	103.54	23.07	33.50	8.77	133.96	7.72	2.15	207.93	1621.00	277.86	1.53	2.40	13.75	7.73	323.82
6	35.02	132.90	128.79	18.47	143.00	108.60	23.33	29.00	9.03	145.32	7.97	2.27	189.59	1629.00	267.74	1.34	2.41	13.37	6.91	369.82
7	32.25	136.10	123.41	19.93	130.00	108.25	23.13	30.33	9.70	149.08	7.80	2.12	161.48	1657.33	264.72	1.72	2.45	13.76	8.44	373.14
8	32.56	137.11	116.18	18.87	143.67	100.95	23.73	28.17	9.47	133.45	7.83	2.08	146.38	1621.67	241.44	1.74	2.25	12.26	7.98	353.28
9	34.10	147.68	125.74	19.67	153.00	98.60	24.13	30.33	9.67	132.12	7.88	2.08	162.86	1634.33	253.30	1.95	2.74	15.22	8.94	350.18
10	37.05	153.77	123.49	19.10	160.00	100.17	24.07	30.00	9.23	137.06	8.58	2.27	184.37	1702.67	263.88	1.77	2.74	15.05	8.18	394.78
11	43.70	161.64	124.89	19.07	166.33	112.08	27.57	25.93	10.30	147.31	10.19	2.73	239.65	1785.67	286.02	1.77	2.73	14.98	8.67	489.41
12	41.96	160.23	114.79	19.60	143.67	107.01	26.27	24.83	10.23	135.48	9.17	2.40	237.48	1693.00	284.65	1.86	2.20	11.29	8.54	416.26
13	38.29	144.75	115.58	20.03	143.00	107.58	26.63	26.17	9.80	129.08	8.97	2.52	201.06	1609.00	276.24	1.74	2.47	12.21	8.01	384.56
14	33.84	155.17	117.81	19.33	157.00	102.07	25.27	24.83	9.17	130.27	7.72	2.16	178.28	1496.00	294.98	1.67	2.29	11.29	7.76	327.73
15	34.56	148.67	119.07	16.67	164.33	116.45	25.87	24.50	9.33	130.48	8.55	2.43	188.79	1511.67	288.41	1.32	2.33	11.52	6.68	363.85
16	32.46	156.00	121.01	20.10	162.67	115.48	25.50	22.50	9.07	138.49	8.38	2.25	202.79	1547.33	307.51	1.44	2.05	10.52	7.22	371.94
17	34.01	156.99	138.86	21.83	144.00	117.92	25.07	27.40	9.10	151.50	10.06	2.54	216.33	1662.67	316.73	1.42	2.66	14.93	7.46	505.86
18	34.58	162.46	145.61	23.90	149.33	104.79	25.17	29.67	9.20	144.95	10.46	2.60	220.73	1685.33	313.12	1.49	2.58	14.54	7.77	502.73
19	34.22	166.22	153.17	22.73	155.67	108.22	24.67	33.83	9.07	150.81	10.01	2.56	208.66	1738.67	288.29	1.85	3.05	17.96	9.29	495.12
20	33.13	162.80	143.02	23.10	144.00	109.56	24.03	30.60	8.93	131.19	8.64	2.34	168.35	1668.33	247.84	1.55	2.40	13.12	7.58	359.93
21	34.27	168.36	135.34	21.20	163.33	115.47	23.13	26.50	12.93	132.39	7.50	1.98	131.32	1711.33	239.02	2.07	2.19	12.33	8.40	325.47
22	31.85	163.56	129.03	20.53	154.67	102.78	24.60	29.00	12.87	117.91	7.89	2.02	117.85	1653.00	234.97	1.81	2.02	10.43	7.41	298.67
23	29.36	160.86	129.24	21.17	193.67	99.50	20.80	37.60	12.50	117.78	7.99	2.11	158.37	1701.67	211.97	2.13	2.38	12.65	8.86	299.57
24	35.01	154.62	140.06	21.53	171.00	90.33	22.53	42.50	9.50	129.01	8.64	2.37	218.03	1692.00	244.12	1.66	2.60	13.66	8.31	356.87
25	35.71	161.44	145.13	22.40	181.00	98.85	21.87	36.73	9.67	131.64	8.67	2.62	263.96	1764.00	289.06	1.87	2.88	15.42	9.13	372.53
26	38.24	159.29	144.58	19.73	161.67	93.00	24.57	37.50	10.57	140.35	9.84	2.84	223.82	1744.67	312.47	1.96	2.65	14.10	9.53	453.42
27	35.36	147.69	144.01	21.93	182.33	104.51	24.37	36.50	10.87	135.28	10.06	2.70	216.55	1747.33	289.87	1.71	2.53	13.53	8.67	450.74
28	33.81	139.24	145.32	22.97	178.00	105.69	23.20	34.50	9.90	137.24	10.00	2.45	197.75	1698.67	286.04	1.49	2.02	10.29	7.83	449.22
29	34.44	139.44	146.65	22.63	193.00	112.40	23.90	29.50	8.50	155.76	8.94	2.08	238.45	1624.00	333.34	1.39	2.01	10.28	7.65	457.62
30	36.05	151.53	154.25	23.53	194.00	103.77	24.10	31.17	8.73	149.60	8.58	2.25	236.20	1542.33	306.56	1.79	2.04	10.45	8.28	420.00
31	37.34	157.67	151.84	21.63	195.33	102.57	24.90	33.50	10.90	161.70	8.42	2.20	232.85	1582.33	291.33	1.76	2.05	10.76	8.12	442.93
32	33.13	151.16	143.56	24.53	191.00	98.26	25.00	29.50	11.03	135.25	7.69	2.20	203.45	1588.33	250.95	2.15	1.80	9.63	7.10	331.91
33	30.20	144.68	133.70	21.73	176.00	95.76	22.50	35.83	10.50	140.91	6.96	1.94	154.35	1611.00	277.85	1.89	2.00	10.19	7.25	321.00
34	30.26	140.64	119.93	21.93	178.67	94.23	20.77	37.67	9.30	128.65	7.12	2.13	159.38	1601.00	286.54	1.89	1.99	10.35	7.28	283.70
35	31.27	141.59	123.96	19.37	180.67	105.51	18.00	35.33	8.93	133.87	6.65	1.91	205.68	1648.00	302.31	1.74	2.23	12.42	8.15	276.98
36	33.17	150.82	131.18	18.80	178.67	107.36	17.37	31.17	9.23	143.59	6.68	1.92	206.11	1634.00	306.61	2.08	2.12	11.87	9.09	300.94
37	32.15	151.23	137.13	22.37	179.67	112.17	20.13	30.83	8.90	143.36	6.83	1.81	203.41	1650.33	304.86	2.12	2.15	11.52	9.27	295.61

Step 1: Transform the data/open/paste the data in SPSS editor.

Step 2: Activate "Analyze"–"Data Reduction"–Factor submenus, as shown below.

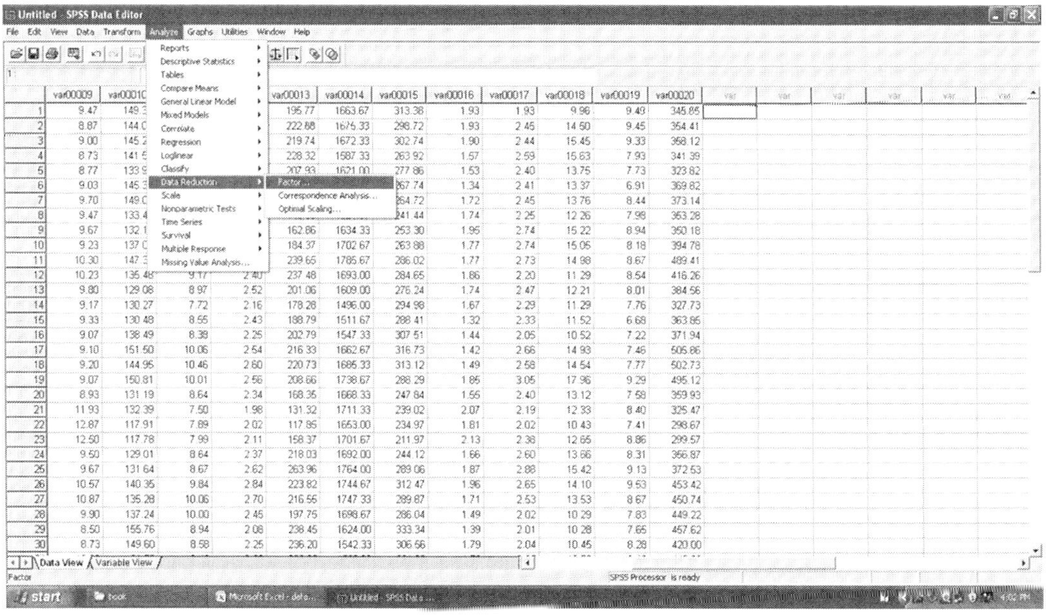

Slide 12.20: Data structure and selection of appropriate submenu of analysis menu for Factor Analysis using SPSS

Step 3: Select the variables for Factor Analysis, as shown below.

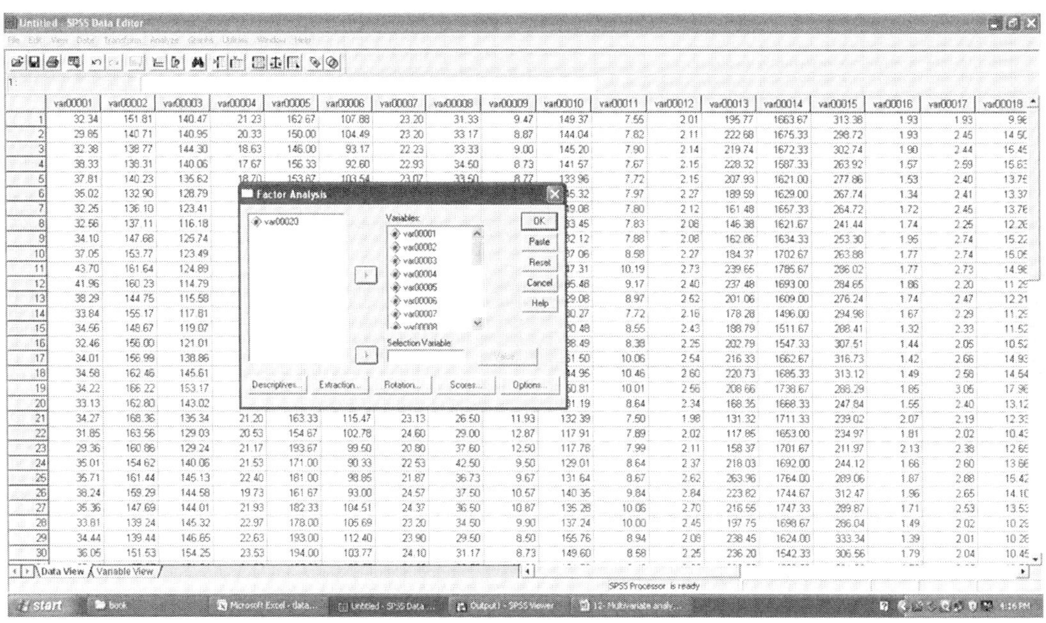

Slide 12.21: Data structure and selection of appropriate variables for Factor Analysis using SPSS

Step 4: Select descriptive to get a subwindow as shown below.

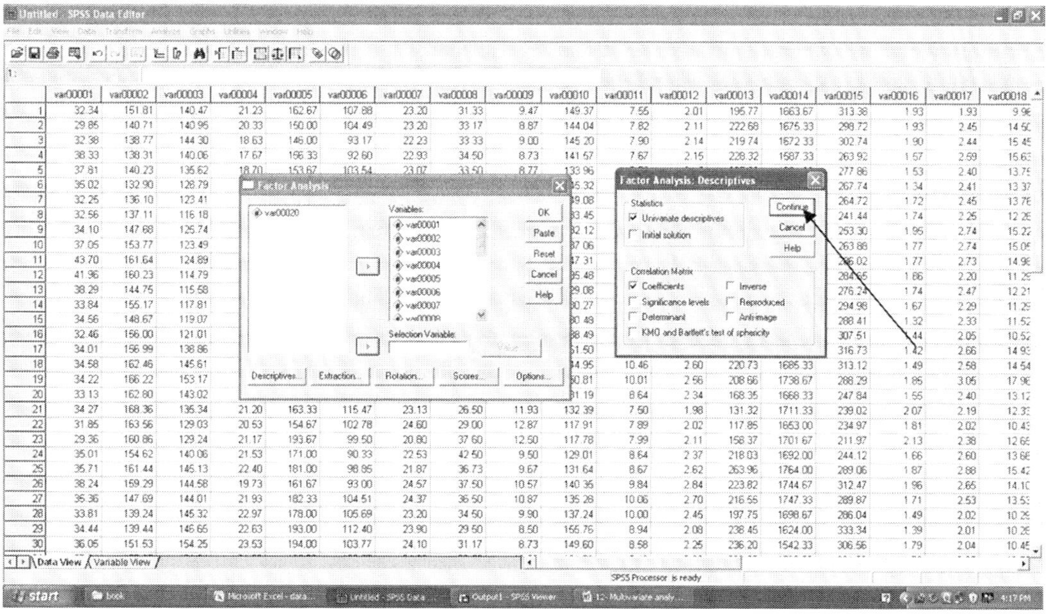

Slide 12.22: Data structure, selection of appropriate variables, and descriptives for Factor Analysis using SPSS

Step 5: Select the required statistics in descriptive window, as shown above.

Step 6: Go to Extraction submenu to get a window at the right-hand side, as shown above, and select the appropriate method of extraction among the available options. If we select the Step 6(a), then PC analysis will continue.

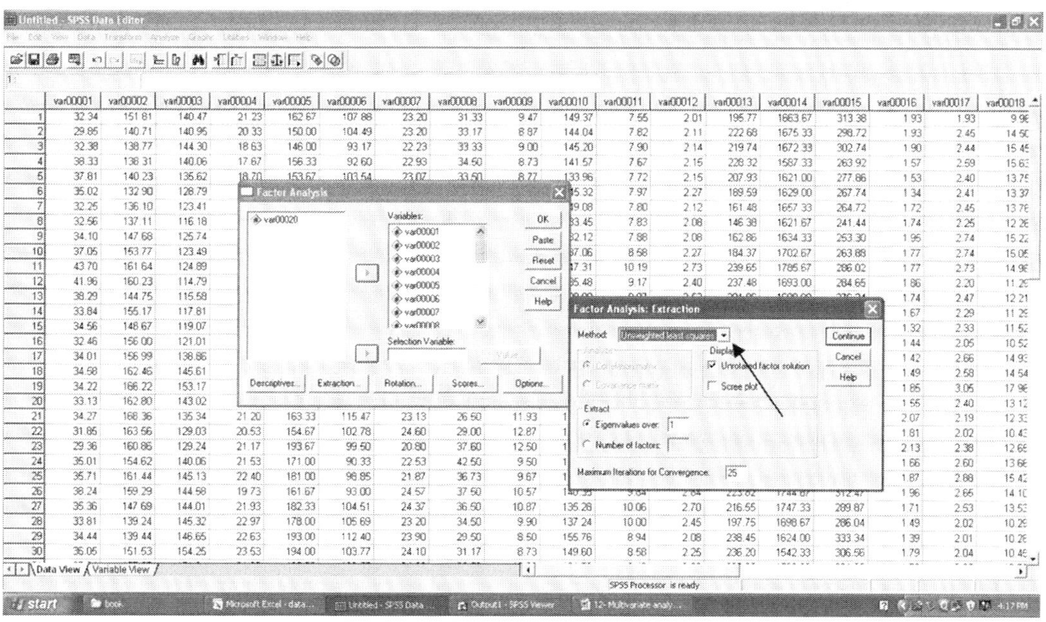

Slide 12.23: Data structure, selection of appropriate variables, and factor extraction method for Factor Analysis using SPSS 6(a)

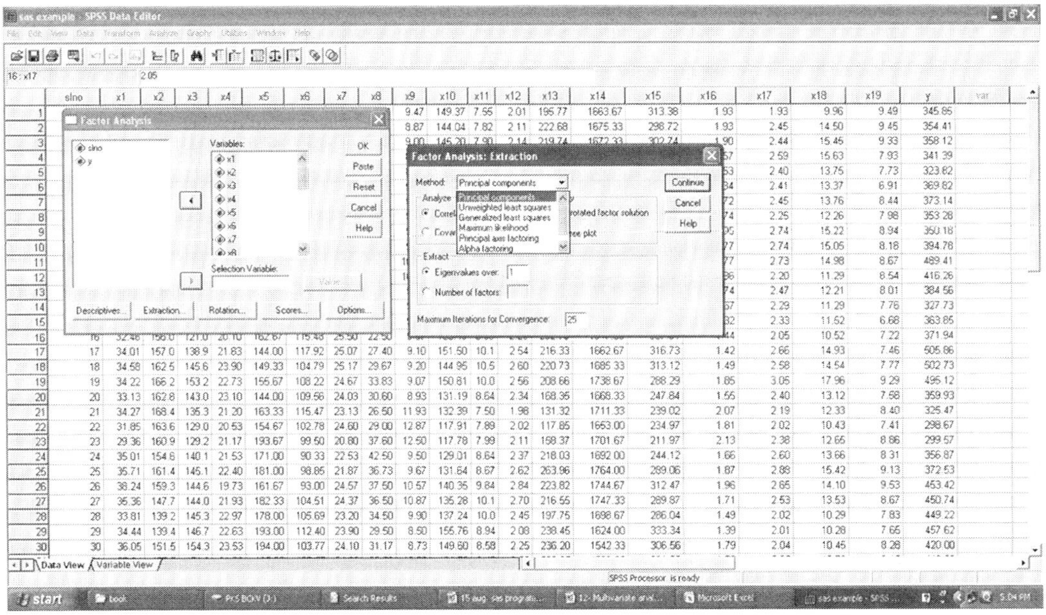

Slide 12.24: Data structure, selection of appropriate variables, and factor extraction method for Factor Analysis using SPSS

Step 7: Activate rotation option and get the window as given, and select appropriate rotation.

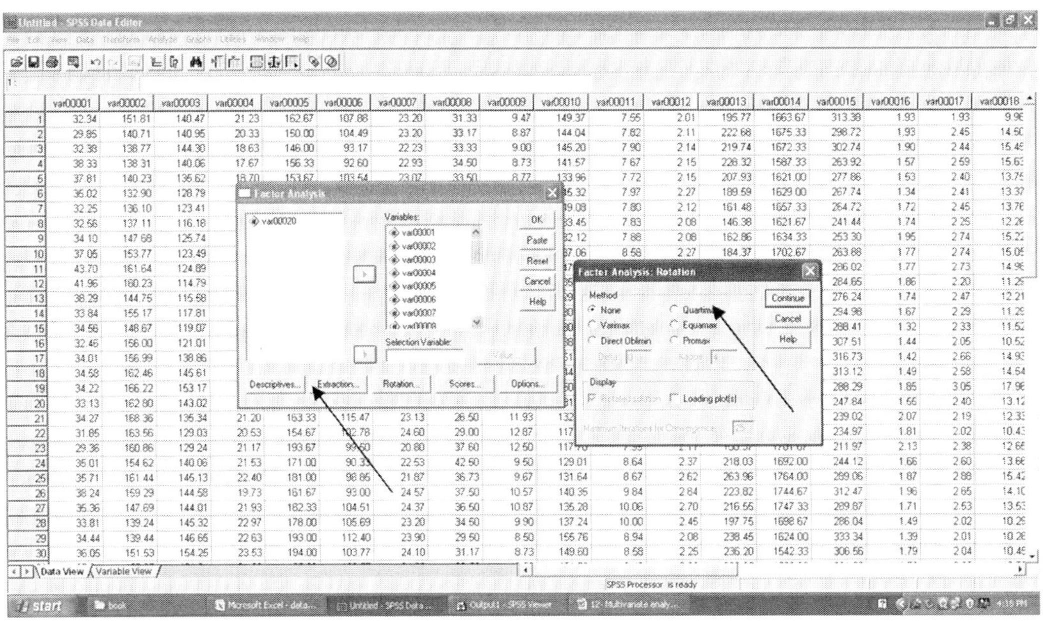

Slide 12.25: Data structure, selection of appropriate variables, and factor rotation method for Factor Analysis using SPSS

Step 8: Activate score option and get the window as given, select appropriate box, and continue.

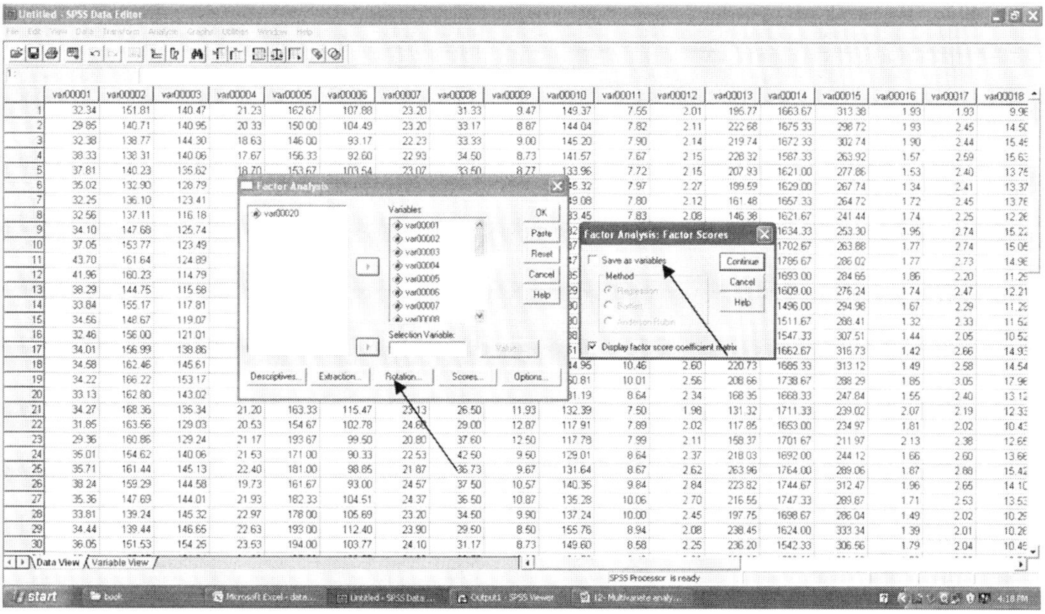

Slide 12.26: Data structure, selection of appropriate variables, factor rotation method, and factor score option for Factor Analysis using SPSS

Step 9: Activate option and get the window as given, select appropriate box, and continue.

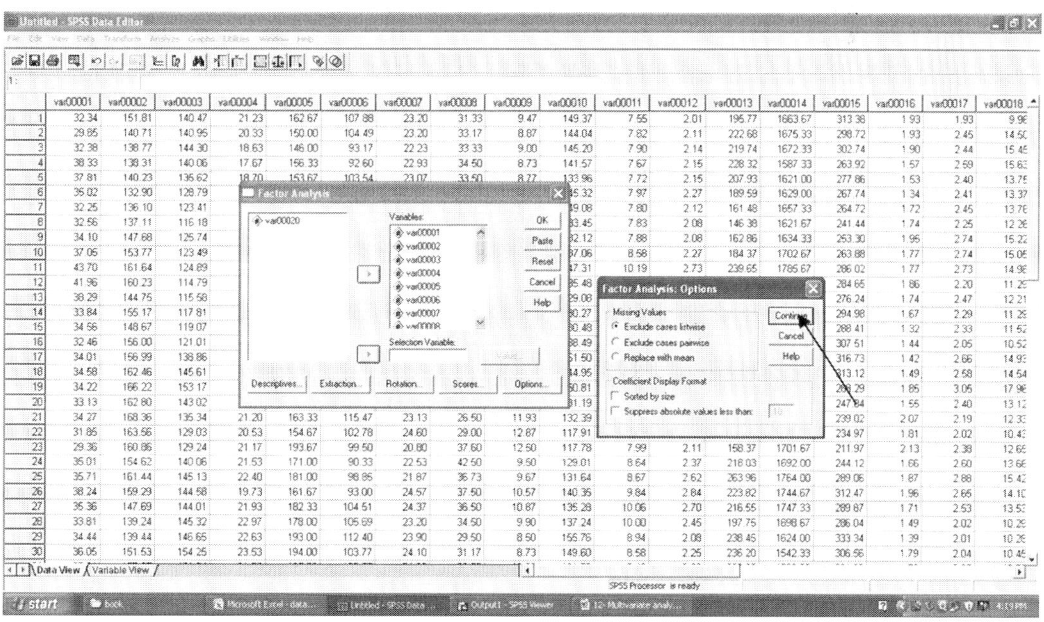

Slide 12.27: Data structure, selection of appropriate variables, and factor analysis option for Factor Analysis using SPSS (Table 12.6)

Table 12.6 Table of output of principal component analysis using SPSS

PC analysis

Communalities

	Initial	Extraction
X1	1.000	.854
X2	1.000	.727
X3	1.000	.802
X4	1.000	.843
X5	1.000	.772
X6	1.000	.812
X7	1.000	.796
X8	1.000	.929
X9	1.000	.824
X10	1.000	.732
X11	1.000	.918
X12	1.000	.885
X13	1.000	.887
X14	1.000	.754
X15	1.000	.796
X16	1.000	.879
X17	1.000	.908
X18	1.000	.917
X19	1.000	.887

Extraction method: principal component analysis

Total variance explained

	Initial eigenvalues			Extraction sums of squared loadings		
Component	Total	% of variance	Cumulative %	Total	% of variance	Cumulative %
1	4.665	24.552	24.552	4.665	24.552	24.552
2	3.439	18.101	42.653	3.439	18.101	42.653
3	2.894	15.231	57.884	2.894	15.231	57.884
4	2.364	12.443	70.327	2.364	12.443	70.327
5	1.434	7.548	77.875	1.434	7.548	77.875
6	1.125	5.921	83.795	1.125	5.921	83.795
7	.685	3.607	87.402			
8	.568	2.989	90.391			
9	.509	2.680	93.071			
10	.339	1.784	94.855			
11	.259	1.362	96.217			
12	.210	1.104	97.320			
13	.145	.763	98.083			
14	.105	.552	98.635			
15	.096	.505	99.140			
16	.069	.365	99.505			
17	.060	.316	99.821			
18	.026	.139	99.960			
19	.008	.040	100.000			

Extraction method: principal component analysis

Component matrix([a])

	Component					
	1	2	3	4	5	6
X1	.657	−.243	−.135	.068	.019	.583
X2	.326	.225	−.338	.587	.328	−.053
X3	.373	.513	.514	.232	−.140	−.251
X4	.088	.397	.332	.633	-.118	−.391
X5	−.223	.447	.459	.449	−.141	.300
X6	.129	−.555	.119	.274	.535	−.334
X7	.512	−.529	−.263	.408	−.083	.105
X8	−.004	.804	.135	−.210	−.469	−.012
X9	−.211	.323	−.559	.584	.043	.142
X10	.399	−.096	.692	−.072	.272	−.071
X11	.859	−.109	−.085	.340	−.187	−.106
X12	.851	−.056	−.181	.208	−.276	.074
X13	.665	.082	.545	−.073	−.028	.367
X14	.540	.565	−.307	.008	.176	−.134
X15	.351	−.178	.760	−.075	.233	.071
X16	−.289	.701	−.174	.080	.456	.245
X17	.723	.158	−.390	−.427	−.036	−.155
X18	.654	.222	−.305	−.542	.026	−.229
X19	.307	.651	.005	−.272	.527	.131

Extraction method: principal component analysis

[a]Six components extracted

Component score coefficient matrix

	Component					
	1	2	3	4	5	6
X1	.141	−.071	−.047	.029	.013	.518
X2	.070	.065	−.117	.248	.229	−.047
X3	.080	.149	.177	.098	−.097	−.223
X4	.019	.116	.115	.268	−.082	−.348
X5	−.048	.130	.159	.190	−.098	.267
X6	.028	−.161	.041	.116	.373	−.297
X7	.110	−.154	−.091	.172	−.058	.094
X8	−.001	.234	.047	−.089	−.327	−.011
X9	−.045	.094	−.193	.247	.030	.126
X10	.086	−.028	.239	−.030	.190	−.063
X11	.184	−.032	−.029	.144	−.130	−.094
X12	.182	−.016	−.063	.088	−.193	.066
X13	.142	.024	.188	−.031	−.019	.326
X14	.116	.164	−.106	.003	.123	−.119
X15	.075	−.052	.262	−.032	.162	.063
X16	−.062	.204	−.060	.034	.318	.218
X17	.155	.046	−.135	−.180	−.025	−.138
X18	.140	.065	−.105	−.229	.018	−.204
X19	.066	.189	.002	−.115	.367	.117

Extraction method: principal component analysis

12.10 Discriminant Analysis

One of the most important multivariate statistical tools to distinguish among the individual elements of a population is the discriminant analysis. The essence of this analysis lies in providing specific group to each and every element of a population based on its difference from other elements of the population considering multiple characters measured on all elements. Discriminant analysis has got various uses in agricultural and allied sectors, finance, business, etc. To a breeder in the field of agriculture, it is important to know which of the genetic stocks are prospective with respect to certain characters of interest in a particular crop, so that these can be selected and exploited in future breeding improvement program. Thus, the object of such selection is to place a particular genotype into homogeneous group or otherwise. On the basis of information on multiple parameters, a bank manager may wish to know whether a loan applicant is good or otherwise. Before launching any product into the market, a manufacturer may wish to know the areas where the company should concentrate on the basis of certain common characteristics of the consumers of the product. The objectives of the entire discriminant analysis are (1) to test the existence or otherwise of any significant differences among the groups presumed, (2) to test whether the variables under consideration are contributing towards intergroup discrimination, (3) to work out a linear combination of a variables which maximize the ratio of the squared differences between the group means and variance within the group, and (4) given any object or individual to assign a particular group or class. It is essentially a statistical technique to discriminate the individuals or objects belonging to two or more groups based on multiple characters/variables simultaneously.

The idea of discriminant analysis was first proposed by Fisher in 1936. Discriminant analysis has got various uses in agricultural and allied sectors, finance, business, etc. In all these cases, discriminant analysis comes into play in taking decisions. The existence of two or more groups with respect to several characters/variables measured at interval or ratio level is presumed. The groups are mutually exclusive. Given any object, it can be placed in any one of the groups presumed. As because the functional form used to combine the group characteristics towards classifying an object into a particular class or group is linear, it is also known as linear discriminant analysis (LDA). The basic assumption on which the entire LDA is based on is that each and every group or class belongs to multivariate normal population. This assumption, though not necessary, warrants for precise estimation of probabilities and subsequent test of significance.

Let there be k variables, X_1, $X_2, \ldots, X_k$, measured on all the individual elements. One writes linear combination of all these variables to make

$\Psi = \alpha_1 X_1 + \alpha_2 X_2 + \alpha_3 X_3 + \alpha_4 X_4 + \cdots + \alpha_{k-1} X_{k-1} + \alpha_k X_k$, where there are k number of variables combined linearly, α_i's are called discriminant coefficients, and Ψ is the value of the discriminant function of a particular individual/object. The α_i's are estimated in such a way that based on Ψ values the ratio of variance between population to that of the within population is maximized. The technique is to frame k sets of simultaneous equations so that solutions of these equations provide the estimates of α_i's. Without losing generality, one can measure the variables from their respective means.

If we consider the simplest case of two populations with k variables, then the discriminant function can be written in the form

$$\Psi = \alpha_1 X_1 + \alpha_2 X_2 + \alpha_3 X_3 + \alpha_4 X_4 + \cdots + \alpha_{k-1} X_{k-1} + \alpha_k X_k.$$

Thus, for group 1

$$\bar{\Psi}_1 = \alpha_1 \bar{X}^1_1 + \alpha_2 \bar{X}^1_2 + \alpha_3 \bar{X}^1_3 + \alpha_4 \bar{X}^1_4 + \cdots + \alpha_{k-1} \bar{X}^1_{k-1} + \alpha_k \bar{X}^1_k,$$

and for group 2

$$\bar{\Psi}_2 = \alpha_1\bar{X}_1^2 + \alpha_2\bar{X}_2^2 + \alpha_3\bar{X}_3^2 + \alpha_4\bar{X}_4^2 + \cdots + \alpha_{k-1}\bar{X}_{k-1}^2 + \alpha_k\bar{X}_k^2,$$

where $\bar{X}_i^1$ and $\bar{X}_i^2$ are the sample means of ith variable for group 1 and 2, respectively.

Let us define $\underset{\sim}{d} = (\bar{X}_1^1 - \bar{X}_1^2, \bar{X}_2^1 - \bar{X}_2^2, \ldots \bar{X}_k^1 - \bar{X}_k^2)'$ and $\underset{\sim}{W} = \frac{1}{n_1+n_2-2}S$, where $S = S_1 + S_2$, S_1 and S_2 are the sum of squares and products matrices for the first and second with n_1 and n_2 number of observations, respectively, and W is the pooled dispersion matrix for the k characters based on two samples.

These α_i's are so chosen that $M = \frac{\text{Variance between} \bar{\Psi}_1 \text{ and } \bar{\Psi}_2}{\text{Total variance with in groups}}$ is maximized.

Then $\underset{\sim}{\alpha} = (\alpha_1, \alpha_2, \ldots \alpha_k)'$ is given as a solution of the equation $WA = d$. Let $\bar{\Psi}_1$ and $\bar{\Psi}_2$ be the sample means of the values for both samples and let $\bar{\Psi}_1 < \bar{\Psi}_2$. Then a unit is assigned to the first population if $\Psi \le \frac{\bar{\Psi}_1+\bar{\Psi}_2}{2}$ and to the second population if $\Psi = \frac{\bar{\Psi}_1+\bar{\Psi}_2}{2}$.

The $\Psi = \alpha_1X_1 + \alpha_2X_2 + \alpha_3X_3 + \alpha_4X_4 + \cdots + \alpha_{k-1}X_{k-1} + \alpha_kX_k$ score obtained from any set of values for a given object on the above p variables is used to determine the place of the particular object in group 1 or 2. A critical score Ψ^* is calculated (generally the average of $\bar{\Psi}_1$ and $\bar{\Psi}_2$); if the score of the new object lies between the Ψ^* and $\bar{\Psi}_1$, then it belongs to group 1; on the other hand, if it lies between the Ψ^* and $\bar{\Psi}_2$, then it belongs to group 2. The effectiveness of the discriminant function is measured with the help of probability of misclassification. The probability of misclassification is defined as the proportion of objects/individuals that are placed in a group but are actually belonging to other groups. Thus, for a two-group problem, the probability of misclassification is the proportion of objects belonging to group 1 but placed in group 2 and similarly the proportion of objects/individuals belonging to group 2 but placed in group 1.

To test for the equality of two population mean vector, that is, to test $H_0 : \mu_1 = \mu_2$,

we calculate statistic T^2 analogous to Fisher's t statistic, defined as

$T^2 = \frac{n_1n_2}{n_1+n_2}\underset{\sim}{d}'\underset{\sim}{W}^{-1}\underset{\sim}{d}$. This is known as Hotelling's T^2 statistic. This T^2 is distributed as $\frac{(n_1+n_2-2)k}{n_1+n_2-k-1}F_{k,n_1+n_2-k-1}$, assuming that the variance–covariance matrices of the two populations are identical but unknown.

$$D^2 = \underset{\sim}{d}'\underset{\sim}{W}^{-1}\underset{\sim}{d}, = \underset{\sim}{A}'\underset{\sim}{d} = \sum_{i=1}^{k}\alpha_i d_i,$$

where α_i's are the linear discriminant coefficients and d_i's are as defined above.

This D^2 is known as Mahalanobis generalized D^2 statistic.

The relationship between Hotelling's T^2 statistic and Mahalanobis generalized D^2 statistic is $T^2 = \frac{n_1n_2}{n_1+n_2}D^2$.

To test H_0 we have

$$F = \frac{n_1+n_2-k-1}{k} \times \frac{n_1n_2}{(n_1+n_2)(n_1+n_2-2)}D^2, \text{ with } k, n_1 + n_2 - k - 1 \text{ d.f.}$$

Contribution of different characters towards group discrimination is worked out as follows: $\frac{\alpha_i d_i \times 100}{D^2}$.

Test for the equality of p (>2) population mean vectors:

In this problem, our objective is to test the null hypothesis $H_0 : \mu_1 = \mu_2 = \cdots = \mu_k$ vectors based on k characters. In this situation, we use Λ statistic, defined as $\Lambda = \frac{|W|}{|W+B|}$, where W and B are the sum of squares and sum of product matrices for within population and between population. This is known as Wilk's Λ (lambda) criterion, which is asymptotically distributed as χ^2 distribution

$$\chi^2_{k(p-1)} = \left[(n-1) - \frac{k+(p-1)+1}{2}\right]\log_e\Lambda.$$

Example 12.5. The following table gives the information on 20 yield and yield components for 37 varieties distributed in four groups. Justify whether grouping was made correctly on not, contribution of different yield components in group discrimination.

Variety	X1	X2	X3	X4	X5	X6	X7	X8	X9	X10	X11	X12	X13	X14	X15	X16	X17	X18	X19	X20	Group
1	31.27	141.59	123.96	19.37	180.67	105.51	18.00	35.33	8.93	133.87	6.65	1.91	205.68	1648.00	302.31	1.74	2.23	12.42	8.15	276.98	1
2	30.26	140.64	119.93	21.93	178.67	94.23	20.77	37.67	9.30	128.65	7.12	2.13	159.38	1601.00	286.54	1.89	1.99	10.35	7.28	283.70	1
3	32.15	151.23	137.13	22.37	179.67	112.17	20.13	30.83	8.90	143.36	6.83	1.81	203.41	1650.33	304.86	2.12	2.15	11.52	9.27	295.61	1
4	31.85	163.56	129.03	20.53	154.67	102.78	24.60	29.00	12.87	117.91	7.89	2.02	117.85	1653.00	234.97	1.81	2.02	10.43	7.41	298.67	1
5	29.36	160.86	129.24	21.17	193.67	99.50	20.80	37.60	12.50	117.78	7.99	2.11	158.37	1701.67	211.97	2.13	2.38	12.65	8.86	299.57	1
6	33.17	150.82	131.18	18.80	178.67	107.36	17.37	31.17	9.23	143.59	6.68	1.92	206.11	1634.00	306.61	2.08	2.12	11.87	9.09	300.94	1
7	30.20	144.68	133.70	21.73	176.00	95.76	22.50	35.83	10.50	140.91	6.96	1.94	154.35	1611.00	277.85	1.89	2.00	10.19	7.25	321.00	1
8	37.81	140.23	135.62	18.70	153.67	103.54	23.07	33.50	8.77	133.96	7.72	2.15	207.93	1621.00	277.86	1.53	2.40	13.75	7.73	323.82	1
9	34.27	168.36	135.34	21.20	163.33	115.47	23.13	26.50	11.93	132.39	7.50	1.98	131.32	1711.33	239.02	2.07	2.19	12.33	8.40	325.47	1
10	33.84	155.17	117.81	19.33	157.00	102.07	25.27	24.83	9.17	130.27	7.72	2.16	178.28	1496.00	294.98	1.67	2.29	11.29	7.76	327.73	1
11	33.13	151.16	143.56	24.53	191.00	98.26	25.00	29.50	11.03	135.25	7.59	2.20	203.45	1588.33	250.95	2.15	1.80	9.63	7.10	331.91	1
12	38.33	138.31	140.06	17.67	156.33	92.60	22.93	34.50	8.73	141.57	7.57	2.15	228.32	1587.33	263.92	1.57	2.59	15.63	7.93	341.39	1
13	32.34	151.81	140.47	21.23	162.67	107.88	23.20	31.33	9.47	149.37	7.55	2.01	195.77	1663.67	313.38	1.93	1.93	9.96	9.49	345.85	1
14	34.10	147.68	125.74	19.67	153.00	98.60	24.13	30.33	9.67	132.12	7.38	2.08	162.86	1634.33	253.30	1.95	2.74	15.22	8.94	350.18	1
15	32.56	137.11	116.18	18.87	143.67	100.95	23.73	28.17	9.47	133.45	7.33	2.08	146.38	1621.67	241.44	1.74	2.25	12.26	7.98	353.28	1
16	29.85	140.71	140.95	20.33	150.00	104.49	23.20	33.17	8.87	144.04	7.32	2.11	222.68	1675.33	298.72	1.93	2.45	14.50	9.45	354.41	1
17	35.01	154.62	140.06	21.53	171.00	90.33	22.53	42.50	9.50	129.01	8.54	2.37	218.03	1692.00	244.12	1.66	2.60	13.66	8.31	356.87	1
18	32.38	138.77	144.30	18.63	146.00	93.17	22.23	33.33	9.00	145.20	7.90	2.14	219.74	1672.33	302.74	1.90	2.44	15.45	9.33	358.12	1
19	33.13	162.80	143.02	23.10	144.00	109.56	24.03	30.60	8.93	131.19	8.54	2.34	168.35	1668.33	247.84	1.55	2.40	13.12	7.58	359.93	2
20	34.56	148.67	119.07	16.67	164.33	116.45	25.87	24.50	9.33	130.48	8.55	2.43	188.79	1511.67	288.41	1.32	2.33	11.52	6.68	363.85	2
21	35.02	132.90	128.79	18.47	143.00	108.60	23.33	29.00	9.03	145.32	7.97	2.27	189.59	1629.00	267.74	1.34	2.41	13.37	6.91	369.82	2
22	32.46	156.00	121.01	20.10	162.67	115.48	25.50	22.50	9.07	138.49	8.38	2.25	202.79	1547.33	307.51	1.44	2.05	10.52	7.22	371.94	2
23	35.71	161.44	145.13	22.40	181.00	98.85	21.87	36.73	9.67	131.64	8.67	2.62	263.96	1764.00	289.06	1.87	2.88	15.42	9.13	372.53	2
24	32.25	136.10	123.41	19.93	130.00	108.25	23.13	30.33	9.70	149.08	7.80	2.12	161.48	1657.33	264.72	1.72	2.45	13.76	8.44	373.14	2
25	38.29	144.75	115.58	20.03	143.00	107.58	26.63	26.17	9.80	129.08	8.97	2.52	201.06	1609.00	276.24	1.74	2.47	12.21	8.01	384.56	3
26	37.05	153.77	123.49	19.10	160.00	100.17	24.07	30.00	9.23	137.06	8.58	2.27	184.37	1702.67	263.88	1.77	2.74	15.05	8.18	394.78	3
27	41.96	160.23	114.79	19.60	143.67	107.01	26.27	24.83	10.23	135.48	9.[illegible]7	2.40	237.48	1693.00	284.65	1.86	2.20	11.29	8.54	416.26	3
28	36.05	151.53	154.25	23.53	194.00	103.77	24.10	31.17	8.73	149.60	8.58	2.25	236.20	1542.33	306.56	1.79	2.04	10.45	8.28	420.00	3
29	37.34	157.67	151.84	21.63	195.33	102.57	24.90	33.50	10.90	161.70	8.42	2.20	232.85	1582.33	291.33	1.76	2.05	10.76	8.12	442.93	4
30	33.81	139.24	145.32	22.97	178.00	105.69	23.20	34.50	9.90	137.24	10.00	2.45	197.75	1698.67	286.04	1.49	2.02	10.29	7.83	449.22	4
31	35.36	147.69	144.01	21.93	182.33	104.51	24.37	36.50	10.87	135.28	10.06	2.70	216.55	1747.33	289.87	1.71	2.53	13.53	8.67	450.74	4
32	38.24	159.29	144.58	19.73	161.67	93.00	24.57	37.50	10.57	140.35	9.84	2.84	223.82	1744.67	312.47	1.96	2.65	14.10	9.53	453.42	4
33	34.44	139.44	146.65	22.63	193.00	112.40	23.90	29.50	8.50	155.76	8.54	2.08	238.45	1624.00	333.34	1.39	2.01	10.28	7.65	457.62	4
34	43.70	161.64	124.89	19.07	166.33	112.08	27.57	25.93	10.30	147.31	10.19	2.73	239.65	1785.67	286.02	1.77	2.73	14.98	8.67	489.41	4
35	34.22	166.22	153.17	22.73	155.67	108.22	24.67	33.83	9.07	150.81	10.01	2.56	208.66	1738.67	288.29	1.85	3.05	17.96	9.29	495.12	4
36	34.58	162.46	145.61	23.90	149.33	104.79	25.17	29.67	9.20	144.95	10.46	2.60	220.73	1685.33	313.12	1.49	2.58	14.54	7.77	502.73	4
37	34.01	156.99	138.86	21.83	144.00	117.92	25.07	27.40	9.10	151.50	10.06	2.54	216.33	1662.67	316.73	1.42	2.66	14.93	7.46	505.86	4

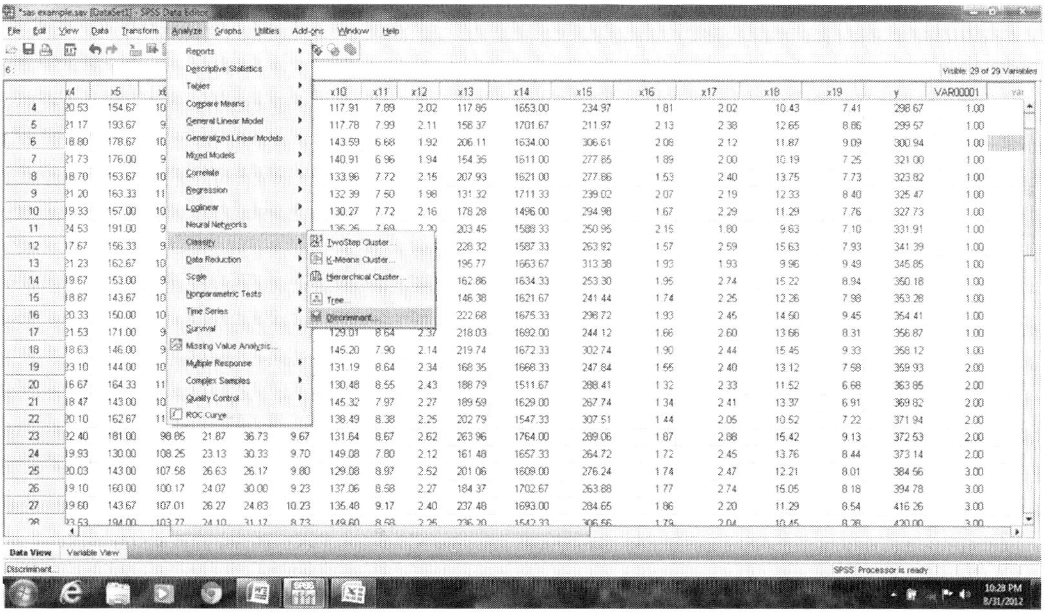

Slide 12.28: Data structure and selection of appropriate analysis submenu of discriminant analysis using SPSS

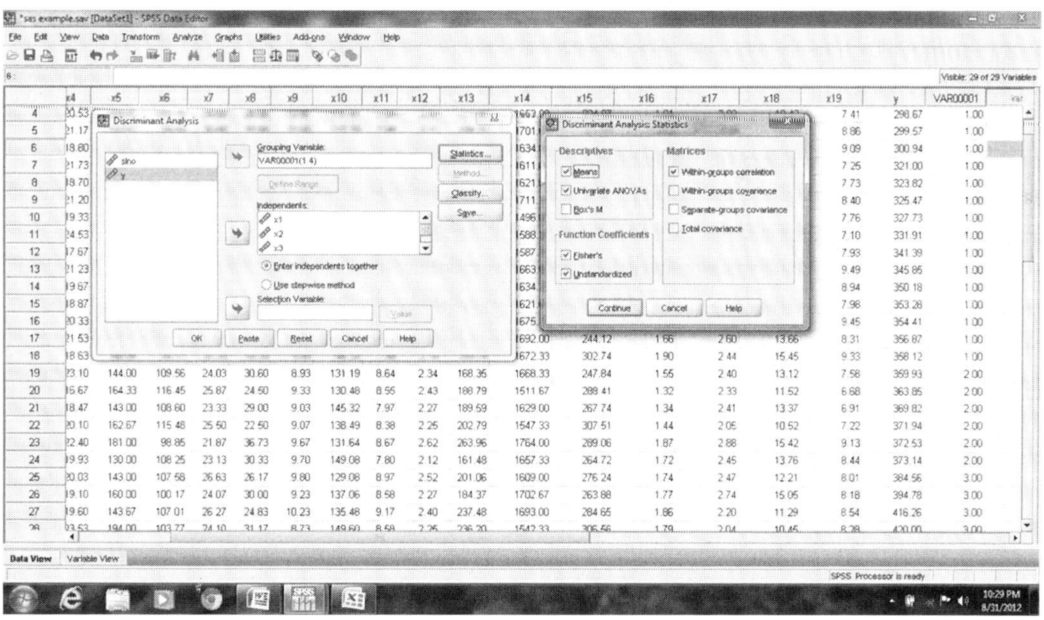

Slide 12.29: Data structure, selection of appropriate variables, and descriptives in discriminant analysis using SPSS

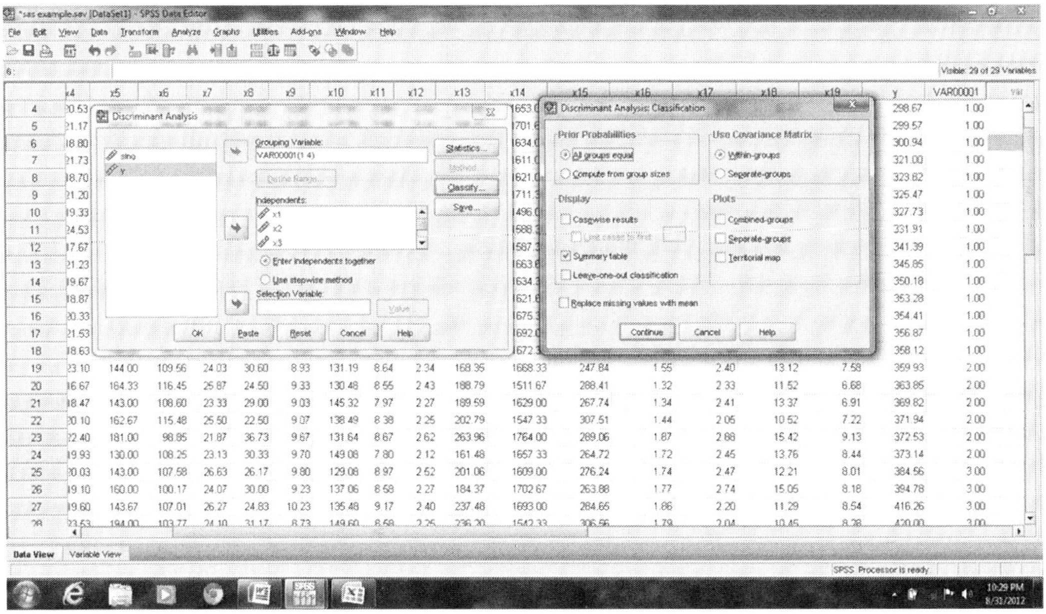

Slide 12.30: Data structure, selection of appropriate variables, and classification option in discriminant analysis using SPSS (Table 12.7)

Table 12.7 Discriminant analysis output using SPSS

Analysis Case Processing Summary			
Unweighted Cases		N	Percent
Valid		37	100.0
Excluded	Missing or out-of-range group codes	0	.0
	At least one missing discriminating variable	0	.0
	Both missing or out-of-range group codes and at least one missing discriminating variable	0	.0
	Total	0	.0
	Total	37	100.0

By using same data with SPSS, one gets the following output:

Group statistics					
				Valid N (listwise)	
VAR00001		Mean	Std. deviation	Unweighted	Weighted
1	x1	32.8822	2.46721	18	18.000
	x2	1.4874E2	9.29225	18	18.000
	x3	1.3246E2	8.86043	18	18.000
	x4	20.4217	1.70423	18	18.000
	x5	1.6609E2	15.40695	18	18.000
	x6	1.0137E2	6.84825	18	18.000
	x7	22.3661	2.19903	18	18.000
	x8	32.5050	4.36185	18	18.000
	x9	9.8800	1.32207	18	18.000
	x10	1.3515E2	8.81418	18	18.000
	x11	7.5578	.51970	18	18.000
	x12	2.0706	.12982	18	18.000

(continued)

(continued)

Group statistics

VAR00001		Mean	Std. deviation	Valid N (listwise) Unweighted	Weighted
	x13	1.8444E2	33.87301	18	18.000
	x14	1.6368E3	50.82311	18	18.000
	x15	2.7253E2	30.68083	18	18.000
	x16	1.8756	.19318	18	18.000
	x17	2.2539	.25597	18	18.000
	x18	12.3950	1.94749	18	18.000
	x19	8.3183	.81692	18	18.000
2	x1	33.8550	1.43826	6	6.000
	x2	1.4965E2	12.78573	6	6.000
	x3	1.3007E2	11.34534	6	6.000
	x4	20.1117	2.39749	6	6.000
	x5	1.5417E2	18.48086	6	6.000
	x6	1.0953E2	6.32227	6	6.000
	x7	23.9550	1.51515	6	6.000
	x8	28.9433	5.03115	6	6.000
	x9	9.2883	.33457	6	6.000
	x10	1.3770E2	7.99281	6	6.000
	x11	8.3350	.36687	6	6.000
	x12	2.3383	.17198	6	6.000
	x13	1.9583E2	36.64345	6	6.000
	x14	1.6296E3	90.56326	6	6.000
	x15	2.7755E2	21.40818	6	6.000
	x16	1.5400	.21900	6	6.000
	x17	2.4200	.26758	6	6.000
	x18	12.9517	1.72697	6	6.000
	x19	7.6600	.94786	6	6.000
3	x1	38.3375	2.58296	4	4.000
	x2	1.5257E2	6.38617	4	4.000
	x3	1.2703E2	18.56861	4	4.000
	x4	20.5650	2.01287	4	4.000
	x5	1.6017E2	23.88554	4	4.000
	x6	1.0463E2	3.41556	4	4.000
	x7	25.2675	1.37337	4	4.000
	x8	28.0425	3.02432	4	4.000
	x9	9.4975	.65541	4	4.000
	x10	1.3780E2	8.58697	4	4.000
	x11	8.8250	.29445	4	4.000
	x12	2.3600	.12570	4	4.000
	x13	2.1478E2	26.37621	4	4.000
	x14	1.6368E3	75.70713	4	4.000
	x15	2.8283E2	17.97179	4	4.000
	x16	1.7900	.05099	4	4.000
	x17	2.3625	.30794	4	4.000
	x18	12.2500	2.00027	4	4.000
	x19	8.2525	.22172	4	4.000
4	x1	36.1889	3.21260	9	9.000
	x2	1.5452E2	9.98084	9	9.000
	x3	1.4388E2	8.28200	9	9.000
	x4	21.8244	1.54563	9	9.000
	x5	1.6952E2	18.62794	9	9.000
	x6	1.0680E2	7.10421	9	9.000
	x7	24.8244	1.19958	9	9.000
	x8	32.0367	4.06234	9	9.000

(continued)

(continued)

Group statistics

VAR00001		Mean	Std. deviation	Valid N (listwise) Unweighted	Weighted
	x9	9.8233	.88571	9	9.000
	x10	1.4721E2	8.72016	9	9.000
	x11	9.7756	.65643	9	9.000
	x12	2.5222	.24702	9	9.000
	x13	2.2164E2	13.83608	9	9.000
	x14	1.6966E3	65.38209	9	9.000
	x15	3.0191E2	17.28947	9	9.000
	x16	1.6489	.20551	9	9.000
	x17	2.4756	.36715	9	9.000
	x18	13.4856	2.59200	9	9.000
	x19	8.3322	.74328	9	9.000
Total	x1	34.4341	3.11995	37	37.000
	x2	1.5071E2	9.75440	37	37.000
	x3	1.3426E2	11.50760	37	37.000
	x4	20.7281	1.85799	37	37.000
	x5	1.6435E2	17.65243	37	37.000
	x6	1.0437E2	7.09265	37	37.000
	x7	23.5354	2.13098	37	37.000
	x8	31.3311	4.46036	37	37.000
	x9	9.7289	1.05023	37	37.000
	x10	1.3878E2	9.65722	37	37.000
	x11	8.3603	1.04694	37	37.000
	x12	2.2551	.25335	37	37.000
	x13	1.9862E2	33.04842	37	37.000
	x14	1.6502E3	67.12036	37	37.000
	x15	2.8160E2	27.36225	37	37.000
	x16	1.7568	.22761	37	37.000
	x17	2.3465	.29686	37	37.000
	x18	12.7349	2.06500	37	37.000
	x19	8.2078	.79184	37	37.000

Tests of equality of group means

	Wilks' lambda	F	d.f.1	d.f.2	Sig.
x1	.618	6.812	3	33	.001
x2	.936	.758	3	33	.526
x3	.747	3.725	3	33	.021
x4	.880	1.498	3	33	.233
x5	.912	1.062	3	33	.379
x6	.793	2.874	3	33	.051
x7	.678	5.222	3	33	.005
x8	.851	1.927	3	33	.144
x9	.953	.544	3	33	.656
x10	.736	3.954	3	33	.016
x11	.227	37.379	3	33	.000
x12	.420	15.206	3	33	.000
x13	.759	3.495	3	33	.026
x14	.840	2.088	3	33	.121
x15	.803	2.692	3	33	.062
x16	.654	5.817	3	33	.003
x17	.894	1.310	3	33	.287
x18	.945	.635	3	33	.598
x19	.904	1.169	3	33	.336

Pooled within-groups matrices

		x1	x2	x3	x4	x5	x6	x7	x8	x9	x10	x11	x12	x13	x14	x15	x16	x17	x18	x19
Correlation	x1	1.000	.109	−.157	−.488	−.173	−.100	.345	−.151	−.001	−.004	.205	.396	.369	.136	−.115	−.086	.295	.264	.066
	x2	.109	1.000	.105	.284	.093	.149	.193	−.134	.479	−.323	.306	.242	-.129	.290	−.261	.440	.162	.074	.199
	x3	−.157	.105	1.000	.551	.329	−.255	−.236	.470	−.127	.354	−.024	−.062	.391	.094	.113	.206	−.027	.089	.277
	x4	−.488	.284	.551	1.000	.402	−.039	−.130	.248	.118	−.021	.019	−.121	−.035	.064	−.067	.316	−.316	−.314	−.047
	x5	−.173	.093	.329	.402	1.000	−.089	−.396	.334	.174	−.021	−.399	−.226	.253	−.165	.088	.267	−.437	−.492	−.067
	x6	−.100	.149	−.255	−.039	−.089	1.000	.060	−.705	−.045	.163	−.195	−.446	−.198	−.148	.206	−.102	−.276	−.246	−.098
	x7	.345	.193	−.236	−.130	−.396	.060	1.000	−.533	.230	−.161	.509	.341	−.195	−.236	−.272	−.203	.031	−.029	−.295
	x8	−.151	−.134	.470	.248	.334	−.705	−.533	1.000	−.011	−.111	.011	.241	.249	.380	−.127	.191	.249	.224	.292
	x9	−.001	.479	−.127	.118	.174	−.045	.230	−.011	1.000	−.522	.133	.128	−.487	.278	−.649	.451	−.166	−.244	−.046
	x10	−.004	−.323	.354	−.021	−.021	.163	−.161	−.111	−.522	1.000	−.538	−.526	.392	−.246	.548	.016	−.167	.010	.204
	x11	.205	.306	−.024	.019	−.399	−.195	.509	.011	.133	−.538	1.000	.778	.012	.344	−.370	−.131	.479	.400	.071
	x12	.396	.242	−.062	−.121	−.226	−.446	.341	.241	.128	−.526	.778	1.000	.214	.353	−.226	.084	.549	.432	.151
	x13	.369	−.129	.391	−.035	.253	−.198	−.195	.249	−.487	.392	.012	.214	1.000	.029	.544	.003	.156	.206	.342
	x14	.136	.290	.094	.064	−.165	−.148	−.236	.380	.278	−.246	.344	.353	.029	1.000	−.320	.494	.505	.508	.610
	x15	−.115	−.261	.113	−.067	.088	.206	−.272	−.127	−.649	.548	−.370	−.226	.544	−.320	1.000	−.125	−.224	−.165	.162
	x16	−.086	.440	.206	.316	.267	−.102	−.203	.191	.451	.016	−.131	.084	.003	.494	−.125	1.000	.072	.071	.633
	x17	.295	.162	−.027	−.316	−.437	−.276	.031	.249	−.166	−.167	.479	.549	.156	.505	−.224	.072	1.000	.954	.447
	x18	.264	.074	.089	−.314	−.492	−.246	−.029	.224	−.244	.010	.400	.432	.206	.508	−.165	.071	.954	1.000	.468
	x19	.066	.199	.277	−.047	−.067	−.098	−.295	.292	−.046	.204	.071	.151	.342	.610	.162	.633	.447	.468	1.000

Analysis 1

Summary of canonical discriminant functions

Eigenvalues

Function	Eigenvalue	% of variance	Cumulative %	Canonical Correlation
1	28.240[a]	84.6	84.6	.983
2	4.258[a]	12.8	97.3	.900
3	.893[a]	2.7	100.0	.687

a. First 3 canonical discriminant functions were used in the analysis

Wilks' Lambda

Test of Function(s)	Wilks' Lambda	Chi-square	df	Sig.
1 through 3	.003	139.000	57	.000
2 through 3	.100	56.299	36	.017
3	.528	15.633	17	.550

Standardized Canonical Discriminant Function Coefficients

	Function		
	1	2	3
x1	1.189	1.256	-.354
x2	−.561	−.322	.213
x3	−.554	−.048	−.248
x4	.551	−.648	.086
x5	.941	−.455	.866
x6	.265	.109	.001
x7	−.925	.271	.532
x8	.541	1.773	.597
x9	1.304	−.430	.712
x10	1.627	−1.162	.017
x11	2.678	2.685	−.431
x12	−.497	−3.419	.282
x13	−1.444	−.101	−.760
x14	−.561	−.816	.178
x15	1.556	1.370	.792
x16	.145	2.966	−.965
x17	.515	−.126	−2.887
x18	.339	.451	3.530
x19	−.404	−1.524	.493

Structure Matrix

	Function		
	1	2	3
x11	.342*	−.106	−.244
x10	.111*	−.027	.096
x15	.093*	−.005	.035
x2	.049*	.009	−.041
x16	−.078	.285*	.114
x6	.055	−.192*	−.141
x19	.010	.148*	.106
x17	.057	−.077*	−.051
x1	.114	.060	−.518*
x8	−.006	.092	.394*
x7	.109	−.060	−.377*

(continued)

(continued)

Structure Matrix

	Function		
	1	2	3
x3	.088	.020	.365*
x12	.206	−.172	−.268*
x5	.021	.106	.201*
x13	.102	−.004	−.164*
x14	.076	.027	.159*
x9	.000	.080	.157*
x4	.065	.041	.113*
x18	.038	−.049	.081*

Pooled within-groups correlations between discriminating variables and standardized canonical discriminant functions
Variables ordered by absolute size of correlation within function.
*. Largest absolute correlation between each variable and any discriminant function

Canonical Discriminant Function Coefficients

	Function		
	1	2	3
x1	.464	.491	−.138
x2	−.057	−.033	.022
x3	−.053	−.005	−.024
x4	.303	−.356	.047
x5	.053	−.026	.049
x6	.040	.017	.000
x7	−.504	.148	.290
x8	.126	.412	.139
x9	1.218	−.401	.665
x10	.188	−.134	.002
x11	5.135	5.150	−.827
x12	−2.902	−19.944	1.645
x13	−.048	−.003	−.025
x14	−.009	−.013	.003
x15	.061	.053	.031
x16	.753	15.424	−5.017
x17	1.757	−.431	−9.851
x18	.162	.215	1.683
x19	−.513	−1.938	.627
(Constant)	−82.362	.234	−21.981

Unstandardized coefficients

Functions at Group Centroids

	Function		
VAR00001	1	2	3
1	−3.918	.930	.417
2	−1.784	−4.266	−.446
3	1.525	1.981	−2.382
4	8.348	.104	.521

Unstandardized canonical discriminant functions evaluated at group means

12.10.1 Classification Statistics

Classification Processing Summary

Processed		37
Excluded	Missing or out-of-range group codes	0
	At least one missing discriminating variable	0
Used in Output		37

Prior Probabilities for Groups

VAR00001	Prior	Cases Used in Analysis Unweighted	Weighted
1	.250	18	18.000
2	.250	6	6.000
3	.250	4	4.000
4	.250	9	9.000
Total	1.000	37	37.000

Classification Function Coefficients

	VAR00001 1	2	3	4
x1	41.072	39.633	44.501	46.345
x2	−3.488	−3.459	−3.893	−4.157
x3	−6.062	−6.131	−6.290	−6.715
x4	30.574	33.027	31.716	34.585
x5	6.158	6.363	6.284	6.840
x6	11.206	11.206	11.441	11.684
x7	16.385	14.288	12.982	10.105
x8	31.385	29.391	32.115	32.602
x9	96.797	100.908	101.143	112.135
x10	13.404	14.501	14.281	15.822
x11	289.057	273.974	324.740	347.709
x12	−215.171	−119.170	−256.539	−234.126
x13	−4.836	−4.899	−5.030	−5.424
x14	.019	.064	−.050	−.078
x15	6.895	6.720	7.195	7.599
x16	4.049	−70.146	38.408	.027
x17	−102.865	−88.375	−66.184	−81.989
x18	62.076	59.851	58.472	64.059
x19	−47.096	−38.665	−53.682	−51.727
(Constant)	−4.940E3	−5.100E3	−5.324E3	−5.979E3

Fisher's linear discriminant functions

Classification Results[a]

		VAR00001	Predicted Group Membership 1	2	3	4	Total
Original	Count	1	18	0	0	0	18
		2	0	6	0	0	6
		3	0	0	4	0	4
		4	0	0	0	9	9
	%	1	100.0	.0	.0	.0	100.0
		2	.0	100.0	.0	.0	100.0
		3	.0	.0	100.0	.0	100.0
		4	.0	.0	.0	100.0	100.0

a. 100.0% of original grouped cases correctly classified.

Example 12.6. Using the SPAR1 software for the same example and taking the first ten characters into consideration with the following four groups, the output would be as given below:

Fisher's Discriminant Analysis

char means group 1

32.97 150.55 131.38 20.61 171.95 102.44 21.96 32.19
10.15 133.29 7.37 2.04 179.54 1625.25 270.99 1.89
2.18 11.84 8.02

char means group 2

32.99 147.77 133.72 20.00 154.33 102.68 23.61 31.74
9.28 136.86 8.10 2.19 190.32 1642.42 273.74 1.75
2.39 13.21 8.47

char means group 3

36.10 149.59 128.31 20.40 157.17 106.21 24.36 28.84
9.43 139.47 8.52 2.34 209.62 1643.08 282.55 1.69
2.41 12.76 8.09

char means group 4

36.19 154.52 143.88 21.82 169.52 106.80 24.82 32.04
9.82 147.21 9.78 2.52 221.64 1696.59 301.91 1.65
2.48 13.49 8.33

chars included in dis. function

1 2 3 4 5 6 7 8 9 10

groups 1 2

group diff. d(i),s

−0.02 2.78 −2.34 0.61 17.61 −0.24
−1.65 0.45 0.87 −3.57

t-values for D(I),S

−0.02 0.66 −0.50 0.68 3.18 −0.07
−1.95 0.20 1.90 −0.94

determinant of pooled var/cov matrix= 0.43857635E+11

discriminant function coefficients L(I),S

0.9938 −0.1752 −0.0033 0.7923 0.1135 −0.1393 −1.5224
−0.6471 1.6698 −0.1320

L(I)xD(I) values

0.0211 −0.4869 0.0078 0.4810 1.9994 0.0336 2.5132
−0.2893 1.4611 0.4705

(continued)

L(I)xD(I)x100/DSQ values

−0.3423 −7.8917 0.1270 7.7969 32.4094
0.5448 40.7367 −4.6896 23.6830 7.6258

d-square= 0.61692864E+01 hotelling t-square= 0.29612570E+02

F value for testing T-SQ= 1.481 WITH 10 AND 9 D.F.

centroid discriminant scores for groups 1 and 2 are -27.3625 and -33.5317

groups 1 3

group diff. D(I), S

−3.13 0.96 3.07 0.22 14.78 −3.78 −2.40
3.35 0.72 -6.18

t-values for d(i),s

−2.27 0.20 0.56 0.26 1.70 −1.38 −2.53
1.70 1.50 −1.65

DETERMINANT OF POOLED VAR/COV MATRIX = 0.31849705E+11

DISCRIMINANT FUNCTION COEFFICIENTS L(I),S

−0.2534 −0.2416 0.2667 0.0395 −0.0181 −0.5544
−2.3654 −1.4696 0.8977 −0.4154

L(I)xD(I) VALUES

0.7929 −0.2322 0.8198 0.0085 −0.2677 2.0936
5.6731 −4.9189 0.6486 2.5654

L(I)xD(I)x100/DSQ VALUES

11.0382 −3.2321 11.4131 0.1186 −3.7266 29.1462
78.9766 −68.4772 9.0293 35.7139

D-SQUARE= 0.71832473E+01 HOTELLING T-SQUARE= 0.34479590E+02

F VALUE FOR TESTING T-SQ= 1.724 WITH 10 AND 9 D.F.

CENTROID DISCRIMINANT SCORES FOR GROUPS 1 AND 3 ARE -214.2960 AND -221.4792

GROUPS 1 4

GROUP DIFF. D(I),S

−3.22 −3.96 -12.50 −1.21 2.43 −4.36 −2.86
0.15 0.33 −13.92

(continued)

(continued)

T-VALUES FOR D(I),S

−2.39 −0.90 −3.50 −1.60 0.33 −1.40 −3.37
0.08 0.62 −3.61

DETERMINANT OF POOLED VAR/COV MATRIX
= 0.56254955E+11

DISCRIMINANT FUNCTION COEFFICIENTS L(I),S

−0.4532 −0.1486 0.0266 −0.7149 −0.0276 −0.7006
−2.5632 −1.7477 1.1586 −0.2437

L(I)xD(I) VALUES

1.4587 0.5893 −0.3328 0.8676 −0.0670 3.0547
7.3314 −0.2651 0.3843 3.3920

L(I)xD(I)x100/DSQ VALUES

8.8877 3.5903 −2.0275 5.2858 −0.4080 18.6114
44.6678 −1.6149 2.3412 20.6663

D-SQUARE= 0.16413136E+02
HOTELLING T-SQUARE= 0.84410420E+02

F VALUE FOR TESTING T-SQ=4.443 WITH 10
AND 10 D.F.

CENTROID DISCRIMINANT SCORES FOR
GROUPS 1 AND 4 ARE -258.3327 AND -274.7458

GROUPS 2 3

GROUP DIFF. D(I),S

−3.11 −1.82 5.42 −0.39 −2.83 −3.54 −0.75
2.90 −0.15 −2.61

T-VALUES FOR D(I),S

−2.48 −0.37 0.84 −0.42 −0.33 −0.98 −1.06
1.21 −0.74 −0.66

DETERMINANT OF POOLED VAR/COV
MATRIX= 0.37497077E+09

DISCRIMINANT FUNCTION COEFFICIENTS L(I),S

−2.5936 0.1834 0.6440 −2.8162 −0.1129 −0.4099 4.6301
0.2521 7.2228 −0.2808

L(I)xD(I) VALUES

8.0595 −0.3335 3.4888 1.1018 0.3198 1.4490
−3.4610 0.7311 −1.1015 0.7332

L(I)xD(I)x100/DSQ VALUES

73.3534 −3.0355 31.7530 10.0284 2.9109 13.1877
−31.5002 6.6544 −10.0250 6.6729

(continued)

D-SQUARE= 0.10987192E+02
HOTELLING T-SQUARE= 0.43948760E+02

F VALUE FOR TESTING T-SQ=1.570 WITH 10
AND 5 D.F.

CENTROID DISCRIMINANT SCORES FOR
GROUPS 2 AND 3 ARE 57.7524 AND 46.7652

GROUPS 2 4

GROUP DIFF. D(I), S

−3.20 −6.74 −10.16 −1.82 −15.18 −4.12 −1.21 −
0.30 −0.54 −10.35

T-VALUES FOR D(I), S

−2.63 −1.49 −2.07 −2.08 −2.10 −1.06 −2.14
−0.13 −1.73 −2.56

DETERMINANT OF POOLED VAR/COV
MATRIX= 0.36381531E+09

DISCRIMINANT FUNCTION COEFFICIENTS L(I),S

−1.8258 1.4017 −0.7546 −4.2540 0.3174 −0.2351
−14.6768 −1.8397 −2.1950 −0.3366

L(I)xD(I) VALUES

5.8383 −9.4535 7.6646 7.7452 −4.8192 0.9682
17.7508 0.5435 1.1926 3.4851

L(I)xD(I)x100/DSQ VALUES

18.8845 −30.5784 24.7920 25.0528 −15.5882
3.1318 57.4170 1.7580 3.8577 11.2730

D-SQUARE= 0.30915662E+02
HOTELLING T-SQUARE= 0.13093690E+03

F VALUE FOR TESTING T-SQ= 5.237 WITH 10
AND 6 D.F.

CENTROID DISCRIMINANT SCORES FOR
GROUPS 2 AND 4 ARE -485.6838 AND -516.5995

GROUPS 3 4

GROUP DIFF. D(I),S

−0.09 −4.93 −15.57 −1.43 −12.35 −0.58 −0.46
−3.20 −0.39 −7.74

T-VALUES FOR D(I),S

−0.06 −0.98 −2.73 −1.80 −1.25 −0.19 −0.65
−1.55 −1.14 −1.94

(continued)

(continued)

DETERMINANT OF POOLED VAR/COV MATRIX= 0.12996363E+10

DISCRIMINANT FUNCTION COEFFICIENTS L(I),S
0.0915 0.0325 −0.1818 −0.1363 0.0288
−0.6955 −4.1911 −1.7772 −0.7058 −0.1388

L(I)xD(I) VALUES
−0.0082 −0.1601 2.8312 0.1948 −0.3557 0.4062 1.9360
5.6790 0.2758 1.0746

L(I)xD(I)x100/DSQ VALUES
−0.0695 −1.3483 23.8443 1.6403 −2.9954
3.4209 16.3054 47.8289 2.3231 9.0502

D-SQUARE= 0.11873580E+02
HOTELLING T-SQUARE= 0.50288100E+02

F VALUE FOR TESTING T-SQ=2.012 WITH 10 AND 6 D.F.

CENTROID DISCRIMINANT SCORES FOR GROUPS 3 AND 4 ARE −266.6579 AND −278.5314

12.11 Cluster Analysis

Researchers, particularly in the field of social science, breeding, genetics, medical, and other fields, always want to have groups of homogeneous elements so that there exist maximum variation among the groups and minimum variation among the individual elements of particular group. This will facilitate further action-oriented research with the elements of these groups. In conventional breeding, the possibility of success in a breeding program increases with the crossing between two parents in two different groups having maximum genetic distance between them. The possibility of in breeding depression increases as the distance between the parents decreases. Clustering based on genetic distances may help in identifying the parents for breeding improvement program. In social sciences, an extension specialist may try to adopt different extension strategies for different groups of people depending upon their multiple characteristics. The extension personal may like to have groups of homogeneous respondents which may fit well to specific extension strategy. In medical sciences, classification of diseases may provide a structural basis towards combating a group of diseases. Cluster analysis is a multivariate statistical technique used to form group of relatively homogeneous elements/individuals/entities together. The main emphasis in cluster analysis remains in exploring the possibility of formation of homogeneous groups of individuals or objects based on multiple characteristics. In all the clustering activities, the main objective remains in bringing together the individuals or the objects or the entities within a group having minimum variance among them and maximum variance between the groups.

The main activities of cluster analysis encompass three major steps: (1) to measure the distance between any two objects based on multiple characteristics, (2) the procedure for formation of clusters, that is, amalgamation technique, and (3) to form the clusters based on the distances calculated taking multiple characters into consideration and following the appropriate amalgamation technique. This will help in further utilization of information generated through cluster analysis.

12.11.1 Distance Measures

Let there be n number of individuals, all measured for p number of characters. The first task is to find out a suitable measure which can be used to have an idea about the distances between any two individuals based on the p characters. These will be used to form the distance matrix among the n individuals.

In literature several distance measures are found; we discuss some of these below:

1. *Minkowski distance* is defined as $M_{ij} = \sqrt[r]{\sum_{k=1}^{p} \left|X_{ik} - X_{jk}\right|^{r}}$, where M_{ij} denotes the distance between two objects i and j, $k = 1,2,3\ldots p$ the number of characters, and r is the parameter chosen suitably.
2. If we select $r = 2$, then we get the *Euclidean* distance measure. Thus, Euclidean distance measure between ith and jth individual is given as $E_{ij} = \sqrt{\sum_{k=1}^{p} \left(X_{ik} - X_{jk}\right)^2}$.
3. *Squared Euclidean* distance $= E_{ij}^2 = \sum_{k=1}^{p} \left(X_{ik} - X_{jk}\right)^2$ is also used when anyone wants to put greater weights on objects that are further apart.
4. *City block distance*: In Minkowski distance measure, if we put $r = 1$ and take the absolute value, we get the measure $\mathrm{CB}_{ij} = \sum_{k=1}^{p} \left|X_{ik} - X_{jk}\right|$. This is nothing but the sum of the distance across the p dimensions between any two objects. This distance measure is also known as Manhattan distance measure.
5. *Chebychev's distance* measure: According to this measure, two individuals are different if they differ in anyone of the characteristics, and as such this distance is calculated as $C_{ij} = \text{Maximum}\left|X_{ik} - X_{jk}\right|$. Thus, in this distance measure dissimilarity is the major point of interest rather than the similarity.
6. *Power distance* measure: The power distance measure is defined as $M_{ij} = \sqrt[r]{\sum_{k=1}^{p} \left|X_{ik} - X_{jk}\right|^{q}}$, where q and r are the user-defined parameters. In fact the same weight r is placed to each and every dimension of Minkowski measure.
7. *Percent disagreement* $= \mathrm{PD}_{ij} = \frac{\text{Number of } X_{ik} \neq X_{jk}}{p}$; when the information is of categorical in nature, then this type of measure becomes useful. In taxonomic studies, the difference in banding pattern in response to an enzymatic action is used to differentiate two samples.
8. *Mahalanobis D^2* value as described in discriminant analysis is also used in measuring the distance between two individuals based on multiple characters.

It must clearly be noted that some of the above-mentioned distances are based on similarity while the others are based on dissimilarity; each of the above-mentioned distances has got merits and demerits; not all the measures are equally useful in all situations. Hence, proper selection of distance measure is one of the most important points in cluster analysis.

12.11.2 Clustering Technique

Amalgamation of the similar individuals into different groups on the basis of appropriate similarity or dissimilarity distance measure constitutes the clustering technique. Among the different clustering techniques, the hierarchical technique, partitioning technique, graphical method, etc., are use mostly.

12.11.2.1 Hierarchical Technique

In this method of clustering, each n individual is assumed to be in n individual clusters initially. Successive fusions take place by linking the most similar to more similar cases or objects and so on in the successive steps. The reverse process of successive divisions of the cases into the most dissimilar to more dissimilar and so on also takes place. The special feature of this technique is that once an object is allocated to a cluster, it is never removed or combined with other objects belonging to some other clusters. Among the different methods of linking the objects in different clusters, single linkage (nearest neighbor), complete linkage (furthest neighbor), unweighted pair group average (UPGMA), weighted pair group average (WPGMA), unweighted pair group centroid (UPGMC), weighted pair group centroid (WPGMC), Ward's method, etc., are important and mostly used.

(a) *Single linkage*: In this method, two objects having minimum distance (nearest neighbor) form the first cluster. In the next step, either a third object will join them or another two closest unclustered objects are joined to form a second cluster. This depends on whether the distance from one of the unclustered objects to the first cluster is shorter than the distance between the two closest unclustered objects or not. This process continues to present a long chain.

(b) *Complete linkage*: In this method, first, the two objects having maximum distance (furthest neighbor) between them constitute two groups. Next the object either joins one of the previous two clusters or forms its own group following the above rule as in the case of first two objects. The process continues till all the objects have become member of any cluster or form their own groups.

(c) *Unweighted pair group average (UPGMA)*: The distance between a cluster and an object is calculated as the average distance between all the objects in the cluster and the objects supposed to enter into the cluster.

(d) *Weighted pair group method (WPGMA)*: Use of weights as the number of objects in the clusters to which another object is sought to be included or excluded in the unweighted pair group method is weighted pair group method.

(e) *Centroid method*: Cluster centers are identified for different clusters, and the distances (closest or furthest) are calculated from this centroid to the object (being considered for inclusion or exclusion) following either unweighted or weighted method. This process results in unweighted pair group centroid or weighted pair group centroid linkage distance, respectively.

(f) *Ward's method*: This is a method of using the technique of analysis of variance (ANOVA). The error sum of squares (ESS) serves as objective function which is minimized during amalgamation. This method assumes n groups of objects with one object per group initially. Then the first group is formed by selecting two of these n groups. This $(n - 1)$ set of groups is examined to determine the objective function. The process is followed systematically for initial n groups to $n - 1$ to $n - 2$ to $n - 3\ldots$, to 1 group, and in each stage, the objective function is assessed and the stage at which the objective function is minimum is taken. This method has the tendency to form clusters of small sizes.

(g) *Tocher method of clustering*: D^2 values are arranged in ascending order. Two objects having smallest distance from each other (nearest neighbor) are considered to form the initial cluster. A third object having smallest average D^2 value from the first two objects is added. The process continues till one comes across with an abrupt change in the average D^2 value. A new cluster is started and the same procedure is followed. Care should be taken that intercluster distance should be greater than the intracluster distance.

12.11.2.2 *k*-means Clustering Technique

Instead of forming unknown number of clusters with n objects, it becomes useful to the researchers or users to cluster the whole n objects into some prespecified number of clusters. While making the clusters, one of the distance measures discussed and deemed suitable for the situation may be adopted. So far as amalgamation technique is concerned, it starts with k random classes and uses ANOVA technique to compare within and between group variances to compute significance test to test the null hypothesis that the group means are different from each other. In the process, the objects are moved in and out of groups to get the most significant ANOVA result so as to minimize within cluster variance and maximize between cluster variance.

12.11.3 Graphical Method (Glyph and Metroglyphs)

This method does not require extensive formal computational algorithm in forming the clusters and is especially useful in the absence of efficient computer knowledge or facility. The essence of this method lies on placing the objects/individuals on a two-dimensional space taking any two of the p characters. Next other characters for each and every individual are presented with the help of lines/rays or some other geometric figures proportional to some suitably transformed index/scales/scores (if required). From the placement and the cumulative values (or scores) of the individuals, one can have distinct idea about the different groups. This method is also objected for its subjective nature.

12.11.4 Dendrogram (Tree Diagram)

Graphical representation linking the objects or variables based on the criteria discussed is known as dendrogram or tree diagram.

Readers must note that clustering is a powerful multivariate technique helpful in grouping the variables as well as individuals of any population, but it requires extensive knowledge about the theory of the method to be used, its usefulness, and appropriateness under specific situation. Researchers should be guided by the conceptual framework about the subject concerned and the requirement and should not be tempted to be guided by the availability of the options in the statistical packages. While dealing with cluster analysis, the following points may be noted:

1. Cluster analysis is a useful technique and can be applied for univariate as well as multivariate data.
2. Presence of outliers should be dealt with caution as most of the clustering techniques are sensitive to the presence of outliers in the data.
3. Selection of appropriate similarity or dissimilarity measure should be taken with utmost care.
4. Similarly, the linkage method to be adopted is also not exclusive for a particular situation, though average linkage is preferred over the others.
5. As most of the clustering analyses are to be performed using computer software and in standard packages there are a number of options in each stage, one should be very careful in selecting the appropriate one. Knowledge of the technique and methods is essential to have meaningful conclusion from cluster analysis.

Example 12.7. The table in the next page gives data on 19 yield component characters for 37 varieties. Using the data cluster the varieties.

Variety	X1	X2	X3	X4	X5	X6	X7	X8	X9	X10	X11	X12	X13	X14	X15	X16	X17	X18	X19
1	32.34	151.81	140.47	21.23	162.67	107.88	23.20	31.33	9.47	149.37	7.55	2.01	195.77	1663.67	313.38	1.93	1.93	9.96	9.49
2	29.85	140.71	140.95	20.33	150.00	104.49	23.20	33.17	8.87	144.04	7.82	2.11	222.68	1675.33	298.72	1.93	2.45	14.50	9.45
3	32.38	138.77	144.30	18.63	146.00	93.17	22.23	33.33	9.00	145.20	7.90	2.14	219.74	1672.33	302.74	1.90	2.44	15.45	9.33
4	38.33	138.31	140.06	17.67	156.33	92.60	22.93	34.50	8.73	141.57	7.67	2.15	228.32	1587.33	263.92	1.57	2.59	15.63	7.93
5	37.81	140.23	135.62	18.70	153.67	103.54	23.07	33.50	8.77	133.96	7.72	2.15	207.93	1621.00	277.86	1.53	2.40	13.75	7.73
6	35.02	132.90	128.79	18.47	143.00	108.60	23.33	29.00	9.03	145.32	7.97	2.27	189.59	1629.00	267.74	1.34	2.41	13.37	6.91
7	32.25	136.10	123.41	19.93	130.00	108.25	23.13	30.33	9.70	149.08	7.80	2.12	161.48	1657.33	264.72	1.72	2.45	13.76	8.44
8	32.56	137.11	116.18	18.87	143.67	100.95	23.73	28.17	9.47	133.45	7.83	2.08	146.38	1621.67	241.44	1.74	2.25	12.26	7.98
9	34.10	147.68	125.74	19.67	153.00	98.60	24.13	30.33	9.67	132.12	7.88	2.08	162.86	1634.33	253.30	1.95	2.74	15.22	8.94
10	37.05	153.77	123.49	19.10	160.00	100.17	24.07	30.00	9.23	137.06	8.58	2.27	184.37	1702.67	263.88	1.77	2.74	15.05	8.18
11	43.70	161.64	124.89	19.07	166.33	112.08	27.57	25.93	10.30	147.31	10.19	2.73	239.65	1785.67	286.02	1.77	2.73	14.98	8.67
12	41.96	160.23	114.79	19.60	143.67	107.01	26.27	24.83	10.23	135.48	9.17	2.40	237.48	1693.00	284.65	1.86	2.20	11.29	8.54
13	38.29	144.75	115.58	20.03	143.00	107.58	26.63	26.17	9.80	129.08	8.97	2.52	201.06	1609.00	276.24	1.74	2.47	12.21	8.01
14	33.84	155.17	117.81	19.33	157.00	102.07	25.27	24.83	9.17	130.27	7.72	2.16	178.28	1496.00	294.98	1.67	2.29	11.29	7.76
15	34.56	148.67	119.07	16.67	164.33	116.45	25.87	24.50	9.33	130.48	8.55	2.43	188.79	1511.67	288.41	1.32	2.33	11.52	6.68
16	32.46	156.00	121.01	20.10	162.67	115.48	25.50	22.50	9.07	138.49	8.38	2.25	202.79	1547.33	307.51	1.44	2.05	10.52	7.22
17	34.01	156.99	138.86	21.83	144.00	117.92	25.07	27.40	9.10	151.50	10.06	2.54	216.33	1662.67	316.73	1.42	2.66	14.93	7.46
18	34.58	162.46	145.61	23.90	149.33	104.79	25.17	29.67	9.20	144.95	10.46	2.60	220.73	1685.33	313.12	1.49	2.58	14.54	7.77
19	34.22	166.22	153.17	22.73	155.67	108.22	24.67	33.83	9.07	150.81	10.01	2.56	208.66	1738.67	288.29	1.85	3.05	17.96	9.29
20	33.13	162.80	143.02	23.10	144.00	109.56	24.03	30.60	8.93	131.19	8.64	2.34	168.35	1668.33	247.84	1.55	2.40	13.12	7.58
21	34.27	168.36	135.34	21.20	163.33	115.47	23.13	26.50	11.93	132.39	7.50	1.98	131.32	1711.33	239.02	2.07	2.19	12.33	8.40
22	31.85	163.56	129.03	20.53	154.67	102.78	24.60	29.00	12.87	117.91	7.89	2.02	117.85	1653.00	234.97	1.81	2.02	10.43	7.41
23	29.36	160.86	129.24	21.17	193.67	99.50	20.80	37.60	12.50	117.78	7.99	2.11	158.37	1701.67	211.97	2.13	2.38	12.65	8.86
24	35.01	154.62	140.06	21.53	171.00	90.33	22.53	42.50	9.50	129.01	8.64	2.37	218.03	1692.00	244.12	1.66	2.60	13.66	8.31
25	35.71	161.44	145.13	22.40	181.00	98.85	21.87	36.73	9.67	131.64	8.67	2.62	263.96	1764.00	289.06	1.87	2.88	15.42	9.13
26	38.24	159.29	144.58	19.73	161.67	93.00	24.57	37.50	10.57	140.35	9.84	2.84	223.82	1744.67	312.47	1.96	2.65	14.10	9.53
27	35.36	147.69	144.01	21.93	182.33	104.51	24.37	36.50	10.87	135.28	10.06	2.70	216.55	1747.33	289.87	1.71	2.53	13.53	8.67
28	33.81	139.24	145.32	22.97	178.00	105.69	23.20	34.50	9.90	137.24	10.00	2.45	197.75	1698.67	286.04	1.49	2.02	10.29	7.83
29	34.44	139.44	146.65	22.63	193.00	112.40	23.90	29.50	8.50	155.76	8.94	2.08	238.45	1624.00	333.34	1.39	2.01	10.28	7.65
30	36.05	151.53	154.25	23.53	194.00	103.77	24.10	31.17	8.73	149.60	8.58	2.25	236.20	1542.33	306.56	1.79	2.04	10.45	8.28
31	37.34	157.67	151.84	21.63	195.33	102.57	24.90	33.50	10.90	161.70	8.42	2.20	232.85	1582.33	291.33	1.76	2.05	10.76	8.12
32	33.13	151.16	143.56	24.53	191.00	98.26	25.00	29.50	11.03	135.25	7.69	2.20	203.45	1588.33	250.95	2.15	1.80	9.63	7.10
33	30.20	144.68	133.70	21.73	176.00	95.76	22.50	35.83	10.50	140.91	6.96	1.94	154.35	1611.00	277.85	1.89	2.00	10.19	7.25
34	30.26	140.64	119.93	21.93	178.67	94.23	20.77	37.67	9.30	128.65	7.12	2.13	159.38	1601.00	286.54	1.89	1.99	10.35	7.28
35	31.27	141.59	123.96	19.37	180.67	105.51	18.00	35.33	8.93	133.87	6.65	1.91	205.68	1648.00	302.31	1.74	2.23	12.42	8.15
36	33.17	150.82	131.18	18.80	178.67	107.36	17.37	31.17	9.23	143.59	6.68	1.92	206.11	1634.00	306.61	2.08	2.12	11.87	9.09
37	32.15	151.23	137.13	22.37	179.67	112.17	20.13	30.83	8.90	143.36	6.83	1.81	203.41	1650.33	304.86	2.12	2.15	11.52	9.27

Solution. The following slides demonstrate the data transformation to SAS data editor, the commands for execution of cluster analysis, the distance matrix, and the dendrogram. It may be noted that in each and every step, there are alternatives/options and related commands. The following method is neither unique nor exhaustive.

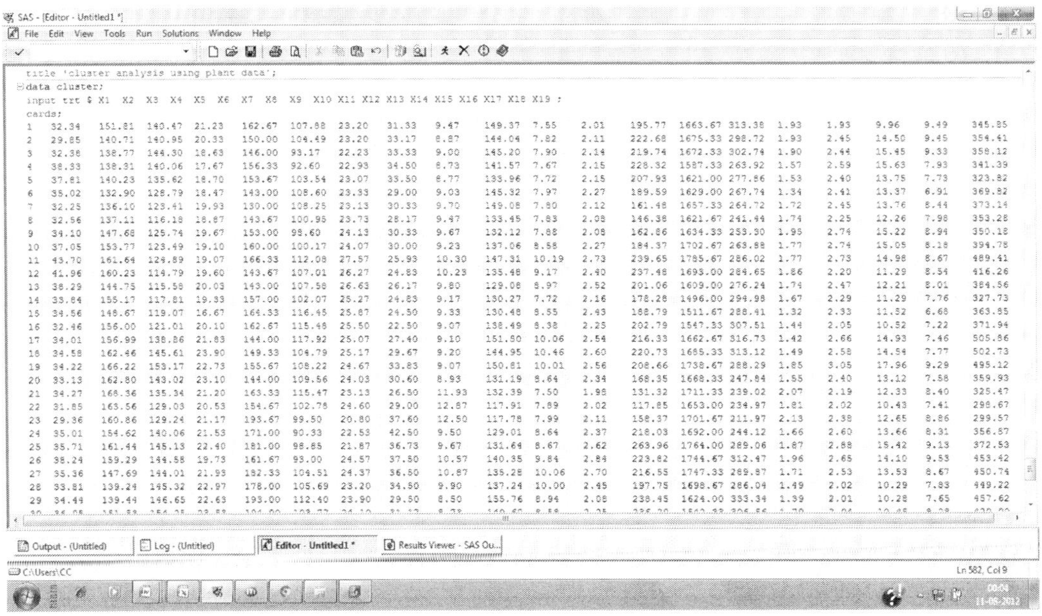

```
title 'cluster analysis using plant data';
data cluster;
input trt $ X1  X2  X3  X4  X5  X6  X7  X8  X9  X10 X11 X12 X13 X14 X15 X16 X17 X18 X19 ;
cards;
1  32.34  151.81  140.47  21.23  162.67  107.88  23.20  31.33  9.47   149.37  7.55   2.01  195.77  1663.67  313.38  1.93  1.93  9.96   9.49  345.85
2  29.85  140.71  140.95  20.33  150.00  104.49  23.20  33.17  8.87   144.04  7.82   2.11  222.68  1675.33  298.72  1.93  2.45  14.50  9.45  354.41
3  32.38  138.77  144.30  18.63  146.00  93.17   22.23  33.33  9.00   145.20  7.90   2.14  219.74  1672.33  302.74  1.90  2.44  15.45  9.33  358.12
4  38.33  138.31  140.06  17.67  156.33  92.60   22.93  34.50  8.73   141.57  7.67   2.15  228.32  1587.33  263.92  1.57  2.59  15.63  7.93  341.39
5  37.81  140.23  135.62  18.70  153.67  103.54  23.07  33.50  8.77   135.96  7.72   2.15  207.93  1621.00  277.86  1.53  2.40  13.75  7.73  323.82
6  35.02  132.90  128.79  18.47  143.00  108.60  23.33  29.00  9.03   145.32  7.97   2.27  189.59  1629.00  267.74  1.34  2.41  13.37  6.91  369.82
7  32.25  136.10  123.41  19.93  130.00  108.25  23.13  30.33  9.70   149.08  7.80   2.12  161.48  1657.33  264.72  1.72  2.45  13.76  8.44  373.14
8  32.56  137.11  116.18  18.87  143.67  100.95  23.73  28.17  9.47   133.45  7.83   2.08  146.38  1621.67  241.44  1.74  2.25  12.26  7.98  353.28
9  34.10  147.68  125.74  19.67  153.00  98.60   24.13  30.33  9.67   132.12  7.88   2.08  162.86  1634.33  253.30  1.95  2.74  15.22  8.94  350.18
10 37.05  153.77  123.49  19.10  160.00  100.17  24.07  30.00  9.23   137.06  8.58   2.27  184.37  1702.67  263.88  1.77  2.74  15.05  8.18  394.78
11 43.70  161.64  124.89  19.07  166.33  112.08  27.57  25.93  10.30  147.31  10.19  2.73  239.65  1785.67  286.02  1.77  2.73  14.98  8.67  489.41
12 41.96  160.23  114.79  19.60  143.67  107.01  26.27  24.83  10.23  135.48  9.17   2.40  237.48  1693.00  284.65  1.86  2.20  11.29  8.54  416.26
13 38.29  144.75  115.58  20.03  143.00  107.58  26.63  26.17  9.80   129.08  8.97   2.52  201.06  1609.00  276.24  1.74  2.47  12.21  8.01  384.56
14 33.84  155.17  117.81  19.33  157.00  102.07  25.27  24.83  9.17   130.27  7.72   2.16  178.28  1496.00  294.98  1.67  2.29  11.29  7.76  327.73
15 34.56  148.67  119.07  16.67  164.33  116.45  25.87  24.50  9.33   130.48  8.55   2.43  188.79  1511.67  288.41  1.32  2.33  11.52  6.68  363.85
16 32.46  156.00  121.01  20.10  162.67  115.48  25.50  22.50  9.07   138.49  8.38   2.25  202.79  1547.33  307.51  1.44  2.05  10.52  7.22  371.94
17 34.01  156.99  138.86  21.83  144.00  117.92  25.07  27.40  9.10   151.80  10.06  2.54  216.33  1662.67  316.73  1.42  2.66  14.93  7.46  505.86
18 34.58  162.46  145.61  23.90  149.33  104.79  25.17  29.67  9.20   144.95  10.46  2.60  220.73  1685.33  313.12  1.49  2.58  14.54  7.77  502.73
19 34.22  166.22  153.17  22.73  155.67  108.22  24.67  33.83  9.07   150.81  10.01  2.56  208.66  1738.67  288.29  1.85  3.05  17.96  9.29  495.12
20 33.13  162.80  143.02  23.10  144.00  109.56  24.03  30.60  8.93   131.19  8.64   2.34  168.35  1668.33  247.84  1.55  2.40  13.12  7.58  359.93
21 34.27  168.36  135.34  21.20  163.33  115.47  23.13  26.50  11.93  132.39  7.50   1.98  131.32  1711.33  239.02  2.07  2.19  12.33  8.40  325.47
22 31.85  163.56  129.03  20.53  154.67  102.78  24.60  29.00  12.87  117.91  7.89   2.02  117.85  1653.00  234.97  1.81  2.02  10.43  7.41  298.67
23 29.36  160.86  129.24  21.17  193.67  99.50   20.80  37.60  12.50  117.78  7.99   2.11  158.37  1701.67  211.97  2.13  2.38  12.65  8.86  299.57
24 35.01  154.62  140.06  21.53  171.00  90.33   22.53  42.50  9.50   129.01  8.64   2.37  218.03  1692.00  244.12  1.66  2.60  13.66  8.31  356.87
25 35.71  161.44  145.13  22.40  181.00  98.85   21.87  36.73  9.67   131.64  8.67   2.62  263.96  1764.00  289.06  1.87  2.88  15.42  9.13  372.53
26 38.24  159.29  144.58  19.73  161.67  93.00   24.57  37.50  10.57  140.35  9.84   2.84  223.82  1744.67  312.47  1.96  2.65  14.10  9.53  453.42
27 35.36  147.69  144.01  21.93  182.33  104.51  24.37  36.50  10.87  135.28  10.06  2.70  216.55  1747.33  289.87  1.71  2.53  13.53  8.67  450.74
28 33.81  139.24  145.32  22.97  178.00  105.69  23.20  34.50  9.90   137.24  10.00  2.45  197.75  1698.67  286.04  1.49  2.02  10.29  7.83  449.22
29 34.44  139.44  146.65  22.63  193.00  112.40  23.90  29.50  8.50   155.76  8.94   2.08  238.45  1624.00  333.34  1.39  2.01  10.28  7.65  457.62
```

Slide 12.31: Data structure for cluster analysis using SAS

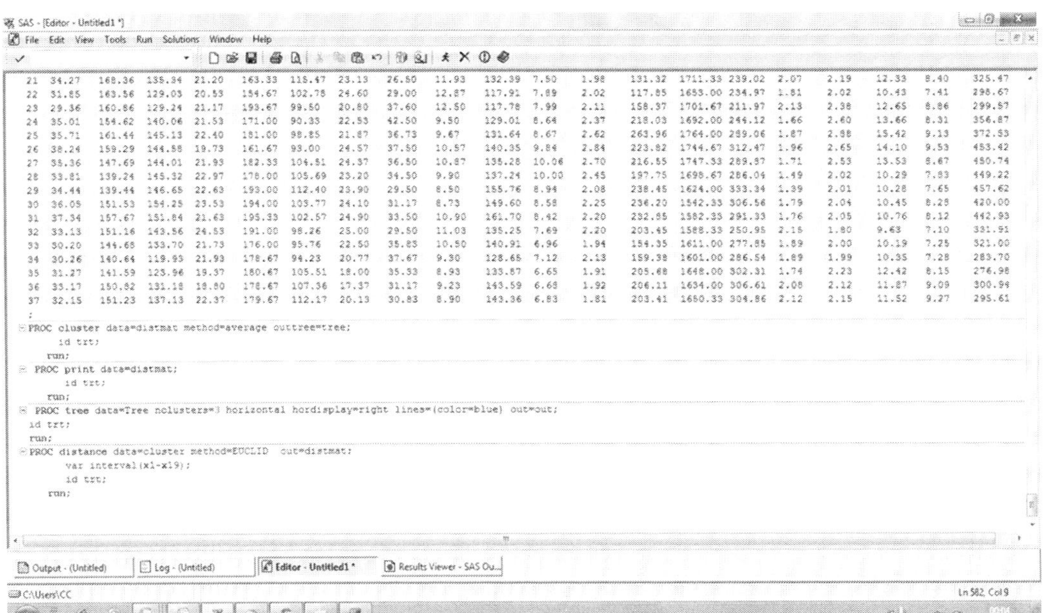

```
21 34.27  168.36  135.34  21.20  163.33  115.47  23.13  26.50  11.93  132.39  7.50   1.98  131.32  1711.33  239.02  2.07  2.19  12.33  8.40  325.47
22 31.85  163.56  129.03  20.53  154.67  102.78  24.60  29.00  12.87  117.91  7.89   2.02  117.85  1653.00  234.97  1.81  2.02  10.43  7.41  298.67
23 29.36  160.86  129.24  21.17  193.67  99.50   20.80  37.60  12.50  117.78  7.99   2.11  158.37  1701.67  211.97  2.13  2.38  12.65  8.86  299.57
24 35.01  154.62  140.06  21.53  171.00  90.33   22.53  42.50  9.50   129.01  8.64   2.37  218.03  1692.00  244.12  1.66  2.60  13.66  8.31  356.87
25 35.71  161.44  145.13  22.40  181.00  98.85   21.87  36.73  9.67   131.64  8.67   2.62  263.96  1764.00  289.06  1.87  2.88  15.42  9.13  372.53
26 38.24  159.29  144.58  19.73  161.67  93.00   24.57  37.50  10.57  140.35  9.84   2.84  223.82  1744.67  312.47  1.96  2.65  14.10  9.53  453.42
27 35.36  147.69  144.01  21.93  182.33  104.51  24.37  36.50  10.87  135.28  10.06  2.70  216.55  1747.33  289.87  1.71  2.53  13.53  8.67  450.74
28 33.81  139.24  145.32  22.97  178.00  105.69  23.20  34.50  9.90   137.24  10.00  2.45  197.75  1698.67  286.04  1.49  2.02  10.29  7.83  449.22
29 34.44  139.44  146.65  22.63  193.00  112.40  23.90  29.50  8.50   155.76  8.94   2.08  238.45  1624.00  333.34  1.39  2.01  10.28  7.65  457.62
30 36.05  151.53  154.25  23.53  194.00  103.77  24.10  31.17  8.73   149.60  8.58   2.25  236.20  1542.33  306.56  1.79  2.04  10.45  8.28  420.00
31 37.34  157.67  151.84  21.63  195.33  102.57  24.90  33.50  10.90  161.70  8.42   2.20  232.85  1582.33  291.33  1.76  2.05  10.76  8.12  442.93
32 33.13  151.16  143.56  24.53  191.00  98.26   25.00  29.50  11.03  135.25  7.69   2.20  203.45  1588.33  250.95  2.15  1.80  9.63   7.10  331.91
33 30.20  144.65  133.70  21.73  176.00  95.76   22.50  35.83  10.50  140.91  6.96   1.94  154.35  1611.00  277.85  1.89  2.00  10.19  7.25  321.00
34 30.26  140.64  119.93  21.93  178.67  94.23   20.77  37.67  9.30   128.65  7.12   2.13  159.38  1601.00  286.54  1.89  1.99  10.35  7.28  283.70
35 31.27  141.59  123.96  19.37  180.67  105.51  18.00  35.33  8.93   133.87  6.65   1.91  205.88  1648.00  302.31  1.74  2.23  12.42  8.15  276.98
36 35.17  150.82  131.18  18.80  178.67  107.36  17.37  31.17  9.23   143.59  6.68   1.92  206.11  1634.00  306.61  2.08  2.12  11.87  9.09  300.94
37 32.15  151.23  137.13  22.37  179.67  112.17  20.13  30.83  8.90   143.36  6.83   1.81  203.41  1650.33  304.86  2.12  2.15  11.52  9.27  295.61
;
PROC cluster data=distmat method=average outtree=tree;
     id trt;
   run;
 PROC print data=distmat;
      id trt;
   run;
 PROC tree data=Tree nclusters=3 horizontal hordisplay=right lines=(color=blue) out=out;
 id trt;
 run;
PROC distance data=cluster method=EUCLID  out=distmat;
      var interval(x1-x19);
      id trt;
   run;
```

Slide 12.32: Data structure and commands for cluster analysis using SAS

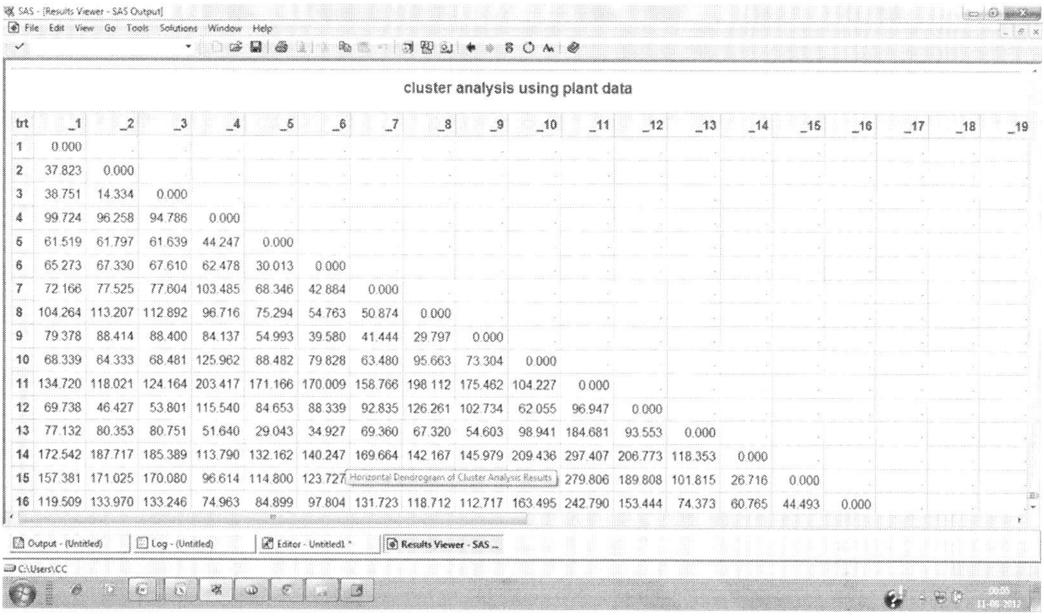

cluster analysis using plant data

trt	_1	_2	_3	_4	_5	_6	_7	_8	_9	_10	_11	_12	_13	_14	_15	_16	_17	_18	_19
1	0.000	.	.	.	.	.	.	.	.	.	.	.	.	.	.	.	.	.	.
2	37.823	0.000	.	.	.	.	.	.	.	.	.	.	.	.	.	.	.	.	.
3	38.751	14.334	0.000	.	.	.	.	.	.	.	.	.	.	.	.	.	.	.	.
4	99.724	96.258	94.786	0.000	.	.	.	.	.	.	.	.	.	.	.	.	.	.	.
5	61.519	61.797	61.639	44.247	0.000	.	.	.	.	.	.	.	.	.	.	.	.	.	.
6	65.273	67.330	67.610	62.478	30.013	0.000	.	.	.	.	.	.	.	.	.	.	.	.	.
7	72.166	77.525	77.604	103.485	68.346	42.884	0.000	.	.	.	.	.	.	.	.	.	.	.	.
8	104.264	113.207	112.892	96.716	75.294	54.763	50.874	0.000	.	.	.	.	.	.	.	.	.	.	.
9	79.378	88.414	88.400	84.137	54.993	39.580	41.444	29.797	0.000	.	.	.	.	.	.	.	.	.	.
10	68.339	64.333	68.481	125.962	88.482	79.828	63.480	95.663	73.304	0.000	.	.	.	.	.	.	.	.	.
11	134.720	118.021	124.164	203.417	171.166	170.009	158.766	198.112	175.462	104.227	0.000	.	.	.	.	.	.	.	.
12	69.738	46.427	53.801	115.540	84.653	88.339	92.835	126.261	102.734	62.055	96.947	0.000	.	.	.	.	.	.	.
13	77.132	80.353	80.751	51.640	29.043	34.927	69.360	67.320	54.603	98.941	184.681	93.553	0.000	.	.	.	.	.	.
14	172.542	187.717	185.389	113.790	132.162	140.247	169.664	142.167	145.979	209.436	297.407	206.773	118.353	0.000	.	.	.	.	.
15	157.381	171.025	170.080	96.614	114.800	123.727	Horizontal Dendrogram of Cluster Analysis Results				279.806	189.808	101.815	26.716	0.000	.	.	.	.
16	119.509	133.970	133.246	74.963	84.899	97.804	131.723	118.712	112.717	163.495	242.790	153.444	74.373	60.765	44.493	0.000	.	.	.

Slide 12.33: Distance matrix as obtained by cluster analysis using SAS

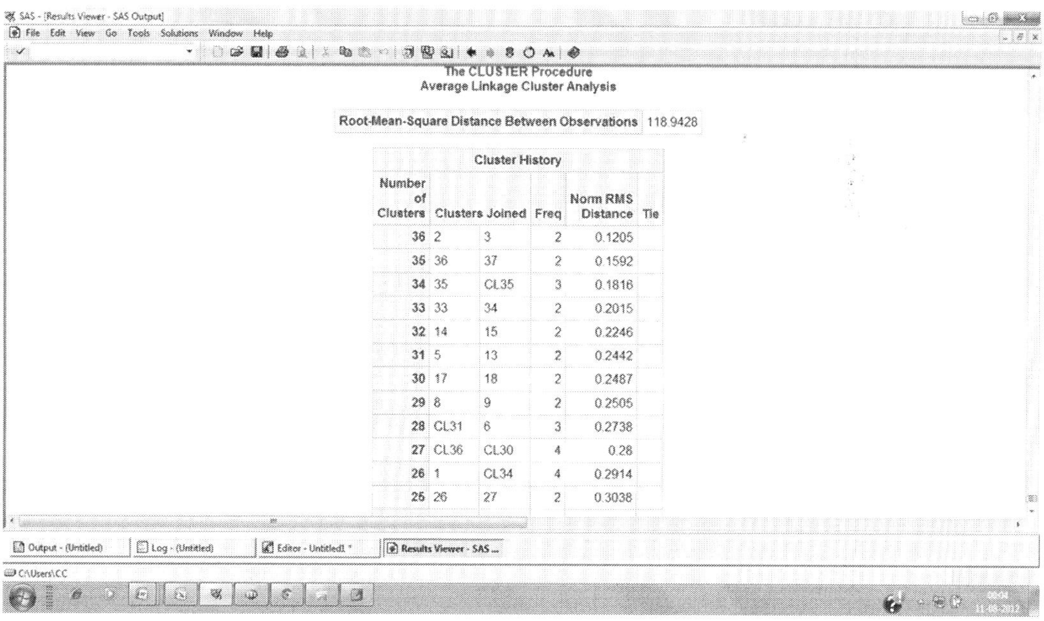

The CLUSTER Procedure
Average Linkage Cluster Analysis

Root-Mean-Square Distance Between Observations	118.9428

Cluster History

Number of Clusters	Clusters Joined		Freq	Norm RMS Distance	Tie
36	2	3	2	0.1205	
35	36	37	2	0.1592	
34	35	CL35	3	0.1816	
33	33	34	2	0.2015	
32	14	15	2	0.2246	
31	5	13	2	0.2442	
30	17	18	2	0.2487	
29	8	9	2	0.2505	
28	CL31	6	3	0.2738	
27	CL36	CL30	4	0.28	
26	1	CL34	4	0.2914	
25	26	27	2	0.3038	

Slide 12.34: Portion of the output as obtained by cluster analysis using SAS

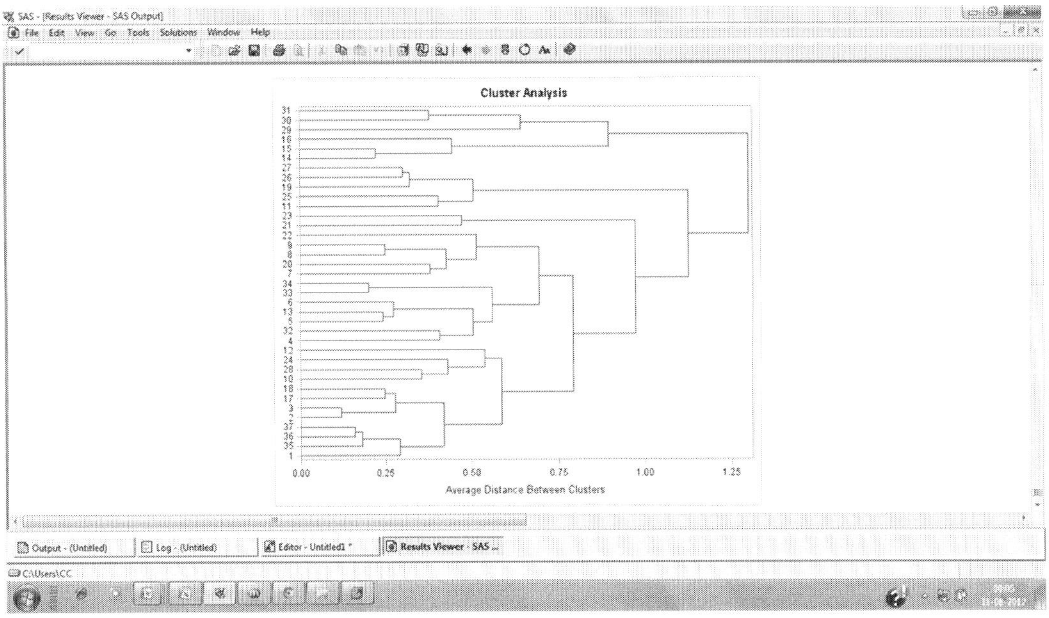

Slide 12.35: Dendrogram of 37 varieties as obtained by cluster analysis using SAS

Taking the same example of 37 varieties for 20 characters (along with yield) and using STATISTICA software, one can have the cluster analysis of the varieties, as shown in the following slides. Readers may kindly note that each software has its own way of presentation and having varied options, the researchers must be very careful in using the appropriate one befitting with the requirement of the research program.

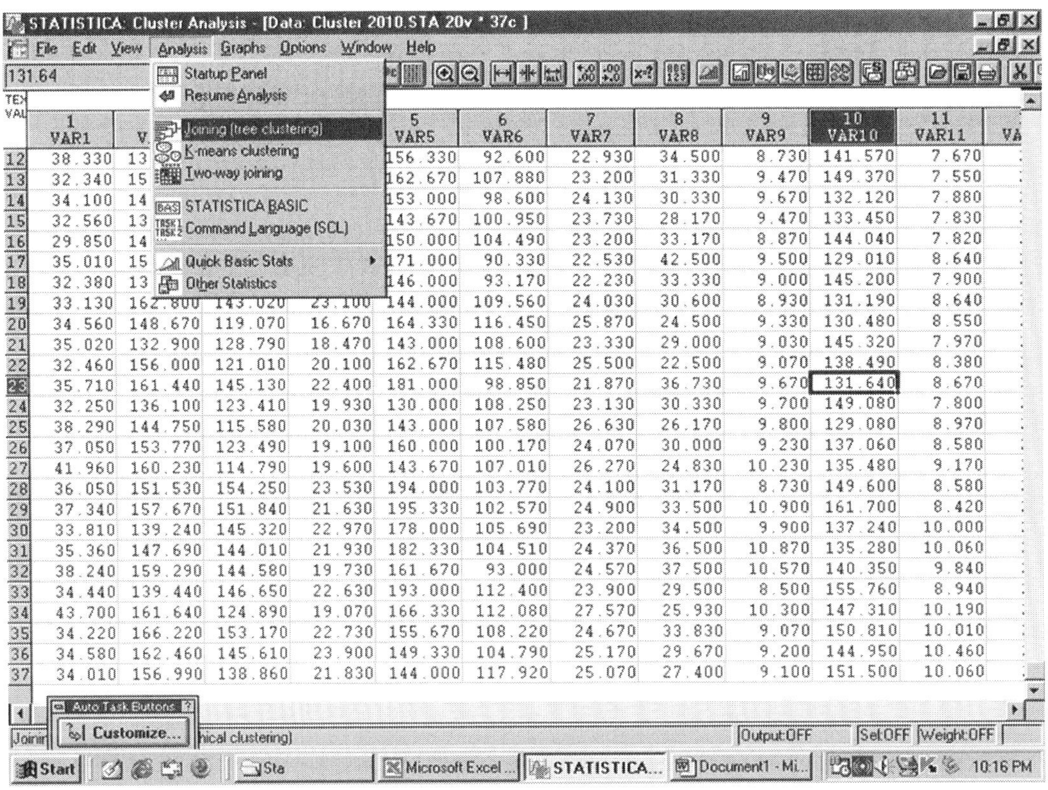

Slide 12.36: Data along with menu bar for cluster analysis using STATISTICA

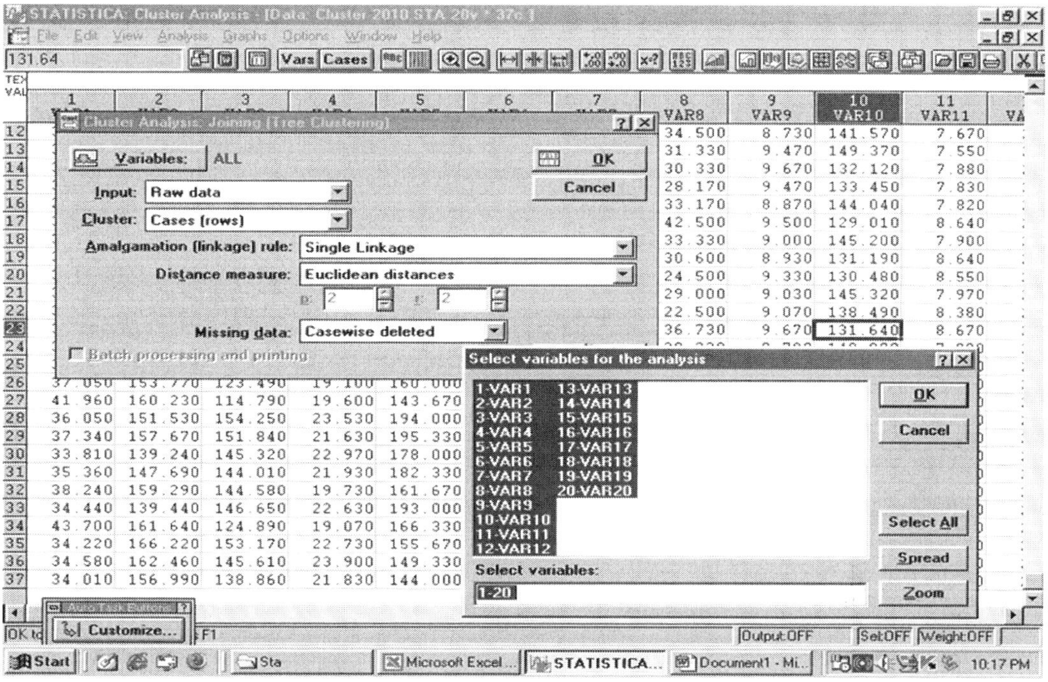

Slide 12.37: Data along with selected variables for cluster analysis using STATISTICA

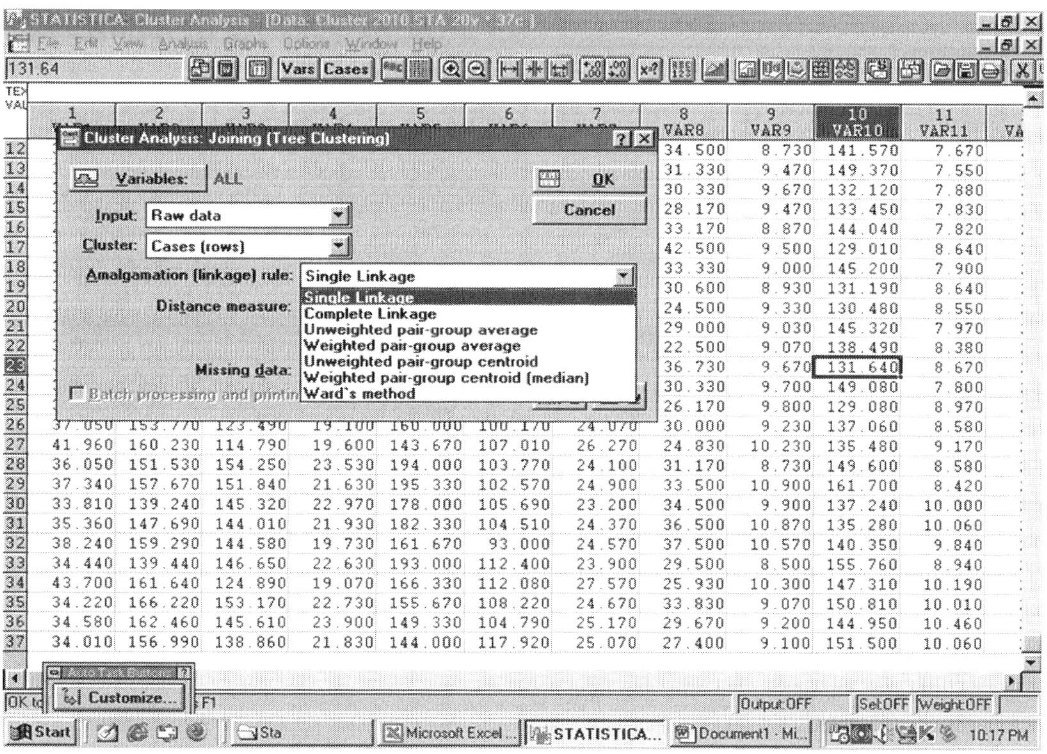

Slide 12.38: Data along with selected linkage method for cluster analysis using STATISTICA

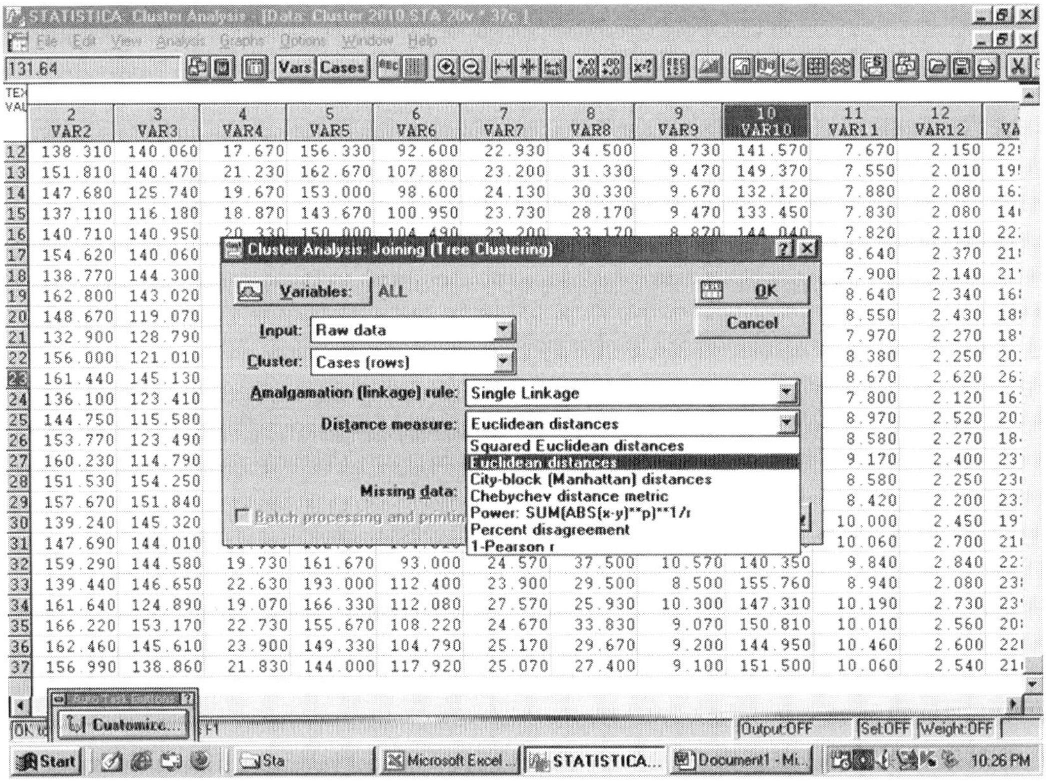

Slide 12.39: Data along with selected linkage and distance measure for cluster analysis using STATISTICA

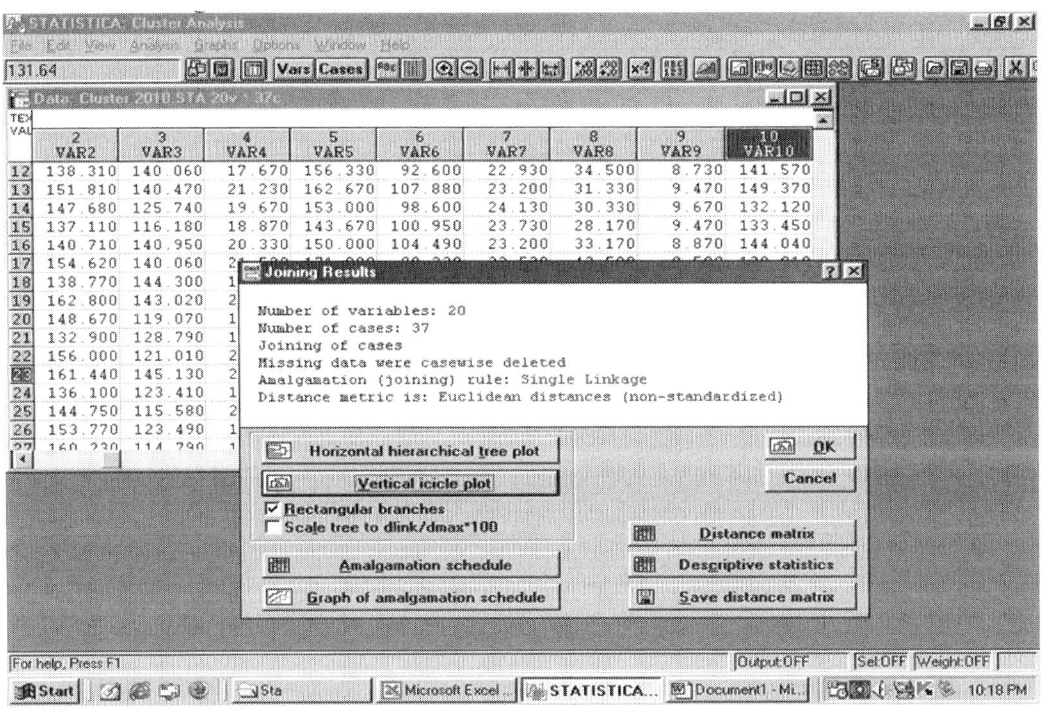

Slide 12.40: Menu bar for getting output for cluster analysis using STATISTICA

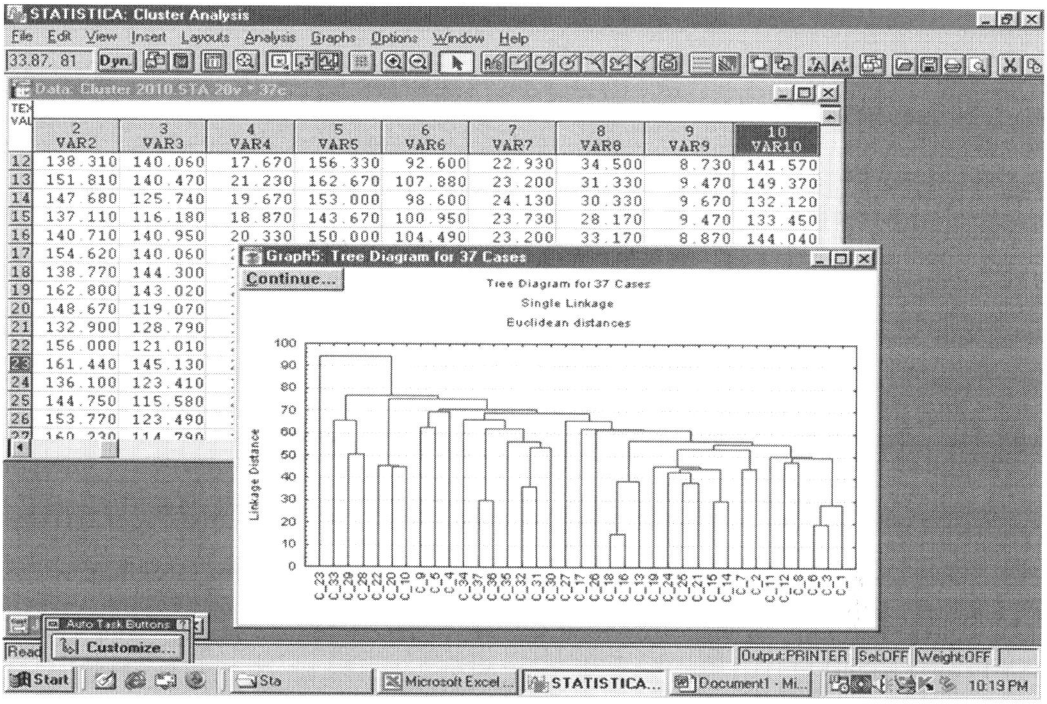

Slide 12.41: Data along with dendrogram for 37 varieties (cases) through cluster analysis using STATISTICA

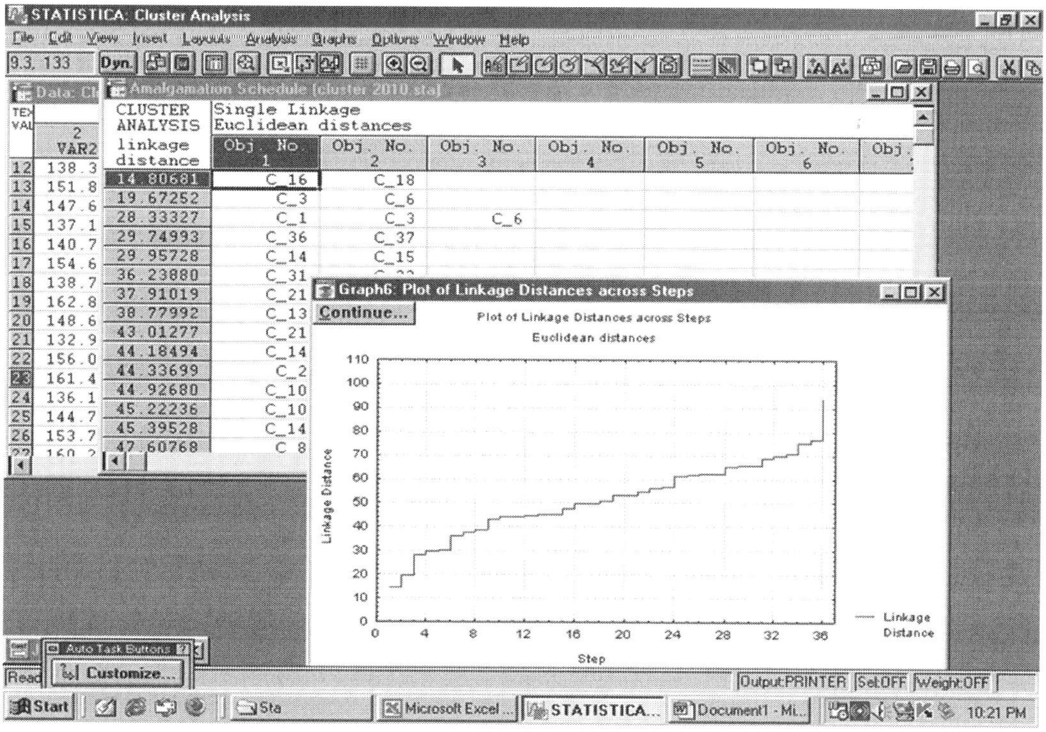

Slide 12.42: Amalgamation schedule along with linkage distance plot through cluster analysis using STATISTICA

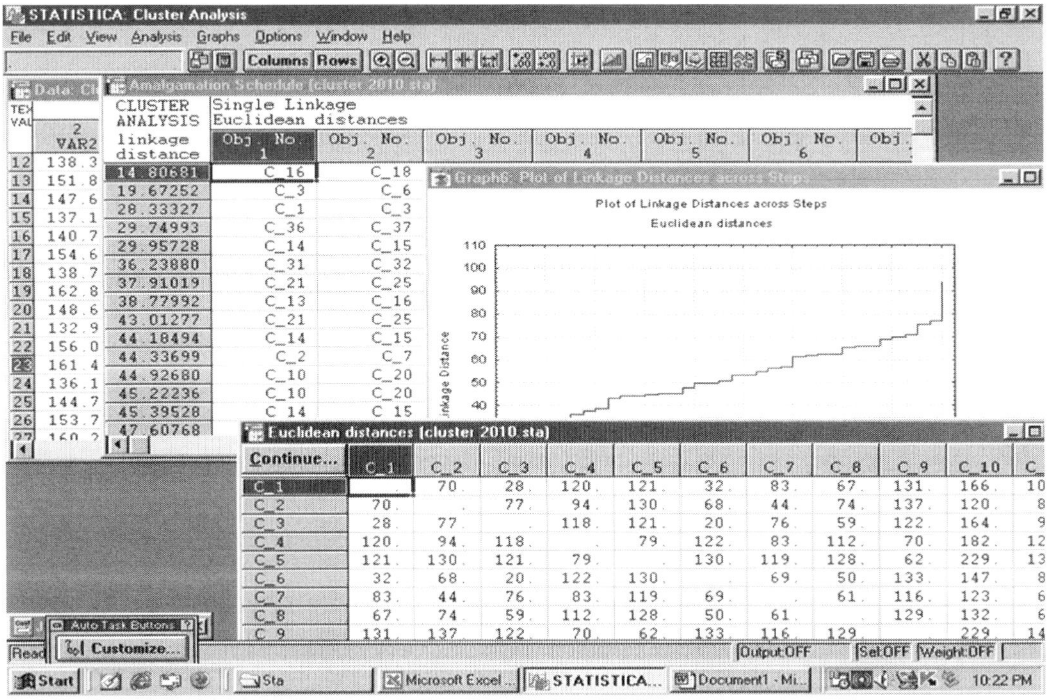

Slide 12.43: Amalgamation schedule along with linkage distance plot and distance matrix among the varieties through cluster analysis using STATISTICA

13 Instrumentation and Computation

13.1 Instruments

Instruments have become an indispensible part of modern scientific research; may it be laboratory/ field research, use of one or more instruments is inevitable. The knowledge of instruments about their function, principle, and safety measure during use is an essential part of research activities. Thousands of instruments with their variants are being put into use in research. Everyday a number of new instruments are being invented and dedicated for the betterment of the humanity. One can hardly provide description of all these instruments at any point of time. In this chapter, our attempt will remain to provide a brief description of some of the mostly used instruments in agriculture and allied fields. It must clearly be noted that this attempt is neither exhaustive nor any effort to list the instruments. The following table gives a glimpse of few instruments used.

P.K. Sahu, *Research Methodology: A Guide for Researchers in Agricultural Science, Social Science and Other Related Fields*, DOI 10.1007/978-81-322-1020-7_13, © Springer India 2013

Instrument	Feature	Image
pH meter	pH is the measure of the dissolved hydrogen ion concentration in a solution. The pH range of 0–14 accounts for hydronium activities from 10 to 1E-14 mol/l. A pH meter measures the pH of a solution utilizing a glass electrode. A pH meter measures essentially the electrochemical potential between a known liquid inside the glass electrode (membrane) and an unknown liquid outside. Because the thin glass bulb allows mainly the agile and small hydrogen ions to interact with the glass, the glass electrode measures the electrochemical potential of hydrogen ions or the *potential of hydrogen*. A pH meter must not be used in moving liquids of low conductivity (thus, measuring inside small containers is preferable).	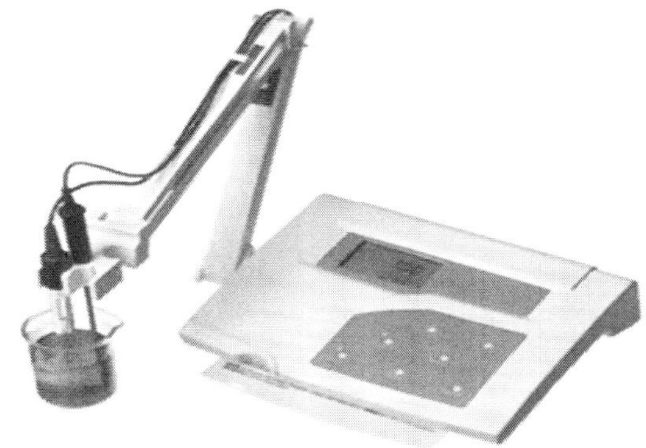 Instant pH Meter 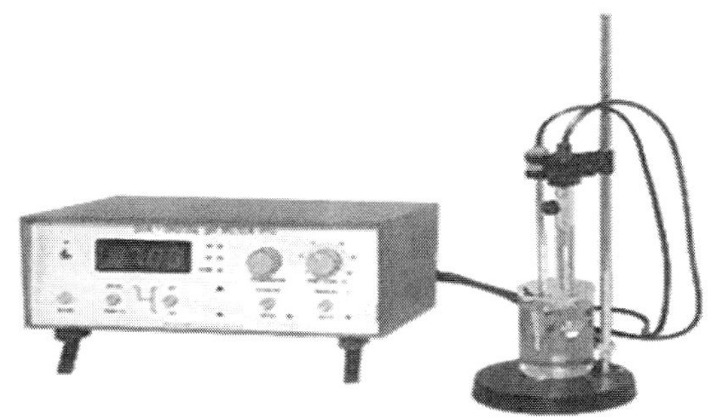 Digital pH meters

Caring for a pH meter depends on the types of electrode in use. Modern pH meters do not mind their electrodes drying out provided they have been rinsed thoroughly in deionized water or potassium chloride. Remember that a liquid of pH = 4 has 10,000 more hydrogen ions than a liquid of pH = 8. Thus, a single drop of pH = 4 in a vial measuring 400 drops of pH = 8 really upsets measurements! Remember also that the calibration solutions consist of chemical buffers that "try" to keep pH levels constant, so contamination of your test vial with a buffer is really serious. pH meters have got extensive use in the field of pharmaceutical, agricultural, wine, and food industry such as manufacturing of soft drinks, butter, and yogurt.	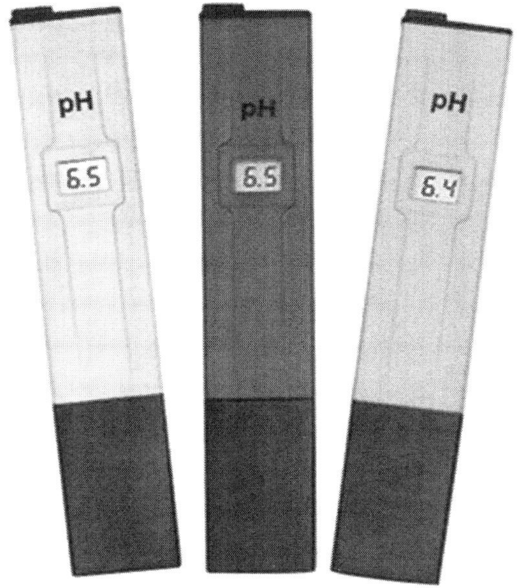 Pen type digital pH meters
There are different types of pH meters like instant pH meter, digital pH meter, electronic digital pH meter, and pH indicators. In instant pH meter one can see the results instantly; the instrument also does not require any calibration with buffer solutions. Digital pH meters are handy and accurate for pH measuring of waste water, chemical, and food and for some laboratory applications. pH indicators are widely used in various industrial applications, and also they are technically designed to meet the industrial requirements	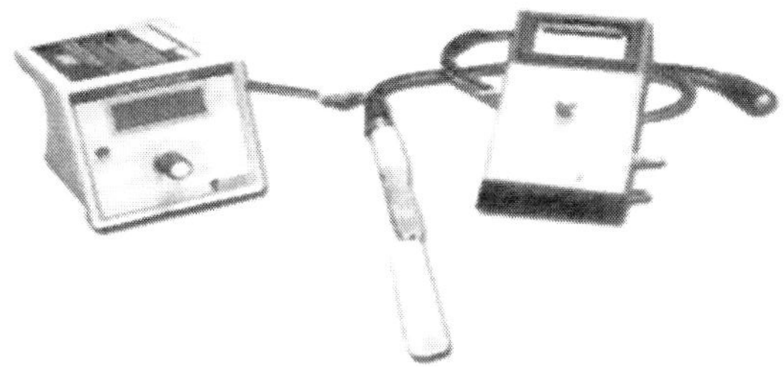 Electronic Digital pH meters

(continued)

(continued)

Instrument	Feature	Image
	While using pH meters depending upon the type, one should be careful on the following points: (a) stabilization of solution is required before taking reading, (b) better to keep the electrodes deep in buffer of 7.00 pH solution when not in use, (c) the pH meter must be calibrated using standard solution before unknown solution is measured, and (d) temperature of the solution also plays an important role in accurate measurement of pH	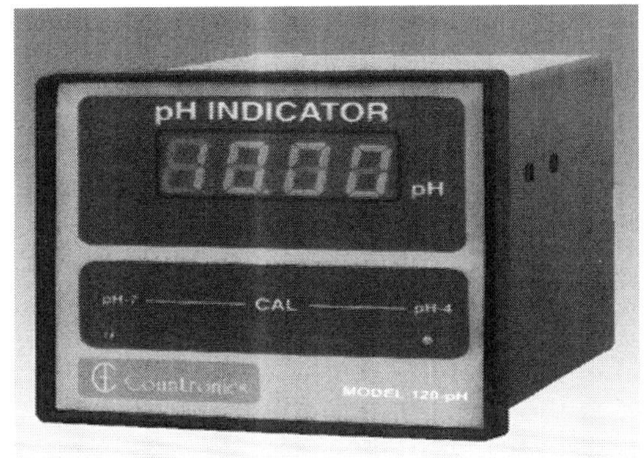 pH Indicators
Laminar flow system	Likewise to that of medical and other research laboratories, agriculture laboratories also require sterile working environments in order to carry out specialized work. Laminar airflows can maintain a working area free from contaminants. Laminar flow cabinets create particle-free working environments by projecting air through a filtration system and exhausting it across a work surface in a laminar or unidirectional air stream. They provide an excellent clean air environment for a number of laboratory requirements. Among different types of cabinets with a variety of airflow patterns for different purposes, the vertical laminar flow cabinets, horizontal laminar flow cabinets, laminar flow cabinets and hoods, and laminar flow benches and booths are widely used	

BOD incubator	An incubator is an instrument used mostly in biological studies which can maintain optimal temperature, humidity, and other conditions for growth of microbiological or cell culture. BOD incubator is an incubator designed to maintain 20°C temperature necessary to perform a test called biochemical oxygen demand (BOD). It involves incubating samples saturated with oxygen at 20°C temperature (usually) for 5 days. There are other incubators designed to maintain temperatures of 5°C or more above ambient to as high as 100°C to study the growth of organisms of under temperate or tropical condition. Incubators designed to maintain temperatures below ambient to as low as about 10°C are generally called low-temperature incubators	
Orbital shaking incubator	This is a special type of incubator useful for life sciences applications, fermentation studies, aging tests, growth studies, and biological cultures under various controlled temperature conditions. Advanced shaking mechanism provides quiet shaking and precise speed control with digital display	

(continued)

(continued)

Instrument	Feature	Image
Digital colony counter	Digital colony counter is an indispensable benchtop tool for the busy microbiologist designed for quick and accurate counting of bacterial and mold colonies in petri dishes or similar experimental units. Simply place the petri dish on the illuminated pad and touch the dish with the pen provided to mark each colony in turn. This causes a count to be registered on the digital display and audible tone confirms each count made. Marking the dish with the pen avoids missing colonies or double counting. The digital count on the display can be reset manually any time by pressing the reset key provided. Glare-free illumination is essential for optimum viewing of colonies	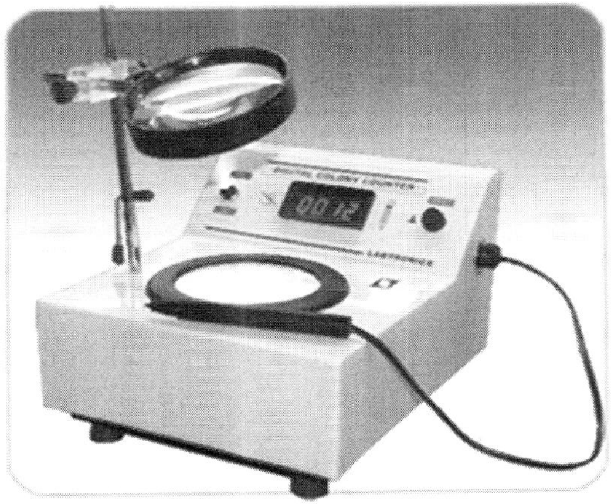
Rotary shaking machine	For proper mixing or making of solution, continuous shaking is necessary. The shaking becomes difficult when preparing a number of such solutions/ mixtures. Rotary shaking machine is there to help the researchers. Continuous shaking at variable speed from 10 to 250 RPM single/double or 3-tier platform unit is the feature of this type of instrument. Speed can be adjusted digitally. Rotary swirling agitation is used extensively in tissue culture work and other chemical mixing procedures. D.C. motor drives the shaker mechanism through mechanical transmission with v-belt drive. Minimum precautionary measures should be taken to maintain the speed, time, and proper placement of the samples	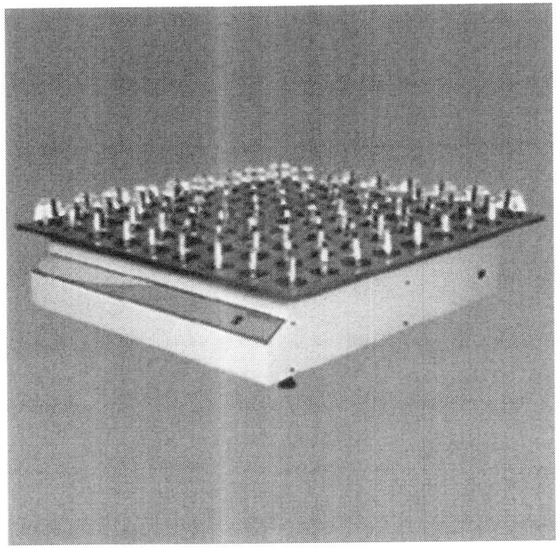

Flame photometer	Reliable and accurate concentration of Na, K, Li, Ca, etc., in solutions remains important in clinical and soil laboratories, and flame photometer plays a great role in this direction. Flame photometry (also known as flame atomic emission spectrometry) is a branch of atomic spectroscopy in which the species examined in the spectrometer are in the form of atoms. The other two branches of atomic spectroscopy are atomic absorption spectrophotometry (AAS) and inductively coupled plasma–atomic emission spectrometry (ICP-AES), a relatively new and very expensive technique. In all cases the atoms under investigation are excited by light. Flame photometry is simple and relatively inexpensive and used for clinical, biological, and environmental analysis. Flame photometry is suitable for qualitative and quantitative determination of several cations, especially for metals that are easily excited to higher energy levels (namely, Na, K, Rb, Cs, Ca, Ba, Cu) at a relatively low flame temperature. Careful and frequent calibration is necessary for good results, and it is very important to measure the emission from the standard and unknown solutions under conditions that are as nearly identical as possible. The processes in a flame photometer include the following stages: desolvation (drying), vaporization, atomization, and ionization. Each of these stages includes the risk of interference in case the degree of phase transfer is different for the analyte in the calibration standard and in the sample	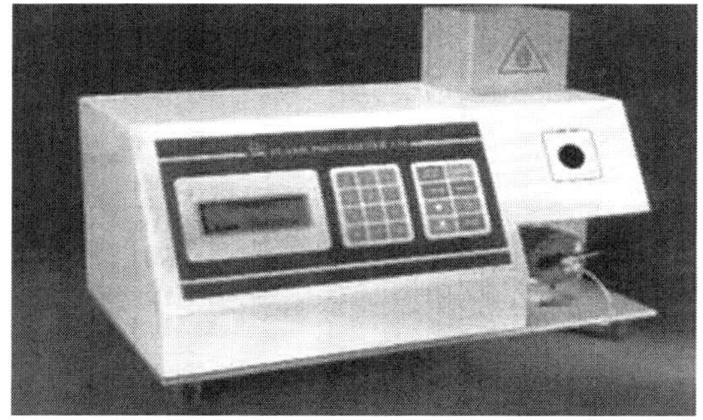
Hot air oven and hot air sterilizer	An electrical instrument used for sterilization. The oven uses dry heat 50–300°C (generally) to sterilize articles. These are widely used to sterilize articles that can withstand high temperatures and not get burned, like glassware and powders. Linen gets burned and surgical sharps lose their sharpness. Generally, a digitally controlled thermostat maintains the temperature. Double walled insulation separated by an air-filled space in between keeps the heat in and conserves energy. An air circulating fan helps in uniform distribution of the heat. The capacities of these ovens vary. Power supply needs vary from country to country, depending on the voltage and frequency (hertz) used. Temperature-sensitive tapes or other devices like those using bacterial spores can be used to work as controls, to test for the efficacy of the device in every cycle. They do not require water and there is not much pressure buildup within the oven, unlike an autoclave, making them safer to work with. A complete cycle involves heating the oven to the required temperature, maintaining that temperature for the proper time interval for that temperature, turning the machine off, and cooling the articles in the closed oven till they reach room temperature	
	If the door is opened before time, heat escapes and the process becomes incomplete. Thus, the cycle must be properly repeated all over	

(continued)

(continued)

Instrument	Feature	Image
Autoclave	An autoclave is a device used to sterilize equipment and supplies by subjecting them to high-pressure saturated steam at 121 °C for around 15–20 min depending on the size of the load and the contents. The name comes from Greek *auto-*, ultimately meaning self, and Latin *clavis* meaning key—a self-locking device. Autoclaves are widely used in microbiology, medicine, tattooing, body piercing, veterinary science, mycology, dentistry, chiropody, and prosthetics fabrication. They vary in size and function depending on the media to be sterilized. Typical loads include laboratory glassware, surgical instruments, medical waste patient pair utensils, and animal cage bedding. A notable growing application of autoclaves is the predisposal treatment and sterilization of waste material, such as pathogenic hospital waste. Machines in this category largely operate under the same principles as conventional autoclaves in that they are able to neutralize potentially infectious agents by utilizing pressurized steam and superheated water. Autoclaves are also widely used to cure composites and in the vulcanization of rubber. The high heat and pressure that autoclave creates allow to ensure the best possible physical properties of the material used for sterilization	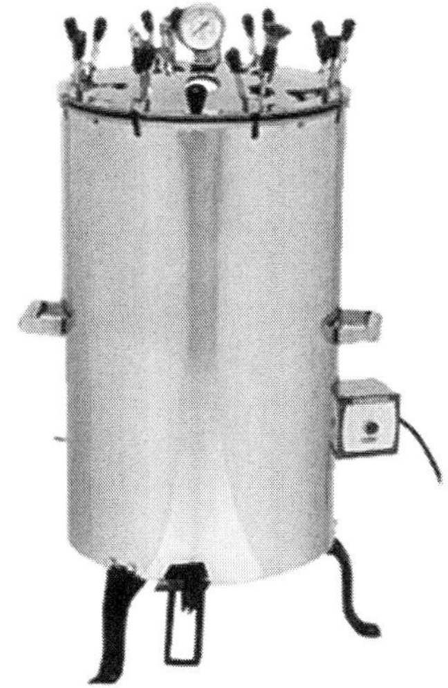
	While using this instrument in the laboratory, care must be taken such that the pressure and temperature do not cross beyond the capacity of the instrument; otherwise, there is every possibility of meeting accidents. Maintenance of appropriate time, temperature, and pressure is the key point to be noted for highest effectivity	 Stovetop autoclaves—the simplest of autoclaves

Analytical balance	Analytical balance is used to measure mass to a very high degree of precision and accuracy. The measuring pan(s) of a high precision (0.1 mg or better) analytical balance is inside a transparent enclosure with doors so as to avoid dust particles and airflow in affecting the balance's operation. Analytical precision is achieved by maintaining a constant load on the balance beam, by subtracting mass on the same side of the beam to which the sample is added. Precaution should be taken to avoid air flow during recording of weights and under dust-free condition	
Compound microscope	Microscope is a tool used to view small tiny elements, which are mostly not clearly visible by the naked eye. In life science, medical science, and other branches, microscope is an essential instrument in research. As light from a source passes through the object, the lens nearest the object, the objective lens, produces an enlarged image of the object in the primary image angle. The lens that you look into, the eyepiece, acts as a magnifier and produces an enlarged image of the image produced by the objective lens. The magnification is the product of the eyepiece magnification by the magnification of the objective lens, usually 4×, 10×, 40×, and 100×. For example, a 10× eyepiece in conjunction with a 40× objective will give you a magnification factor of 400; that means the object will be magnified 400 times larger than you can view it with the naked eye. Viruses, molecules, and atoms are beyond the capabilities of today's compound microscopes and can be viewed only with an electron microscope. Always use immersion oil while viewing any object using 100× eyepiece to prevent the damage of the lens. Wipe off the lens gently with tissue paper soaked in xylene to remove the oil after use. Never use a dry cloth or paper towel to wipe any optical surface as you could scratch a lens. Use an air blower or a camel hair brush to whisk away dust. If there is dirt on the eyepiece that can't be removed with air or the brush, gently wipe it with a piece of clean cotton	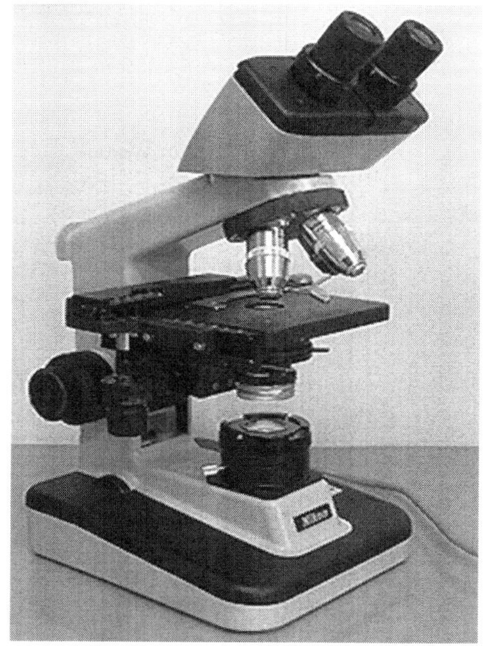

(continued)

(continued)

Instrument	Feature	Image
Stereoscopic microscope	The stereo or dissecting microscope is an optical microscope variant designed for low-magnification observation of a sample using incident light illumination. It uses two separate optical paths with two objectives and two eyepieces to provide slightly different viewing angles to the left and right eyes. In this way it produces a three-dimensional visualization of the sample being examined. Stereomicroscopy overlaps macrophotography for recording and examining solid samples with complex surface. The stereo microscope should not be confused with a compound microscope equipped with double eyepieces and a binoviewer. In such a microscope, both eyes see the same image, but the binocular eyepieces provide greater viewing comfort. However, the image in such a microscope is no different from that obtained with a single monocular eyepiece	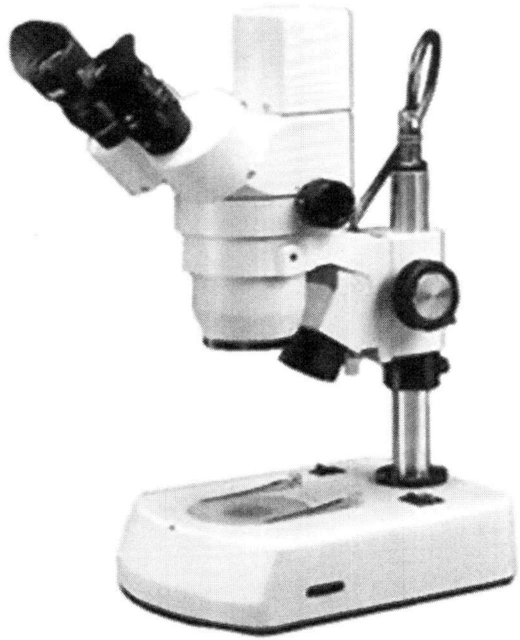

Electron microscope	An electron microscope is the modern variant of microscope. It uses a beam of electrons to illuminate a specimen and produce a magnified image. An electron microscope has greater resolving power than a light-powered optical microscope because electrons have wavelengths about 100,000 times shorter than visible light (photons). Magnifications of up to about 10,000,000× can be achieved through these microscopes. Electron microscopes are used to observe a wide range of biological and inorganic specimens including microorganisms, cells, large molecules, biopsy samples, metals, and crystals. Industrially, the electron microscope is often used for quality control and failure analysis. As such these are used in all types of research activities. Electron microscope may be (1) transmission microscope (TEM), (2) scanning electron microscope, (3) reflection electron microscope, (4) scanning transmission electron microscope, and (5) low-voltage electron microscope	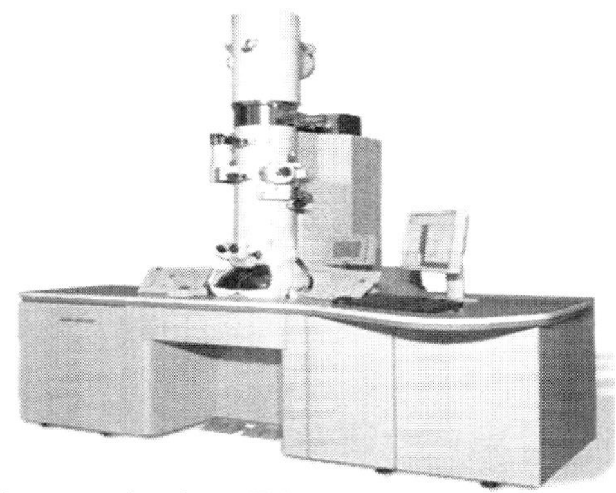 Transmission Electron Microscope 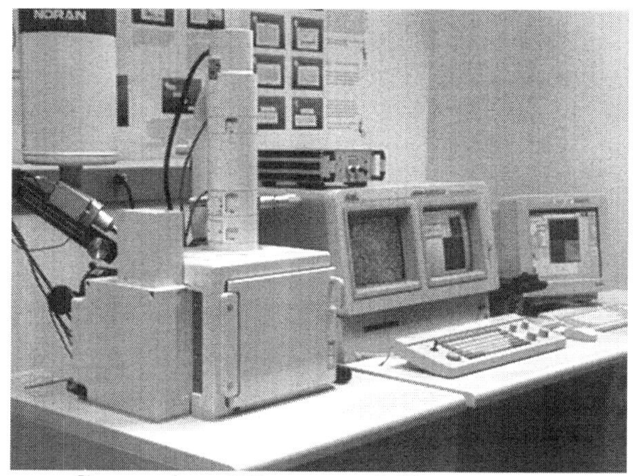 Scanning Electron Microscope

(continued)

(continued)

Instrument	Feature	Image
Weather station	Weather parameters have become indispensible part of modern research, particularly in the field of life science and agriculture. Because of tremendous development associated with other advantages, nowadays, *automatic weather stations* (AWS) are preferred over manual weather station. Measurements on parameters can also be made from remote areas through the use of these stations. Most automatic weather stations have thermometer for measuring temperature, anemometer for measuring wind speed, wind vane for measuring wind direction, hygrometer for measuring humidity, and barometer for measuring atmospheric pressure. Some of them even have ceilometer for measuring cloud height, rain gauge along with data logger, rechargeable battery, and telemetry, and the meteorological sensors with an attached solar panel or wind turbine are mounted upon a mast. The specific configuration may vary due to the purpose of the system. The system may report in near real time via the Argos Systems and the Global Telecommunication System or save the data for later recovery	 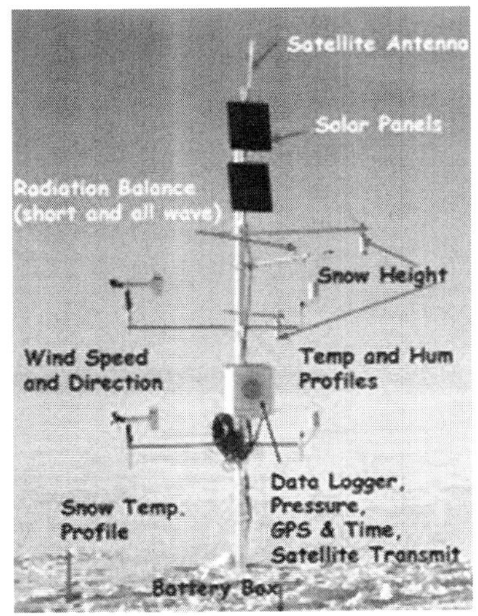

Spectroscopy	Spectroanalytical procedures are nowadays used for determination of concentration of elements in solutions. Atomic absorption spectroscopy (AAS) is one of such quantitative procedures employing the absorption of optical radiation (light) by free atoms in the gaseous state. AAS can be used to determine over 70 different elements in solution or directly in solid samples	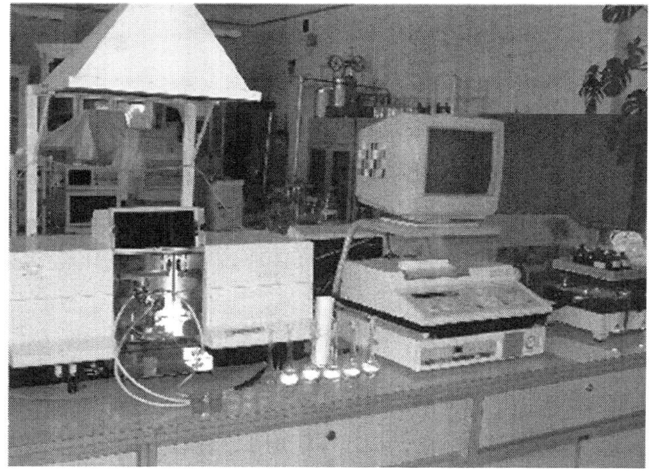
	Based on Beer–Lambert law, it requires standards with known analyte content to establish the relation between the measured absorbance and the analyte concentration. The principle of AAS lies on promoting the electrons of the atoms into higher orbital (excited state) for a few nanoseconds by absorbing a defined quantity of energy (i.e., radiation of a given wavelength) which is specific to a particular electron transition in a particular element because, in general, each wavelength corresponds to only one element	

(continued)

(continued)

Instrument	Feature	Image
PCR unit	One of the major tools in molecular biology is the PCR. The *polymerase chain reaction* (PCR) unit is used to amplify a single or a few copies of a piece of DNA across several orders of magnitude, generating thousands to millions of copies of a particular DNA sequence. PCR is now a common and often indispensible technique used in medical and biological research labs for DNA cloning for sequencing, diagnosis of hereditary diseases, identification of genetic fingerprints (in forensic sciences and paternity testing), detection and diagnosis of infectious diseases, etc. A basic PCR setup requires several components and reagents like DNA template, two primers, Taq polymerase deoxynucleoside triphosphates, buffer solution, divalent cations, and monovalent cation. The researchers must be careful because PCR can fail for various reasons, in part due to its sensitivity to contamination causing amplification of spurious DNA products. According to the variations on the basic PCR technique, it can be allele-specific PCR, asymmetric PCR, dial-out PCR, hot start PCR, intersequence-specific PCR (ISSR), inverse PCR, ligation-mediated PCR, methylation-specific PCR, miniprimer PCR, multiplex ligation-dependent probe amplification (MLPA), multiplex PCR, nested PCR, overlap extension PCR, splicing by overlap extension (SOE), quantitative PCR (qPCR), reverse transcription PCR (RT-PCR), solid phase PCR, thermal asymmetric interlaced PCR (TAIL-PCR), touchdown PCR (step-down PCR), PAN-AC, universal fast walking, and in silico PCR (digital PCR, virtual PCR, electronic PCR, e-PCR)	 A thermal cycler 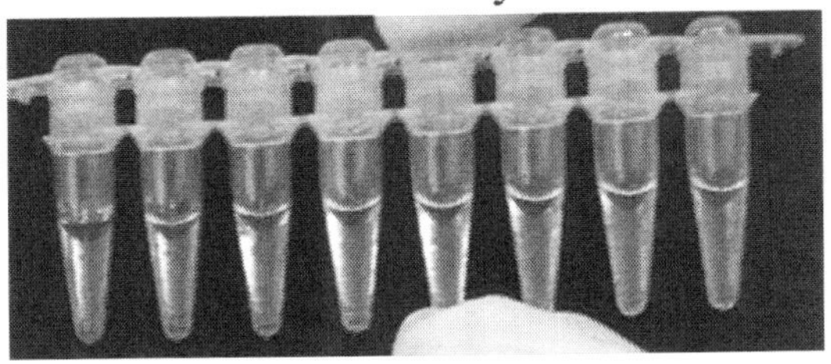

Chromatography	Separation of constituents from the mixture has remained the main focal point in chromatography. Chromatography is the collection of a set of laboratory techniques. There are two phases—the *mobile phase*, which carries, and the *stationary phase*. The various constituents of the mixture travel at different speeds, causing them to separate. Chromatography may be preparative (separate the components of a mixture) or analytical (measuring the relative proportions of analytes in a mixture). Chromatography may be categorized by (a) *chromatographic bed shape* (column chromatography, planar chromatography, paper chromatography, thin-layer chromatography) and (b) *physical state of mobile phase* (gas chromatography, liquid chromatography, etc.)	 Gas chromatograph
HPLC	One of the important chromatographic techniques in analytical chemistry and biochemistry is the high-performance liquid chromatography, also known as high-pressure liquid chromatography (HPLC), used to separate a mixture of compounds for identifying, quantifying, and purifying the individual components of the mixture. HPLC has many uses in medical, purifying substances from a complex biological sample, manufacturing, pharmaceutical, quality assurance, etc.	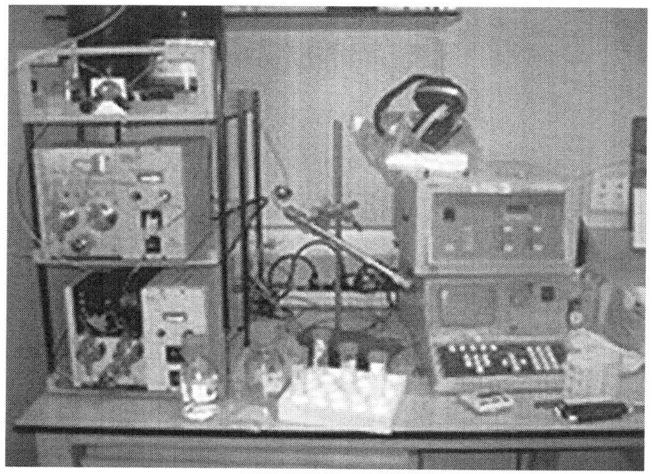 HPLC unit

(continued)

(continued)

Instrument	Feature	Image
Computer	A landmark in the history of science is the invention of computer. Computer is an electronic device that is designed to work with information. The term "computer" *is* derived from the Latin term *computare*; this means to calculate. Computer cannot do anything without a program. Computers not only help in calculation but also have become integrated components of many modern instruments. A general-purpose computer has four main components: the arithmetic logic unit (ALU), the control unit, the memory, and the input and output devices (collectively termed I/O). These parts are interconnected by busses, often made of groups of wires. Inside each of these parts are thousands to trillions of small electrical circuits which can be turned off or on by means of an electronic switch. The circuits are arranged in logic gates so that one or more of the circuits may control the state of one or more of the other circuits. The control unit, ALU, registers, and basic I/O (and often other hardware closely linked with these) are collectively known as a central processing unit (CPU). A *peripheral* is a device connected to a host computer, but not part of it. It expands the host's capabilities but does not form part of the core computer architecture. It is often, but not always, partially or completely dependent on the host	 Computer and peripheral

13.2 Laboratory Safety Measures

Safety measures can broadly be categorized into two groups—*the human safety* and the *instrumental or laboratory safety*. Almost all laboratory equipments are sensitive and can become dangerous if safety precautions are not taken properly. Many equipments are something that one can't just drag around and store anywhere because most equipments have their conditions and specifications in installing, use, and storing. Most of the instruments achieve high levels of performance due to carefully designed interface between external connectors and internal components. As a result, specific handling precautions must be observed for device reliability and optimum performance. An appropriate safety precaution to use with a laboratory instrument depends on the type of instrument and analysis to be carried out. For example, if one is going to use an instrument having furnace, then one is required to be careful about fire, overheating, burning, etc. On the other hand, if the instrument uses compressed gas, one needs to be aware of how to make and maintain tight connections to the gas source and may need to look at adequacy of ventilation. While using the instrument, one should be careful about the following points, though these are not exhaustive:

1. The equipment should be placed away from heat sources. Do not block airflow around equipment. If equipment be operated at high ambient temperatures, mount with a good thermal connection to a large thermal mass.
2. Do not allow foreign material into the enclosure. Do not allow contamination to be introduced into the connectors.
3. If applicable, always use the provided AC adaptor. Do not power the unit with a different adaptor. Do not modify the power plug or wall outlet to remove the third (ground) pin.
4. Do not drop or shake the equipment. Minimize vibration and handle with care.
5. Make sure to read the instructions in the lab equipment and do not do experiments on your own.
6. If you don't know how to use the lab equipment, it's better to ask the expert to avoid mistakes caused by hunches.

13.2.1 Precautionary Measures

Given below are some of the precautionary measures one should follow while using the instruments:

1. *Autoclaves*, heat sterilizers, and pressure cookers should be run only by experts or professionals. In using a pressure cooker, check the safety valve before pressure is built up. The equipment should be turned off and allowed to cool before the stopcock is opened to equalize pressure.
2. *Incubators* may be useful components of a biology laboratory; care should be taken to keep incubators safe and well maintained. Unwanted growth of organisms should be restricted by regular cleaning.
3. *Microscope*: if it has a cover or case, always put it in place when you aren't using the device. Wet or dirty slides should never be put on the stage, which should always be kept dry. When cleaning your microscope, unplug it first, if applicable, and then clean the outside using a damp, soft cloth only. Never use a dry cloth or paper towel to wipe any optical surface as it could scratch lens. Use an air blower or a camel hair brush to whisk away dust. If there is dirt on the eyepiece that can't be removed with air or the brush, gently wipe it with a piece of clean cotton.
4. *Chromatography*
   (a) Dissolving and developing solvents give off toxic vapors. They must be stored in closed containers and the room should properly be ventilated.
   (b) Solvents are highly flammable and must not be used near an open flame.
   (c) Avoid skin contact when spraying the developing solvents.
   (d) Use a fume hood when appropriate.

5. *Biotechnology*
   (a) Handle all microorganisms and DNA carefully. Treat them as if they could cause infections.
   (b) Hands should be washed with soap and water before and after handling microorganisms and before leaving the laboratory regardless of what materials were used. When handling microorganisms or other living materials, wear rubber gloves to protect against infection.
   (c) Use only mechanical pipetting devices for transferring any material. Do not allow mouth pipetting.
6. *Greenhouse maintenance and operation*
   (a) Check waterlines, heating system, fans, and temperature control. These are usually routine procedures but must be done every time.
   (b) Make sure all automatic equipments are functional and accurate.
   (c) Clean tools after use and store them appropriately.

13.2.2 Human Safety Measures

(a) Use safety equipment in performing tests and experiments such as safety goggles to avoid anything from entering your eyes, which could be anything from chemicals to shrapnel.
(b) Don't eat or drink in the laboratory or near lab equipment that has chemicals or sample in them. While you may want to feel refreshed and relaxed in the lab, it's not worth to risk in case you swallow something by mistake.
(c) Many people are allergic to pollen, mold spores, or other plant exudates. When using flowers, mushrooms, fungi, etc., in the laboratory, adequate ventilation is essential. Pollen and mold spores should be displayed in closed glass petri dishes.
(d) Do not apply cosmetics in the laboratory. Keep fingers and writing instruments away from your face and mouth.

13.2.3 Overall Laboratory Condition

(a) Keep the laboratory clean.
(b) Disinfect the work area before and after each laboratory procedure. Use of a commercial disinfectant to wipe down the area is acceptable.
(c) While working the researcher should use gloves, chemical splash safety goggles, and aprons as would be found necessary.
(d) Containers should be cleaned before and after use.
(e) Do not leave laboratory materials unattended or ill maintained. They should be cleaned out regularly to prevent unwanted growth of organisms.
(f) Laboratory should be well equipped with proper lighting, adequate heat, and water supply along with fire extinguisher.
(g) Commercial potting mixtures are recommended over garden soil because they are relatively sterile.

13.3 Computer, Computer Software, and Research

By this time, the importance of statistics to a researcher is clear. Statistics are best used when it is supported by strong computing facilities. The need for developing computing facility was felt time and again. Many of the statistical tools would not have been used at a large scale if there was no effective development in the computing facilities. Development of computers and statistical software has made it possible to have wider application of statistics in unearthing and/or explaining the so long hidden truth of this universe. But this development is not unidirectional or flawless. For best application, understanding the theories and situations where actually the specific statistical tools are required to be used is essential. Statistical theories are used best by the subject matter specialists in consultation with an efficient statistician. Understanding of both the specialists towards the field of each other to a certain degree is essential for efficient use of statistical theories towards advancement of human civilization. Statistics is just like a molded clay; one can make God or devil out of it as per the choice of the user. Unfortunately, in many of the cases, statistical calculations have been made with the help of the computer packages, without knowing the logic and utilities. Misuse of statistical concept is increasing day by day. We must be cautious about garbage in, garbage out (GIGO). The researcher must have a clear idea about what are the requirements, what are inputted to the computer, and what are the commands to be given to the computers to get the required information analyzed properly befitting to the requirement of the objectives of the research program. A computer does not have any brain; the only thing it can do is the use of stored program as per the direction of the user. Using a set of data, one can get innumerable types of output just by changing instruction to the computer. But definitely all of these outputs are neither relevant nor correct under the given situation. For best use of the computer, it is feasible to get from the information fed to the computer using statistical theories, what should be the direction to the computer and what is the output generated by the computer. In the following example, we shall demonstrate how many types of statistical analysis could be taken up. While using statistical software, the following points are required to be noted:

1. The background information about the data
2. Nature of the data
3. Objectives of the experiment
4. Hypothesis to be tested
5. Appropriate statistics to be calculated for 3 and 4 using the specific type of data
6. Background knowledge about the statistical package to be used w.r.t. its algorithm, techniques used in performing the task
7. Testing the accuracy of the software using solved identical example
8. Knowledge about the explanation of the output after running the statistical package and linking these with the objective of the study

Example 13.1 The following table gives yield (q/ha) of a particular paddy variety. The experiment was conducted with three types of manure and three doses of nitrogen to standardize the supplement of inorganic nitrogen by manure, and the experiment was conducted under field condition for two consecutive seasons. The problem is to find out the best dose of nitrogen and best manure to get the best yield.

	Nitrogen	N1	N1	N1	N2	N2	N2	N3	N3	N3
	Rep	R1	R2	R3	R1	R2	R3	R1	R2	R3
Manure 1	Season 1	57.80	65.06	67.78	68.77	70.37	76.47	76.07	64.67	75.37
	Season 2	58.43	65.69	68.41	71.37	61.01	68.07	68.81	69.31	70.10
Manure 2	Season 1	76.27	66.37	69.17	59.47	71.47	72.77	69.77	77.63	71.84
	Season 2	70.11	66.29	70.14	68.31	74.30	73.72	70.82	71.56	72.87
Manure 3	Season 1	69.07	87.74	81.70	83.07	90.48	72.66	71.86	62.92	61.80
	Season 2	73.71	75.38	72.34	73.81	73.12	73.26	72.61	71.76	68.54

Before analyzing the data information on the type design followed is a must. But most frequently it is found that after completion of the experiment, the researcher searches for appropriate method of data analysis. Actually in designs of experiments, once the experimental design is fixed, its analysis is also fixed. In the absence of full information, the above data could be analyzed in various forms, and the results of all these analyses are not identical. Here, some of the possible analyses along with experimental design assumed in each stage are provided. Readers may note that these are not exhaustive; one can have other types of analysis with the same data. We have presented the ANOVA tables only, from which it is clear that there exist differences among the analyses and as such the interpretation will also vary. Thus, appropriate analysis of the data is required to extract the information. Otherwise, misleading

Table 13.1 Analysis of data using three-factor randomized complete block design

Analysis of variancetable

K value	Source	Degrees of freedom	Sum of squares	Mean square	*F*-value	Prob
1	Replication	2	25.020	12.510	0.4745	
2	Season	1	36.787	36.787	1.3955	0.2457
4	Manure	2	351.847	175.923	6.6735	0.0036
6	SM	2	28.528	14.264	0.5411	
8	Nitrogen	2	53.707	26.853	1.0187	0.3718
10	SN	2	32.885	16.442	0.6237	
12	MN	4	454.189	113.547	4.3073	0.0063
14	SMN	4	190.808	47.702	1.8095	0.1498
−15	Error	34	896.296	26.362		
	Total	53	2070.066			

Table 13.2 Analysis of data using randomized complete block design for season, with manure and nitrogen as split plots on season

Analysis of variancetable

K value	Source	Degrees of freedom	Sum of squares	Mean square	*F*-value	Prob
1	Replication	2	25.020	12.510	1.4902	0.4016
2	Season	1	36.787	36.787	4.3822	0.1714
−3	Error	2	16.789	8.395		
4	Manure	2	351.847	175.923	6.4008	0.0046
6	SM	2	28.528	14.264	0.5190	
8	Nitrogen	2	53.707	26.853	0.9770	
10	SN	2	32.885	16.442	0.5982	
12	MN	4	454.189	113.547	4.1313	0.0082
14	SMN	4	190.808	47.702	1.7356	0.1665
−15	Error	32	879.507	27.485		
	Total	53	2070.066			

Table 13.3 Analysis of data using randomized complete block design for season and manure with nitrogen as a split plot on season and manure

Analysis of variancetable

K value	Source	Degrees of freedom	Sum of squares	Mean square	*F*-value	Prob
1	Replication	2	25.020	12.510	0.5800	
2	Season	1	36.787	36.787	1.7055	0.2208
4	Manure	2	351.847	175.923	8.1563	0.0079
6	SM	2	28.528	14.264	0.6613	
−7	Error	10	215.691	21.569		
8	Nitrogen	2	53.707	26.853	0.9469	
10	SN	2	32.885	16.442	0.5798	
12	MN	4	454.189	113.547	4.0040	0.0126
14	SMN	4	190.808	47.702	1.6821	0.1869
−15	Error	24	680.605	28.359		
	Total	53	2070.066			

conclusion may be drawn. In fact in statistical software, there are a varied range of options for execution of analysis; one must be sure about the exact procedure to be adopted for that knowledge on analytical tools and the specifications/command of the statistical software are essential (Tables 13.1, 13.2, 13.3, 13.4, 13.5, and 13.6).

Table 13.4 Analysis of data using randomized complete block design for season, with manure as a split plot on season and nitrogen as a split plot on manure

Analysis of variancetable						
K value	Source	Degrees of freedom	Sum of squares	Mean square	*F*-value	Prob
1	Replication	2	25.020	12.510	1.4902	0.4016
2	Season	1	36.787	36.787	4.3822	0.1714
−3	Error	2	16.789	8.395		
4	Manure	2	351.847	175.923	7.0758	0.0170
6	SM	2	28.528	14.264	0.5737	
−7	Error	8	198.902	24.863		
8	Nitrogen	2	53.707	26.853	0.9469	
10	SN	2	32.885	16.442	0.5798	
12	MN	4	454.189	113.547	4.0040	0.0126
14	SMN	4	190.808	47.702	1.6821	0.1869
−15	Error	24	680.605	28.359		
	Total	53	2070.066			

Table 13.5 Analysis of data using two-factor randomized complete block design combined over seasons

Analysis of variancetable						
K value	Source	Degrees of freedom	Sum of squares	Mean square	*F*-value	Prob
1	Season	1	36.787	36.787	1.3384	0.2559
3	R(S)	4	41.809	10.452	0.3803	
4	Manure	2	351.847	175.923	6.4008	0.0046
5	SM	2	28.528	14.264	0.5190	
8	Nitrogen	2	53.707	26.853	0.9770	
9	SM	2	32.885	16.442	0.5982	
12	MN	4	454.189	113.547	4.1313	0.0082
13	SMN	4	190.808	47.702	1.7356	0.1665
−15	Error	32	879.507	27.485		
	Total	53	2070.066			

Table 13.6 Analysis of data using two-factor randomized complete block design with split plot combined over seasons

Analysis of variancetable						
K value	Source	Degrees of freedom	Sum of squares	Mean square	*F*-value	Prob
1	Season	1	36.787	36.787	1.4796	0.2585
3	R(S)	4	41.809	10.452	0.4204	
4	Manure	2	351.847	175.923	7.0758	0.0170
5	SM	2	28.528	14.264	0.5737	
−7	Error	8	198.902	24.863		
8	Nitrogen	2	53.707	26.853	0.9469	
9	SN	2	32.885	16.442	0.5798	
12	MN	4	454.189	113.547	4.0040	0.0126
13	SMN	4	190.808	47.702	1.6821	0.1869
−15	Error	24	680.605	28.359		
	Total	53	2070.066			

14 Research Proposal and Report Writing

14.1 Research Proposal

Research is a systematic process and research proposal is the documentation of the process in a stepwise manner. Discussions have been made on the type of research from a qualitative point of view. For successful implementation of research program, funding is a major aspect; at the same time there are various established and reputed research organizations to carry out different types of researches. Each and every research organization has clear-cut guidelines for framing research proposals. Besides, the above researches are also carried out by the individual researcher. Moreover, there are certain researches which do not require dependence on financial support for carrying out research program. However, a research program should always be documented in the form of research proposals. The content, steps, and procedure of writing a research proposal may vary to some extent, but there are certain common points to be documented in any research proposal. A research proposal is a standalone document which clarifies what the proposed project is about, what it is trying to do and achieve, how it will go about doing that, what we will learn from it, and why it is worth learning. It is a document written to convince funding agencies and academic bodies that the project is worth their attention. If the research proposal is asking for financial support, then there are several components to a strong grant application. First, the subject must be creative, exciting, and worthy of funding. Second, the project must have been developed through a rigorous, well-defined experimental plan. One of the most important points to consider when presenting a research proposal for funding is presenting the information in crystal clear language with the application following the rules and guidelines of the funding authority. The research idea, questions, or problems must be very clearly stated and persuasive and address a demonstrable gap in the existing literature. One must be sure that the departmental staffs are interested in the subject area and available for the project. One must also ensure that the scope of the project is reasonable and must remember that there are significant limits to the size and complexity of a project that can be completed and written up within a given period of time. Assessment of proposals takes place not only for their intellectual ambition and significance but also for the likelihood that the researcher can complete this project. International agencies, particularly the agencies in the Western countries, use objective criteria in screening the proposals. In India, the evaluation is subjective, that is, by peer review. Experts review the projects, and few promising projects are approved directly; some are accepted with modifications, and a fairly large number of projects are rejected. Most rejections are mainly attributed to inappropriate presentation style.

P.K. Sahu, *Research Methodology: A Guide for Researchers in Agricultural Science, Social Science and Other Related Fields*, DOI 10.1007/978-81-322-1020-7_14, © Springer India 2013

While reviewing a project proposal, the following points are generally given importance along with other factors: (a) how best are the intellectual quality and merit of the study; (b) what is its potential impact; (c) how holistic is the proposal, whether the research proposal is likely to produce new data and concepts or confirm existing hypotheses; (d) are the hypotheses valid and whether these have been presented with supporting evidences; (e) whether the aims are logical; (f) whether the procedures proposed are appropriate, adequate, and feasible for the research; (g) whether the investigators/proposers are qualified and competent enough as shown by their credentials and experience; (h) are the facilities adequate and the environment conducive to the research; and (i) is there any other organization where the similar types of work are being conducted, if so how the present proposal is different from that, and so on.

In the following section, an attempt is made to note down the possible steps in research proposal.

14.1.1 Title

The title is the stepping stone to a project proposal. It gives first impression of a proposal and as such is required to be catchy, small, and informative. The title should not be the same as the objective. It should be shorter and at the same time must indicate broadly what is being attempted to.

14.1.2 Introduction and Rationale

In this part of the research proposal, the researcher makes an attempt to put forward a research problem, which he/she has conceived. This section concerns with background information, urgency, critical gaps in knowledge, and need for the present study. The background part of the introduction deals with a context having a set of problems. In the process, the researcher should emphasize on the origin of the problems, its importance, and its impact on society. A researcher should try to make clear why the problem was selected and why it has an importance to the society and thereby justify the research approach.

14.1.3 Review of Literature

Once after the synthesis of the problem, the researcher should try to search whether the same types of work are already taken up in the same area, if any, their short falls, and how previous work could be improvised or thought afresh to solve the present problem. This will help the researcher to concretize the research idea and the methodology to be adopted or developed.

14.1.4 Objectives and Specific Objectives of the Study

Based on the above two stages, the researcher is now better placed to formulate the objective and specific objective of the research program proposed. The objectives of the study should clearly be spelled out. Generally there are objective(s), and to each and every objective, specific objectives are there, which are generally followed by hypothesis to be tested (not in all cases). In all these efforts, the researchers should try to clear out what he/she wants to achieve, for whom the objectives are valuable, whether the objectives are in measurable form or not, and of course how far the objectives are realistic and achievable under the given situation. It should be clearly noted that for mandatory researches, the objectives, specific objectives, and the hypotheses are mostly stipulated by the research organizations or the funding agencies. The researcher may have options to slightly modify or reorient keeping the mandate in mind.

14.1.5 Materials and Methodology

This is purely a technical section. This section clearly answers to questions on how to realize the objectives. Approaches with details and references wherever possible must be provided. Facilities available and additional resources needed and the method of acquiring resources must be specified. To fulfill the objectives, any research proposal should clearly spell out the information/data required and to be used, how the data are to be collected, and the methodology for analysis of the data. Depending on the type of research, whether the primary data or secondary data are to be used must be clearly mentioned. The analytical tools to be used or to be developed must also be cited clearly so that the reviewer of the project can understand that using the information and methodology mentioned, the objectives of the project would be achieved.

14.1.6 Time Trend

In a project proposal, the researcher should draw up the time trend for the completion of different steps of the project along with the output expected. This will help monitor the progress of the research project. Activity-wise time frame and investigators who will be held responsible for carrying out the specific work must clearly be spelled out. Scheduling of works provided in the methodology in a sequence is to be provided. Generally standard project management techniques like flow chart and PERT are used to illustrate this. The proposer must specify facilities available and additional resources needed and the method of acquiring resources. Specify activity-wise time frame. Give milestones for each objective. Elaborate how the work will be managed.

14.1.7 Financial Outlay

Other than the institutional research project, all types of research project proposals should contain the budgetary requirement. Even institutional researches need budgetary allocation for smooth functioning of the project. The researcher must provide year-wise and activity-wise budget in detail and justify costly equipments and other facilities. Try to include reasonable inflation while calculating year-wise budget. The breakup of budget requirement for different components of research and justifications thereof must be provided. Unless and otherwise stipulated by the funding agency, the researcher should have such a budgetary provision so as to complete the research process effectively. Generally, there are three parts in budgetary provisions: (1) *the recurring contingency*, (2) *the nonrecurring contingency*, and (3) *salaries and wages*. In ad hoc projects, provisions for salaries are generally avoided. Under the recurring contingency, generally the consumable, the expenses for day-to-day functioning, traveling allowances, etc., are included. In nonrecurring contingency, the expenditures on fixed items like purchase of equipments and facility development are included. To each and every research project, supporting staff are required, may it be in the form of research scholars, research fellows, research assistance, technical staff, etc. To meet the expenditures on staff, salary and wages form the budgetary provision. The financial outlay should clearly mention the amount of money required at different stages of the research process. The various budgets should match the details given in the work program. Avoid overbudgeting, as underutilization may invite negative remark. Once all the project activity costs are listed, split the costs as per the funding agencies' format.

14.1.8 Information About the Research Organizations

Not all types of research works could be taken up in all types of research institutions. Capacity and credential of the research organization play a vital role in deciding whether the research program could be taken up at the particular organization or not. Details of institutional capabilities/credential may help in sanctioning the research proposals.

14.1.9 Information About the Research Persons

Whether a particular researcher (or a team) is competent enough to carry out the proposed research work required to be ascertained. A brief biodata of the researcher and the associated personals should be provided to assess the capabilities and the expertise of the group. This is specially required in competitive mode of research funding.

14.1.10 Monitoring and Evaluation

To judge the success of the research program, periodic evaluation report at various stages is needed and if required midterm corrections may also be taken up. Participation of stakeholders and interest groups in planning, monitoring, and evaluation is the most widely used method in internationally reputed research organizations. To assess the progress, key indicators are specified in each of the objectives. The indicators should be as far possible as quantitative in nature.

14.1.11 Layout of the Research Proposal

A research proposal should be framed and designed meticulously. There should be different parts of the research proposals. Though not exhaustive, along with the above-mentioned sections, there should be one cover page and another inner cover page. The cover page should contain the title of the project, address of the implementing organizations, name of the person who proposed the research, name of person the project is submitted to, etc. The inner cover page will also contain similar information. The cover page is generally colorful and lucrative in nature contrary to the simple presentation of the inner cover page.

14.2 Research Report Writing

Research is an endeavor towards the betterment of civilization. All research activities are directed to having a better world. Unless and otherwise people come to know about the research activities and its findings, the whole process results in a futile exercise and the whole process of carrying out research gets vitiated. So, passing on the findings and inferences of the research work carried out is one of the most important tasks of the researcher. Writing of research reports is the last and most probably the most difficult step of the research process. Research reports have many purposes. The report informs the rest of the world about what a researcher has done, what has been invented or discovered, what are the conclusions a researcher have drawn from the research findings, how these findings or information is going to enrich the knowledge bank, how the results/findings/recommendations are going to help the society, etc. In mandatory research or sponsored researches, the reports are of great importance. It is also mandatory on the part of the researchers to satisfy the sponsors in regard to the extent of commitment fulfilled or otherwise towards the objectives. A research is of least importance and value unless it is being communicated effectively to others.

Research reports maybe of different types: (a) *printed form*, (b) *audio form*, (c) *electronic form, and* (d) *documented as audio–video film*, etc. Most of the research reports are presented in printed form. Additionally, other forms may also come into existence. One can find very few or rare examples where the report of the research projects has not been documented in typed or printed form. As such, in this section we shall consider the printed version of the research report along with general features of the research reports in brief. The layout, content, arrangement, and chronology of the reports vary from types of research, purpose of research report, etc. However, there are certain guidelines and norms with regard to the preparation of the research report.

Before writing a research report, (1) *the researcher must satisfy himself or herself* (a) about the conceptualization of the research project, of course, that has been checked and rechecked during the preparation of the project proposal, (b) that the data generated for the project are trustworthy, appropriate, and adequate to draw inference befitting to the objectives of the research program, and (c) that proper analytical/statistical tools have been used; (2) *the researcher must be cautious about the possible errors which may creep into the inference*; (3) *the researcher must keep a strong vigil about the fact that the interpretations are supported by facts and analysis of the facts and figures*; (4) *the researcher must be aware about the fact that it is the task of the researcher to not only state the striking observations but also identify and unreveal the factors so long remained hidden*; and (5) *there should be consistent interaction between the initial hypothesis or presumption (if any), the empirical observations, and the theoretical background/conceptions.*

14.2.1 Steps in Research Report Writing

Writing of the report is the last stage of research process. It requires a set of great skills from different allied departments/sections. A researcher is always advised to be most careful and to take help from the experts in relevant field while writing reports so that the whole process could be accomplished in a holistic and faithful manner. The whole process of report writing is completed in a steady and in no hurry condition involving the following steps: (1) *logical analysis of the whole work*, (2) *preparation of outline of the report*, (3) *preparation of draft report*, (4) *reviewing and fine-tuning the draft report*, and (5) *preparation of bibliography and writing the final report*. The activities to be performed in different steps may vary from project to project, but by and large, one has to follow the above steps with modification, alteration, and/or addition.

In the first step of report preparation, the researcher needs to take the whole work into his or her mind. Then, sequencing is required to be done either logically or chronologically or both, such that the problem with which the research work initially started and took place gets its solution. Once the researcher has taken the whole work in his or her mind, the outline of the presentation is required to be framed. In doing so, utmost care should be taken to inform the rest of the world in its simplest form. Soon after conceptualization of the outline of the report, a draft report is required to be prepared accordingly. In this phase, the researcher writes chronologically or logically or both on why the research was undertaken; what has been done; what are the procedures adopted; what are the overall observations, analysis, findings, and inferences that could be drawn; what are the limitations and suggestions; etc. Ideally, there should be a gap of reasonable time period between the preparation of the draft report and rewriting and fine-tuning of the same. The step of rewriting and fine-tuning of the draft report is the most vital, tedious job and requires ample patience and time. Consistency in every sphere is required to be checked during this period. Grammar, spelling check, important omissions, or deletions are required to be taken care of with utmost sincerity. Lastly, the lists of literature consulted are required to be acknowledged in the form of a bibliography.

14.2.2 Components of Research Report

A research report is meant for the rest of the world and is an authentic unique documentation so as to convey the general scientific context of the research work, the adequacy of the method, and the findings for the betterment of humanity. Thus, the layout of the report has great importance in reaching the world. In general, a research report should have three main parts: *the preliminary or initial part, the main part, and the end part or the last part*. The initial part is devoted to introducing the title of the work, location of the work, preface or the foreword, acknowledgement, and content of the reports, that is, table of contents and table of figures/illustrations. This part facilitates the readers to locate the portion of his or her interest. Sometimes, an executive summary is also presented in between the initial pages and the main part of the report, whereas in the main part, the reader understands the introduction of the research problem; its rationale; sources of finance along with financial layout, timeline, objectives, and specific objectives; the materials and method section; experimental part (if any); and observations made and analysis of data, followed by statement of the findings, interpretations and implications of results and findings, recommendations (if any), summary and conclusion, etc. The end part or the last part is consisting mainly of the bibliography, questionnaires, mathematical deductions, annexure or appendix, etc. Thus, it is clear that though the report is divided in to three parts, each part is again constituted of subparts. By and large, a research report should have the following sections: (1) *the cover page*, (2) *the inner cover page*, (3) *the foreword*, (4) *the preface*, (5) *acknowledgement*, (6) *the table of contents*, (7) *the table of figures*, (8) *the financial outlay of the project (optional)*, (9) *the executive summary*, (10) *the introduction and rationale of the project*, (11) *the objectives and specific objectives of the study along with hypotheses to be tested (if any)*, (12) *location/implementation area of the project (optional)*, (13) *materials and methodology*, (14) *results and findings along with discussion*, (15) *conclusion and recommendation*, (16) *future scope of research*, (17) *the reference or bibliography*, and (18) *the annexures/appendices*. It may emphatically be noted that these sections may change or get modified as per the mandate or need of the research project. Though these are self-explanatory, in the following sections these will be discussed in brief.

1. *The Cover Page*: The cover page is generally a colored, attractive, high-quality page consisting of the title of the project, the name (s) of the researcher(s) submitting the report, the year of submission, the agency to whom the report is being submitted (optional), and the name and address of the implementing agency.
2. *The Inner Cover Page*: It is almost the replica of the cover page but mostly on the paper of the same quality as that of the inner pages of the report.
3. *The Foreword*: These pages are generally to introduce the readers about the necessity and usefulness of the research program. A scientist of eminence in the concerned field or an authority in the area of research generally writes this portion. Appreciation of the work done in the report is generally provided in the foreword pages. It may be noted that this portion is not a compulsory one.
4. *The Preface*: In the preface, the authors/ researchers writing the report try to provide, in a very short form, about the synthesis, the execution, and the outcomes of the research program under report. During the process, the researcher acknowledges the help, cooperation, and assistance received from various corners.
5. *Acknowledgement*: This is the portion carefully written by the researcher to acknowledge the name of the persons, the institutions, the agencies, etc., from which the researcher has been benefitted in various ways during the process of research program. Acknowledgement at personal level and official level also comes under this section. If otherwise not stipulated, the financial sources or financial assistances received from different sources are to be mentioned in this section.
6. *Table of Contents*: In the table of contents, page-wise distributions of different sections of the research report are mentioned. A reader, going through this table of contents, can have an idea what are the contents of the research report and where lies the chapter(s) of his/her interest and the page concerned. The table of contents helps the readers in getting the overall view of the research report at a glance.
7. *Table of Figures/Illustrations*: Likewise to that of the table of contents, additionally there may be table of figures and illustrations. This will help the readers who are in a hurry to search and get a glimpse of the chapter or portions of his/her interest. Going through the pages of tables or figures of interest, the readers can have an idea, in no time, about his/her point of interest.
8. *Financial Statement*: The financial outlay portion is not obligatory to each and every research report. Generally the ad hoc research project reports funded by some

agencies for a short period of time may contain these financial statement pages. In the financial statement, there would be revelations of the amount sanctioned for the project, details of breakup of the allocation of fund, and utilization thereof, and it may also include the auditor's report.

9. *Executive Summary*: During the fast-moving era of the present days, people of high profile get very little time to go through all the tidbits of any huge research report. This is particularly more applicable for the topmost administrator, policy makers, and other prominent persons. The executive summary provides the bird's eye view of the entire research program carried out, vis-a-vis the output and its implications for the development of the society thereof. The researcher should be very careful while writing the executive summary so that all essential information is put together in a nutshell.
10. *Introduction and Rationale*: The purpose of writing the introduction and rationale of the project is to make the readers understand the purpose, the importance, and the usefulness of the research program taken up. In this section, the brief background upon which the project has been conceptualized and the status of the research vis-à-vis the shortcoming of the already taken research program are clearly spelled out. A clear definition or the statement of the problem must be focused with backup from review of literature on recent works done in the area of the proposed research.
11. *Objectives*: As has already been mentioned, clear statement of the objective of any research program lays the foundation of the whole program. In writing the objective and specific objective of the program, one should be very careful to state these in a clear-cut and meaningful manner which are operationalized and achieved. The objective of the research program may be hypothesis testing type or otherwise. In research program requiring testing of hypothesis, the hypothesis tested under the given situations must be clearly explained.
12. *Site of Implementation*: Different research programs have got different areas of operation, which vary over the types of research. If a research is to be carried out under laboratory conditions, then the locations and facilities used are required to be mentioned. If a research program is field oriented, survey type, or experimental type, the locations of the experimental fields or the survey fields must clearly be mentioned. But location specifications are not absolutely necessary for all types of research program; these are specially required for field experiment or field research.
13. *Materials and Methods*: Experimental materials and method constitute a very important part of the research report. Synchronization of these two factors is essential for a fruitful research outcome. Absence or fault in any one of these factors is bound to have tremendous impact on the whole research process. A good methodology accompanied by a faulty or bad material is not going to help the researcher in fulfilling the objective of the research program. Similarly, good material with unjustified methodology also results in the same output. In this section the researcher clearly spells out the materials, information and data generated, technique, and field of data collections used in the research program. The researcher also clearly indicates the methodology used in analyzing the information to extract the hidden truth from the data. If the researcher develops some new methods or methodology that also required to be clearly discussed. In fact, the material and methodology section induces the future researchers to take up similar studies in other fields. Various limitations of the materials and method used (if any) are also discussed.
14. *Results and Discussion*: Statement of the observations made and findings obtained upon analysis of the information constitute the most vital part of the research report. The presentations of the results and findings generally have two parts: *nontechnical part and technical part*. At the initial stage, the

statement of the observations and findings should be made in such a way that any person, even a nontechnical one, is also able to easily understand the findings of the program. In the second half, the results and findings are to be presented with scientific logic, references, etc., to justify the conclusions that the researcher is going to draw in the next section. To present the findings in a meaningful and lively manner, it should be made in tabular, graphical, pictorial, etc., forms wherever possible along with explanations. The discussion and inference part of the research report should be made unambiguously and in a straightforward manner. Too much generalization of the findings or inferences drawn over ambitiously, not supported by the methodology or analytical tool, may be a drawback of the report. During the end of this section, the researcher should again write the results clearly and precisely followed by implications of each and every result and finding.

15. *Conclusion and Recommendation*: Researches are for the betterment of human life. The conclusion and recommendation part of any research program, therefore, is the most important part of the research report in furthering the betterment of human life. In the conclusion part of the research report, a very brief summary of the research problem vis-à-vis the findings obtained using specific methodologies followed by the recommendations is provided. The policy makers and the executives at the highest levels are more interested on these aspects rather than the whole research report.
16. *Future Scope*: A good research program encourages or lays the foundations for many research problems. Moreover, particularly in exploratory type of research, all the aspects may not be covered in a single research program. As such the younger researchers or other researchers look for the avenues and possibilities of newer research programs/problems from any research report. The future scope of research, in a research report, thus gives birth to many research problems. This also enlightens the researchers about the other possible ways of solving a particular problem.
17. *Reference/Bibliography*: During the entire research process, a researcher needs to survey the literature, may it be printed source, e-source and/or other sources. In each and every step, this helps in the efficient execution of the whole research process. So, one section is devoted in noting down the resources used for the purpose under reference or bibliography section. There are different methods of writing the bibliography or reference section. Different organizations have set different standards and types of writing references; unless and otherwise stipulated, the researcher should follow any one of the standard methods. The reference section is very useful for future researchers.
18. *Annexure/Appendix*: Starting from the conceptualization of the problem to the ultimate presentation of the research report, a researcher is required to go through a host of essential supporting materials which cannot be provided or mentioned in the text. These supporting materials like questionnaires and schedules are presented in the form of appendix or annexure at the end of the report.

So long we have discussed the presentations of the research report in written/printed form and the steps to be followed in writing the report. But the reader should note that the written form of the report is not the only form of presentation of report. Nowadays, reports are also presented in audio, audiovisual, and other formats. Audio recording for the research report is very much helpful in reaching the research report to the ultimate stakeholders who may be illiterate or sitting at distant places or otherwise. "Seeing is believing"—following this truth, sometimes research reports are presented in audiovisual form like video cassette/documentary film/movie.

14.2.3 Qualities of a Good Research Report

Depending upon the objective, mandate of the funding agency reports may be of various types. Mainly, reports can be of *technical* and/or *popular type*. Whatever may be the form of the report, a good research report should have certain qualities, and some precautions are required to be taken up during the preparation of the research report. The following are some of the qualities of a good research report:

1. A research report should not be too lengthy
2. A research report should not be dull and must be attractive, eye catching, neat, and clean.
3. The layout of the research report should be well planned and in accordance with the objective of the study.
4. A research report should be written in simple and clear sentences. Crystal clear sentences to the readers should be used.
5. A research report should be devoid of any grammatical mistake.
6. A research report should be framed like a short story such that the interest of the readers continues till the end of the report.

Suggested Readings

Abrams M (1951) Social survey and social action. Heinemann, London

Ackoff RL (1953) The design of social research. University of Chicago Press, Chicago

Agarwal BL (2006) Basic statistics. New Age International Publishers, New Delhi

Aigner DJ (1971) Basic econometrics. Prentice-Hall, Englewood Cliffs

Allen RGD (1951) Statistics for economics. Hutchinson Universal Library, London

Anderson TW (1958) An introduction to multivariate analysis. Wiley, New York

Anderson TW (1963) An introduction to multivariate statistical analysis. Wiley, New York

Annual Report (1994) International Centre for Agricultural Research in the Dry Areas (ICARDA), PB 5466, Aleppo, Syria, pp 29–30

Anonymous (1984) Linear probability, logit and probit models. Sage, Bevery Hills

Arnold SJ (1979) A test for clusters. J Market Res 16:545–551

Bailey KD (1978) Methods of social research. The Free Press, London

Baker LT (1988) Doing social research. McGraw Hill, New York

Barnett V, Lewis T (1978) Outliers in statistics. Wiley, New York

Berndt ER (1991) The practice of econometrics: classic and contemporary. Addison and Wesley, Reading

Bhattacharya GK, Johnson RA (1977) Statistical concepts and methods. Wiley, New York

Black JA, Champion DJ (1976) Methods and issues in social research. Wiley, New York

Blackwell D, Girshick MA (1954) Theory of games and statistical decision. Wiley, New York

Blalock HM (1972) Social statistics. McGraw Hill, New York

Bridge JI (1971) Applied econometrics. North Holland, Amsterdam

Broadbeck M (ed) (1968) Readings in the philosophy of science. The McMillan, New York

Campbell DT, Stanley JC (1963) Experimental and quasi experimental designs for research. Houghton Mifflin, Boston

Chapin FS (1974) Experimental design in sociological research. Harper, New York

Chatterji S, Price B (1991) Regression analysis by example. Wiley, New York

Chiang C (1984) Fundamental methods of mathematical economics, 3rd edn. McGraw-Hill, New York

Child D (1970) The essentials of factor analysis. Holt, Rinehart and Winston Inc, New York

Chow GC (1960) Test of equality between sets of coefficient in two linear regressions. Econometrica 28(3):591–605

Chow GC (1983) Econometric methods. McGraw-Hill, New York

Christ C (1966) Economic models and methods. Wiley, New York

Christopher AH (1982) Interpreting and using regression. Sage, Beverly Hills

Chung KL (1968) A course in probability theory. Harcourt, Brace & World, New York

Cochran WG (1985) Sampling technique. Wiley Eastern Limited, New Delhi

Coombs CH (1950) The concepts of reliability and homogeneity. Educ Psychol Meas 10:43–56

Cronbach LJ (1964) Essentials of psychological testing. Harper and Row, International Education, New York

Croxton FE, Cowden DJ (1964) Applied general statistics. Prentice-Hall, Englewood Cliffs

D' Amato MR (1970) Experimental psychology: methodology psychophysics and learning. McGraw-Hill Kogakusha Ltd, Tokyo

Dabholkar AR (1992) Elements of biometrical genetics. Concept Publishing Co, New Delhi

Darlington RB, Weinberg S, Walberg H (1973) Canonical variate analysis and related techniques. Rev Educ Res 43(3):433–454

Das SK (1998) An inventory of local flora. Project work submitted to the Indira Gandhi National Open University for Certificate in Environmental Studies, Coochbehar, West Bengal

Das NG (2002a) Statistical methods, vol 1. M Das and Co, Salt Lake

Das NG (2002b) Statistical methods, vol 2. M Das and Co, Salt Lake

P.K. Sahu, *Research Methodology: A Guide for Researchers in Agricultural Science, Social Science and Other Related Fields*, DOI 10.1007/978-81-322-1020-7, © Springer India 2013

Das J, Mandal TK, Basu D (1997) Content analysis of farmer education programme in All India Radio, Kolkata. J Educ Rabindra Bharati University 2(1):52–61
Department of Agricultural Statistics (2002) Manual on computational statistics in agricultural sciences. Bidhan Chandra Krishi Viswavidyalaya, Mohanpur
Department of Agricultural Statistics (2004) Manual on recent advances in computational statistics in agricultural sciences. Bidhan Chandra Krishi Viswavidyalaya, Mohanpur
Des Raj, Chandhok P (1999) Sample survey theory. Narosa Publishing House, New Delhi
Dey PK (1968) Relative effectiveness of radio and television as mass communication media in dissemination of agricultural information. M. Sc. thesis, Division of Agricultural Extension, Indian Agricultural Research Institute, New Delhi
Dillon WR, Goldstein M (1984) Multivariate analysis: methods and applications. Wiley, New York
Dixon WJ, Massey FJ (1957) Introduction to statistical analysis. McGraw-Hill Book Company Inc, New York
Doby JT, Suchman EA, Mckineey JC, Dean JP (eds) (1954) An introduction to social research. Stackpole, Harrisburg
Draper NR, Smith H (2003) Applied regression analysis, 3rd edn. Wiley, New York
Durbin J (1960) Estimation of parameters in time series regression model. J R Stat Soc Ser B 22:139–153
Dutta M (1975) Econometric methods. South Western Publishing Company, Cincinnati
Eberhart SA, Russell WL (1966) Stability parameters for comparing varieties. Crop Sci 6:36–40
Edwards AL (1969) Techniques of attitude scale construction. Vakils, Feffer and Simons Private Ltd, Mumbai
Eide WB (1992) Nutrition research in developing countries- 'Data Imperialism'- or a tool in the fight against hunger and malnutrition. Paper for 5th Nordic Congress of Nutrition, Reykjavik, Iceland, 14–17 June
Engelman L, Hartigan JA (1969) Percentage points of a test for clusters. J Am Stat Assoc 64:1647–1648
English HB, English AC (1961) A comprehensive dictionary of psychological and psychoanalytical terms. Longman Green and Co, New York
Epstein I, Tripodi T (1974) Research techniques for program planning, monitoring and evaluation. Columbia University Program, New York
Everitt B (1980) Cluster analysis. Wiley, New York
Ezekiel M, Fox KA (1959) Methods of correlation and regression analysis. Wiley, New York
Farrar DE, Glauber RR (1967) Multicollinearity in regression analysis: the problem revisited. Rev Econ Stat 49:92–107
Feller W (1968) An introduction to probability theory and its applications, vol I, 3rd edn. Wiley, New York
Feller W (1971) An introduction to probability theory and its applications, vol II, 2nd edn. Wiley, New York
Ferguson TS (1967) Mathematical statistics. Academic, New York
Festinger L, Katz D (eds) (1953) Research methods in behavioural sciences. Holt, Rinehart and Winston Inc, New York
Finley KW, Wilkinson GM (1963) The analysis of adaptation in plant breeding programme. Aust J Agric Res 14:742–757
Finney DJ (1981) Probit analysis. S Chand and Company Ltd, New Delhi
Fisher RA, Frank Y (1979) Statistical tables for biological, agricultural and medical research. Longman, London
Fisz M (1963) Probability theory and mathematical statistics, 3rd edn. Wiley, New York
Flanagan JC (1953) The critical incident technique. American Institute for Research, Pittsburgh
Flesch R (1960) How to write, speak and think more effectively. Harper and Row Publishers, New York
Food and Agriculture Organization (1990) Participatory monitoring and evaluation: handbook for training field workers. FAO, Bangkok
Fox K (1968) Intermediate economic statistics. Wiley, New York
Fraser DAS (1965) Nonparametric methods in statistics. Wiley, New York
Freeman FS (1965) Theory and practice of psychological testing. Oxford and IBH Publishing Company Pvt Ltd, New Delhi
Freund JE (1992) Mathematical statistics. Prentice-Hall of India, New Delhi
Galtung J (1970) Theory and methods of social research. George Allen and Unwin, London
Gambhir GD (1979) Labour in small scale industries. Nagpur University, Nagpur
Gangwar B, Katyal V, Anand KV (2003) Productivity, stability and efficiency of different cropping sequences in Maharashtra. Indian J Agri Sci 73 (9):471–477
Garrett HE (1979) Statistics in psychology and education. Vakils, Feffer and Simons Ltd, Mumbai
Gibbons JD (1971) Nonparametric inference. McGraw-Hill, New York
Gibbons JD, Chakrabarty S (1985) Nonparametric methods for quantitative analysis. American Sciences Press, New York
Giles WJ, Hatt PK (1981) Methods in social research. McGraw-Hill Book Company, Singapore
Glejser H (1969) A new test for heteroscedasticity. J Am Stat Assoc 64:316–323
Goldberg S (1960) Probability, an introduction. Prentice-Hall, Englewood Cliffs
Goldberger AS (1964) Econometric theory. Wiley, New York
Goldfield SM, Quandt RE (1972) Nonlinear methods in econometrics. North Holland Publishing Company, Amsterdam
Goode WJ, Hatt PK (1952) Methods of social research. McGraw Hill, New York
Goon AM, Gupta MK, Dasgupta B (1998a) Fundamentals of statistics, vol 1. World Press, Kolkata

Goon AM, Gupta MK, Dasgupta B (1998b) Fundamentals of statistics, vol 2. World Press, Kolkata
Goon AM, Gupta MK, Dasgupta B (1998c) Outline of statistics, vol 1. World Press, Kolkata
Goon AM, Gupta MK, Dasgupta B (1998d) Outline of statistics, vol 2. World Press, Kolkata
Gorsuch RL (1983) Factor analysis. Erlbaum, Hillsdale
Granger CWJ (1969) Investigating causal relations by econometric models and cross-spectral methods. Econometrica 37(3):424–438
Granger CWJ, Mowbold P (1976) R^2 and the transformation of regression variables. J Econom 4:205–210
Graybill FA (1961) Introduction to linear statistical models, vol 1. Mc-Graw Hill Inc, New York
Guilford JP (1954) Psychometric methods. McGraw-Hill Book Company Inc, New York
Gujarati DN (1995) Basic econometrics. McGraw-Hill, Inc, Singapore
Gunning R (1952) The technique of clear writing. McGraw-Hill Book Co Inc, New York
Gupta SC (2001) Fundamentals of statistics. Himalaya Publishing House, Mumbai
Gupta SC, Kapoor VK (2002) Fundamentals of mathematical statistics. Sultan Chand and Sons, New Delhi
Gupta SC, Kapoor VK (2004) Fundamentals of applied statistics. Sultan Chand and Sons, New Delhi
Haque A (1981) Study of some factors related to the adoption of recommended species of fish in composite fish culture. PhD thesis, Department of Agricultural Extension, Bidhan Chandra Krishi Viswavidyalaya, West Bengal
Harman HH (1976) Modern factor analysis. The University of Chicago Press, Chicago
Hartigan JA (1975) Clustering algorithm. Wiley, New York
Hogg RV, Craig AT (1972) Introduction to mathematical statistics. Amerind, New Delhi
Hollander M, Wolfe DA (1973) Nonparametric statistical methods. Wiley, New York
Holsti OR, Loomba JK, North RC (1968) Content analysis. In: Lindzey G, Aronson E (eds) The handbook of social psychology, vol 2. Addison-Wesley Publishing Company Inc, Reading
House ER (1980) Evaluating with validity. Sage, Beverly Hills
Johnston J (1985) Econometric methods. Mcgraw-Hill Book Company, Singapore
Kaiser HF (1958) The varimax criterion for analytic rotation in factor analysis. Psychometrika 23:187–200
Kane EJ (1968) Economic statistics and econometrics. Harper International, New York
Kaplan A, Goldsem JM (1949) The reliability of content analysis categories. In: Lasswell H et al (eds) The language of politics: studies in quantitative semantics. George steward, New York
Kapoor JN, Saxena HC (1973) Mathematical statistics. S Chand and Co (Pvt) Ltd, New Delhi
Katyal V, Sharma SK, Gangwar KS (1998) Stability analysis of rice (*Oryza sativa*)- wheat (*Triticum aestivum*) cropping system in integrated nutrient management. Indian J Agric Sci 68(2):513–516
Katyal V, Gangwar KS, Gangwar B (2000) Yield stability in rice (*Oryza sativa*)- wheat (*Triticum aestivum*) system under long term fertilizer use. Indian J Agric Sci 70(5):277–281
Katz D (1953) Field studies. In: Festinger L, Katz D (eds) Research methods in the behavioural sciences. Holt, Rinehart and Winston Inc, New York
Kaufman R, Thomas S (1980) Evaluation without fear. New Viewpoints, New York
Kendall MG (1962) Rank correlation methods, 3rd edn. Griffin, London
Kendall MG, Stuart A (1968) The advance theory of statistics, vol 3, 2nd edn. Charles Griffin and Company Limited, London
Kendall M, Stuart A (1973) The advance theory of statistics, vol 2. Charles Griffin and Co. Ltd, London
Kendall M, Stuart A (1977) The advance theory of statistics, vol 1. Charles Griffin and Co. Ltd, London
Kerlinger FN (1973) Foundations of behavioural research. Holt, Rinehart and Winston Inc, New York
Khanna KP, Mathew M (1979) Women workers in unorganised sector of coir industries in India. I.C.S.S.R. research abstract
Khaparde MS (1998) Action research process. National Council of Educational Research and Training, New Delhi
Kidder LH (1981) Research methods in social relations. Holt, New York
Klien LR (1962) An introduction to econometrics. Prentice-Hall, Englewood Cliffs
Klien LR, Shinkai Y (1963) An econometric model of Japan, 1930–1959. Int Econ Rev 4:1–28
Kmenta J (1986) Elements of econometrics, 2nd edn. Macmillan, New York
Kochar VK (1970) Strategy and framework for ethnographic research in India. In: Sinha SC (ed) Research programmes on cultural anthropology and allied disciplines. Anthropological Survey of India, Kolkata
Kolmogorov AN, Fomin SV (1961) Elements of the theory of functions and functional analysis, vol 2. Graylock Press, Albany
Koontz H, Weihrich H (1988) Management. McGraw-Hill Book Company, New York
Kothari CR (1996) Research methodology: methods and techniques. Wishwa Prakasam, New Delhi
Koul L (1998) Research methodology: methods and techniques. Wishwa Prakashan, New Delhi
Koutsoyiannis A (1977) Theory of econometrics. Macmillan Press Ltd, London
Kraft CH, Eeden CV (1968) A nonparametric introduction to statistics. Macmillan, New York
Kramer JS (1991) The logit model for economists. Edward Arnold publishers, London
Krejcie M (1970) Determining Sample Size for Research Activities. Educ Psychol Meas 30:607–610
Kumar Somesh (2002) Methods for community participation: a complete guide for practitioners. Vistaar Publications, New Delhi
Kvalseth TO (1985) Cautionary note about R^2. Am Stat 39:279–285

Lal Das DK (2000) Practice of social research: social work perspective. Rawat Publications, Jaipur
Lee KL (1979) Multivariate tests for clusters. J Am Stat Assoc 74:708–714
Lehmann EL (1959) Testing statistical hypotheses. Wiley, New York
Leser C (1966) Econometric techniques and problems. Griffin, London
Li C (1958) Population genetics. The University of Chicago press, Chicago
Lindgren BW (1968) Statistical theory, 2nd edn. The Macmillan Company, New York
Lindquist EF (ed) (1951) Educational measurement. American council of education, Washington
Loeve M (1963) Probability theory, 3rd edn. Van Nostrand, Princeton
Lundberg GA (1946) Social research. Longman, New York
Lush JL (1943) Animal breeding plans. Iowa State College Press, Ames
Madala GS (1983) Limited dependent and qualitative variables in econometrics. Cambridge University Press, New York
Maddaka GS (1983) Limited dependent and qualitative variables in econometrics. Cambridge University Press, Cambridge
Madnani JMK (1988) Introduction to econometrics: principles and applications, 4th edn. Oxford and IBH Publishing Company Pvt Ltd, Calcutta
Mahajan SL (1998) Postgraduate diploma in future studies (PGDFS) course at India Gandhi National Open University (IGNOU) Level (Development of Curriculum Outline through Delphi Method). Indian Psychol Rev 51 (Special Issue):261–271
McClain JO, Rao VR (1975) CLUSTSIZ: a programme to test for the quality of clustering of a set of objects. J Market Res 12:456–460
Mehta P (1958) A study of communication of agricultural information and the extent of distortion occurring from district to village level workers in selected IADP districts. PhD thesis, The University of Udaipur, Rajasthan
Mikkelsen B (1995) Methods for development work and research: a guide for practitioners. Sage, New Delhi
Mill JS (1930) A system of logic, 8th edn. Longmans, New York
Millers DC (1964) Handbook of research design and social measurement. David Mckey Co, New York
Mohsin SM (1984) Research methods in behavioural sciences. Orient and Longman Limited, Calcutta
Monete DR et al (1986) Applied social research: tool for the human services. Holt, Chicago
Montgomery D, Elizabeth P (1982) Introduction to linear regression analysis. Wiley, New York
Monthly Public Opinion surveys (1981) A readership study in metropolitan cities: May 1981. 22(2)
Monthly Public Opinion surveys (1983) Radio and television audience surveys. 28(9)
Monthly Public Opinion surveys (1986) An assessment of preferences of viewers of advertisements and sponsored programmes on Doordarshan. 31(5 and 6)
Mood AM, Graybill FA, Boes DC (1974) Introduction to the theory of statistics. McGraw-Hill, New York
Morrison DF (1990) Multivariate statistical methods. McGraw-Hill, New York
Moser CA, Kalton G (1950) Survey methods in social investigation. Heinemann Educational Books Harper and Brothers, London
Moulik TK (1965) A study of the predictive values of some factors of adoption of nitrogenous fertilisers and the influence of sources of information on adoption behaviour. PhD thesis, Division of Agricultural Extension, Indian Agricultural Research Institute, New Delhi
Mulay S, Sabarathanam VE (1980) Research methods in extension education. Manasayan, New Delhi
Nachmias D, Nachmias C (1981) Research methods in the social sciences. St. Martina Press, New York
Nanjappa D, Ganapathy KR (1987) Content analysis of agricultural information in selected Kannada dailies. Indian J Ext Educ 23(1 and 2):20–23
Narain P, Soni PN, Pandey AK (1990) Economics of long-term fertilizer use and yield sustainability: soil fertility and fertilizer use. Vol. IV nutrient management and supply system for sustaining in agriculture. Indian Farmers Fertilizers Co-operative Limited. Agricultural Services Department, Marketing Division, New Delhi
Osgood CE, George JS, Percly HT (1957) The measurement of meaning. University of Illinois Press, Urbaba
Pal Satyabrata, Sahu PK (2007) On assessment of sustainability of crops and cropping system – some new measures. J Sustain Agric 31(3):43–54
Panse VG, Sukhatme PV (1967) Statistical methods for agricultural workers. Indian Council of Agricultural Research, New Delhi
Pareek U, Trivedi G (1964) Manual of the socio-economic status scale (rural). Manasayan, Delhi
Park RE (1966) Estimation with heteroscedastic error terms. Econometrica 34(4):888
Parzen E (1972) Modern probability theory and its applications. Wiley Eastern, New Delhi
Patel V (1982) "Women and work" (mimeographed). SNDT Women's University, Mumbai
Patton MQ (1980) Qualitative evaluation and research methods, 2nd edn. Sage, New Delhi
Patton MQ (1982) Qualitative evaluation methods. Sage, London
Perry NC, Michael WB (1951) The estimation of a phi-coefficient. Educ Psychol Meas 11:629–638
Pfohl J (1986) Participatory evaluation: a users guide. PACT, New York
Posova EJ, Carey RG (1985) Programme evaluation methods and case studies. Prentice Hall, Englewood Cliffs
Prajneshu (1998) A non-linear statistical model for aphid population growth. J Indian Soc Agric Stat 51:73–80
Prajneshu (2007) Non-linear statistical models and their applications to crops, pests and fisheries. In: A diagnostic study of design and analysis of field experiments. IASRI, New Delhi

Punch KF (1998) Introduction to social research. Sage, New Delhi
Radhakrishna RB, Bowen BE (1991) Agricultural extension problems: perceptions of extension Directors. Indian J Ext Educ 27(3 and 4):7–15
Raifa H, Schlaifer R (1961) Applied statistical decision theory. Division of Research, Harvard Business School, Harvard University, Boston
Ramachandran P (1968) Social work research and statistics. In: History and philosophy of social work in India. Allied Publishers, Bombay
Rangaswamy R (2000) A text book of agricultural statistics. New Age International (P) Limited, Publishers, New Delhi
Rao CR (1952) Advanced statistical methods in biometric research. Wiley, New York
Rao CR (1968) Linear statistical inference and its application. Wiley, London
Rao MP (1998) Scientific literacy-its impact on neoliterates. Indian J Adult Educ 59(1):10–15
Ray GL (1967) A study of agricultural and sociological factors as related to high and low levels of urbanisation of farmers. PhD thesis, Division of Agricultural Extension, Indian Agricultural Research Institute, New Delhi
Ray GL (2003) Extension communication and management. Kalyani Publishers, Ludhiana
Roethlisberger FJ, Dickson WJ (1939) Management and the worker. Harvard University Press, Cambridge
Rogers EM (1995) Diffusion of innovations. The Free Press, New York
Rosenberg M (1968) The logic of survey analysis. Basic Books, New York
Rossi P, Freeman H (1982) Evaluation: a systematic approach. Sage, Beverly Hills
Rubin A, Babbie E (1989) Research methods for social work. Belmont, California
Ruttman L (ed) (1977) Evaluation research methods: a basic guide. Sage, Beverly Hills
Sagar RL (1983) Study of agro-economic, socio-psychological and extension- communication variables related with the farmers productivity of major field crops in Haringhata Block. PhD thesis, Department of Agricultural Extension, Bidhan Chandra Krishi Viswavidyalaya, West Bengal
Sahu PK (2007) Agriculture and applied statistics –I, Kalyani Publisher, Ludhiana, India
Sahu PK, Das AK (2009) Agriculture and applied statistics –II, Kalyani Publisher, Ludhiana, India
Sahu PK, Kundu AL, Mani PK, Pramanick M (2005) Sustainability of different nutrient combinations in a long term rice-wheat cropping system. J New Seeds 7(3):91–101
Scheffe H (1959) The analysis of variance. Wiley, New York
Scheirer MA (1980) Programme evaluation: the organizational context. Sage, Beverly Hills
Seber GAF, Wild CJ (1989) Nonlinear regression. Wiley, New York
Sellitz G et al (1973) Research methods in social relations, 3rd edn. Holt, Rinehart and Winston Inc, New York
Selltiz C, Wrightsman LS, Cook SW (1976) Research methods in social relations. Holt, Rinehart and Winston Inc, New York
Sen Gupta T (1966) Developing job-chart and a rating scale for measuring effectiveness of village level workers in the intensive agricultural district programme. PhD thesis, Division of Agricultural Extension, Indian Agricultural Research Institute, New Delhi
Shah VF (1977) Research design. Rachna Prakashan, Ahmadabad
Sharma JC (1974) Measurement of social concepts: indicators and indexes. Indian J Soc Work 34 (4):359–366
Sharma DD (1998a) Marketing research-principles, applications and cases. Sultan Chand and Sons, New Delhi
Sharma JR (1998b) Statistical and biometrical techniques in plant breeding. New Age International Publishers, New Delhi
Shaughnessy JJ, Zechmeister EB (1950) Research methods in psychology. McGraw-Hill Publishing Company, New York
Shenoy GV, Pant M (1994) Statistical methods in business and social sciences. Macmillan India Limited, New Delhi
Siddaramaiah BS, Jalihal KA (1982) An experimental study of one-sided and two-sided presentation of messages with advance organisers. Indian J Ext Educ 18(1 and 2):45–50
Siegal S (1956) Nonparametric statistics for the behavioural sciences. McGraw-Hill Book Company Inc, New York
Singh SN (1969) A study on adoption of high yielding varieties and investment pattern of additional income by the farmers of Delhi Territory. PhD thesis, Division of Agricultural Extension, Indian Agricultural Research Institute, New Delhi
Singh AK (1981) Study of some agro-economic socio-psychological and extension-communication variables related with the level of fertiliser use of the farmers. PhD thesis, Department of Agricultural Extension, Bidhan Chandra Krishi Viswavidyalaya, West Bengal
Singh Daroga, Chaudhary FS (1989) Theory and analysis of sample survey designs. Wiley Eastern Limited, New Delhi
Singh RK, Chaudhary BD (1995) Biometrical methods in quantitative genetic analysis. Kalyani Publishers, Ludhiana
Sinha AKP, Upadhyaya OP (1962) Eleven ethnic groups on a social distance scale. J Soc Psychol 57:49–54
Sjoberg G, Nett B (1968) A methodology of social research. Harper & Row, New York
Slessinger D, Stevenson M (1930) Social research, in encyclopaedia of the social science, vol 9. The MacMillan Company, New York
Sneath P (1957) The application of computers to taxonomy. J Gen Microbiol 17:201–226
Snedecor GW, Cochran WG (1967) Statistical methods. Iowa State University Press, Ames

Sokal RR, Sneath PHA (1963) Principles of numerical taxonomy. Freeman, London
Soni PN, Sikarwar HS, Moheta DK (1988) Long term effects of fertilizer application on productivity in rice-wheat sequence. Indian J Agron 33:167–173
Spicer EH (ed) (1952) Human problems in technological change: a casebook. Wiley, New York
Spiegel MR (1988) Theory and problems of statistics. McGraw-Hill Book Co, Singapore
Sproull NL (1988) Handbook of research methods; a guide for practitioners and students in the social sciences. The Scarecrow Press Inc, Metuchen
Stouffer SA (1962) Social research to test ideas. Free Press of Glencoe, New York
Sulaiman VR, Sadamate VV (2000) Privatising agricultural extension in India. Policy paper 10. National Centre for Agricultural Economics and Policy Research, New Delhi
Theil H (1970) On the relationships involving qualitative variables. Am J Sociol 76:103–154
Theil H (1972) Principles of econometrics. North Holland, Amsterdam
Theil H (1978) Introduction to econometrics. Prentice-Hall, Englewood Cliffs
Theis J, Grady HM (1991) Participatory rapid appraisal for community development: a training manual based on experience in the Middle East and North Africa. International Institute for Environment and Development and Save the Children Federation, London
Thomas PT (1977) Social research. In: Encyclopaedia of social work India. Government of India, New Delhi
Thompson WA Jr (1969) Applied probability. Holt, Rinehart and Winston Inc, New York
Thurstone LL (1946) Comment. Am J Sociol 52:39
Tintner G (1965) Econometrics. Wiley, New York
Tuckman BW (1978) Conducting educational research. Harcourt Brace Jovanovich Inc, New York
UNICEF (1990) Strategy for improved nutrition of children and women in developing countries. A UNICEF policy review. UNICEF, New York
Walker HM, Lev J (1965) Statistical inference. Oxford and IBH Publishing Company Pvt Ltd, Calcutta
Ward J (1963) Hierarchical grouping to optimize an objective function. J Am Stat Assoc 58:236–244
Watson G (1987) Writing a thesis: a guide to long essays and dissertations. Longman Inc, New York
White H (1980) A heterosedasticity consistent covariance matrix estimator and direct test of heterosedasticity. Econometrica 48:817–898
Wilkinson TS, Bhandarkar PL (1977) Methodology and techniques of social research. Himalaya, Bombay
Wilks SS (1962) Mathematical statistics. Wiley, New York
Wimmer RD, Dominick JR (2000) Mass media research: an introduction. Wadsworth Publishing Company, New York
Wolfe JH (1970) Pattern of clustering by multivariate mixture analysis. Multivar Behav Res 5:329–350
Yamane T (1970) Statistics. Harper International, New York
Yang WY (1980) Methods of farm management investigation. FAO Agricultural Development Paper No. 80. FAO, Rome
Yin RK (1984) Case study research: design and methods. Applied social research methods series, vol 5. Sage, Beverly Hills
Young PV (1996) Scientific social surveys and research. Prentice-Hall of India Pvt. Ltd, New Delhi
Yule GU, Kendall MG (1950) Introduction to the theory of statistics. Charles Griffin, London
Zacks S (1971) The theory of statistical inference. Wiley, New York

Index

P.K. Sahu, *Research Methodology: A Guide for Researchers in Agricultural Science, Social Science and Other Related Fields*, DOI 10.1007/978-81-322-1020-7, © Springer India 2013

Printed by Publishers' Graphics LLC
CISO20130409.15.35.6